Recent Advances in CROP PHYSIOLOGY

VOLUME 2

The Editor

Dr. Amrit Lal Singh, Principal Scientist, Plant Physiology at the Directorate of Groundnut Research, Junagadh, Gujarat obtained his M.Sc. (Botany) from BHU, Varanasi and Ph.D. in Botany from Utkal University, Bhubaneswar. During his Ph.D. Dr. Singh worked on the biological nitrogen fixation by *Azolla* and blue green algal biofertilizers and demonstrated their use in rice. Dr. Singh, joined ICAR as an ARS Scientist in Plant Physiology, at DGR in Jan. 1985, upgraded to Senior and Principal Scientists and even as Director of DGR. Last 35 years he has been working on the various aspects of crop physiology. He is the life member of more than a dozen of scientific societies and published more than 150 research papers, 25 reviews and book chapters, and several technology bulletins. He is Fellow of the Indian Society for Plant Physiology (ISPP), New Delhi and Indian Society for Oilseeds Research, DOR, Hyderabad, and recipient of the J.J. Chinoy Gold medal award 2011 of the ISPP, New Delhi. Dr. Singh served as a Vice-President (2012) of the ISPP New Delhi, visited China, Japan, Tanzania, Turkey and USA and has worked as Referee/Reviewer of the Annals of Applied Biology, UK, Australian J. of Crop Science, Journal of Plant Nutrition, USA and Journal of Plant Nutrition and Soil Science, Germany.

Recent Advances in CROP PHYSIOLOGY

VOLUME 2

— Editor —

Dr. Amrit Lal Singh

Principal Scientist,
Plant Physiology
Directorate of Groundnut Research,
Junagadh, Gujarat

2016

Daya Publishing House®

A Division of

Astral International Pvt. Ltd.

New Delhi – 110 002

Cataloging in Publication Data–DK
Courtesy: D.K. Agencies (P) Ltd. <docinfo@dkagencies.com>

Recent advances in crop physiology / editor, Dr. Amrit Lal Singh.
  volume 2   cm
Includes bibliographical references and index.
ISBN 978-93-5130-744-0 (International Edition)

  1. Crops–Physiology. 2. Crops–Physiology–India.   I. Singh, Amrit Lal, editor.
    DDC 571.2   23

Published by   :   **Daya Publishing House®**
                            A Division of
                            Astral International Pvt. Ltd.
                            – ISO 9001:2008 Certified Company –
                            4760-61/23, Ansari Road, Darya Ganj
                            New Delhi-110 002
                            Ph. 011-43549197, 23278134
                            E-mail: info@astralint.com
                            Website: www.astralint.com

Laser Typesetting   :   **Classic Computer Services,** Delhi - 110 035

Printed at   :   **Thomson Press India Limited**

प्रो॰ स्वपन कुमार दत्ता
उपमहानिदेशक (फसल विज्ञान)
Prof. Swapan Kumar Datta
Deputy Director General
(Crop Science)

भारतीय कृषि अनुसंधान परिषद्
कृषि भवन, डा. राजेन्द्र प्रसाद मार्ग, नई दिल्ली-110001
**INDIAN COUNCIL OF
AGRICULTURAL RESEARCH**
KRISHI BHAWAN, DR.
RAJENDRA PRASAD ROAD,
NEW DELHI-110001

भाकृअनुप
I C A R

Foreword

Agriculture plays a pivotal role for food and nutritional security, and in alleviation of poverty. But, agriculture sector has been confronted with numerous challenges linked to food and energy crisis, climate change and natural resources. With beginning of 21st century, India is being recognized as the global power in the key economic sectors with high economic growth, but its slow growth in agriculture sector is major concerns for the future food and nutritional security, as one-third of the country's population lives below poverty line, and about 80 per cent of our land mass is highly vulnerable to drought and floods. Indian agriculture, with only 9 per cent of world's arable land, contribute 8 per cent to global agricultural gross domestic product to support 18 per cent of the world population. Also, India has nearly 8 per cent of the world's biodiversity and many of these are crucial for livelihood security of poor and vulnerable population. Thus, acceleration of agricultural growth along with natural resources conservation is of supreme importance.

As the Global food demand is expected to be doubled by 2050, world must learn to produce more food with less land, less water and less labour by devising more efficient and profitable production systems that are resilient to climate change. Thus, more than ever, we need to produce more food with less land. Also looking to the demand of 2050 all the institutions and agricultural universities need to redesign their research and teaching programmes for harnessing power of science and bringing excellence in agricultural research and education that ensures food, nutrition and livelihood security for all.

The ICAR with the help of SAUs has brought green revolution in agriculture in India through its research and technology development in past and its subsequent efforts have enabled the country to increase the production of food grains by 4-fold, horticultural crops by 6-fold since 1950-51 which made a visible impact on the national food and nutritional security. Using cutting edge technologies, there is tremendous development in agriculture during the last two decades and it is hoped that with ingenuity, determination and innovative partnerships among everyone working in the agricultural sector, we can meet the food needs of 9 billion people by 2050 without irreparably harming our planet. However, all these informations are scattered and need to be compiled and circulated widely.

This series on "Recent Advances in Crop Physiology" is a timely effort in this direction, which will act as a reference for directly implementing the available technologies and to help the researchers for planning their future research programme.

Swapan Kumar Datta

Preface

"Food security exists when all people, at all times, have physical, social and economic access to sufficient safe and nutritious food that meets their dietary needs and food preferences for an active and healthy life."

Global food demand is expected to be doubled by 2050, while production environment and natural resources are shrinking and deteriorating. World cereal production has gone 2525 million tonnes (mt) during 2013-14 and is expected to be 2535 mt in 2014-15. Same time, world cereal utilization which was 2416 mt in 2013-14 is put 2464 mt in 2014-15. To feed the world in 2050, yields on maize, rice, wheat, and soybeans will have to rise by 60-110 per cent, but the present projections show an increase of only 40-65 per cent and most rice and wheat had very low rates of increase in crop yields. In other places, the trajectories of population growth and food production are heading in different directions. The rice, is the central to existence in many nations, feeds the world, and provides more calories to humans than any other food, and more than a billion people depend on rice cultivation for their livelihoods. Changes in the price and availability of rice have caused social unrest in developing countries and in 2008, when rice prices tripled, 100 million people were pushed into poverty. About 90 per cent of the world's rice is grown in Asia, on more than 200 million small scale farms (about 1 acre), where additionally 8-10 m t of rice need to be produced every year to keep prices affordable with population increase. However, the International Food Policy Research Institute estimates that by 2050 rice prices may increase 35 per cent because of yield losses due to climate change.

Malnutrition in form of under nutrition, micronutrient deficiencies and obesity imposes unacceptably high economic costs and improving nutrition requires a multisectoral approach that begins with food and agriculture. A total of 842 million people in 2011–13, or around one in eight people in the world, are estimated to be suffering from chronic hunger, regularly not getting enough food for an active life.

The agriculture play its fundamental role in producing food and its processing, storage, transport and consumption contribute to the eradication of malnutrition. Because of better agriculture the total number of undernourished during 2013 has fallen by 17 percent since 1990–92. Agricultural policies and research must continue to support productivity growth for staple foods with greater attention to nutrient-dense foods and more sustainable production systems. Traditional and modern supply chains can enhance the availability of a variety of nutritious foods and reduce nutrient waste and losses.

Recently the Intergovernmental Panel on Climate Change (IPCC) predicted that global food production due to climate change will decline 2 per cent per decade for the remainder of this century compared to food production without climate change even as food demand increases 14 per cent per decade. In 2007, the panel was hopeful that gains in agricultural productivity would more than make up for losses due to climate change. But later research revealed in greater detail the impacts of climate change on sensitive crops and raised questions about how much elevated carbon dioxide levels could increase productivity.

The organic material decays without oxygen, in water-logged rice paddies, soil microbes generate methane, a greenhouse gas with 25 times more warming potential than CO_2. In India, rice methane emission accounts for about 10 per cent of the nation's total greenhouse gas (GHG) emissions. Also, nitrous oxide emissions from rice grown under dryer and aerated conditions, can be as significant as methane emissions which has about 300 times more warming potential than CO_2. It has not yet been estimated what percentage of nitrous oxide emissions come from rice cultivation in India, and other rice growing regions in Asia.

If we are unable to double yields on existing cultivated lands, due to food insecurity pressure, we are likely to clear more land for agriculture leaving environmental concerns and efficiency measures a side. This will have a ripple effect, putting additional pressure on already stressed water resources and wildlife habitat, accelerates climate change. This cycle, left unchecked, can only end with farmers competing for increasingly scarce water and arable land in the face of ever more extreme weather – from floods to droughts – brought on by climate change.

These colliding trends indicate that the world must learn to produce more food with less land, less water and less labour by devising, climate resilient more efficient and profitable production systems. Thus, more than ever, we need to produce more food with less land. Farmers must seek out crop production technologies that will be highly productive and have a smaller impact on water quality and quantity, climate and habitat. To do this, we have the tools and technologies that reduce the need for inputs like fertilizer, pesticides and herbicides; innovative irrigation methods that reduce water demand; and methods that reduce greenhouse gas emissions. Using improved technologies, there has been tremendous development in agriculture and productivity during the last two decades and it is hoped that with ingenuity, determination and innovative partnerships among everyone working in the agricultural sector, we can meet the food needs of 9 billion people by 2050 without irreparably harming our planet on which we all depend. However, all these informations are scattered and need to be compiled and circulated widely. This series on Recent Advances in Crop Physiology is an effort in this direction, which will act as a reference to the farmers for directly implementing the technologies and also to help the researchers for planning their future research to improve crop productivity.

This second volume of 'Recent Advances in Crop Physiology' encompasses 13 chapters written by the experts in the field describing production physiology, drought and salinity stresses, nutrient efficiencies particularly P and N, radiotracer and their use in mineral nutrition, nutritional quality of potato and wheat and role of bioregulators in increasing productivity through amelioration of abiotic stressed. Abiotic stresses are the major factors limiting crop productivity worldwide. The chapter one on 'Drought management in pulses and their diversification under new niches' and chapter seven on 'Can water deficit be useful in potato? – Some issues', widey covers the physiological behavior of these crops under water stresses and how best the water stress could be managed to increase productivity and quality of pulses and potato in India. Chapter eight on 'Bioregulators ameliorate water deficit stress in wheat' is an effort on water stress management through bioregulators and new molecules altogether a different approach.

There are plenty of acid soils and the soil salinity problem is increasing in India and worldwide due to faulty irrigation and drainage practices. A comprehensive chapter three on 'Salinity Management in Vertisols: Physiological Implications' and chapter six on 'Physiological basis of Iron toxicity and its management in crops' takes care of soil and crop management in saline soil and iron- toxicity in crops in acid soils and provide a guidelines how to manage these crops under these stresses. The nutrients and fertilizers are the driving force in increasing the productivity of any crop, but in recent years there is an indiscriminate use of nitrogen and phosphorus inspite of the fact that there is limited P sources on the planet. The use of nutrient efficient crop varieties are the best alternative for managing both deficiencies and excess of these nutrients and in chapter two on 'Role of phosphorus efficient genotypes in increasing crop production' and chapter 12 on 'Nitrogen-use efficiency and productivity of wheat crop' discuss these issues in depth with solutions.

The precise study of mineral nutrition in crop plants require use of radiotracer and hence chapter 10 on the 'Radiotracer use in understanding mineral nutrition of crop plants' is fully devoted on the same.

India is emerging as an export hub of several horticultural crops and chapter four on 'Physiological basis for maximizing yield potentials in coffee' extensively covers the major hurdles and list the ways to increase production and quality of coffee for domestic consumption as well as export. Similarly the chapter five on 'Bioregulators improve the productivity and quality of Indian table grapes' list the best practices and use of bioregulators to increase the productivity of indian grapes.

The forest cover majority of the geographical areas of India and world and play an important role in the climate management and environmental protection, but there are no systematic studies on the productivity of forest. The chapter nine on 'Phenology and productivity of forest flora of Gujarat' is an effort in this direction to highlight the issues how the phenological studies can help to increase forest productivity of Gujarat and reduce the carbon dioxide concentrations on earth through carbon sequestration by forest plant species.

Finally, seed, which is the primary requirement for enhancing crop productivity, plays a vital role in ensuring food security, and a chapter on 'Quality seed- a mega factor in enhancing crop productivity' are well composed by the renowned scientists in the field.

I would like to express my gratitude to all the stalwarts of agriculture and plant biology from various disciplines who has contributed in enhancing agricultural production. Thanks are also due to all the staffs of plant physiology at DGR Junagadh for their help in the various ways. Finally, I would like to express my sincere thanks to Mr. Prateek Mittal for coming forward to take up the responsibility of publishing the series and Mr. Anil mittal and the staff of Astral International (P) Ltd, New Delhi for their care and diligence in producing the book timely.

Dr. Amrit Lal Singh

Contents

Recent Advances in Crop Physiology Vol. 2 (2016)

Pages 1–18

Editor: Dr. Amrit Lal Singh

Published by: DAYA PUBLISHING HOUSE, NEW DELHI

Chapter I

Drought Management in Pulses and their Diversification under New Niches

P.S. Basu and Jagdish Singh

ICAR-Indian Institute of Pulses Research,
Kanpur – 208 024, Uttar Pradesh, India

1. Introduction

India is the largest producer accounting for about 25 per cent of the global share as well as consumer of pulses being an inseparable ingredient in the diet. About a dozen of pulse crops, namely chickpea, pigeonpea, greengram, blackgram, lentil, field pea, lathyrus, cowpea, common bean, moth bean, horse gram, and ricebean are cultivated on about 25 million ha area under varied agro-ecological conditions. Pulses are mainstay of sustainable crop production, an important component of the rainfed agriculture contributed significantly in providing nutritionally balanced food to the people of India since millennia (UNICEF, 2009; WHO 2010). The nutritional benefits of pulses are enormous, contain 18-25 per cent protein and comprise one of the cheapest sources of protein. The per capita availability of pulses has been declining continuously, from 62 g in 1990 to 37 g in 2012. Most of the times their demand had exceeded the domestic production and the additional demand was met through imports. It is estimated that in order to overcome protein deficiency through pulses alone India will require about 20-24 million tons of pulses by 2020; and to produce this quantity domestically it would be essential to either double its area at current yield levels or double the productivity keeping the acreage constant. Though substantial progress has been made in evolving techniques to obtain high yields of pulses, their production has stagnated for the last several decades. About 90 per cent of the global pigeonpea, 75 per cent of chickpea and 37 per cent of lentil area falls in India, (FAOSTAT, 2009). The production of total pulses in India is presently about 17 -18 million tons covering an area of about 25 million hectare majority of which falling under rainfed, resource poor and harsh environments frequently prone to drought and other abiotic stresses. Pulses are least preferred by farmers because of high risk and less remunerative than cereals; consequently, the production of the pulses is sufficiently low. Pulses being a leguminous crop, makes better use of atmospheric nitrogen and fixes it in its root nodules, which in turn enriches the soil and reduces requirement of fertilizers by succeeding crops. Its root system also checks soil erosion and improves soil structure.

Assuming a moderate requirement of 37 g pulses/head/day, the projected pulse requirement during 2030 is 32 million tons which necessitates annual growth rate of 4.2 per cent in production. To meet the projected requirement the productivity needs to uplift at 1361 kg per ha and about 3.0 million ha additional area has to be brought under pulses (Ali and Kumar, 2000). But in the face of shrinking natural resources and high population growth, enhancing production of pulses had been a major concern for the nation. The area under pulses is unlikely to increase in the irrigated areas. Thus, the focus should be on improving yields and increasing area wherever possible, such as new niches like rice fallow, foothills of northern plains, diara land, Bhal and Tal areas of Gujarat and Bihar and other unexplored areas.

Water scarcity is being increasingly considered as a major constraint limiting agricultural production. In adequate rainfall, its erratic distribution and faster decline in the water table are some of the factors causing low availability of irrigation water for diverse crops under different agroclimatic zones. These water crises remain to be the major issues and challenge for sustainable agriculture and food production in future. Hence, introduction of drought tolerant genotypes with improved grain yield through physiological and breeding approaches could be very pertinent areas of research in the present context which enables us to use available water judiciously for longer.

2. Drought, a major Challenge for Pulses

Water scarcity imposes huge reduction in crop yield and is one of the greatest limitations to crop expansion outside present-day agriculture areas. Present trend of global environmental change suggests a future increase in aridity leading to extreme events in many areas of the earth. Therefore, irrigation and use of appropriate crops are important issues worldwide. Approximately 70 per cent of the global available water is presently employed in agriculture and 40 per cent of the world food is produced in irrigated soils. Some irrigation (around 10 per cent) uses water from aquifers, leading to many underground water tables being exploited unsustainably. Factors such as erratic distribution, delayed onset or early withdrawal of monsoon and intermittent long dry spell contribute towards development of drought like situation with various magnitudes. The probability of occurrence of drought in India in majority of the states could be about 2-5 years. Drought stress is a complex phenomenon involving several climatic, edaphic and agronomic factors, and is characterized by timing of occurrence, duration and intensity. The drought problems are often intensified under the Semi-arid tropics by erratic and unpredictable rainfall and by the occurrence of high temperatures/solar radiation and poor soil characteristics. In India, droughts occur mainly due to failure of south-west monsoon which accounts for 70-80 per cent of the total annual precipitation.

Plant water deficit may occur as a result of seasonal decline in soil water availability developed in long-term, or may be due to drought spells. An increased atmospheric evaporative demand affects total carbon gain by the crops. The timings, intensity and duration of stress episodes are pivotal to determine the effects produced by drought. Plant strategies to control water status and resist drought are numerous. Besides numerous metabolic activities those are affected adversely due to drought, reduction in the photosynthesis and biomass is the first and foremost targeted component that results from the water stress and eventually yield is drastically reduced.

The response of cultivated crops to drought varies from species to species. The variation in the sensitivity of crop plants to drought is considered to largely depend upon the water retention capacity of the tissue which involves water and osmotic potential, bulk modulus of elasticity, relative water content etc., and to a greater extent the root capacity to draw water from soil.

2.1. Drought Tolerant Features

As compared to other crops, pulses are relatively tolerant to drought thus providing opportunities to adapt under wide range of soil moisture regimes. Pulse crops include chickpea (*Cicer arietinum* L.), pigeonpea, lentil, green gram, blackgram, soybean (*Glycine max* L.), dry pea (*Pisum sativum* L.), lentil (*Lens culinaris* Medik.), dry bean (*Phaseolus vulgaris* L.), cowpea (*Vigna unguiculata*), mothbean (*Phaseolus aconitifolius*), horse gram (*Dolichos biflorus*) etc.

☆ Pulses have relatively higher water-use efficiency than other crops: Example: Dry pea and chickpea have high WUE similar to spring wheat.

☆ Pulses have the ability to grow at lower relative water contents than other crops: Example; drypea and chickpea (Angadi *et al.*, 1999)

☆ Pulse crops possess deep tap root system for efficient water extraction from soil

☆ Pulse crops, drypea had the highest WUE, however, water use has less correlation with grain yield in pulses compared to wheat. This difference may reflect differences in growth habit (determinate vs. indeterminate).

☆ Under the same growing conditions, pulse crops use less water than other crops e.g wheat

☆ As leaf water content decreased, the leaf water potential of the pulse crops did not decrease as rapidly as that of wheat. This suggests pulse cell walls have additional elasticity, which helps maintain turgor under water-stressed conditions.

☆ Pulses in general have higher dehydration tolerance compared to cereals (Table 1.1). Among grain legumes, pigeonpea, have very high dehydration tolerance compared to other minor pulses and cereals.

Table 1.1: Lethal Leaf Water Potential for a Range of Grain Legumes

Species	Crops	Lethal Water Potential (M Pa)	Dehydration Tolerance
Pigeonpea	Legume	−7.0 to -8.2	Very high
Groundnut	Legume	−3.4 to -8.2	Very high
Soybean	Legume	−5.0	High
Mungbean	Legume	−1.9	Moderate
Cowpea	Legume	−1.8 M Pa	Moderate
Sorghum	Cereal	−3.0 M Pa	Moderate
Wheat	Cereal	0 to -2.0 MPa	Majority sensitive except few

☆ Major grain legumes such as pigeon pea, chickpea and soybean have very high level of osmotic adjustment thus enabling them to maintain positive turgor under severe soil moisture deficit (Lecoeur *et al.*, 1992). Other minor pulses also possess moderate degree of osmotic adjustment and delay desiccation under moderate water stress. On the contrary cereals and vegetables are relatively sensitive to drought due to low or lack of osmotic adjustment. The range of osmotic adjustment in pulses, cereal and vegetable is shown below (Table 1.2).

Table 1.2: Range of Osmotic Adjustment in Grain Legumes

Species (Pulses)	Range of Osmotic Adjustment (M Pa) in Leaves	Degree of Dehydration Postponment	Species (Cereals/ vegetables)	Range of Osmotic Adjustment (M Pa)	Degree of Dehydration Postponment
Groundnut	0.2 to 1.6	Very high	Fieldpea	0.0 to 0.4	Moderate
Pigeonpea	0.1 to 1.3	High	Faba bean	0.0 to 0.2	Low/sensitive
Soybean	0.3 to 1.0	High	Lathyrus	0.0 to 0.1	Low/sensitive
Chickpea	0.0 to 1.3	High	Sorghum	0.8 to 1.7	Very high
Lentil	0.0 to 0.6	Moderate	Wheat	0.2 to 1.5	High
Greengram	0.3 to 0.4	Moderate	Barley	0.2 to 0.5	Moderate
Black gram	0.1 to 0.5	Moderate	Maize	0.1 to 0.4	Moderate
Cowpea	0.0 to 0.4	Moderate	Potato	0.0 to 0.25	Low /sensitive
Lupin	0.1 to 0.5	Moderate			

☆ The relative water content (RWC) in chickpea, a drought tolerant pulse crop, decreased very slowly in respect to decline in the soil moisture. About 10 per cent decrease in the RWC results in about a drop of leaf water potential of −0.6 MPa in chickpea whereas with same value of decline in RWC causes about −1.0 MPa reduction in the leaf water potential of potato, a highly drought sensitive crop, which is much faster. The LWP in potato rarely falls below −1.2 Mpa when leaves attain this threshold level, the leaves loss complete turgor whereas, the corresponding turgor loss in chickpea does occur at sufficiently lower LWP, suggesting that pulses have the ability to withstand to high level of water stress. The slow decrease in the RWC is brought about by the inherent ability of chickpea to adjust osmotically during dehydration.

2.2 Improving Crop-Water Use Efficiency

Due to increasing scarcity and competition for water resources, irrigation is generally not a possible option for alleviating drought problems in the semi-arid tropics. Therefore, for increasing biomass and seed yield in agricultural crops, drought management strategies, whether agronomic or genetic, need to focus on maximum extraction of available soil moisture and its most efficient use in both crop establishment and maximum crop growth.

3. Integrated Drought-Management

The integrated approach for managing drought and producing desired yield involve genetic improvement of the crops through conventional breeding, marker-assisted selection (MAS) of the targeted traits conferring resistance to drought, molecular techniques to incorporate specific genes and finally exploring a very efficient agronomic base to reduce the overall water-requirement of the crops as well as atmospheric demand of water. Most of the molecular tools are presently in the experimental stage. The integrated drought management is a holistic approach comprising of a multiple interaction of various components such as genetic improvement of crops, agronomic options for efficient water use, plant protection measures and Government policies, all combined together to manage the problems of drought and increase the productivity. There are two major approaches to combat uncertainties caused by drought and increase stable production of our agricultural crops.

First consideration is of course, the strategic planning by the Government which involves vigorous monitoring of weather pattern through Geographical Information System (GIS) tool and to identify the agroecological areas-prone to recurrence of drought, launching of the effective programmes such as rural work group, drought-prone area and desert development programmes, technology mission on drought, integrated watershed management etc to combat drought and provide employment opportunity to rural mass. The second, the most important component is the development of scientific technologies to address the drought problems through integrated manner. Integrated approaches to tackle drought problems broadly deals with:

1. Genetic options to modify plants suitable for dry areas and produce better economic yield
2. Agronomic options which involve choice of crops/varieties, water harvesting, moisture conservations, crop management practices and soil management etc.

The following steps are essential for planning improvement programmes for crop yields for a given target drought-prone area:

☆ Characterize the major patterns of drought stress and their frequency of occurrence in the target environment.

☆ Evaluate crop response to the major drought patterns (simulation modelling).

☆ Match crop phenology (growth period, sowing, flowering, seed filling) with the most favourable period of soil moisture and climatic regimes.

☆ Develop a strategy for the optimal use of supplementary irrigation, when available.

☆ Increase the soil water available to crops through agronomic management practices.

☆ Identify plant traits that would maximize (i) the use of available soil moisture in transpiration; (ii) the production of biomass per unit water transpired; and (iii) partitioning into seed, thereby conferring enhanced crop water productivity.

3.1 Genetic-Enhancement

Major genetic-enhancement approaches adopted worldwide to improve the crop yields under drought-prone environments are as follows:

☆ Development of short-duration genotypes that can escape terminal drought;

☆ Development of genotypes with superior yield performance through conventional breeding;

☆ Development of drought-resistant genotypes following the physiological breeding approach;

☆ Identification of quantitative trait loci (QTL) for drought tolerance and their use in marker-assisted breeding.

A large germplasm collection of chickpea and pigeon pea crops, available in the ICRISAT gene bank, provides the base material for implementing the above four approaches of genetic enhancement in drought resistance.

3.2 Development of Short-Duration Genotypes

Short-duration varieties that mature before the onset of severe terminal drought have proved successful in increasing yield under drought-prone conditions. Significant progress has been made in developing improved chickpea varieties of short duration that mature in 70–90 days in mild-winter chickpea-growing conditions, as prevailing in peninsular India. Even extra-short-duration chickpea varieties, termed super-early *e.g.,* ICCV 2, have now been developed. The development of these new varieties has expanded options to include chickpea as a crop in many prevailing and evolving new production systems, such as rice fallows. Similarly, a large spectrum of genotype duration is now available in pigeon pea which is matching with likely period of soil-water availability. This is the first line of defense against terminal-drought stress.

3.3 Screening Tools and Breeding for Drought Tolerance

In order to identify sources of drought tolerance, it is necessary to develop screening methods that are simple and reproducible under the target environmental conditions. Several field- and laboratory-screening methods have been used to screen the pulses crops for drought tolerance, including line-source-sprinkler irrigation, rain-out shelters and measurement of the drought-susceptibility index (DSI) at ICRISAT. Both line-source and DSI methods have been found to be very effective in identifying sources of tolerance to terminal drought in chickpea. Promising drought-tolerant germplasm, such as the line ICC 4958 in chickpea and hybrids ICPH 8 and ICPH 9 in pigeon pea were identified as the most drought-tolerant.

3.4 Crop-Improvement Strategies

A first approach to minimizing yield losses due to terminal drought has been to breed for earliness. The empirical breeding approach is based on the selection for yield and its components under a given drought environment. On the other hand, physiological traits can help greatly in establishing screening protocols, which allow better management of G x E interactions. Physiological responses associated with drought tolerance can be identified and incorporated into elite high-yielding genotypes of appropriate crop duration. The QTL-mapping approach can identify individual genetic factors associated with a specific response and monitor the incorporation of the identified factors into the breeding programmes. An ideotype approach was followed for genetic improvement of drought tolerance in chickpea. Using ICC 4958 (drought-tolerant parent), Annigeri (a high-yielding parent) and ICC 12237 (a wilt- and root-rot-resistant parent), a three-way cross was made. Following a diversified bulk method of breeding, generations were advanced and nine yield- and root-trait-based selections were made. Yield-based selections were effective in producing varieties with high yield and trait based selections in producing varieties with a greater degree of drought tolerance. Promising drought-tolerant, *Fusarium*-wilt-resistant lines with high yields are ICCVs 94916–4, 94916–8, 94920–3, 94924–2 and 94924–3 have been identified. A backcross programme was also initiated at ICRISAT, with the objective of incorporating drought-tolerant traits in elite cultivars and of combining drought-tolerant traits.

Seven varieties that combine the traits of large roots and fewer pinnules were developed (ICCV 98901 to ICCV 98907). A few of these recombinants showed a greater degree of drought tolerance than and a yield similar to that of the high-yielding parent. While the empirical approach was partly successful, it was concluded that a more efficient breeding approach required the selection of traits associated with drought resistance. There has been significant progress in understanding the physiological basis of genotypic variability of drought response in number of crops. Significant genotypic variation in total amount of water transpired (T) and transpiration efficiency (TE) has been shown under field conditions. These studies made it possible to analyse the yield variation under drought conditions using the physiological framework proposed by Passioura (1977).

Where: pod yield = T x TE x harvest index (HI)

Research has also shown that TE and carbon isotope discrimination in leaves are well correlated in groundnut, suggesting the possibility of using isotope discrimination technique as a rapid, non-destructive tool for selection of TE in groundnut. However, further research has shown that specific leaf area (SLA, cm^2 g^{-1}) is well correlated with using isotope discrimination and TE in groundnut. Further studies are currently in progress to compare the efficiency of the trait-based selection approach *vis-à-vis* the empirical approach.

3.5 Improving Drought Tolerance: A Most Suitable Option under Resource Poor Lands

Pulses are water-efficient crops and are known for their drought tolerance. Pulses improve soil fertility, require less water in comparison to cereals, and improve moisture-holding capacity of the soils. Among various crops, pulses have drawn special attention by virtue of their unique adaptation to dry land agro-ecosystem, hence enormous scopes are still left to improve the productivity, extend the cultivation of pulses and diversifying the cropping system with different legumes that can substantially reduce the water requirement as a whole. Consistent efforts have been made in this direction to further improve the yield and at the first stage, some potential morphophysiological traits conferring resistance to drought such as high water use efficiency measured by carbon-isotope descrimination ratio, deeper root length, high biomass and harvest index, osmotic adjustment have been identified (Basu *et al.*, 2007 a, b). Use of QTLs, AFLPs are being used to speed up breeding programme for improving drought resistance. The osmotic stress regulated genes including DREB-1A and 2A and ALFINI Zinc finger transcription factors are being used to increase the drought tolerance. Rooting behaviour of different genotypes and their recombinant inbred lines (RILs) had shown drought tolerance in addition to other desired characters such as early vigour and membrane stability. This holds new promise for development of drought tolerant genotypes based on the physiological trait identified.

4. Strategies to Expand Areas of Pulses

With growing population and receding agricultural land, the food and nutritional security of the country remain uncertain and has become a major challenge. Pulses are being consistently pushed towards harsher environment as other competitive crops with higher remunerative values are replacing at faster rates. These factors eventually led to low or stagnated production of pulses in the country for the past several decades. Under this situation, only marginal increase in the productivity could be possible with most improved technologies. Therefore, expansion of pulse area would be the only option to enhance total production of pulses to meet the future demand. The rainfed rice fallows offer significant opportunities for the intensification of agricultural production in some states. Chickpea, lentil and lathyrus are the important pulses that can be successfully grown in these fallows on residual moisture, and can escape terminal drought. Insufficient irrigation facilities and low residual soil moisture are the main limiting factors for utilization of rainfed rice fallows for crop production in *rabi*. Drought alone may reduce crop yield by 50 per cent. A quantum jump in productivity can be achieved by applying life-saving irrigation especially in *rabi* pulses grown on residual moisture. Extraction and use of ground and surface water for irrigation is difficult and costly.

4.1. Limited Option for Horizontal Expansion of Agricultural Crops

Agricultural land is limited and there is little scope to bring additional area under cultivation because of increasing demand for land for non-agricultural purposes. The only opportunities to increase food grain production are by increasing cropping intensity and/or growing more than one crop in a year on the same piece of land. The rainfed rice-fallow lands offers some scope to address the problems of food and nutrition insecurities (Joshi *et al.*, 2002). About 12-14 million ha of rainfed rice land remains uncultivated in the postrainy *rabi* season. If the existing rice-fallows are brought under cultivation, it may usher another green revolution in the predominantly rice-fallow states, benefiting millions of small landholders.Lack of irrigation is the main limiting factor to rainfed rice-fallows productivity of *rabi* crops. Extraction and use of ground and surface water for irrigation is difficult and costly and public irrigation infrastructure involves a huge investment. Therefore, introduction of crops that can escape terminal drought is one of the plausible options to harness the potential opportunities in rainfed rice-fallows.

4.2. Pulses under Conservation Agriculture (Rainfed Rice Fallows)

It has been projected that water would be a major limiting factor for agricultural crops in larger part of the world. Therefore the conservation agriculture with drought tolerant pulses, both in terms of saving soil nutrients and water is most urgent need under present scenario to sustain the overall agricultural productivity. Pulses and coarse cereals are the best substitutes to conserve the presently available irrigation water. The potential areas of rice fallows have long been considered to be exploited for pulses as succeeding crop after rice harvest primarily because of availability of conserved moisture in deep layer where low input responsive and hardy crops like pulses can be best grown without any further tillage practices. Rice fallow cultivation is considered as Conservation Agriculture (CA). The resource conservation technology (RCT) is a vital component of the strategy for food security and poverty alleviation, better health, rural development, enhancing productivity, improve environmental quality and preserve natural resources. RCTs/CA are also strategies to mitigate and adapt to climate change. Generic elements of Conservation Agriculture are significant reduction in tillage, retention of crop residues on soil surface, reduce compaction and economically feasible, diversified crop rotations, zero tillage and raised bed planting, timely crop establishment and improve water productivity. Chickpea amongst pulses is one of the most important pulses that can be successfully grown in rainfed rice fallows on residual moisture and escape terminal drought. Evidence indicates that pulses can be grown in a cost efficient manner because of their low input requirements (Joshi *et al.*, 2002). Growing pulses such as chickpea, lentil, greengram, blackgram and lathyrus could be economically beneficial crops in the rainfed rice fallows spreading over Bihar, Meghalaya, Jharkhand, Odisha, Chattisgarh, Madhya Pradesh, Andhra and Tamil Nadu states of India. Madhya Pradesh and Chattisgarh hold about 40 per cent of the total 40 million ha rainfed rice fallows in India and have great potential of introducing chickpea as a *rabi* crop. The productivity and production of chickpea can be significantly enhanced using the improved pulse production and protection technology.

The rice fallows are one of the potential areas of expansion of pulse cultivation in India which has been little exploited so far. Approximately 12 million ha of the 41 million ha rice area cultivated during the rainy season (*kharif*) remains uncultivated during *rabi*. About 40 per cent of the total rice fallow area lies in the states of MP and Chhattisgarh which together control about 5.0 million ha of rainfed rice fallow, just about half of the country's total chickpea area (Subbarao *et al.*, 2001). The *rabi* rice-fallow area in these states, is as high as 80 per cent in Chhattisgarh and 85 per cent in Madhya Pradesh. Together, the rainfed rice fallows in these two states accounts for about 85 per cent of the total *kharif* rice area. This indicates that only 16 per cent of the *kharif* rice area is utilized for cultivation of *rabi* crops and the rest of the land remains fallow. It is a common practice. Farmers usually either leave the rice area vacant during *rabi* after rice harvest or to cultivate traditional low-yielding varieties of chickpea. Rainfall during the *kharif* in these fallows over the country is usually much adequate to grow rice. Hence, there is tremendous opportunity for cultivation of a second crop on available soil moisture after harvest of rice with a little modification of agronomic practices. Lack of irrigation facilities is the main impediment to production of another crop in the rice fallows. Chickpea is one of the crops that have better tolerance to moisture stress. Thus short-duration wilt-resistant varieties of chickpea can be successfully expanded following improved pulse production and protection technology.

4.3. Opportunities for Chickpea Production in Rainfed Rice Fallows

The rainfed rice fallows offer enormous opportunities for intensification of pulse production in India. Pulses have several unique characteristics, which make them one of the most eligible crops for production in rainfed rice fallows. Among pulses, chickpea has a better prospect due to better adaptability to the harsh weather conditions, soils, and moisture stress. Besides, it has a great economic and nutritional value. The analysis reveals that chickpea production in rice-fallow areas of Madhya Pradesh and Chhattisgarh has opened several new opportunities to the farmers in terms of increased farm income and employment.

The traditional varieties of chickpea often have very low production potentials and are also vulnerable to a number of biotic and abiotic constraints. These constraints inflict a huge gap between the average attainable yield and the average yield of the traditional chickpea varieties in rainfed rice fallows. Average attainable yield is just the average yield of the improved chickpea varieties cultivated by the farmers in the village. It is obvious that there is tremendous scope for increasing chickpea production by bridging these yield gaps in different regions. It is obvious that on an average chickpea yield in both Chhattisgarh and Madhya Pradesh can be almost doubled just by replacing all the local chickpea varieties with improved varieties in the regions.

4.4. The Potential Areas to be Targeted for Expansion

The major future expansion of area under pulse crops may take place in rice fallows, where there is no other crop to compete. A considerable area remains fallow after rice harvest in north eastern and peninsular regions of the country, where rice is grown as the principal crop. Majority of this fallow lies in the states like Assam, Bihar, Chhattisgarh, Jharkhand, Madhya Pradesh, Odisha and West Bengal and remaining area in the states like Tamil Nadu, Karnataka and Andhra Pradesh. Pulses may prove as ideal crops for rice fallow cultivation. Lentil, chickpea and lathyrus are the candidate crops that can occupy the area under rice fallows of northern India. Production of chickpea in rainfed rice fallows is economically viable and technically feasible. Chickpea fits in rotation with the *kharif* rice and it can be successfully produced on less puddle heavy vertisol soils with residual soil moisture after the rice harvest. Similarly, blackgram and greengram are successful in rice fallows of peninsular region (Satyanarayana *et al.*, 1988; Kumar *et al.*, 2000).

4.5. Pulses Cultivation under New Niches

In India, over 10 lakh hectares of fertile land lie fallow in the *rabi* season after the monsoon which remails fallow after rice harvest and farmers traditionally avoid growing crops. Majority of the people in these areas are poor and malnourished. However, planting pulse crops on fallow lands are some of the simple low-risk options that could help these people grow more food and make more money. During *rabi* season, rains are often erratic causing frequent droughts leading to crop failure. Often seed germination is uncertain as the paddy fields dry out and hard layers of soil (hard pans) form and crops don't mature before drought sets in. Early-maturing varieties of hardy pulses, low-cost seed treatments and simple farming practices would be beneficial and proved less risk and good returns.

Pulses provide opportunities for crop diversification and intensification of cropping under rainfed production systems. Evidence indicates that pulses are the most ideal crops that can be successfully cultivated in rainfed rice fallows. Of the various pulses, chickpea offers the best option for *rabi* cultivation in rainfed rice fallows. Chickpea fits in rotation with the *kharif* rice and it can be successfully produced on less puddle heavy vertisol soils with residual soil moisture after the rice harvest. It has better adaptability to moisture stress conditions and can thrive well on residual soil moisture after the harvest of rice.

Table 1.4: Potential Areas of Pulses under Projected Expansion

Potential Areas	Crop	States	Additional Area (million ha)
Rice fallow	Chickpea	Eastern UP, Bihar, Odisha, Jharkhand, Chhattisgarh, WB	4.0
	Urd/Mungbean	AP, TN, Odisha, Karnataka	5.0
	Lentil	E.UP, Bihar,WB	1.0

4.6. Pulses in Rice Fallows Need Mechanization

Vast areas of rice fallow exist in the north east states and Tamil Nadu, Karnataka and Andhra Pradesh where pulses like lentil, chickpea, black gram and greengram are grown in considerable areas. In Krishna district of Andhra Pradesh, black gram as relay crop is grown in large areas in deep black soil with excess soil moisture at sowing and the crop continues to grow in residual soil moisture. Farmers have adopted 'LBG 752' successfully in rice fallow using excess amount of seed, spraying insecticide and liquid fertilizer that leads to over plant population and increased cost of production. In west Godavari district the similar situation exists. A strong research advisory programme is needed to educate farmers to avoid use of excess inputs. At the same time planting techniques for maintaining optimum plant population need to be standardized. In Aduthurai, Tamil Nadu about 3 lakhs ha area is covered with black gram as relay crop with rice in black soil. There is a possibility to bring about additional 4 lakhs ha area of black gram in rice fallow situation in the state. Farmers traditionally broadcast black gram seed in the standing rice field 3-4 days before rice harvest in waxy soil moisture situation. The terminal drought and low plant population of black gram are common phenomena in these areas. This situation needs critical intervention of implements (either for line sowing separately immediately after harvest of rice or simultaneous harvesting and sowing of black gram in waxy soil moisture situation), the post emergence herbicide and soil moisture conservation practices. Line sowing of pulses in combine harvested rice fields with uniform spreading of residue by mounting a device at the rear of combine is one possibility for better crop establishment. To achieve simultaneous sowing of pulses and harvesting of rice in single operation, suitable attachment to existing rice harvester should be developed. Mobile sprinkler in combination with residue management is one of the options to mitigate terminal drought. The non-traditional pulse crop area in north-east, particularly Meghalaya, Manipur, Nagaland have the potential of promoting pulses in rice fallows both in upland and lowland situations. The ideal crops in these areas are lentil and field pea. The variety developed by IIPR (DPL-62) and field pea (Prakash) performed well in this region. To reduce the drudgery and for timely sowing, the manually operated no-till drill designed by IIPR as well as by ICAR research complex for rice fallow is found useful.

5. Constraints under Rice Fallows

The major constraints under rice fallow include abiotic stress such as drought caused by low residual moisture and erratic rainfall and biotic stress due to insect-pest and disease attacks. The pulses are most suitable crops for cultivation under rice fallows as they require little moisture and most hardy crop with deep root penetration ability. The country is under deficit of about 2-3 million tons of pulses which is being imported every year to meet the growing demand of pulses. Rainfed farming suffers great risks and uncertainties. The rainfed rice fallows of Chhattisgarh and Madhya Pradesh consist of varied range of soil types from sandy-loam (Entisols) to heavy textured deep vertisols soils. Deep vertisol soils are more suitable for chickpea in rainfed rice fallows. On the other hand, deep vertisol soils provide a better environment for the root zone and retain moisture for a longer period. However, low organic matter and humus in the soils of rainfed rice fallow is common problem. In sandy-loam soils, cultivation of chickpea or any other pulses is difficult with limited irrigation because of their poor moisture holding capacity.

Farmers depend mostly on traditional or self produced seeds that have low genetic potential and poor seed replacement rates. Non-availability of short duration varieties of *kharif* rice as well as of chickpea is another serious problem. Most of the existing rice varieties are of long duration (about 130-150 days). This delays the recommended sowing of chickpea from mid-October to early-November after rice harvesting. Late sowing of chickpea often leads to poor seed germination, poor crop stand and suffering from terminal drought. It is high time to develop some short-duration high-yielding varieties of rice and chickpea to promote *rabi* cropping in rice fallows and to escape terminal drought. Non availability of resource conserving machineries and implements to farmers limits their work efficiency and income. Farmers cannot afford to buy zero-till machines for sowing of seeds of different crops and other farm machines that perform multiple tasks of inter cultivation and sowing. Short duration chickpea varieties with multiple resistances to important insects and diseases may be helpful in the expansion of chickpea in rice fallows. In addition, there is a need to develop some economical and effective management practices to minimize the losses due to insect-pests and diseases. The most serious problem is non-availability of quality seeds in desired quantity (Bantilan and Parthasarathy, 1998) at appropriate time at reasonable cost. Chickpea productivity and production in rainfed rice fallows can be significantly enhanced using the Improved Pulse Production and Protection Technology.

A number of abiotic factors limit the utilization of rainfed rice fallows for *rabi* pulses. Among all, water scarcity is the most critical constraint. Adequate soil moisture is required for seed germination, crop establishment, efficient conversion of soil nutrients for proper crop growth and grain filling. Low soil moisture after rice harvest, low and depleting water table due to exploitation of ground water for production of crops such as wheat, mustard and vegetable, and terminal drought towards flowering and harvest stages limit the crop productivity. Lack of irrigation coupled with low residual soil moisture after rice harvest is the main factor for prevalence of fallow during the *rabi*. During the *kharif* water tables are raised high, but as the monsoon rains cease, the water tables recede swiftly. The water table at many places is beyond accessible limits. Post-monsoon rainfall is uncertain and sparse. A drought-like condition during advanced stages of the *rabi* crops adversely affects the productivity and sometimes also leads to crop failure. Drought alone may reduce seed yields by 50 per cent in the tropics. A quantum jump in productivity can be achieved by applying life-saving irrigation, especially in *rabi* pulses grown on residual moisture. Other problems are related with quality, texture and type of soils having a significant bearing on pulse production.

A very high proportion of the farmers perceived the problem of soil cracking as one of the most serious constraints. Deep vertisol soils become hard and compact after puddling. Under prolonged moisture stress conditions these soils develop cracks that facilitate rapid escape of available moisture from the field and prove fatal for the standing *rabi* crops. Low organic matter content and humus in the soils of rainfed rice fallow are another constraints. Cow dung is one the major sources of soil organic matter and the availability of dung is unlikely to increase in future. The other soil related constraints are development of soil salinity, particularly in some of the canal-irrigated areas.

Chickpea, wheat, mustard, lentil and lathyrus are the most preferred crops for *rabi* production.

However, there is an obvious preference for chickpea everywhere in rainfed rice fallows. Chickpea, lentil and lathyrus are well suited to rainfed conditions, but chickpea has the best adaptability of all. Though cultivation of lathyrus is preferred by the farmers of Chhattisgarh, its detrimental effect on human health discourages its cultivation. Cultivation of lentil is preferred in Madhya Pradesh, but it is more susceptible to some of the diseases and pests. Chickpea is relatively hardy. Since cultivation of rice is subject to the extent of monsoon rains, and most of the existing rice varieties are of long duration (about 130 to 150 days) there remains a very short growing period for chickpea production. If the harvest of rice is delayed, chickpea sowing also gets delayed. The recommended sowing time for chickpea is mid-October to mid-November but it is often sown up to last week of December or well into the first fortnight of January. This leads to reduced seed germination and a poor crop stand. Since the chickpea is gown on residual soil moisture after the harvest of rice, farmers perceive that better germination of chickpea occurs if sowing is done immediately after the harvest of rice. Perhaps the soil moisture, temperature and micro-environment in the field after the harvest of rice favour chickpea. Unfortunately, the sowing of chickpea is further delayed due to threshing and preparation of the field for *rabi* sowing. Most of the cultivated chickpea varieties are of long duration. These varieties often suffer the worst form of terminal drought or even witness massive failures if there is no rainfall.

The farmers look forward to short-duration, high-yielding varieties of rice and chickpea specifically developed to promote *rabi* cropping in rainfed rice fallows. It would help the region to escape terminal drought and promote the livelihood status of the farmers who are constrained to leave the paddy fields vacant during *rabi*. Another remedy could be the introduction of resource-conserving technologies such as zero-tillage (seed cum fertilizer drills) and sprinkler irrigation.

5.1. Challenges Associated with Rice Fallows

There are lots of challenges associated with rice fallows. After rice harvest, the top layers become hard due to prolonged submergence of the rice field, soil porosity and microbial activities decreased causing poor aeration, The rice fallows are subjected to a number of problems such as poor seed germination or decay, poor crop establishment, restricted root proliferation and nodulation, low nutrient availability and organic matter followed by terminal drought.

5.2. Technology Gap

Mostly after rice harvest, nothing grows in the fallows covering vast areas falling under post-rainy *rabi* season. Such fallows are suitable for hardy pulses like chickpea and lentil providing enormous scope for farmers to increase their returns. After *kharif* season and rice harvest, paddy field soils are still damp. These soils are usually highly fertile. Lack of suitable varieties and appropriate production technologies compel to keep such a large area under fallow after rice cultivation. Besides, there are number of abiotic, biotic and socioeconomic constraints for rice fallow cultivation.

Poor germination after rice harvest and occurrence of drought during flowering to grain filling stages are serious constraints. Soil hardiness along with low organic matter content in the soil adversely affects the soil moisture distribution and root growth of pulses. Lack of appropriate moisture conservation and sowing technique for successful seed germination in the low moisture regime also limit the cultivation in rice fallow. Even if the crop is sown timely and established well, pulses experience high incidence of insect-pests and diseases.

6. Conclusions and Implications

Farmers can use residual moisture after rice harvest to grow pulses but situation varies widely. Pulses are suitable candidate crops by virtue of having deep roots system. Soybean, mungbean, black gram, pigeonpea, groundnut, chickpea, lentil, khesari, faba bean and pea are all suitable for fallow paddies. Including legumes in rice rotations adds nitrogen and organic matter to the soil and improves rice yields. Pulses are more profitable than irrigated *rabi* crops such as wheat because they have low input costs and the grain fetches high prices. Introducing *rabi* pulse crops to rice fallows is low-cost and can raise both yield per hectare and income for the poorest farmers. In addition, rice-pulse rotations conserve both land and water while producing more food and employing more labour.

The early rice harvest ensures early sowing of follow-on crop which allows more moisture remains in the soil for the next crop. It is important to keep as much moisture in the soil as possible. This means minimum tillage with faster preparing the soil for sowing without too much disturbance. Early-maturing varieties are now available that can bring the rice harvest forward by ten days or more. Another way of bringing the harvest forward is direct seeding where, instead of transplanting rice seedlings, rice is sown directly into paddies. In this case, more resources need to be spent on weeding.

Seed Priming, that is soaking seed overnight, helps seed germinate and gives seedlings a head start. This simple practice raises yields of chickpea by over 40 percent and costs practically nothing. Priming seeds with both molybdenum and *Rhizobium* effectively increases yields. After priming, the seeds need to dry just enough to be sown easily. Seed priming is easy, low-cost, low-risk and very effective.

Short-duration pulse varieties, early sowing, minimal tillage and seed priming are set to change this rice-fallows dramatically. Microbial degradation of residue of the preceding crop modifes the physical properties of the soil and conservation of resources like carbon, nitrogen and other trace elements. Suitable matching varieties of chickpea and lentils for rice fallows are primarily required to mitigate the initial moisture stress besides conservation of other nutrients and water resources. Rapid proliferation and deep penetration of root, accelerated canopy growth, higher water and nutrient efficiency, enhanced photosynthesis, prolific nodulation, higher drought tolerance are some of essential plant attributes that have been conceptualized for better adaptation under rice fallow ecosystem. A wide genetic variability in phenology, seed size, plant types, water and nutrient use efficiency and drought, salinity and thermotolerance traits are available in the germplasm stock of chickpea and lentil which can be efficiently exploited for improved productivity under rice fallows.

Therefore, specific programmes need to be developed and implemented for rice fallow ecosystem. The programmes should include development of suitable cultivars and technologies for resource conservation like conservation tillage, stubble mulching and weed management, integrated pest management (IPM) and supported by efficient machines. Thus, introduction of pulses in rice fallows with appropriate production technologies may usher in another green revolution in the backward, poverty-ridden and deprived region of the country.

The major strength of the MP and Chattisgarh is the existence of large *rabi* fallow lands that virtually has zero opportunity cost. Labour is also abundant and cheap, and remains grossly underutilized during the *rabi*. Besides, chickpea has better adaptability to marginal lands. Yields of chickpea in Chhattisgarh and Madhya Pradesh are also higher than the national average. The region has an abundance of Vertisol, which favours chickpea production. The monsoon is generally good in these states and that helps chickpea cultivation on the residual moisture after *kharif* rice. Unavailability of quality seed, poor accessibility to markets, practice of cultivation of long duration rice varieties, lack of short-duration chickpea varieties, weak extension system, drought-like situations at the time of crop maturity and uncertain rainfall are some of the obstacles to chickpea production in rainfed rice fallows. Besides, the farmers lack water-saving/harvesting technologies and improved tools and implements required for crop establishment. The Government of India is committed to introduce the 'Food Security Act (FSA)' and the success of FSA will depend on augmentation of agricultural production by raising agricultural productivity and/ or cropping intensities of mono-cropped, rainfed and marginal lands, apart from other measures. Pulses complement cereals in both production and consumption. These improve soil fertility, require less water in comparison to cereals and control diseases and pests in rotation with cereals. There is a persistent increase in area of chickpea under rice fallow. Because of cultivation of chickpea after rice, farmers could obtain an average additional income. Farmers perceived the positive impact of chickpea cultivation on soil fertility as seen from the increased yield of subsequent rice crops. There is a need to develop seed-cum-fertilizer drills suitable for chickpea crop establishment soon after the harvest of rice. Increase R and D endeavour to develop short-duration chickpea cultivars with multiple resistance to major biotic and abiotic constraints such as Fusarium wilt, collar rot, dry root rot, pod borer, terminal drought, etc. Increase R and D efforts to develop short-duration rice varieties with resistance to diseases and pests suitable for rainfed rice fallows, increase on-station and on-farm R and D with short-duration rice varieties with short duration chickpea in different agro-ecological regions of Chhattisgarh and Madhya Pradesh. Develop low cost and effective insect pest/disease management technologies. Disseminate relevant information on different aspects of production, crop protection, soil and water conservation, markets and prices. Ensure sufficient regulatory and policy mechanisms to regulate role of private sector in seed and input marketing and delivery.

References

Ali M and Kumar, S. 2009. Major technological advances in pulses: Indian scenario. In: Milestones in Food Legumes Research (Ali M and Kumar S, eds.). Indian Institute of Pulses Research, Kanpur. pp: 1-20.

Angadi, S., B. McConkey, D. Ulrich, H. Cutforth, P. Miller, M. Entz, S.Brandt, and K. Volkmar. 1999. Developing viable cropping options for the semiarid prairies. Project Rep. Agric. Agri-Food Canada, Swift Current, SK.

Bantilan, M.C. S, and Parthasarathy, D. 1999. Efficiency a n d sustainability gains from adoption of short -duration pigeonpea in nonlegume -based cropping systems. (In En Summaries in EnFr.) Impact Series no. 5. Patancheru 502 324, Andhra Pradesh, India: International Crops Research Institute for the Semi - Arid Tropics. 28 p p. ISBN 92 - 9066 - 407 -X. Order code ISE 005.

Basu P S, Berger, J.D. Turner N.C., Chaturvedi,S.K., Ali Masood, and Siddique K H M 2007 a.Osmotic adjustment of chickpea (*Cicer arietinum*) is not associated with changes in carbohydrate composition or leaf gas exchange under drought. Annals of Applied Biology 150: 217-225 (Blackwell Publishers, U.K.)

Basu PS, Ali, Masood and Chaturvedi S. K. 2007 b.Osmotic adjustment increases water uptake, remobilization of assimilates and maintains photosynthesis in chickpea under drought. *Indian Journal of Experimental Biology (CSIR)* 45: 261-267.

Joshi PK, Birthal PS and Bourai V A. 2002. Socioeconomic constraints and opportunities in rainfed *rabi* cropping in rice fallow areas of India. Patancheru 502 324, Andhra Pradesh, India.

International Crops Research Institute for the Semi-Arid Tropics. 58 pp.

Kumar P. Joshi PK, Johansen C and Asokan M. 2000. Total factor productivity of rice-wheat cropping systems in India – The role of legumes. Pages 166-175 *in* Legumes in rice and wheat cropping systems of the Indo-Gangetic Plain – Constraints and opportunities (Johansen, C., Duxbury JM, Virmani SM, Gowda CLL, Pande S and Joshi PK eds.) Patancheru 502 324, Andhra Pradesh, India: International Crops Research Institute for the Semi-Arid Tropics; and Ithaca, New York, USA: Cornell University.

Lecoeur J, Wery J, Turc O. 1992. Osmotic adjustment as a mechanism of dehydration postponement in chickpea (*Cicer arietinum* L.) leaves. *Plant and Soil* 144: 177-189.

Passioura, J. B. (1977). Grain yield, harvest index, and water use of wheat. *J. Aust. Inst. Agric.Sci.* 43, 117–121.

Satyanarayana A, Murthy SS, Johansen C, Laxman Singh, Chauhan YS, and Kumar Rao JVDK. 1988. Introducing pigeonpea into rice-fallows of coastal Andhra Pradesh. *International Pigeonpea Newsletter* 7: 11-12.

Subbarao GV, Kumar Rao JVDK, Kumar J, Johansen C, Deb UK, Ahmed I, Krishna Rao MV, Venkataratnam L, Hebbar KR, Sai MVSR and Harris D. 2001. Spatial distribution and quantification of rice-fallows in South Asia - potential for legumes. Patancheru 502 324, Andhra Pradesh, India: International Crops Research Institute for the Semi-Arid Tropics. 316pp. ISBN 92-9066-436-3. Order code BOE029.

UNICEF. 2009. Tracking progress on child and maternal nutrition: A survival and development priority. New York.

WHO. 2010. Global database on child growth and malnutrition: India. http://www.who.int/nutgrowthdb/database/countries/who_standards/ind.pdf.

Recent Advances in Crop Physiology Vol. 2 (2016)   *Pages* **19–50**
Editor: **Dr. Amrit Lal Singh**
Published by: **DAYA PUBLISHING HOUSE, NEW DELHI**

Chapter 2

Role of Phosphorus Efficient Genotypes in Increasing Crop Production

B.C. Ajay[1]*, A.L. Singh[1], Narendra Kumar[1], M.C. Dagla[1], S.K. Bera[1] and R. Abdul Fiyaz[2]

[1]*ICAR-Directorate of Groundnut Research,*
PB 5, Junagadh – 362 001, Gujarat, India
[2]*ICAR-Directorate of Rice Research, Rajendranagar, Hyderabad*

1. Introduction

Today, agricultural sector supports food to nearly 7 billion people in the world; of this nearly 4.5 billion people are living in Asia where the food scarcity, which was repeatedly claimed due to increasing population, has been evaded by the tremendous progress of agricultural technology particularly in India and China during the past four decades (FAO, 2013; Singh, 2014). The world population is predicted to become nearly 10 billion in next 20-25 years, of this more than 60 per cent people will be living in Asia with only 50 per cent of the world production. Thus food shortage may become an important problem in future in Asia where optimization of mineral nutrition holds a key to optimize crop production (Singh and Mann, 2012).

Phosphorus (P) is one of the important element required for crop growth and development and is often applied in the form of fertilizers for obtaining high productivity. The phosphorus fertilizers is derived from inorganic minerals such as phosphate rock and around 90 per cent of the phosphate rock extracted globally is for

* *Corresponding Author:* E-mail: ajaygpb@nrcg.res.in

food production and the remainder is for industrial purposes (Jasinski, 2006). These phosphate rock reserves are distributed in very few countries around the world (Table 2.1). Morocco who controls western Sahara's reserves holds 75 per cent of worlds phosphate rock reserves. Importing Western Saharan P rock via Moroccan authorities is condemned by the UN and has recently been boycotted by several Scandinavian firms (Corell, 2002). Phosphate rock is a non-renewable resource that takes 10-15 million years to form from seabed to uplift and weathering, and current known reserves are likely to be depleted in 50-100 years (Jainski, 2006). It is expected that global P requirement will reach its peak by 2040 (Cordell *et al.*, 2009). Like oil and other natural resources P has no substitute in agriculture and as an element can't be manufactured or synthesized. Hence it becomes very important to conserve and efficiently use these limited natural resources.

Table 2.1: Phosphate Rock Reserve Estimates around the World

Country	Reserves (million metric tons)	Share (per cent) of World Total
Moroco and Western Sahara	50000	74.9
China	3700	5.5
Algeria	2200	3.3
Syria	1800	2.7
South Africa	1500	2.2
Jordan	1300	1.9
Russia	1300	1.9
United states	1100	1.6
Australia	870	1.3
Peru	820	1.2
Iraq	430	0.6
Brazil	270	0.4
Kazakhstan	260	0.4
Saudia Arabia	211	0.3
Israel	130	0.2
Egypt	100	0.1
Tunisia	100	0.1
Senegal	50	0.1
India	35	0.1
Other countries	582	0.9
World total	16758	

Data: U.S. Geological Survey, Mineral Commodity Summary 2014.

Several agronomic practices have been proposed for efficient phosphorus utilization in agriculture system but are cost affected. The development of P-efficient cultivars is regarded as efficient strategy to mitigate the problem of phosphorus limitation. Release of P-efficient genotypes for both high and low input farming systems would reduce the production cost of P fertilizer application in both acidic and calcareous soils (Singh and Basu, 2005a, b), minimize environmental pollution and contribute to the maintenance of P reserve globally (Cakmak, 2002; Vance *et al.*, 2003). Development of P-efficient genotypes with a great ability to grow and yield in -deficient soil is therefore an important goal in plant breeding (Rengel, 1999; Hash *et al.*, 2002; Wissuwa *et al.*, 2002; Yan *et al.*, 1992, 2004).

2. Role of P in Crop Production and its Widespread Deficiency

The phosphorus, known to be a constituent of nucleic acid, phytin and phospholipids in plants is the second most important nutrient for crop growth and development. It plays important role in energy storage and transfer within cells, speeds up root development, facilitates greater N uptake and results in higher grain protein yields. Phosphorus is essential for the formation of chlorophyll and absorption of potassium which is an essential part of life, cell division and development of meristematic tissues, helps in the seed development and maturity of plant and phosphorus has got specific role in nodule formation as a component of ATP and ADP involved in the energy transformations, driving most of the biochemical reactions including respiration and photosynthesis (Marschner, 1995; Singh, 2004; Singh and Basu, 2005a; Singh *et al.*, 2004). Thus high P supply is required for realising high yields (Clark, 1990; Singh and Basu, 2005a). However phosphorous availability in calcareous and low pH soils is very less as it forms complexes with calcium and aluminium making it unavailable to the plants (Singh, 2000, Singh *et al.*, 2004). Hence phosphorus is regarded as most limiting nutrient for plant growth. It is estimated that P availability to plant roots is limited in two thirds of the cultivated soil in the world (Batjes, 1997; Singh, 2004).

While studying the effects of P deficiency on plant growth in a wide range of species of various ecological habitats Atkinson (1973) described three common features of P-deficiency (a) Leaves are the first organs to be affected and their growth reduced most severely, (b) Root growth is the least affected, root/shoot ratio increased with time but increase is proportionate and greater in deficient plants, (c) Leaf development is delayed. Hence it is considered as one of the essential nutrient for plant growth and development.

In order to maintain P availability to the plants external application of phosphorus is required. But concerns are being expressed that due to limited P resources, lasting only a few decades, lack of P fertilizers may become a serious problem in the future (Mucchal *et al.*, 1996). Annually 17.5 million tons of phosphate rock is mined (Cordell *et al.*, 2009) however, these P reserves are finite resources and are concentrated in few places like Morocco and China which together hold 70 per cent of world's reserves (Rosamarin, 2004). Cordell *et al.* (2009) suggested that we may experience "peak" in P supply as early as 2033 a point at which global supply cannot keep pace with demand for P. As the demand increases cost of P fertiliser increase and in long term it affects sustainability and economic viability of agriculture. Thus, these finite P reserves should be used judiciously by adopting suitable crop husbandry practices and by developing P efficient crop varieties.

In most of the crops and in almost all soils right from acidic ferrosols to neutral to acidic andosols and in calcareous soils phosphorus deficiency is a widespread problem. Ferrosols with low pH and Andisols with wider range of pH exhibit a high P fixing capacity. In natural soils, the level of soluble orthophosphate (Pi) is often below many of the minor elements (Epstein, 1972). Tropical soils also frequently have high capacities for P fixation and may require heavy P fertilization to achieve economic yields. Excess of P fertilizer application not only increases the costs to farmers but also creates serious problem of nutrient pollution. Phosphorus is added to the soil as phosphatic fertilizer from where plants acquire their P in soluble ionic forms HPO_4^{2-} and $H_2PO_4^-$. The low P availability to leguminous crops, having nodules responsible for N fixation has a high P requirement (Vance, 2001). The fixation of P in the soil converts most often applied P into insoluble form. Some work has been done to understand the factor associated with P-efficiency in plants (Gerloff and Gabelman, 1983; Blair, 1993). Understanding the P-efficiency at physiological and molecular level (Goldstein, 1991) should assist in developing and refining selection criteria for plant improvement programs.

3. Defining P Efficiency

Several definitions have been proposed for nutrient use efficiency and accordingly criterions used by these definitions also vary. Moll *et al.* (1982) defined N and P use efficiency as grain yield per unit of nutrient supplied (from the soil and / or fertilizer). Fohse *et al.* (1998) defined P efficiency as the ability of plant to produce its certain percentage of its maximum yield at a certain level of soil P. *i.e.* P content in soil required to produce 80 per cent of maximum yield. These definitions may include absolute yield and amount of P absorbed under P–limited conditions, relative shoot dry weight, P acquisition and utilisation efficiency, rate of P absorbed per unit of root weight or root length, relative reduction in shoot dry weight etc. It is of the opinion that relative P-efficiency indices like relative shoot dry weight, would take into account both the acquisition and P utilisation efficiencies (Rengel, 1999).

3.1. P Stress Factor (PSF)

P-stress factor (PSF) is a tolerance index of cultivars under P-starvation. PSF indicates relative reduction in SDM due to P-stress and cultivars exhibiting low PSF values are considered more P-tolerant under P-deprivation (Akhtar *et al.*, 2007). P stress factor (PSF) takes into account shoot dry weight (SDW) under P-deficiency and p-sufficiency conditions and measures relative reduction (per cent) in SDW under P-deficiency conditions (Iqbal *et al.*, 2001). It determines the responsive and nonresponsive behaviour of a crop towards a nutrient. In general, varieties showing smaller PSF values are preferred in screening programs, because they show lesser decrease in SDW production with decreased nutrient supply in root medium (Iqbal *et al.*, 2001).

3.2. P-Efficiency (PE)

Similarly Sepher *et al.* (2009) used another relative P-efficiency (PE) index i.e relative shoot dry weight (shoot dry weight under P–/shoot dry weight under P+). Genotypes with PE value close unity are regarded as efficient.

3.3. Agronomic P Use Efficiency (APE)

Another relative P efficiency index used is "Agronomic P use efficiency (APE)" which is calculated as increase in yield per unit of added P fertilizer. APE will account both the acquisition and P utilisation efficiencies (Hammond *et al.*, 2009). Preferably reduction in yield per unit of P applied under P deficit conditions among genotypes should be low. Hence for a genotype to be efficient lower APE values are preferred.

3.4. P Use Efficiency (PUE)

It is defined as the kg of grain yield produced per kg of soil available P. PUE depends on ability of genotype to acquire P from soil and the way it is being utilised in the plant. Hence PUE is obtained by multiplying two variables *viz*. P acquisition efficiency (PAE) and P internal utilisation efficiency (PUTIL). PAE is defined as the amount of phosphorus accumulated in plant per unit of P available in soil. PUTIL is defined as the grain yield produced per unit of P in plant.

3.5. P Efficiency Index (PEI)

Pan *et al.* (2008) used principal component analysis to calculate PEI. This method not only simplifies parameters into several important principal components but also provides relative weights of different principal components. Higher the PEI more efficient is the genotype (Pan *et al.*, 2008).

The practices proposed for efficient phosphorus utilisation in agricultural system and development of P-efficient cultivars has been regarded as efficient strategy to mitigate the problem of P limitation. Considerable work has been done to understand the complex factor associated with P-efficiency in plants (Gerloff and Gabelman, 1983; Blair, 1993). Understanding P-efficiency at physiological and molecular level (Goldstein, 1991) should assist in developing and refining selection criteria for plant improvement programs. Various other definitions of P-efficiency are listed in Table 2.2. Ranking genotypes for nutrient efficiency can vary according to definition used (McLachlan, 1976; Blair and Cordero, 1978 and Blair, 1993).

3.6. P Efficiency and Response

Genotypes may be classified in two different ways based on efficiency of genotype (Efficient and In-efficient) and based on its response to applied fertilizer (responders and non-responders). Responsive plants would increase uptake and yield as nutrient supply increases whereas P-efficient plants would produce high yields at low levels of P (Randall, 1995). P-efficient plants depending on their responsiveness may or may-not respond to applied nutrients. Hence combining responsiveness and efficiency in one genotype becomes important. By combining both responsiveness and efficiency Blair (1993) proposed four response classes (Figure 2.1).

In order to identify suitable criteria to classify genotypes as "efficient or in-efficient" or "responders and non-responders" all available definitions of phosphorus use efficiency were classified into two groups: efficiency and responders. Definitions listed under "efficiency" will classify genotypes into "efficient" or "in-efficient".

Table 2.2: Definitions of Phosphorus Use Efficiency to Identify P Efficient Genotypes

Name	Abbreviation	Description	Formula	Authors
Phosphorus acquisition efficiency	PAE	kg P in the plant kg⁻¹ of soil available P	Pt/Ps	Parentoni and Junior, 2008
Phosphorus uptake efficiency	PUpE	P in plant/P in soil	Pt/Ps	Kakar et al., 2002
Root P absorption efficiency	RPAE	Capacity of root to absorb P from soil	Pt/RDW	Gerloff and Gabelman, 1983
Root efficiency ratio	RER	Shoot P uptake/Root dry matter	Pso/RDW	Jones et al., 1989
P utilisation efficiency	PUtE	Organ P content/Plant P content	Sto/Pt	Sepehr et al., 2009
P efficiency ratio	PER	Yield/amount of P in the plant (tissue P concentration)	Gr/Pt or Sto/Pt	White et al., 2005; White and Hammond, 2008
Relative P uptake	RP	Organ P content/Plant P content	Gr/Pt or Sto/Pt	Grant and Matthews, 1996
physiological P efficiency index	PPEI	Grain yield/Total P uptake	Gr/Pt	Yaseen and Malhi, 2009
P internal utilisation efficiency	PUtE	kg of grain produced per kg of P in the plant	Gr/Pt	Parentoni and Junior, 2008
Phosphorus utilisation efficiency	PUtE	Yield/P in plant	Gr/Pt	Kakar et al., 2001
		The efficiency of plant to utilize the absorbed P within the plant	(Gr+Sto)/Pt	Gerloff and Gabelman, 1983
Whole Phosphorus utilisation efficiency	WPUE	Rate of whole plant weight and total phosphorus content	(Gr+Sto)/Pt	Su et al., 2006
Phosphorus biological yield efficiency ratio	PBER	Biological yield/Total P uptake	(Gr+Sto)/Pt	Yaseen and Malhi, 2009
P efficiency ratio	PER	dry weight/P content	Sto/Pso or Gr/Pgo	Gerloff and Gablemen, 1983
Physiological P use efficiency	PPUE	Yield/tissue P concentration at a given P concentration	Gr/GPc	White et al., 2005; White and Hammond, 2008
Phosphorus efficiency ratio	PER	Shoot dry matter/shoot P content	Sto/Pso	Jones, 1989
Phosphorus utilisation index	PUI	Shoot dry matter/shoot P content	Sto/Pso	Siddiqi and Glass, 1981
Quotient of P utilisation	QUt	kg of grain dry matter per kg of P in the grain	Gr/Pg	Parentoni and Junior, 2008
Shoot Phosphorus utilisation efficiency	SPUE	Above ground biomass/total P content in shoot	Sto/Pso	Su et al., 2006
Phosphorus harvest index	PHI	P uptake in grain/Total P uptake * 100	Pgo/Pt	Jones et al., 1989
		Shoot P content/Total P content	Pso/Pt	Siddiqi and Glass, 1981

Contd...

Table 2.2–*Contd...*

Name	Abbreviation	Description	Formula	Authors
Phosphorus use efficiency	PUE	Gr/ps Yield/P in soil	Gr/Ps Gr/Ps	Parentoni and Junior, 2008 Kakar et al., 2001
Fertilizer P uptake efficiency	FPUpE	(P in fertilised plants – P in control plants)/P applied	$[(Pt_{high} - Pt_{low})/$P applied$]$	Kakar et al. 2002
P acquisition efficiency	PACE		Pso_{low}/Pso_{high}	Sepehr et al., 2009
P uptake efficiency	PUpE	Increase in plant P content per unit of added P fertilizer	$(Pgo_{high} - Pgo_{low})/\Delta Papp$ or $(Pso_{high} - Pso_{low})/\Delta Papp$	White et al., 2005; White and Hammond, 2008
P response efficiency	PRE	Yield responses/difference between amounts of P supplied	$(Sto_{high} - Sto_{low})/\Delta Papp$ or $(Gr_{high} - Gr_{low})/\Delta Papp$	Pang et al., 2010
P efficiency	PE	Sto_{low}/Sto_{high}	Sto_{low}/Sto_{high}	Sepehr et al., 2009
Fertilizer P Utilisation efficiency	FPUtE	$[(Yield_{high} - yield_{low})/(Pt_{high} - Pt_{low})]$	$[(Gr_{high} - Gr_{low})/(Pt_{high} - Pt_{low})]$	Kakar et al., 2002
P utilization efficiency	PUtE	Increase in yield per unit increase in plant P content	$(Sto_{high} - Sto_{low})/(Pso_{high} - Pso_{low})$	White and Hammond, 2008
Fertilizer P Use efficiency	FPUE	$[(Yield_{high} - yield_{low})/$P applied$]$	$[(Yield_{high} - yield_{low})/$P app$]$	Kakar et al., 2002
Agronomic P use efficiency	APE	Increase in yield per unit of added P fertilizer	$(Sto_{high} - Sto_{low})/\Delta Papp$ or $(Gr_{high} - Gr_{low})/\Delta Papp$	White et al., 2005; White and Hammond, 2008
P efficiency index	PEI	$PEI = \sum_{i=1}^{n} PCi \times RWi$		Pan et al., 2008

Papp: Applied phosphorus; Pgo: Grain/kernel P content; Pso: Shoot P content; Pgc: Grain/kernel P concentration; Pt: Total P content; Ps: P concentration in soil; Gr: Grain/kernel dry weight; Sto: Shoot dry weight; RDW: Root dry weight.

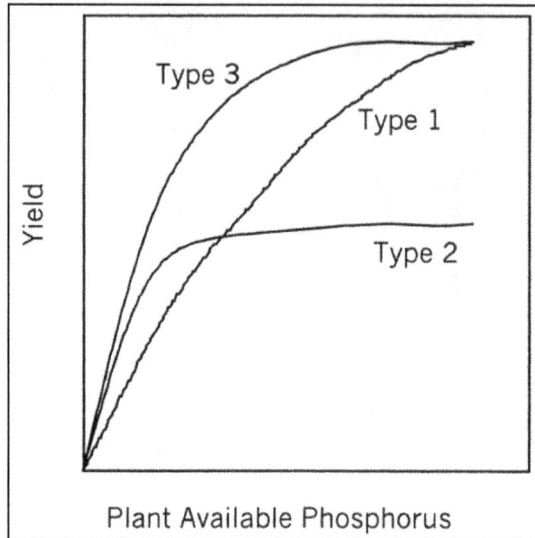

Figure 2.1: Four Classes in Response to P (Blair, 1993): 1) Inefficient responder (type 1); Efficient non-responder (type 2); 3) Efficient responder (type 3); and 4) Inefficient non-responder (not shown).

Type I- Inefficient responder: These genotypes give low yield when nutrient availability is less, but increases their yield as the nutrient availability increases. Thus their ability to respond to applied fertilizer could be used in breeding to develop new efficient responders.

Type II- Efficient non-responder: These genotypes are capable of giving high yield even when nutrient availability is less, but do not respond with increased yield under high input conditions. These can be used in breeding to develop new efficient responding lines.

Type III- Efficient responders: These genotypes show high yield at low level of nutrient supply and their yield level increases as nutrient supply increases. Identifying efficient responders would be an ideal breeding programme for nutrient use efficiency.

Type IV- Inefficient non-responders: These genotypes give low yield irrespective of nutrient availability.

3.7 P Uptake and Utilisation Efficiencies

In order to understand superior performance of efficient genotypes over inefficient genotypes it is important to study their P-uptake and P-utilisation separately. Hence, Parentoni and Junior (2008) obtained P use efficiency by multiplying the means of "uptake" and "utilisation". To study P-uptake pattern four definitions are available which consider rate of P-content of whole plant or P- content in shoot over P available from soil or root dry weight. To further complicate the matter four different acronyms are used for each definition. The P-utilisation pattern explained using 15 different types of definitions in literature, accordingly the criterion used to explain utilisation efficiency also varies and ranking of genotypes for nutrient efficiency also varies according to definition used (McLachlan, 1976; Blair and Cordero, 1978 and Blair, 1993). Thus there is a need of common universal definition to explain utilisation pattern of P in plant.

Most of the available definitions of P-utilisation consider either grain yield or shoot weight over P-content of whole plant. Few other definitions consider grain yield or shoot weight over P content in grain or shoot respectively to explain P-utilisation efficiency. Three definitions proposed by Gerloff and Gabelman, 1983, Su *et al.* (2006) and Yaseen and Malhi (2009) consider both grain yield and shoot biomass weight over P-content of whole plant but with different acronyms (*viz.* PUE, WPUE and PBER). Rose and Wissuwa (2012) are of the view that for valid comparison of genotypes and to improve physiological mechanisms, internal P-use efficiency should be defined as biomass produced per unit of P accumulated in tissue and further this PUE has to be dissected into components such as "shoot PUE", "root PUE" and "grain PUE". "Root PUE" has been defined as P accumulation per unit of dry matter which may be misleading because P uptake from soil depends on volume of soil explored which in turn depends on root length and its surface area. Hence "root PUE" should be interpreted as root surface area per mg P (Rose and Matthews, 2012). In maize P-efficient genotype had lower root P concentration and high root surface area than P in-efficient genotype. Definitions proposed by Gerloff and Gabelman (1983), Su *et al.* (2006) and Yaseen and Malhi (2009) consider only grain and shoot weight and lack root surface area for calculating "root PUE". Adding root surface area to the above proposed formula PUtE can be written as:

PUtE = (Gr+Sto+RSA)/Pt

To study the physiological mechanisms of PUE this formula could be dissected out as:

PUtEg = Gr/Pt (P utilisation efficiency of grain)

PUtEs = Sto/Pt (P utilisation efficiency of shoot)

PUtEr = RSA/Pt (P utilisation efficiency of root), RSA is root surface area.

PUtE = PUtEg+ PUtEs+ PUtEr

Parentoni and Junior (2008) obtained PUE by multiplying the means of PUtE and PUpE. PUE = PUpE * PUtE

= (Pt/Ps) * (Gr+Sto+RSA)/Pt

PUE = (Gr+Sto+RSA)/Ps

This definition would facilitate in improving PUE of shoot, grain and root individually. Interpreting root PUE on the basis of root biomass per unit of P may be misleading because soil explored by roots for P acquisition depends on root length (or surface area) rather than on root biomass (Rose and Matthews, 2012).

In order to identify genotypes responsive to fertilizer application i.e "responders" and "non-responders" several definitions are available. Classification of genotypes as efficient and responders using formulas proposed here would help in identifying four classes of genotypes as proposed by Blair (1993).

As there were no well-defined selection criteria for P-efficiency in groundnut, an effort was made in our laboratory (Singh and Basu 2005b) by growing the crop under control and P fertilized condition and recording the relative pod and haulm yields (RPY, RHY) and relative P uptake by groundnut genotypes calculated as:

Relative yield = 100 x Yield in P-unfertilized plot/Yield of P-fertilized plot.

The groundnut genotypes were sorted based on their high pod and haulm yields, per cent P contents in leaves at 60 DAE, P uptake by Pod, total P uptake by Plant, both under P-fertilized and P-unfertilized conditions, separately as well as combined. The genotypes having high values of these parameters were categorized as P-efficient and the one having low values were categorized as the P-inefficient one.

The data on various parameters, shows genotypic differences under both P-fertilized and P-unfertilized conditions. These genotypic differences were more pronounced on pod yields, P concentrations and uptake. With the data on these parameters, when the groundnut genotypes were arranged in the descending order, no single genotypes could top the list on the basis of all these parameters. However, certain genotypes were common in the top ten in most of the parameters assessed and showed their higher values. In a similar fashion a few genotypes were common in bottom ten showing lower values of these parameters. Accordingly, for demarcating P-efficient and P-inefficient genotypes, the average values of these parameters from top 20, and bottom 20, were taken into consideration. However such study need to be conducted over the years and, only the genotypes fulfilling majority of the criteria during most of the years need to be categorized as P-efficient and P-inefficient.

Among the various parameters, high pod yield followed by high P uptake were the most important for identifying P-efficient genotypes, the relative pod yield and relative P-uptake further strengthen these parameters (Singh, 2004; Singh and Basu, 2005a, b).

4. Mechanisms of PUE

The P-efficient plants can employ a number of potential adaptive mechanisms for better growth on low-P soils.

4.1 Root Morphology and Architecture

Root hair formation, growth of primary root and lateral root formation are particularly sensitive to changes in the internal and external concentration of nutrients increases absorptive area and soil volume explored. Wang et al. (2004) found that root hair density, average root hair length and root hair length per unit root, varied among different genetic materials and that these variations were highly associated with P status.

Root architecture: Indicates the extent to which soil volume is explored and includes, lateral root branching, length and growth angle of basal roots and root growth plasticity. Plants with shallow root architecture have higher P-efficiency attributed to higher nutrient availability in topsoil.

The high P availability in top soil causes shallower growth angles of axial roots, enhanced adventitious rooting, and greater dispersion of lateral roots are associated with foraging of P from top soil and thus P acquisition. Variation in root growth angle among bean contributed 600 per cent increase in P acquisition and 300 per cent increase in yield (Bonser et al., 1996; Liao et al., 2001). The root growth angle (RGA) is influenced by basal roots which appear in distinct nodes/whorls. Basal root whorl number (BRWN) varies among genotypes from one to four (16 basal roots). Shallow basal RGA are found in topmost whorls whereas lower whorls produce steeper basal RGA. The RGA has been successfully used for breeding varieties for low fertility soils (Lynch, 2007). In dicots adventitious roots grow from subterranean portion of hypocotyls horizontally through the top soil and are associated with P acquisition in low P soils. Metabolic cost of soil exploration by these roots is also less.

4.2 Symbiosis

4.2.1. Rhizobium

P addition has considerable impact on rhizobium symbiosis and biological N_2 fixation by increasing nodule formation and nitrogenase activity on the upper parts of the roots (Kuang *et al.*, 2005). The shallow root systems increase P-uptake efficiency and facilitate biological N_2 fixation. Improved N status with resulting enhanced root growth might be the mechanism by which soybean P uptake was increased in plants inoculated with the effective rhizobium strains on low-P acid soils.

4.2.2. Azolla and Blue Green Algae

The blue green algae (BGA) inside Azolla fixes atmospheric nitrogen in symbiotic association. Azolla encompass a BGA *Anabaena azollae* inside its leave where fertilizer P application either as foliar spray or in split doses increases the growth of azolla and BGA and nitrogen fixation and finally the growth and productivity of rice and fertility of rice field (Singh and Singh, 1990; Singh *et al.*, 1988).

4.2.3. Mycorrhizae

Mycorrhizal fungi can increase phosphorus availability by exudating various organic acids themselves, freeing phosphates in the same manner as those exuded from plant roots. Under low P conditions, plants often have higher mycorrhizal infection rate and contribute more to P uptake (Singh and Chaudhari,1996). Plant growth response to arbuscular mycorrhizae (AM) associations (*i.e.* the 'mycorrhizal growth response', MGR) varies widely among plant species and even varieties.

Colonization by these beneficial fungi improved access of phosphorus by extending the crop's root system with mycorrhizal hyphae (Bucher, 1971), indirectly increasing the root surface area for nutrient absorption and crop growth. Mycorrhizal hyphae work to improve nutrient acquisition by increasing their affinity for phosphorus ions and decreasing the concentration gradient required for more energy efficient absorption (Shenoy and Kalagudi, 2005). Additionally, biodiversity of AM fungi is greater in low-input production systems compared to high-input, likely due to the availability of nutrients making microbial symbiotic relationships obsolete and energy expensive to the crop (Oehl *et al.*, 2004).

4.2.4. Root Exudates

Root apices exude a variety of organic acids, which can influence plant nutrition and provide an easily degradable nutrient source for soil microorganisms (Rengel and Marscner, 2005). The roots under phosphorus deficiency exude citrate, malate, and oxalate organic acids which are the most effective at mobilizing soil phosphorus

(Hinsinger, 2001; Ryan *et al.*, 2001). These organic acids release unavailable phosphorus from bound minerals, allowing for the chelation of Al^{3+}, Fe^{3+}, and Ca^{2+} consequently freeing phosphorus and helping to alleviate P stress (Marschner, 1995; Singh, 2000; 2008). Differences in the exudation of organic acids can be seen between crops under P-deficiency or not (Neumann and Romheld, 1999; Yan *et al.*, 2002), In addition to improving access of previously unavailable phosphate via rhizosphere acidification, exuded carboxylates promote microbial growth, and potentially exploit beneficial microbial relationships that correlate with P bioavailability (Rengel and Marscner, 2005). Thus beneficial relationships between crops and mycorrhizal fungi improve availability and uptake of phosphorus (Li *et al.*, 2010).

4.3 Activation of High-Affinity Phosphate (Pi) Transporters

The inorganic phosphate (Pi) concentration within the plant cells is approximately >10 mM (>3 g kg^{-1}, on a biomass dry-weight basis), and yet the concentration in the soil solution is typically <10 μM (Bieleski, 1973; Marschner, 1995). Due to low concentration of soluble form of P and slow rate of diffusion plants have evolved several mechanisms to increase Pi uptake from soil and among them high-affinity Pi transporters (PTs) are assumed to be the predominant role in Pi acquisition by plant roots (Marschner, 1995; Raghothama, 1999). The genes encoding these PTs were first identified in Arabidopsis (Muchhal *et al.*, 1996) followed by identification of similar such genes from other plant species including cereals, legumes and solanaceous species (Chen *et al.*, 2007; Chiou *et al.*, 2001; Glassop *et al.*, 2005; Harrison *et al.*, 2002; Javot *et al.*, 2007a; Leggewie *et al.*, 1997; Liu *et al.*, 1998a,b; Maeda *et al.*, 2006; Mitsukawa *et al.*, 1997; Mudge *et al.*, 2002; Nagy *et al.*, 2005; Paszkowski *et al.*, 2002; Rae *et al.*, 2003; Smith *et al.*, 1997; Xu *et al.*, 2007).

Most of these plant PTs belong to Pht1 family of genes and are believed to be made of 12 transmembrane (TM) domains (Saier, 2000), containing two partially duplicated subdomains of six TM segments (Lagerstedt *et al.*, 2004). Most of these genes expressed predominantly in roots induced by low-Pi supply or by AM fungi (Bucher, 2007) and few of them also expressed in other plant parts such as stem, leaves, cotyledons, tubers and flowers (Karthikeyan *et al.*, 2002; Mudge *et al.*, 2002). Analysis of the Arabidopsis genome has revealed that there are nine genes in the Pht1 family (Mudge *et al.*, 2002), whereas in barley at least eight members have been identified (Rae *et al.*, 2003).

In vascular plants, at least two forms of PTs are known and they are classified based on the Pi absorption kinetics and affinity to target Pi *i.e.*, high-affinity PTs, Km (Pi) = 3–7 μM; low-affinity PTs Km (Pi) = 50–330 μM (Furihata *et al.*, 1992; McPharlin and Bieleski, 1987; Ullrich-Eberius *et al.*, 1984). The high- and low-affinity PTs belong to Pht1 and Pht2 families, respectively (Bucher *et al.*, 2001). The members of Pht1 family are induced under P deficiency often exclusively in the root (Daram *et al.*, 1998; Liu *et al.*, 1998b; Rae *et al.*, 2003). On the contrary, the members of Pht2 family are mostly expressed constitutively in the aerial parts of the plant (Daram *et al.*, 1999; Rae *et al.*, 2003). High-affinity PTs are involved in regulating Pi uptake and transcriptional control of PTs activity (Muchhal and Raghothama, 1999; Raghothama and Karthikeyan, 2005) and post-transcriptional regulatory mechanisms (Bucher *et al.*, 2001). Hence, high-affinity PTs have been suggested as potential targets for improving

Pi uptake (Mitsukawa *et al.*, 1997; Rae *et al.*, 2003; Vance *et al.*, 2003). Studies have indicated that plasma membrane H^+-ATPase is also involved in P uptake (Shen *et al.*, 2006).

4.4. Secretion of Organic Acids and Phosphatases into the Rhizosphere

A major portion of Pi in soil may be present in organic forms. Organic P complexes such as phytic acid may contribute to significant portions (20–80 per cent) of P in soil (Jungk *et al.*, 1993; Richardson, 1994). The organic P complexes need to be broken down by enzymatic activity before the inorganic Pi is released into the rhizosphere (Raghothama and Karthikeyan, 2005). Inoculation of food crops with plant growth-promoting rhizobacteria (PGPR) or mycorrhizae can directly increase plant available P via mechanisms of solubilization and mineralization of fixed P from inorganic and organic forms (Rengel and Marschner, 2005; Hodge *et al.*, 2009). Mechanisms include the release of organic acids, protons and phosphatases into the rhizosphere. Bacteria from the genera Pseudomonas and Bacillus and fungi, primarily Penicillum and Aspergillus are among the most powerful P solubilizers.

Another mechanism which indirectly leads to increased P acquisition by plants is the production of phytohormones (mainly auxins) by rhizobacteria that stimulate root growth (Richardson, 2001; Jacobsen *et al.*, 2005; Richardson *et al.*, 2009a). Inoculation with Azospirillum, known to produce substantial amounts of indole-3-acetic-acid (IAA), increases the length and density of root hairs as well as the appearance and elongation rates of lateral roots in many plant species (Fallik *et al.*, 1994) which increases the surface area for absorption of P.

5. Strategies to Improve P Use Efficiency

In order to improve phosphorus use efficiency (PUE) it is important to understand the source through which P is obtained, its movement in the soil and different ways in which it is deposited in the soil. When the fertilizer is added, phosphate ions are readily available as plant-available P in soil solution. These Pi ions may get adsorbed on to the soil surface or may be lost through leaching. Phosphate ions after entering plant and animal system get exported in different forms. In order to maintain yield levels, it is important that, P removed from the soil is replaced back to maintain the balance. Simpson *et al.* (2011) gave P balance efficiency formula as,

$$P_{fertilizer} = P_{export} + P_{erosion/leaching} + P_{waste\ disposal} + P_{soil\ accum}$$

where,

P_{export} = removal of P in products, $P_{erosion/leaching}$ = P lost by leaching, runoff o r soil movement, $P_{waste\ disposal}$ = P accumulated in small areas of farms as a result of uneven dispersal of animal excreta rendering the P less available and $P_{soil\ accum}$ = P accumulating as sparingly-available phosphate or organic P compounds that are slowly mineralised.

5.1. Improving P Use Efficiency by Minimizing Losses

P fertilizer efficiency can only be achieved when P loss through erosion is reduced, uniform distribution of excreta reducing P fixation in soils. This could be achieved by interventions of fertilizer, agronomic, microbial and plant based technologies.

5.1.1. P Export Loss through Farm Products

The amount of P applied to soil can be reduced by modified deliver products with low P content leading to low P export. In this system P absorbed from soil is less; hence quantity of P fertilizer applied may also be reduced. This is an ideal system for low P fixing soils but leads to accumulation of applied P in moderate to high P fixing soils. Hence in high P fixing soils if P loss by export is reduced it is counteracted by accumulation of unavailable forms of P in soils.

5.1.2. P Loss through Erosion, Runoff and Leaching

For P to cause an environmental problem there must be a source of P (*i.e.*, high soil levels, manure or fertilizer applications, etc.) and P must be transported to a sensitive location (*i.e.*, leaching, runoff, erosion, etc.) (Gburek *et al.*, 2000). A high P source with little opportunity for transport, while it may be a waste of a resource, may not constitute an environmental threat. Likewise, a situation where there is a high potential for transport, but no source of P to move, is also of little threat.

P losses from agricultural fields vary with soil type. In soils with high P fixing ability loss is only 0.4 -5 per cent of applied P, (McCaskill and Cayley, 2000; Ridley *et al.*, 2003; Melland *et al.*, 2008), but on soils with low P fixing capacity losses could range from 40 -90 per cent of applied P (Ozanne *et al.*, 1961; Lewis *et al.*, 1987). P lost from agricultural fields is entering water bodies and is causing serious environmental problems. In agricultural fields P is lost through erosion; runoff and leaching which could be avoided to improve P efficiency using following strategies (Chambers *et al.*, 2000; Uusi-Kamppa *et al.*, 2000):

- ☆ Appropriate forms and placement of obilized,
- ☆ Appropriate timing of obilized in relation to rain and crop growth,
- ☆ Use of soil amendments to reduce nutrient transport,
- ☆ Use of buffer strips and fencing of waterways to capture obilized nutrients and avoid direct contamination,
- ☆ Location of field access points,
- ☆ Attention to cultivation methods
- ☆ Use of minimum tillage and attention to ground cover
- ☆ Conservation tillage and crop residue management
- ☆ Terracing
- ☆ Contour tillage
- ☆ Cover crops

5.2 Agronomic Interventions to Improve P Availability to Plants

5.2.1. Microbial Activity and Organic P Cycling

The rate of mineralization of fixed form of P in soil increases the P balance efficiency of our farming system. Organic manure increases microbial activity in soils. This inturn will increase organic P mineralization in soil. But recovery of P from organic manure was affected when plant available P is high in soil solution. Low uptake of residual P from manure also indicates that, yield response of plant is due to the use of residual P reserves of soil than plant available residual P.

Organic P mineralisation and microbial P pool could be enhanced under different farming systems. Grass-legume and grass only pastures have demonstrated that organic P and microbial activity can increase plant available P in soils (Oberson *et al.*, 1999). Thus managing interactions of residues with soil will also slow down P sorption reaction and P held in microbial biomass is also protected temporarily from absorption.

5.2.2. Minimising Inefficiencies and Constraints Associated with Yield

Reducing Constraints to Yield

P availability also influences yield indirectly through various yield constraints like root growth affected by soil acidity, root diseases, soil compaction and others which in turn affect nutrient uptake and finally yield. Alleviating these constraints will improve P export and P balance efficiency. In general, P fertilizers are applied in excess of requirement to overcome several other constraints and relieving these constraints can increase are relieved P-use efficiency. For example in acid soils when Aluminium (Al^{3+}) concentration in soil solution increases it will affect root elongation and which inturn affects water and nutrient uptake and finally P uptake. The recovery or amelioration of Al-toxicity through the use of lime increased the P availability.

Targeted Use of P-Fertilizer

In efficient farming systems there is accumulation of sparingly available phosphate, losses due to leaching, erosion and runoff, and accumulations due to uneven distribution of excreta will continue to increase beyond critical P level. In these farming systems when soil P is in excess of critical P, lower P inputs and short term cessation of fertilizer is an viable option (Butkitt *et al.*, 2010).

Uneven Dispersal of Excreta

Grazing animals deposit excreta in camps, in shade or close to water bodies and thus results in poor P balance efficiency. For example sheep on low slope areas deposits 25 to 47 per cent of dung in 5 to 15 per cent of area. Over long period of time this uneven disposal of excreta may lead to accumulation of P in small areas and thus making it unavailable for plant growth.

Precision Agriculture

Inefficient P utilisation can also occur due to productivity gradients which may arise due to topography, botanical composition, soil type and depth even when fertiliser is applied uniformly. As the price of fertiliser increases its application as per the fertility or productivity gradient is expected to give economic benefits (Hackney, 2009). Differential fertilizer application would be easy in areas where there are large and easily identified differences, but complex productivity patterns can be identified using ground level, airborne or space borne canopy reflectance sensing devices (Trotter *et al.*, 2010). GPS technology can also be used to develop productivity maps for variable rate fertiliser application.

These agronomic practices would help avoid inefficient use of P-fertilizer and bring farming systems to P balance efficiency. The productive farming systems under low soil P concentrations would be economically efficient and environment friendly also.

5.3 Genetic Interventions to Improve P Availability to Plants

The P uptake has been found to be multigenic with involvement of additive, dominance and epistatic effects (Duncan and Carrow, 1999). Several genes for P uptake and transportation have been identified, cloned and characterized. QTL mapping has been well conducted with P efficiency in Arabidopsis (Ma *et al.*, 2001), rice (Shimizu *et al.*, 2004), maize (Zhu *et al.*, 2006), wheat (Su *et al.*, 2006) and common bean (Beebe *et al.*, 2006). Expression of genes imparting tolerance to Pi starvation have been identified in soybean (Guo, 2008), rice (Wissuwa, 2005), Arabidopsis (Hammond *et al.*, 2003) and in other crops.

When external-P level drops low to micro-molar concentration, the high affinity transporter mRNA transcripts in roots increase leading to enhanced capacity of roots for P uptake. The inorganic P starvation is known to enhance synthesis of these carrier systems, resulting in better PUE (Duncan and Carrow, 1999). The high-affinity transporters are expressed in the cells in close contact with soil solution (*e.g.* epidermal cells with their associated root hairs and outer layer of cortex) and play an important role in acquisition of P. The low-affinity transporters are active in vascular loading and unloading, *i.e.*, in internal distribution and re-mobilization of acquired P in millimolar concentration range (Smith, 2001). The *Pht1* and *Pht2* families of genes are the two well characterized gene families of P transporters in plants. Three major classes of P transporters with partially overlapping specificities and genes for them have been identified (Duncan and Carrow, 1999). Genes for the four major P metabolic enzymes have been identified. The cytosolic GAPDH is coded by the nuclear gene *GapC*, whereas the chloroplastic GAPDH is encoded by the nuclear genes *GapA* and *GapB*. The nuclear *GapN* encodes the cytosolic GAPDHN (Valverde *et al.*, 1999).

5.4. Low P-Farming Systems

5.4.1. Prior Fertilizer Application

Prior application of phosphate lowers the P sorption capacity of soil and increases the availability of subsequent fertilizer application (Bolland and Baker, 1998; Bolland and Allen, 2003; Burkitt *et al.*, 2008). This has got significance in soils where large quantity of fertilizer is added to soil. Bolland and Becker (1998) in a study applied P fertilizer at different rates starting from 0 kg/ha to 599 kg/ha only once 20 years ago and after 20 years he applied fresh fertilizer at different rates for wheat. He observed that the initial P-fertilizer applied increased effectiveness of freshly applied fertilizer in increasing yield.

5.4.2. P mobilisation by Soil Microorganisms

Microorganisms can enhance P mobilisation capacity of plants by increased root growth, alteration in sorption equilibria: Results in release of orthophosphate ion into soil solution and increase mobility of organic P and solubilising and mineralisation of sparingly available forms of P (both organic and inorganic P). Microorganisms decompose organic matter in soil, mineralise organic P and later are incorporated into microbial biomass. Increased microbial activity is observed as P in soil solution decreases (when P is not added or P limiting soils or when P is added) when organic matter (C and N) is added. Hence microorganisms also compete with plants for available P from soil solution and make it temporarily unavailable. This is an important mechanism for regulating P supply as it avoids reaction of P in soil solution with soil particles.

Mineralization of Organic P

Organic P has to be mineralised by phosphatases (plant or microbial origin) before it can be utilised by microorganisms or plants. Phosphatase activity increases when there is a deficiency of P as a part of P starvation response. When soil suspensions were treated with phosphatases orthophosphate was released (George *et al.*, 2007). Bunemann (2008) reported that 60 per cent of the total organic P may be hydrolyzed by phosphatases with highest amounts being released by phytases (monoester phosphatases active against phytate). Grasses and Legumes grown in media inoculated with soil microorganisms showed increased utilisation of phytate P (Richardson *et al.*, 2001). Genetically modified plant with their ability to release extracellular fungal phytase was able to acquire P directly from phytate. The organic acids chelate cations like Fe and Al bound to P, compete with P for reaction sites on cations and thus release P. Thus organic acids also help by preventing Al from entering into plant.

Solubilisation of Inorganic P

The bacteria (*e.g. Actinomycetes, Pseudomonas*, and *Bacillus* spp.) and fungi (*e.g. Aspergillus* and *Penicillium* spp.) acidify growth media, release organic anions like citrate, gluconate, oxalate and succinate. Amount of P solublised depend on type of inorganic P present like Ca, iron, aluminium phosphates and other sources of rock phosphate. Inoculation of plants with P solublising microorganisms' results in improved growth and P nutrition. These P solublisers include *Pseudomonas, Bacillus, Penicillum* and *Aspergillus* (Richardson *et al.*, 2009). The plant growth promoting rhizobacteria (PGPR) or mycorrhizae solublise and mineralise fixed form of phosphorus by releasing organic acids or phosphatases. The Azospirillum produces phytoharmones like auxin (Indole – 3 acetic acid) which increases root hair length and density (Fallik *et al.*, 1994).

5.4.3. Slow Release Fertilizer

When plants efficiency for P uptake is increased it would lower total soil P content. To avoid total exhaustion of P from soil slow release fertilizer like rock phosphate may be added. In low P fixing soils this strategy prevents leaching loss where as in P fixing soils it will prevent P fixation. Hence slow release fertiliser along with P use efficient lines will ensure that added P-fertilizer is utilised efficiently.

5.4.4. Application of Silica

Application of Silica or silicate in the form of rice-husk ash or calcium silicate increase P availability for the plant by competing with phosphate ion for adsorption sites within the soils. But some soils may not respond to silicate application further silicon sources must be water soluble.

5.4.5. Space and Timing of Fertiliser Application

When Plant available P is high in soil, then external application of P may not have any effect on yield. P can also be applied when soil is near to critical soil P level.

5.4.6. Application of Fertilizer Close to Seeds

Small quantity of fertilizer may be placed close to seed instead of broadcasting without affecting yield (Van der Eijk, 2006). Since applied phosphorus is in the vicinity of seedlings major quantity is available for plants uptake and only smaller quantity is fixed in the soil. This will also increase P-use efficiency.

6. Phosphorus Use Efficient Varieties

Several Phosphorus use efficient varieties have been released in several crops globally and examples of such varieties are listed in Table 2.3.

Table 2.3: Phosphorus Use Efficient Cultivars/Breeding Materials Reported in Various Crops

Sl.No.	Crop	Varieties	Reference
1	Wheat	Lovrin	Su *et al.*, 2006
		81 (85) 5-3-3-3, Ji 87-4617	Chun-Jian *et al.*, 2003
		rye Bevy, rye PC00361	Osborne and Rengel, 2002a
		Egret and Durati	Osborne and Rengel, 2002b
		BR10, CPAC89128, and NL459	Fageria and Baligar, 2008
2.	Common bean	G19833	Yan *et al.*, 1995a,b
		Milenio, BAT477, and A785	Mourice and Tryphone, 2009
		BAT477	L'taief *et al.*, 2012
3.	*Brassica napus*	Eyou Changjia.	Yang *et al.*, 2011
		RIL 102	Yao *et al.*, 2011
4.	Maize	Mo17	Kaeppler *et al.*, 2000
		082	Chen *et al.*, 2008
		NY821	Reiter *et al.*, 1991
		Mutant 99038	Li *et al.*, 2007b
		L3, 228-3	Parentoni *et al.*, 2008, 2010
		CB-2, DP x Tromba, HV313 x DEM, Macho III-04, and CIMMYT-1	Bayuelo-Jiménez and Ochoa-Cadavid, 2014
5.	Rice	Kasalath	Wissuwa and Ae, 2001
		Zhangzao 18	Li *et al.*, 2010
		99112	Zai-Hua *et al.*, 2006
		BRA032048, BRA042094, BRA02601, BRA032051, RA032033, BRA052015, BRA042156, BRA01600, BRA01506, BRA052023 and BRA042160	Fageria *et al.*, 2014

Contd...

Table 2.3–*Contd...*

Sl.No.	Crop	Varieties	Reference
6.	Arabidopsis	C24, Co, Cal	Narang *et al.*, 2000
7.	Soybean	Nannong 94-156	Zhang *et al.*, 2009
		BX10	Zhao *et al.*, 2004
		IAC-1, IAC-2, IAC-4, IAC-5, IAC-6, IAC-9, Sta. Rosa and UFV-1	Furlani *et al.*, 2002
8.	Groundnut	SAMNUT 10 and 21	Gabasawa and Yusuf, 2013
		ICGV 86590, ICG 14475, Mutant 68, ICGV 92188	Amit *et al.*, 2009
		GG5, FeESG 10, SP 250A	Singh and Basu, 2005
9.	Cowpea	IT90K-277-2	Singh, 1999
10.	Potato	CGN 17903 and CIP 384321.3	Balemi and Schenk, 2009
11.	Black gram	DBS-7, DBS-13	Shridevi *et al.*, 2009.

Conclusion and Future Prospects for PUE

The P deficiency is most wide spread problem of most of the crop worldwide, where fertilization is of utmost important. However the world P reserves are limited and shrinking. Under such circumstances the role of P efficient genotypes is of utmost importance. All efforts are needed to find out the P efficient and P responsive genotypes for each and every crop.

The nutrient efficiency is the ability of a system to convert inputs into desired outputs, or to minimize the conversion of inputs into waste. The supply or availability of the mineral nutrient is the input and plant growth and yield are the outputs. Thus the efficiency is the relationship of output to inputs and expressed as simple ratio, such as kg yield per kg fertilizer or kg dry weight per g of nutrient supply. However this efficiency depends upon the uptake efficiency (uptake of nutrient per unit supply of nutrients) and utilization efficiency (dry matter production per unit of nutrient taken up). The nutrient responsiveness is the capacity of a plant to increase uptake and yield as nutrient supply increases. The responsive plants are most desirable in fertilized high-input systems, while the nutrient efficient plants, which produce high yields at low levels of nutrients, are most valuable in low-fertility situations.

Breeding for P use efficient lines differs between low P soils and high P fixing soils. In low input agro-ecosystem or in soils with low P content where erosion losses are more, phosphorus availability is more in top soils. Hence in these conditions cultivars with shallow roots would cover the soil surface enabling it to use the available phosphorus and reduce its loss through erosion. Also, plants' internal P use efficiency also can be improved which is defined as capacity to produce a large amount of organic matter per unit of P taken up. These P efficient plants show higher growth potential under same amount of P added.

Luckly in India the peanut varieties released through multi-locational testing under All India Coordinated Research Project on Groundnut (AICRP-G) are as P-efficient as phosphorus fertility status of the most of the soils where these are evaluated are moderate in majority of states. However, if the responsiveness and efficiency are combined in one genotype through breeding program or achieve through natural selection, it is the best and this is high time to put efforts in this direction in the present day modern agriculture.

References

Amit K, Kusuma P, Gowda MVC. 2009. Genotypic variation for root traits in groundnut germplasm under phosphorus stress conditions. *SAT ejournal* 7: 1-4.

Araújo AP, Plassard C, Drevon JJ. 2008. Phosphatase and phytase activities in nodules of common bean genotypes at different levels of phosphorus supply. *Plant Soil* 312: 129–138.

Atkinson, D., 1973. Some general effects of phosphorus deficiency on growth and development. *New Phytology*, 72: 101-111.

Balemi T, Schenk MK. 2009. Genotypic difference of potato in carbon budgeting as a mechanism of phosphorus utilization efficiency. *Plant Soil* 322: 91–99.

Bariola P A, Howard C J, Taylor C P, Verburg M T, Jaglan V D and Green P J 1994 The Arabidopsis ribonuclease gene RNS1 is tightly controlled in response to phosphate limitation. *Plant J*. 6: 673–685.

Batjes N H 1997. A world data set of derived soil properties by FAO-UNESCO soil unit for global modeling. *Soil Use Manage* 13: 9–16.

Bayuelo-Jiménez JS, Ochoa-Cadavid I. 2014. Phosphorus acquisition and internal utilization efficiency among maize landraces from the central Mexican highlands. *Field Crops Research* 156 (2014) 123–134.

Beebe SE, Rojas-Pierce M, Yan X, Blair MW, Pedraza F, Munoz F, Tohme J, Lynch JP (2006). Quantitative trait loci for root architecture traits correlated with phosphorus acquisition in common bean. *Crop Sci* 46: 413–423.

Bieleski, R.L. (1973). Phosphate pools, phosphate transport, and phosphate availability. *Annu. Rev. Plant Physiol. Plant Mol. Biol*. 24, 225–252.

Blair G J 1993. In *Genetic aspects of mineral nutrition*. Eds. P J Randall, E Haize, R A Richards and R Munson. pp 205-213. Kluwer Academic Publishers, Dordrecht, Netherlands.

Blair G J and Cordero S. (1978). The phosphorus efficiency of three annual legumes. *Plant Soil* 50: 387-398.

Bolland MDA, Allen DG (2003). Phosphorus sorption by sandy soils from Western Australia: effect of previously sorbed P on P buffer capacity and single-point P sorption indices. *Aust J Soil Res* 41: 1369–1388.

Bolland MDA, Baker MJ (1998). Phosphate applied to soil increases the effectiveness of subsequent applications of phosphate for growing wheat shoots. *Aust J Exp Agr* 38: 865–869.

Bonser AM, Lynch J, Snapp S (1996). Effect of phosphorus deficiency on growth angle of basal roots in *Phaseolus vulgaris*. *New Phytol* 132: 281–288.

Bucher M, Rausch C, Daram P (2001). Molecular and biochemical mechanisms of phosphorus uptake into plants. *J Plant Nutr Soil Sci* 164: 209–217.

Bucher, M. (2007). Functional biology of plant phosphate uptake at root and mycorrhiza interfaces. *New Phytol*. 173, 11–26.Chen *et al.*, 2007.

Bünemann EK (2008). Enzyme additions as a tool to assess the potential bioavailability of organically bound nutrients. *Soil Biol Biochem* 40: 2116–2129.

Burkitt LL, Donaghy DJ, Smethurst PJ (2010). Low rates of phosphorus fertiliser applied strategically throughout the growing season under rain-fed conditions did not affect dry matter production of perennial ryegrass (*Lolium perenne* L.). *Crop Pasture Sci* 61: 353–362.

Cakmak I (2002). Plant nutrition research: priorities to meet human needs for food in sustainable ways. *Plant Soil* 247: 3–24.

Chambers BJ, Garwood TWD and Unwin RJ. (2000). Controlling soil water erosion and phosphorus losses from arable land in England and Wales. *Journal of Environmental Quality* 29: 145-150.

Chen J, Xu L, Cai Y, Xu J. 2008. QTL mapping of phosphorus efficiency and relative biologic characteristics in maize (*Zea mays* L.) at two sites. *Plant Soil* 313: 251–266.

Chiou, T.J., Liu, H. and Harrison, M.J. (2001). The spatial expression patterns of a phosphate transporter (MtPT1) from *Medicago truncatula* indicate a role in phosphate transport at the root/soil interface. *Plant J.* 25, 281–293.

Chun-Jian L, Xin P, Fu-Suo Z. 2003. Comparison of response of phosphorus-efficient wheat varieties to phosphorus-deficient stress. *Acta Botanic Sinica*. 45: 936.

Clark RB 1990. In *Crops as enhancers of nutrient use*. Eds. VC Baligar and R R Duncan. pp. 131-209. Academic Press, Inc., San Diego, CA.

Cordell D, Drangert J-O, White S. 2009. The story of phosphorus: global food security and food for thought. *Global Environmental Change* 19: 292–305.

Corell, H. (2002). Letter dated 29 January 2002 from the Under-Secretary-General for Legal Affairs, the Legal Counsel, addressed to the President of the Security Council. United National Security Council, Under-Secretary- General for Legal Affairs The Legal Counsel.

Dan Zhang D, Cheng H, Geng L, Kan G, Cui S, Meng Q, Gai J. Yu D. 2009. Detection of quantitative trait loci for phosphorus deficiency tolerance at soybean seedling stage.

Daram P, Brunner S, Persson BL, Amrhein N, Bucher M (1998). Functional analysis and cell-speciWc expression of a phosphate transporter from tomato. *Planta* 206: 225–233.

Duan GL, Zhou Y, Tong YP, Mukhopadhyay R, Rosen BP, Zhu YG (2007). A CDC25 homologue from rice functions as an arsenate reductase. *New Phytol* 174: 311–321.

Duncan RR, Carrow RN. 1999. Turfgrass molecular genetic improvements for abiotic/edaphic stress resistance. *Advances in Agronomy* 67, 233–305.

Epstein, E. (1972). *Mineral Nutrition of Plants: Principles and Perspectives*. John Wiley and Sons, New York.

Fageria NK and Baligar VC. 2008. Phosphorus-use efficiency in wheat genotypes. *Journal of Plant Nutrition* 22: 331-340.

Fageria NK, de Morais OP, dos Santos AB and Vasconcelos MJ. 2014. Phosphorus use efficiency in upland rice genotypes under field conditions. *Journal of Plant Nutrition*, 37: 633–642.

Fallik, E., Sarig, S., Okon, Y., 1994. Morphology and physiology of plant roots associated with *Azospirillum*. In: Okon, Y. (Ed.), *Azospirillum/plant associations*. CRC Press, Boca Raton, pp. 77-86.

Faye I, Diouf O, Guisse A, Se'ne M, and Diallo N. 2006. Characterizing Root Responses to Low Phosphorus in Pearl Millet [*Pennisetum glaucum* (L.) R. Br.]. *Agron. J.* 98: 1187–1194.

Fohse, D., N. Claassen and A. Jung (1998). Phosphorus efficiency of plants. I. External and internal P requirements and P uptake efficiency of different plant species. *Plant Soil*, 110: 101-109.

Furihata T, Suzuki M, Sakurai H (1992). Kinetic characterization of two phosphate uptake systems with different affinities in suspension-cultured Catharanthus roseus protoplast. *Plant Cell Physiol* 33: 1151–1157.

Furlani AMC, Furlani PR, Tanaka RT, Mascarenhas HAA, Delgado MDP. 2002. Variability of soybean germplasm in relation to phosphorus uptake and use efficiency. *Scientia Agricola* 59: 529-536.

Gabasawa AI, Yusuf AA. 2013. Genotypic variations in phosphorus use efficiency and yield of some groundnut cultivars grown on an Alfisol at Samaru, Nigeria. *Journal of Soil Science and Environmental Management* 4(3): 54-61.

Gburek W J, Sharpley A N, Heathwaite A L, and Folmar G J. 2000. Phosphorus management at the watershed scale: a modification of the phosphorus index. *J. Environ. Qual.* 29, 130–144.

George TS, Gregory PJ, Simpson RJ, Richardson AE (2007). Differential interactions of *Aspergillus niger* and *Peniophora lycii* phytases with soil particles affects the hydrolysis of inositol phosphates. *Soil Biol. Biochem* 39: 793-803.

Gerloff, G.C. and W.H. Gabelman. 1983. Genetic basis of inorganic plant nutrition. Vol. 15B. pp. 453-476. In: A. Lauchli and R.L. Bieleski (eds.), *Encyclopedia of Plant Physiology*, New Series. Springer-Verlag, New York, NY.

Gill H.S., Singh A, Sethia S.K, Behl R.K. 2004. Phosphorus uptake and use efficiency in different varieties of bread wheat (*Triticum aestivum* L). *Archives of Agronomy and Soil Science* 50: 563-572.

Glassop, D., Smith, S.E. and Smith, F.W. (2005). Cereal phosphate transporters associated with the mycorrhizal pathway of phosphate uptake into roots. *Planta*, 222, 688–698.

Goldstein, A.H. 1991. Plant cells selected for resistance to phosphate starvation show enhanced P use efficiency. *Theoretical and Applied Genetics* 82: 191-194.

Grant RS, Matthews MA (1996). The influence of phosphorus availability and rootstock on root system characteristics, phosphorus uptake, phosphorus partitioning and growth efficiency. *Am. J. Enol. Vitic.* 47: 403-409.

Guo B, Jin Y, Wussler C, Blancaflor EB, Motes CM, Versaw WK. 2008. Functional analysis of the Arabidopsis PHT4 family of intracellular phosphate transporters. *New Phytologist* 177, 889–898.

Hammond JP, Broadley MR, White PJ, King GJ, Bowen HC, Hayden R, Meacham MC, Mead A, Overs T, Spracklen WP and Greenwood DJ. 2009. Shoot yield drives phosphorus use efficiency in *Brassica oleracea* and correlates with root architecture traits. *Journal of Experimental Botany* 60: 1953-1968.

Hammond, J. P., Bennett, M. J., Bowen, H. C., Broadley, M. R., Eastwood, D. C., May, S. T., Rahn, C., Swarup, R., Woolaway, K. E., and White, P. J. (2003). Changes in gene expression in Arabidopsis shoots during phosphate starvation and the potential for developing smart plants. *Plant Physiol.* 132, 578–596.

Hash C T, Schaffert R E and Peacock J M 2002. Prospects for using conventional techniques and molecular biological tools to enhance performance of 'orphan' crop plants on soils low in available phosphorus. *Plant Soil* 245, 135–146.

Hinsinger, P. 2001. Bioavailability of soil inorganic P in the rhizosphere as affected by root-induced chemical changes: A review. *Plant Soil* 237, 173-195.

Hodge, A., Berta, G., Doussan, C., Merchan, F., Crespi, M., 2009. Plant root growth, architecture and function. *Plant Soil* 321, 153-187.

Isral, D. W., 1987. Investigations of the role of the phosphorus in symbiotic nitrogen fixation. *Plant Physiol.*, 84 : 835-840.

Jacobsen, I., Leggett, M.E., Richardson, A.E., 2005. Rhizosphere microorganisms and plant phosphorus uptake. In: Sims, J.T. and Sharpley, A.N. (Eds.), *Phosphorus, agriculture and the environment*. Soil Sci. Soc. America, Madison, pp. 437-494.

Jasinski, S.M., 2006. Phosphate Rock, Statistics and Information. US Geological Survey.

Javot, H., Penmetsa, R.V., Terzaghi, N., Cook, D.R. and Harrison, M.J. (2007a). A Medicago truncatula phosphate transporter indispensable for the arbuscular mycorrhizal symbiosis. *Proc. Natl Acad. Sci. USA*, 104, 1720–1725.

Johnson JF, Allan DL, Vance CP, Weiblen G. 1996a. Root carbon dioxide fixation by phosphorus-deficient *Lupinus albus*, Contribution to organic acid exudation by proteoid roots. *Plant Physiology* 112: 19–30.

Jones, G. P. D., Blair, G. J., and Jessop, R. S. (1989). Phosphorus efficiency in wheat: A useful selection criterion? *Field Crop Res.* 21, 257–264.

Jungk A, Seeeling B and Gerke J 1993. Mobilization of different phosphate fractions in the rhizosphere. *Plant Soil* 155/156: 91–94.

Kaeppler S M, Parke J L, Mueller S M, Senior L, Stuber C and Tracy W F 2000. Variation among maize inbred lines and detection of quantitative trait loci for growth at low P and responsiveness to arbuscular mycorrhizal fungi. *Crop Sci.* 40, 358–364.

Kakar, K. M., Tariq, M., Taj, F. H., and Nawab, K. (2002). Phosphorus use efficiency of soybean as affected by phosphorus application and inoculation. *Pak. J. Agron.* 1, 49–50.

Karthikeyan, A.S., Varadarajan, D.K., Mukatira, U.T., D'Urzo, M.P., Damsz, B. and Raghothama, K.G. (2002). Regulated expression of Arabidopsis phosphate transporters. *Plant Physiol.* 130, 221–233.

Kuang R, Liao H, Yan X, Dong Y. 2005. Phosphorus and nitrogen interactions in field-grown soybean as related to genetic attributes of root morphological and nodular traits. *Journal of Integrative Plant Biology* 47: 549-559.

LLtaief B, Sifi B, Zaman-Allah M, Horres R, Molina C, Beebe S, Winter P, Kahl G, Drevon J. and Lachaâl M. 2012. Genotypic variability for tolerance to salinity and phosphorus deficiency among N_2-dependent recombinant inbred lines of Common Bean (*Phaseolus vulgaris*). *African Journal of Microbiology Research* 6(20): 4205-4213.

Leggewie, G., Willmitzer, L. and Riesmeier, J.W. (1997). Two cDNAs from potato are able to complement a phosphate uptake-deficient yeast mutant: identification of phosphate transporters from higher plants. *Plant Cell*, 9, 381–392.

Lewis DC, Clarke AL, Hall WB (1987). Accumulation of plant nutrients and changes in soil properties of sandy soils under fertilised pastures in south-eastern South Australia. I. Phosphorus. *Aust J Soil Res* 25: 193–202.

Li L, Liu C, Lian X. 2010. Gene expression profiles in rice roots under low phosphorus stress. *Plant Mol Biol* 72: 423–432.

Li, K., Xu, Z., Zhang, K., Yang, A. and Zhang, J. (2007b). Efficient production and characterization for maize inbred lines with low phosphorus tolerance. *Plant Sci.* 172, 255–264.

Li, L., Liu, C., Lian, X. 2010. Gene expression profiles in rice roots under low phosphorus stress. *Plant Mol. Biol.* 72, 423-432.

Liao H, Ge Z, Yan X. 2001a. Ideal root architecture for phosphorus acquisition of plants under water and phosphorus coupled stresses: from simulation to application. *Chinese Science Bulletin* 46: 1346–1351.

Liu L, Liao H,Wang X, Yan X. 2008. Regulation effect of soil P availability on mycorrhizal infection in relation to root architecture and P efficiency of Glycine max. *Chinese Journal of Applied Ecology* 19: 564–568.

Liu, C.M., Muchhal, U.S., Uthappa, M., Kononowicz, A.K. and Raghothama, K.G. (1998a). Tomato phosphate transporter genes are differentially regulated in plant tissues by phosphorus. *Plant Physiol.* 116, 91–99.

Liu, H., Trieu, A.T., Blaylock, L.A. and Harrison, M.J. (1998b). Cloning and characterization of two phosphate transporters from *Medicago truncatula* roots: regulation in response to phosphate and to colonization by arbuscular mycorrhizal (AM) fungi. *Mol. Plant Microbe Interact.* 11, 14–22.

Lynch J (1995). Root architecture and plant productivity. *Plant Physiol* 109: 7–13.

Lynch J. 2007. Roots of the second green revolution. *Australian Journal of Botany* 55, 493–512.

Ma Z, Walk TC, Marcus A, Lynch JP (2001). Morphological synergism in root hair length, density, initiation and geometry for phosphorus acquisition in Arabidopsis thaliana: A modeling approach. *Plant Soil* 236: 221–235.

Maeda, D., Ashida, K., Iguchi, K., Chechetka, S.A., Hijikata, A., Okusako, Y., Deguchi, Y., Izui, K. and Hata, S. (2006). Knockdown of an arbuscular mycorrhiza-inducible phosphate transporter gene of Lotus japonicus suppresses mutualistic symbiosis. *Plant Cell Physiol.* 47, 807–817.

Marschner, H. (1995). *Mineral Nutrition of Higher Plants*. London: Academic Press.

McCaskill MR, Cayley JWD (2000). Soil audit of a long-term phosphate experiment in South-Western Victoria: total phosphorus, sulfur, nitrogen and major cations. *Aust J Agr Res* 51: 737–748.

McLachlan KD 1976. Comparative phosphorus responses in plants to a range of available phosphorus situations. *Aust. J. Agric. Res.* 27: 323-341.

McPharlin J, Bieleski R (1987). Phosphate uptake by Spirodela and Lemna during early phosphate deficiency. *Aust J Plant Physiol* 14: 561–572.

Melland AR, McCaskill MR, White RE, Chapman DF (2008). Loss of phosphorus and nitrogen in runoff and subsurface drainage from high and low input pastures grazed by sheep in Australia. *Aust J Soil Res* 46: 161–172.

Mitsukawa N, Okamura S, Shirano Y, Sato S, Kato T, Harashima S, Shibata D (1997). Overexpression of an Arabidopsis thaliana high-affinity phosphate transporter gene in tobacco cultured cells enhances cell growth under phosphate-limited conditions. *Proc Natl Acad Sci USA* 94: 7098–7102.

Moll RH, Kamprath EJ, Jackson WA (1982). Analysis and interpretation of factors which contribute to efficiency of nitrogen utilization. *Agron. J.* 74: 562-564.

Mourice SK and Tryphone GM. 2009. Evaluation of Common Bean (*Phaseolus vulgaris* L.) Genotypes for Adaptation to Low Phosphorus. *ISRN Agronomy* 2012 doi: 10.5402/2012/309614.

Muchhal US, Pardo JM, Raghothama KG (1996). Phosphate transporters from the higher plant *Arabidopsis thaliana*. *Proc Natl Acad Sci USA* 93: 10519–10523.

Muchhal US, Raghothama KG (1999). Transcriptional regulation of plant phosphate transporters. *Proc Natl Acad Sci USA* 96: 5868–5872.

Mudge, S.R., Rae, A.L., Diatloff, E. and Smith, F.W. (2002). Expression analysis suggests novel roles for members of the Pht1 family of phosphate transporters in Arabidopsis. *Plant J.* 31, 341–353.

Nadeem M, Mollier A, Morel C, Vives A, Prud'homme L, Pellerin S (2011a). Relative contribution of seed phosphorus reserves and exogenous phosphorus uptake to maize (*Zea mays* L.) nutrition during early growth stages. *Plant Soil* 346: 231–244.

Nagy, R., Karandashov, V., Chague, W., Kalinkevich, K., Tamasloukht, M., Xu, G.H., Jakobsen, I., Levy, A.A., Amrhein, N. and Bucher, M. (2005). The characterization of novel mycorrhiza-specific phosphate transporters from *Lycopersicon esculentum* and *Solanum tuberosum* uncovers functional redundancy in symbiotic phosphate transport in solanaceous species. *Plant J.* 42, 236–250.

Narang RA, Bruene A, Altmann T. 2000. Analysis of Phosphate Acquisition Efficiency in Different Arabidopsis Accessions. *Plant Physiology* 124: 1786-1799.

Neumann, G.; Romheld, V. 1999. Root excretion of carboxylic acids and protons in phosphorus-deficient plants. *Plant Soil*, 211, 121-130.

Oberson A, Friensen D K, Tiessen H, Morel C and Stahel W 1999. Phosphorous status and cycling in native savanna and improved pastures on an acid low-P. *Colombian Oxisol.* 77–88.

Oehl, F.; Sieverding, E.; Mader, P.; Dubois, D.; Ineichen, K.; Boller, T.; Wiemken, A. 2004. Impact of long-term conventional and organic farming on the diversity of arbuscular mycorrhizal fungi. *Oecologia*, 138, 574-583.

Osborne LD. and Rengel Z. 2002a. Genotypic differences in wheat for uptake and utilisation of P from iron phosphate. *Australian Journal of Agricultural Research* 53 (7) 837 – 844.

Osborne LD. and Rengel Z. 2002b. Screening cereals for genotypic variation in efficiency of phosphorus uptake and utilization. *Australian Journal of Agricultural Res* 53 (3) 295 - 303.

Ozanne PG, Kirton DJ, Shaw TC (1961). The loss of phosphorus from sandy soils. *Aust J Agr Res* 11: 927–924.

Pan XW, Li WB, Zhang QY, Li YH, Liu MS. Assessment on Phosphorus Efficiency Characteristics of Soybean Genotypes in Phosphorus-Deficient Soils. *Agricultural Sciences in China* 7(8): 958-969.

Pang J, Tibbett M, Denton MD, Lambers H, Siddique KHM, Bolland MDA, Revell CK. (2010). Variation in seedling growth of 11 perennial legumes in response to phosphorus supply. *Plant Soil* 328: 133–143.

Parentoni, S.N., C.L. Souza Junior, V.M.C. Alves, E. E.G. Gama, A.M. Coelho, A.C. de Oliveira, C.T. Guimaraes, M.J.V. Vasconcelos, C.A.P. Pacheco, W.F. Meirelles, J.V. Magalhaes, L.J.M. Guimarães, A.R. da Silva, F.F. Mendes, and R.E. Schaffert. 2010. Inheritance and breeding strategies for phosphorus efficiency in tropical maize (*Zea mays* L.). *Maydica* 55: 1–15.

Parentoni, S.N., C.L. Souza Junior. 2008. Phosphorus acquisition and internal utilization efficiency in tropical maize genotypes. *Pesq. agropec. bras., Brasília,* 43: 893-901.

Park MR, Baek SH, de los Reyes BG, Yun SJ. 2007. Overexpression of a high-aYnity phosphate transporter gene from tobacco (NtPT1) enhances phosphate uptake and accumulation in transgenic rice plants. *Plant Soil* 292: 259–269.

Paszkowski, U., Kroken, S., Roux, C. and Briggs, S.P. (2002). Rice phosphate transporters include an evolutionarily divergent gene specifically activated in arbuscular mycorrhizal symbiosis. *Proc. Natl Acad. Sci. USA*, 99, 13324–13329.

Pathak H. 2010. Trend of fertility status of Indian soils. *Current Advances in Agricultural Sciences* 2(1): 10-12.

Rae, A.L., Cybinski, D.H., Jarmey, J.M. and Smith, F.W. (2003). Characterization of two phosphate transporters from barley; evidence for diverse function and kinetic properties among members of the Pht1 family. *Plant Mol. Biol.* 53, 27–36.

Raghothama KG (1999). Phosphate acquisition. *Ann Rev Plant Physiol Plant Mol Biol* 50: 665–693.

Raghothama KG, Karthikeyan AS (2005). Phosphate acquisition. *Plant Soil* 274: 37–49.

Randall, PJ. 1995. Genotypic differences in phosphorus uptake. In: C. Johansen, K.K. Lee, K.K. Sharma, G.V. Subbarao, E.A. Kueneman eds. *Genetic Manipulation of crop Plants to Enhance Integrated Nutrient Management in Cropping Systems - 1. Phosphorus*. Patancheru, India: ICRISAT, pp. 31-47.

Reiter, R. S., Coors, J. G., Sussman, M. R. *et al.*, 1991. Genetics analysis of tolerance to low-phosphorus stress in maize using RFLP, *Theor. Appl. Genet.*, 82: 561.

Rengel Z, Marschner P (2005). Nutrient availability and management in the rhizosphere: Exploiting genotypic differences. *New Phytol* 168: 305–312.

Rengel Z. 1999. Physiological mechanisms underlying differential nutrient efficiency of crop genotypes. In: Rengel Z, ed, *Mineral Nutrition of Crops: Fundamental Mechanisms and Implications*. Haworth Press, New York. pp. 227-265.

Richardson AE 1994. Soil microorganisms and phosphorus availability. *Soil Biota*, 50–62.

Richardson AE, Barea JM, McNeill AN, Combaret CP (2009). Acquisition of phosphorus and nitrogen in the rhizosphere and plant growth promotion by microorganisms. *Plant Soil* 321: 305–339.

Richardson AE, Hadobas PA, Hayes JE (2001). Extracellular secretion of Aspergillus phytase from Arabidopsis roots enables plants to obtain phosphorus from phytate. *Plant J* 25: 641–649.

Ridley AM, Christy BP, White RE, McLean T, Green R (2003). North-east Victoria SGS National Experiment site: water and nutrient losses from grazing systems on contrasting soil types and levels of inputs. *Aust J Exp Agr* 43: 799–815.

Rose TJ and Matthews W. 2012. Chapter five – Rethinking Internal Phosphorus Utilization Efficiency: A New Approach Is Needed to Improve PUE in Grain Crops. *Advances in Agronomy* 116: 185-217.

Rosmarin, A. (2004). The Precarious Geopolitics of Phosphorous Down to Earth (Science and Environment Fortnightly), (June 30, 2004): p. 27-31.

Ryan, P.R.; Delhaize, E.; Jones, D.L. 2001. Function and mechanism of organic anion exudation from plant roots. *Annu. Rev. Plant Physiol. Plant Mol. Biol.*, 52, 527-560.

Saier, M.H. (2000). Families of transmembrane transporters selective for amino acids and their derivatives. *Microbiol. Sgm*, 146, 1775– 1795.

Schachtman DP, Reid RJ, Ayling SM (1998). Phosphorus uptake by plants: from soil to cell. *Plant Physiol* 116: 447–453.

Sepehr, E., Malakouti, M. J., Kholdebarin, B., Samadi, A., and Karimian, N. (2009). Genotypic variation in P efficiency of selected Iranian cereals in greenhouse experiment. Int. J. Plant Prod., 3: 17–28.

Shen H, Chen J, Wang Z, Yang C, Sasaki T, Yamamoto Y, Matsumoto H, Yan X (2006). Root plasma membrane H+-ATPase is involved in the adaptation of soybean to phosphorus starvation. *J Expt Bot* 57: 1353–1362.

Shenoy, V.V.; Kalagudi, G.M. 2005. Enhancing plant phosphorus use efficiency for sustainable cropping. *Biotechnol. Adv.* 23, 501-513.

Shimizu A, Kato K, Komatsu A, Motomura K, Ikehashi H. 2008. Genetic analysis of root elongation induced by phosphorus deficiency in rice (*Oryza sativa* L.): Fine QTL mapping and multivariate analysis of related traits. *Theor Appl Genet* 117: 987–996.

Shimizu A, Yanagihara S, Kawasaki S, Ikehashi H (2004). Phosphorus deficiency-induced root elongation and its QTL in rice (*Oryza sativa* L.). *Theor Appl Genet* 109: 1361–1368.

Shridevi AJ, Kajjidoni ST, Koti RV. 2009. Genotypic variation for root traits to phosphorus deficiency in blackgram (*Vigna mungo* L. Hepper). *Karnataka J. Agric. Sci.*, 22(5): 946-950.

Siddiqui, M.Y. and A.D.M. Glass, 1981. Utilization index: A modified approach to the estimation and comparison of nutrient utilization efficiency in plants. *J. Plant Nutr.*, 4: 289–302.

Singh A.L. 1999. Mineral Nutrition of Groundnut. In: *Advances in Plant Physiology* (Ed. A. Hemantranjan), Vol II pp. 161-200. Scientific Publishers (India), Jodhpur, India.

Singh, A.L. 2000. Mechanism of Tolerance and Crop Production in acid Soils. In: *Advances in Plant Physiology* Vol. III Plant Physiology, Biochemistry and Plant Molecular Biology in 2000 (Ed. A. Hemantranjan), pp. 353-394. Scientific Publishers (India), Jodhpur, India.

Singh, A.L. 2004. Mineral nutrient requirement, their disorders and remedies in Groundnut. pp. 137-159. In: Groundnut Research in India (Eds. M.S. Basu and N. B. Singh) National Research Center for Groundnut (ICAR), Junagadh, India.

Singh, A.L. 2006. Macronutrient stresses and their management in crop plants. In: *Advances in Plant Physiology* (Ed. P.C. Trivedi), pp.198-234. I.K. International Publishing House Pvt. Ltd. New Delhi, India.

Singh, A.L. 2008. Mineral Stresses and Crop Productivity. In *International Symposium on Natural Resource management in Agriculture*. Dec 19-20 2008, Agricultural Research Station, RAU, Durgapura, Jaipur. pp. 9-23. Invited Lead paper.

Singh, A.L. 2014. Role of nutrient efficient crop genotypes in modern agriculture. In: Proc National Conference of Plant Physiology (NCPP 2014) on *"Frontiers of Plant Physiology Research: Food Security and Environmental Challenges"* 23-25 Nov 2014, Odisha University of Agriculture and Technology, Bhubaneswar, India pp. 135-142.

Singh A.L. and M.S. Basu 2005a. Integrated Nutrient Management in Groundnut: A Farmer's Manual. National Research center for groundnut, Junagadh, India. 54 p.

Singh, A. L., and M. S. Basu. 2005b. Screening and selection of P efficient genotypes for calcareous soils in India. In: Plant nutrition for food security, human health and environmental protection (Eds C.J. Li *et al.*), pp. 1004-1005 (Plant and Soil Series). Proceedings 15th International Plant Nutrition Colloquium, China Agricultural University, Beijing, China 14-19 Sept. 2005. Tsinghua Univ. Press Beijing, China.

Singh, A. L., Basu, M. S. and Singh, N. B. (2004). Mineral Disorders of Groundnut. National Research Center for Groundnut (ICAR), Junagadh, India p. 85.

Singh A.L. and V. Chaudhari 1996. Use of zincated and boronated super phosphate and mycorrhizae in groundnut in calcareous soil. *J. Oilseed Research*, 13: 61-65.

Singh, A.L. and V. Chaudhari 1996. Interaction of sulphur with phosphorus and potassium in groundnut nutrition in calcareous soil. *Indian Journal of Plant Physiology* (New Series), 1: 21-27.

Singh, A.L. and V. Chaudhari 2006. Macronutrient requirement of groundnut: Effects on the growth and yield components. *Indian J. Plant Physiology*, 11: 401-409.

Singh, A.L. and V. Chaudhari 2007. Macronutrient requirement of groundnut: Effects on uptake of macronutrients. *Indian J. Plant Physiology*, 12: 72-77.

Singh, A.L. and Anita Mann 2012. Recent Advances in Plant Nutrition. In: Proc National Seminar of Plant Physiology on *"Physiological and molecular approaches for development of climatic resilient crops"* 12-14 Dec. 2012, ANGRAU, Hyderabad, India pp. 6-22.

Singh, A.L. and P.K. Singh 1990. Phosphorus fertilization and the growth and N_2-fixation of Azolla and blue-green algae in rice field. *Indian J. Plant Physiol.* 33: 21-26.

Singh, A.L., P.K. Singh and Pushp Lata 1988. Effect of split application of phosphorus fertilizer on the growth of Azolla and low land rice. *Nutrient Cycling in Agroecosystems (Fertilizer Research)*, 16: 109-117.

Singh, A.L., Vidya Chaudhari and V.G. Koradia 1991. Foliar nutrition of Nitrogen and Phosphorous in groundnut. In: *Physiological Strategies For Crop Improvement: Proceeding of the International Conference of Plant Physiology* (Eds. D.N. Tyagi, B. Bose, A. Hemantranjan and T. Meghabati Devi), pp. 129-133. B.H.U., Varanasi, India.

Singh BB (1999). Improved drought tolerant cowpea varieties for the Sahel. Project II. Cowpea cereal system improvement for the savannas, p. 36. International Institute of Tropical Agriculture, Ibadan, Nigeria. Annual Report.

Smith, F.W., Ealing, P.M., Dong, B. and Delhaize, E. (1997). The cloning of two Arabidopsis genes belonging to a phosphate transporter family. *Plant J*. 11, 83–92.

Smith, FW. 2001. S and P transport systems in plants. *Plant Soil* 232: 109–118.

Su, J. Y., Xiao, Y., Li, M., Liu, Q., Li, B., Tong, Y., Jia, J., and Li, Z. (2006). Mapping QTLs for phosphorus-deficiency tolerance at wheat seedling stage. *Plant Soil* 281, 25–36.

Teng W, Deng Y, Chen XP, Xu XF, Chen RY, Lv Y, Zhao YY, Zhao XQ, He X, Li B, Tong YP, Zhang FS, Li ZS. 2013. Characterization of root response to phosphorus supply from morphology to gene analysis in field-grown wheat. *Journal of Experimental Botany* 64: 1403-1411.

Trotter MG, Lamb DW, Donald GE, Schneider DA (2010). Evaluating an active optical sensor for quantifying and mapping green herbage mass and growth in a perennial grass pasture. *Crop Pasture Sci* 61: 389–398.

Ullrich-Eberius CI, Novacky A, van Bel A (1984). Phosphate uptake in Lemna gibba G1: energetics and kinetics. *Planta* 161: 46–52.

Uusi-Kamppa J, Braskerud B, Jansson H, Syversen N, Uusitalo R.2000. Buffer zones and constructed wetlands as fillers for agricultural phosphorus. *Journal of Environmental Quality* 29: 151-158.

Valverde F, Losada M, Serrano A. 1999. Engineering a central metabolic pathway: glycolysis with no net phosphorylation in an *Escherichia coli* gap mutant complemented with a plant Gap N gene. *FEBS Lett* 449: 153– 8.

van der Eijk, D., Janssen, B.H. and Oenema, O. 2006. Initial and residual effects of fertilizer phosphorus on soil phosphorus and maize yields on phosphorus fixing soils. A case study in south-west Kenya. *Agr. Ecosyst. Environ*., 116(1–2): 104–120.

Vance CP, Uhde-Stone C, Allan DL (2003). Phosphorus acquisition and use: critical adaptation by plants for securing a non-renewable resource. *New Phytol* 157: 423–447.

Vance, C. P 2001. Symbiotic nitrogen fixation and phosphorus acquisition.Plant nutrition in a world of declining renewable resources. *Plant Physiol* 127: 390-397.

Wang LH, Duan GL, Williams PN, Zhu YG. 2008. Influences of phosphorus starvation on OsACR2.1 expression and arsenic metabolism in rice seedlings. *Plant Soil* 313: 129–139.

Wang S, Basten CJ, Zeng Z-B. 2004. Windows QTL Cartographer 2.0. Department of Statistics, North Carolina State University. Raleigh, NC, USA.

Wang X, Wang Y, Tian J, Lim B, Yan X, Liao H. 2009. Over expressing AtPAP15 enhances phosphorus efficiency in soybean. *Plant Physiol* 151: 233–240.

White PJ, Broadley MR, Greenwood DJ, Hammond JP. 2005. Genetic modifications to improve phosphorus acquisition by roots. Proceedings 568. York, UK: International Fertiliser Society.

White, P. J., and Hammond, J. P. (2008). Phosphorus nutrition of terrestrial plants. In *"The Ecophysiology of Plant–Phosphorus Interactions"* (P. J. White and J. P. Hammond, Eds.), pp. 51–81. Springer, The Netherlands.

Wissuwa M (2005). Combining a modelling with a genetic approach in establishing associations between genetic and physiological effects in relation to phosphorus uptake. *Plant Soil* 269: 57–68.

Wissuwa M, Ae N. 2001. Further characterization of two QTLs that increase phosphorus uptake of rice (*Oryza sativa* L.) under phosphorus deficiency. *Plant and Soil* 237: 275–286.

Wissuwa M, Wegner J, Ae N, Yano M (2002). Substitution mapping of Pup1: a major QTL increasing phosphorus uptake of rice from a phosphorus-deficient soil. *Theor Appl Genet* 105: 890–897.

Wissuwa, M., Gamat, G., and Ismail, A. (2005). Is root growth under phosphorus deficiency affected by source or sink limitations? *J. Exp. Bot.* 56, 1943–1950.

Xu, G.H., Chague, V., Melamed-Bessudo, C., Kapulnik, Y., Jain, A., Raghothama, K.G., Levy, A.A. and Silber, A. (2007). Functional characterization of LePT4: a phosphate transporter in tomato with mycorrhiza-enhanced expression. *J. Exp. Bot.* 58, 2491– 2501.

Yan X, Beebe S E and Lynch J P 1995a. Genetic variation for P efficiency of common bean in contrasting soil types: II. Yield response. *Crop Sci.* 35, 1094–1099.

Yan X, Liao H, Beebe SE, Blair MW, Lynch JP (2004). QTL mapping of root hair and acid exudation traits and their relationship to phosphorus uptake in common bean. *Plant Soil* 265: 17–29.

Yan X, Lynch J P and Beebe S E 1995b. Genetic variation for P efficiency of common bean in contrasting soil types: I. Vegetative response. *Crop Sci.* 35, 1086–1093.

Yan, F., Zhu, Y.Y., Muller, C., Zorb, C., Schubert, S. 2002. Adaptation of H+-pumping and plasma membrane H+ ATPase activity in proteoid roots of white lupin under phosphate deficiency. *Plant Physiol.* 129, 50-63.

Yang M, Ding G, Shi L, Xu F, Meng J. 2011. Detection of QTL for phosphorus efficiency at vegetative stage in *Brassica napus*. *Plant Soil* (2011) 339: 97–111.

Yao Y, Sun H, Xu F, Zhang H, Liu S. 2011. Comparative proteome analysis of metabolic changes by low phosphorus stress in two *Brassica napus* genotypes. *Planta* 233: 523–537.

Yaseen, M., and Malhi, S. S. (2009a). Variation in yield, phosphorus uptake, and physiological efficiency of wheat genotypes at adequate and stress phosphorus levels in soil. *Commun. Soil Sci. Plant Anal.* 40, 3104–3120.

Yaseen, M., and Malhi, S. S. (2009b). Differential growth response of wheat genotypes to ammonium phosphate and rock phosphate phosphorus sources. *J. Plant Nutr.* 32, 410–432.

Yoshida S, Forno DA, Cook JH, Gomez KA (1976). Laboratory manual for physiological studies of rice. International Rice Research Institute, Manila, pp 61–67.

Yun SJ and Keppler SM. 2001. Induction of maize acid phosphatase activities under phosphorus starvation. *Plant and Soil* 237: 109–115,.

Zai-hua G, Ping D, Li-Yuan H, Cai-Guo X. 2006. Genetic Analysis of Agricultural Traits in Rice Related to Phosphorus Efficiency. *Acta Genetica Sinica,* 33: 634–641.

Zhao J, Fu J, Liao H, He Y, Nian H, Hu Y, Qiu L, Dong Y, Yan X. 2004. Characterisation of root architecture in an applied core collection for phosphorus efficiency of soybean germplasm. *Chinese Science Bulletin* 49, 1611–1620.

Zhu JM, Kaeppler SM, Lynch JP (2006) Mapping of QTLs for lateral root branching and length in maize (*Zea mays* L.) under differential phosphorus supply. *Theor Appl Genet* 111((4): 688–695.

Zhu, J., and Lynch, J. P. (2004). The contribution of lateral rooting to phosphorus acquisition efficiency in maize (*Zea mays*) seedlings. *Funct. Plant Biol.* 31, 949–958.

Recent Advances in Crop Physiology Vol. 2 (2016)   *Pages* **51–105**
Editor: **Dr. Amrit Lal Singh**
Published by: **DAYA PUBLISHING HOUSE, NEW DELHI**

Chapter 3

Salinity Management in Vertisols: Physiological Implications

G. Gururaja Rao*

ICAR–Central Soil Salinity Research Institute, Regional Research Station, Bharuch – 392 012, Gujarat

1. Introduction

Global demand for food, fiber and bio-energy are growing at rapid rate, and growth rate in agriculture in most developing countries has failed to catch up with the increase in population growth. One of the principal constraints in achieving the desired growth rate in food grain production is land and water degradation mainly resulting from anthropogenic activities. A major factor contributing to human induced land degradation is soil salinisation. Water logging and salinity problems result in degradation of 20000-30000 ha of irrigated land each year. Salinity development in the country charts a parallel path with irrigation development. As inadequate attention has been paid in the planning stage of irrigation projects, the problems of water logging and salinity have increased at an alarming rate.

Vertisols are dark montmorillonite-rich clays with characteristic shrinking/ swelling properties. Vertisols are churning clay soils with a high proportion of swelling clays (>30 per cent to at least 50 cm from the surface) and in dry state with typical cracks which are at least 1 cm wide and reach a depth of 50 cm or more, are often also called heavy cracking clay soils or swell-shrink soils. These soils form deep wide cracks from the surface downward during the dry season which happens

_____
* *Author:* E-mail: ggrao54@yahoo.com

in most years. These cracks usually show a slight curvature. These soils have a vertic horizon within 100 cm from the surface; possess 30 per cent or more clay in all horizons to a depth of 100 cm or more and possess cracks which open periodically.

Vertisols owe their specific properties to the presence of swelling clay minerals, mainly montmorillonite. As a result of wetting and drying, expansion and contraction of the clay minerals take place. During expansion of clay minerals, high pressures are developed within these soils, causing a characteristic soil structure with wedge-shaped aggregates in the surface soil and planar soil blocks in the subsoil. The slippage of one soil block over the other leads to the formation of typical polished grooved surfaces, "slickensides" on the blocks. Expansion and contraction also cause the formation of micro-topographic features known as "gilgai", a distinctive microrelief of knolls and basins that develops by internal mass movements in the soil and heaving of the underlying material to the surface.

The pedo-climatic environment supporting these soils varies from semi-arid to sub-humid tropics in India. Such climates are characterized by hot and dry pre-monsoon summer months (March to May) followed by well-expressed summer monsoon months (June to September). Vertisols occur principally in the soils of hot environments with marked alternating wet and dry seasons. The mean annual rainfall ranges from 500 to 1500 mm of which 80 to 90 per cent is received during monsoon months. The soil moisture control system (SMCS) remains dry either completely or in part for 4 to 8 months in a year suggesting an Ustic moisture regime.

2. Vertisols, their Characteristics and Production Constraints

2.1 Distribution

Vertisols are generally found on sedimentary plains, both on level land and in depressions. These soils cover 335 million hectares worldwide. While most Vertisols occur in the semi-arid tropics covering some 200 m ha of which one-fourth is considered to be useful land. The largest Vertisol areas are on sediments that have a high content of smectitic clays or produce such clays upon post-depositional weathering (Sudan) and on extensive basalt plateaus (India).

Vertisols and associated soils are found to occur in nearly four agro-climatic conditions of India *i.e.*, arid, semi-arid, dry sub-humid and moist sub-humid conditions that prevail in the states of Maharashtra, Gujarat, Madhya Pradesh, Karnataka, Andhra Pradesh and Tamil Nadu (Figure 3.1). About 76.4 million hectares of Vertisols occur in India which is 23.2 per cent of the total geographical area of the country (Table 3.1, Murthy *et al.*, 1982). Vertisols and associated soils are predominant in Maharashtra, Madhya Pradesh, Gujarat, Andhra Pradesh and Karnataka (6.9 m ha). Of these salt affected Vertisols measuring 0.54 m ha in Maharashtra, 0.12 m ha in Gujarat state and 0.034 m ha in Madhya Pradesh.

The development of salinity and sodicity in black soils region is generally associated with poor drainage and water logging. Vertisols are widespread in India, often have low fertility and are difficult to cultivate. The current level of crop production in these harsh environments is inadequate to meet the needs of rapidly increasing

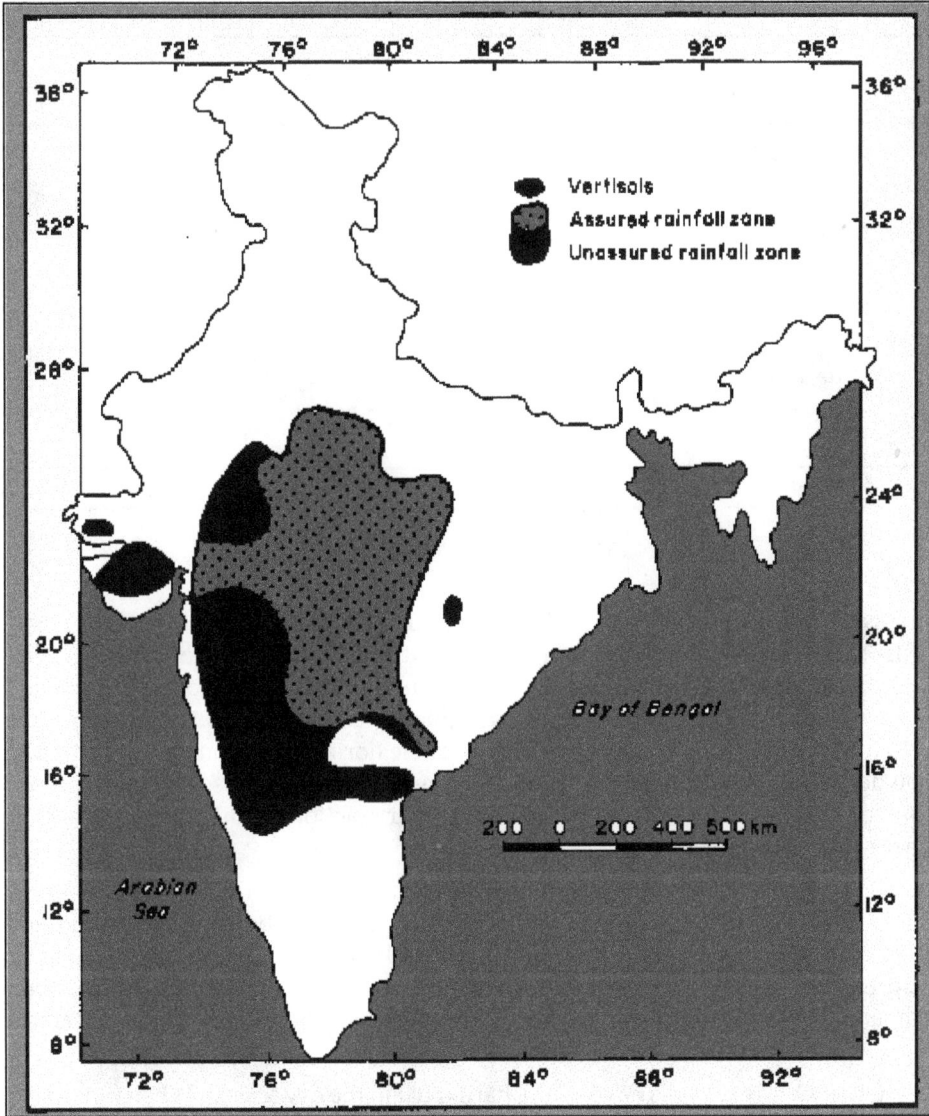

Figure 3.1: Distribution of Vertisols in India.

population. Vertisols have great potential for productive cropping because their stored water that can sustain crops through drought/stress periods.

2.2 Characteristics

The morphological characteristics comprise the expression, shape and orientation of the structural aggregates, depth and width of cracks developing on drying. The structural arrangement together with the wide cracks is probably the most striking morphological feature of Vertisols. In most cases the surface horizons exhibit large, well developed angular blocky or prismatic structures, while in the sub-soil wedge

Table 3.1: Distribution of Black Cotton Soils in India

Sl.No.	State	Area (M ha)	Per cent of Total Black Soil Area of the Country	Per cent of Total Geographical Area of Country
1.	Maharashtra	29.9	39.1	9.1
2.	Madhya Pradesh	16.7	22.0	5.0
3.	Gujarat	8.2	10.7	2.5
4.	Andhra Pradesh	7.2	9.4	2.2
5.	Karnataka	6.9	9.0	2.1
6.	Tamil Nadu	3.2	4.2	1.0
7.	Rajasthan	2.3	3.0	0.7
8.	Odisha	1.3	1.7	0.4
9.	Bihar	0.7	0.9	0.2
	Total	**76.4**	**100.0**	**23.2**

Source: Murthy *et al.*, 1982.

shaped structural elements of all sizes do occur. A typical profile of Vertisols has A (B)C- horizons, the A-horizon comprises both the surface mulch (or crust) and the underlying structured horizon that changes only gradually with depth. Important morphological characteristics such as soil colour, texture, element composition etc are all uniform throughout the solumn. A calcic horizon or a concentration of soft powdery lime may be present in or below the vertic horizon. Vertisols differ in surface characteristics and these strongly influence their reaction to soil tillage operations. There are two broad groups.

Self-mulching Vertisols have a fine (granular or crumb) surface soil, 2-30 cm thick, during the dry season. This fine tilth is produced by desiccation and soil shrinkage. When such soils are ploughed, the clods, after being subjected to repeated wetting and drying, disintegrate. When this mulch is well developed, seedbed preparation is hardly necessary, crust in the dry season. Such soils require mechanical tillage if they are to be cultivated.

Crusty Vertisols soils have a thin, hard crust in the dry season. When ploughed, crusty Vertisols produce large, hard clods that may persist for 2 to 3 years before they have crumbled enough to permit the preparation of a good seedbed. The self-mulching versus crusting characteristic is related to the tensile stress of the soil. One of the factors that influences this stress is soil texture. Soils are also strongly self-mulching when they contain appreciable amounts of fine, sand-sized calcareous concretions: these apparently disturb the continuity of the clayey soil material. Other observations have shown that under higher rainfall, Vertisols do not generally contain calcareous concretions. Very high amounts of sodium favour the formation of a hard surface crust.

Dry Vertisols have a very hard consistence; wet Vertisols are very plastic and sticky. It is generally true that Vertisols are friable only over a narrow moisture range but their physical properties are greatly influenced by soluble salts and/or adsorbed sodium. The most important physical characteristics of Vertisols are as low hydraulic conductivity which varies among different groups, salt content and bulk density. The clay content recorded normally vary from 40 to 95 per cent but little difference exists within the profile. Infiltration of water in dry Vertisols with surface mulch or fine tilth is initially rapid. However, once the surface soil is thoroughly wetted and cracks have closed, the rate of water infiltration becomes almost nil. If, at this stage, the rains continue (or irrigation is prolonged), Vertisols flood readily. Vertisols, are soils with good water holding properties. However, a large proportion of all water held between the basic crystal units is not available to plants. When these soils are irrigated, the high seepage leads to a shallow water table build-up causing secondary salinisation or sodiumisation.

Most Vertisols have a high cation exchange capacity (CEC) and a high base saturation percentage (BS). Dominant cations are Ca^{2+} and Mg^{2+} while Na^+ plays an important role. pH values are in the range of 6.0 to 8.0. Higher pH values (8.0-9.5) are seen in Vertisols with much ESP. Salinity in Vertisols may be inherited from the parent materials or may be caused by over-irrigation. In coastal regions Vertisols with high soluble salts and/or with low sulphates are seen. Leaching of excess salt is hardly possible. It is possible to flush salts that have precipitated on the wall of cracks. Surface leaching of salts from the paddy fields in India was achieved by evacuating the standing water at regular intervals.

The structural stability of Vertisols remains low. They are therefore very susceptible to water erosion. Slopes above 5 per cent should not be used for arable cropping, and on gentler slopes contour cultivation with a groundcover crop is advisable. When terracing, sufficient surface drainage must be provided to avoid slumping.

Studies conducted to understand the effect of electrolyte concentration and Sodium Adsorption Ratio (SAR) and/or Exchangeable sodium percentage (ESP) on flocculation and hydraulic conductivity (Ks) indicated that the ESP of both the soils increased with electrolyte concentration and SAR. The data also indicated that with increase in the ESP of soil, the critical coagulation concentration increases. An electrolyte concentration of 20 meq l^{-1} is necessary to cause flocculation of clayey soil at ESP of 6 and silty clayey soil at ESP of 10, beyond which, these soils undergo structural degradation. It is inferred from the study that at salinity of < 2 dS m^{-1}, the Vertisols can be grouped as sodic if the ESP is > 6 and >10 in clayey and silty clayey soils, respectively. Similarly, at salinity of < 4 dS m^{-1}, the Vertisol can be grouped as sodic if the ESP is > 13 and >21 in clayey and silty clayey soils, respectively. At higher salinity *i.e.*, > 6 dS m^{-1} even at fairly high ESP also, the soil Ks and dispersion are not affected adversely. It can be fairly concluded that the coupled salinity and ESP values be considered as the limit for sodicity classification.

2.3. Production Constraints of Vertisols

Vertisols are difficult to work - as these are of very hard consistence when dry and very plastic and sticky when wet. Therefore, the workability of the soil is often limited to very short periods of medium (optimal) water status. However, tillage operations can be performed in the dry season with heavy machinery. While mechanical tillage in the wet season causes serious soil compaction, wet land are really impassable.

Vertisols are imperfectly to poorly drained, leaching of soluble weathering products is limited, the contents of available calcium and magnesium are high and pH is above 7. This is due to very low hydraulic conductivity of a Vertisol *i.e.*, once the soil has reached its field capacity, practically no water movement occurs. Flooding can be a major problem in areas with higher rainfall. Surface water may be drained by open drains. Mole drainage is virtually impossible.

The adverse physical properties and poor workability of Vertisols are major obstacles to agricultural land use, especially in low-technology societies. Vertisols have a considerable potential for agricultural production but special management practices (tillage and water management) are required to secure sustained production. Main parts of the Vertisol regions are used for low intensity grazing. Where irrigation is possible or in more humid environments, arable land use is possible and shows a considerable potential for agricultural production.

Vertisols are base rich soils and are capable of sustaining continuous cropping. They do not necessarily require a rest period for recovery; because the pedo-turbation brings subsoil to the surface. However, the overall productivity normally remains low, especially where no irrigation water is available. Nitrogen is normally deficient as well as phosphorus. Phosphate fixation (as tricalcium phosphate) may occur but is not a major problem. Potassium contents are variable. Secondary elements and micronutrients are often deficient. In semi-arid areas free carbonate and gypsum accumulations are common. Saline and sodic Vertisols may develop under irrigation, but they are rare under natural conditions. The major production constraints of Vertisols are:

☆ Reduced permeability in swollen state so that both infiltration and internal drainage are very low

☆ Poor aeration of wet soils and related root development

☆ Narrow optimum moisture range for tillage and seeding operations

☆ Germination and difficulties associated with rapid drying of granular surfaces and scaling and crusting

☆ Salinity hazards associated with rising groundwater table and use of poor quality irrigation water

☆ Salinity developed due to irrigation under canal command areas

3. Agricultural Salinity in Vertisols

Salt-affected soils that contain considerable amounts of soluble salts, occur where evapotranspiration greatly exceeds precipitation, *i.e.* in arid and semi-arid areas. Cations and anions accumulate in these soils and cause high salinity or alkalinity. The accumulated ions include sodium, potassium, magnesium, calcium, chloride, sulphate, carbonate and bicarbonate. A distinction can be made between primary and secondary salinisation processes. Primary salinisation involves accumulation of salts through natural processes due to high salt contents in parent materials or groundwater. Secondary salinisation is caused by human interventions such as inappropriate irrigation practices, *e.g.* with salt-rich irrigation water or insufficient drainage.

The state-wise distribution of salt affected soils in India is presented in Table 3.2. Mapping of salt affected soils of India by the NRSA and Associates, (1996) indicated occurrence of about 3.77 m ha of sodic soils in India, out of which about 1.77 m ha represents the sodic nature among the irrigated/unirrigated black soil region (Vertisols) occurring in Rajasthan, Gujarat, Madhya Pradesh, Maharashtra, Andhra Pradesh, Karnataka and Tamil Nadu states.

3.1 Saline Vertisols

Saline soils contain excess neutral soluble salts like chlorides and sulphate of sodium, calcium and magnesium with ECe > 4 dS m^{-1}, pH <8.2 and ESP <15. Saline black soils due to their inherent physico-chemical properties, high clay content, low hydraulic conductivity and narrow workable moisture range are very difficult to manage. Osmotic effect of salt, toxic concentration of soluble ions like Na, Cl, B and reduced availability of essential nutrients due to competitive uptake affect plant growth in this type of soil. Excess salinity in these black soils results into delay germination, poor crop stand, stunted growth and reduced yield.

Table 3.2: State-wise Extent of Salt Affected Soils in India (ha)

State	Saline soils	Sodic soils	Total
Andhra Pradesh	77598	196609	274207
Andaman and Nicobar Islands	77000	0	77000
Bihar	47301	105852	153153
Gujarat	1680570	541430	2222000
Haryana	49157	183399	232556
Karnataka	1893	148136	150029
Kerala	20000	0	20000
Madhya Pradesh	0	139720	139720
Maharashtra	184089	422670	606759
Odisha	147138	0	147138
Punjab	0	151717	151717
Rajasthan	195571	179371	374942
Tamil Nadu	13231	354784	368015
Uttar Pradesh	21989	1346971	1368960
West Bengal	441272	0	441272
Total	2956809	3770659	6727468

Source: NRSA and Associates, 1996.

3.2 Sodic Vertisols

The distinguishing characteristics of sodic soils are high ESP, electrical conductivity (ECe) less than 4 dS m^{-1}, pH more than 8.2 and presence of higher amount of carbonate and bicarbonates of sodium. Some sodic soils are also termed as saline sodic as they contain large quantity of soluble salts and ECe is more than 4 dS m^{-1}. In sodic soils, the dominant cation on the exchange complex is sodium which disperses clay and imparts adverse soil physical conditions such as low permeability, crusting and hardening of the surface soils upon drying. Dense, slowly permeable sodic sub-soils reduce supplies of water, oxygen and nutrients necessary to obtain optimum yield (Rengasamy and Olsson, 1991). Besides, high Na content is often toxic to many plants, which exhibits poor growth and yield.

3.3 Saline Vertisols in Gujarat

The Vertisols of Bhal area and Bara tract in Gujarat (Figure 3.2) are generally very deep (150-200 cm), fine textured with clay content ranging from 45-68 per cent with montmorillonite as the dominant clay mineral. These soils are calcareous in nature with calcium carbonate (2 to 12 per cent $CaCO_3$) occurring in the form of nodules, kankar and powdery form and exhibit alkaline reaction. While the soils of Bhal area are highly saline, the soils of Bara tract have significant concentration of soluble salts in sub-soils, although the concentration in surface layer is low. In Bara tract, it was found that only 39.6 per cent of surface soils are free from salinity (< 2 dS m^{-1}), 49.3 per cent soils are saline (2-4 dS m^{-1}) and only 11.1 per cent soils are having salinity greater than 4.0 dS m^{-1}. Whereas 10 per cent of the sub-soil are having salinity

Figure 3.2: Salt Affected Black Soil in Gujarat.

less than 2 dS m^{-1}, 15 per cent between 2 - 4 dS m^{-1}and 75 per cent greater than 4 dS m^{-1} (Gururaja Rao *et al.*, 2001c, 2013b).

Sodic Vertisols are characterized by high pH (>8.2), high exchangeable sodium, high CaCO$_3$, very low organic matter content, poor physical conditions. The critical values of ESP depend on the electrolyte concentration of soil solution and range from 5 to 15 (Shainberg, 1984). Loveday and Pyle, (1973) working on a range of soil types from two regions of Australia reported that the differences in critical ESP are due to the differences in EC of the soil solution. They put forward a critical value of ESP 8 and above which the surface soils are found to disperse. Northcote and Skene, (1972) suggested that above ESP 6, the soils begin to disperse. Kadu *et al.* (1993) reported that drainage in these soils gets completely impaired due to higher dispersion of clay even at an ESP of 5. Robinson, (1971) working on Vertisols of Sudan reported highest cotton yields were associated with ESP of 8-16. The hydraulic conductivity of the sodic soils is affected by the initial swelling followed by clay dispersion (Gupta and Verma, 1984, 1985). Swelling reduces the pore size and dispersion clogs the soil pores. The sodicity increases the bulk density, which limits the root perforation. The plant cannot take up all of the water remaining in the root zone as rapidly as needed because it is held too tightly by the soil particles. Even at sufficient moisture content, the survival of crops in such soil is very difficult in the absence of sub-soil contribution due to low hydraulic conductivity and diffusivity. The hydraulic conductivity is an important property affected by the salinity and sodicity and can serve as the basis for classification for the degree of degradation. Chaudhary, (2001) reported that an increase of EC from 0.5 to 5 dS m^{-1} resulted in more than threefold increase in the hydraulic conductivity irrespective of SAR. A threshold electrolyte concentration (TEC) in the soil solution is necessary for flocculation. Nayak *et al.* (2004) reported that at electrolyte concentration (EC) < 2 dS m^{-1}, Vertisols can be grouped as sodic if the ESP is > 6 and > 10 in clayey and silty clay soils respectively and at EC of < 4 dS m^{-1}, Vertisols can be grouped as sodic if the ESP is >13 and >21 in clayey and silty clayey soils respectively.

In south Gujarat (Bharuch, Surat, Valsad and Navsari districts) sodic soils occupy about 55000 ha area. Typical soil profile characteristics of sodic Vertisols of Bharuch district is given in Tables 3.3 and 3.4. In Ukai-Kakrapar command area about 40 per cent of soils are affected by sodicity (Patel *et al.*, 2000). Paddy and sugarcane are the prominent crops in this command area due to availability of perennial irrigation facility and productivity of these crops is declining due to salinity/sodicity and water logging problems (Rana and Raman, 1999).

The low organic carbon, high exchangeable sodium, pH and calcium carbonate, toxic concentration of CO$_3$ and HCO$_3$ affect adversely the solubility, transformation and availability of the native and applied nutrient elements, especially N, K, Ca, Zn, Fe and Mn, their uptake and cationic balance in plants hindering crop growth and sustainable high productivity in these soils. The absence of carbonate ions in the saturation extract suggests that during high evaporative demands for soil-water in the semi-arid climatic conditions, maintenance of a proper Ca/Mg ratio in the soil solution becomes difficult because Ca^{2+} ions get precipitated as CaCO$_3$ resulting in an increase in the SAR of the soil solution and the ESP of the soil. Although the correlation

Table 3.3: Physico-chemical Properties of Sodic Haplusterts at Village Sadathala, District-Bharuch (Gujarat)

Horizon	Depth (m)	Sand (per cent)	Silt (per cent)	Clay (per cent)	WHC (per cent)	CEC (cmol/kg)	ESP (per cent)	CaCO$_3$ (per cent)	O.C. (per cent)
Ap	0.00-0.21	13.3	17.4	69.3	58.1	49.5	19.0	9.9	0.35
Bw1n	0.21-0.53	10.0	18.0	72.0	69.2	50.3	30.2	9.7	0.39
Bss1n	0.53-0.96	7.4	16.8	75.8	72.1	44.0	26.2	11.3	0.41
Bss2n	0.96-1.32	9.0	15.0	76.0	83.2	35.5	26.5	13.1	0.39
BC	1.32-1.70	15.0	25.0	60.0	70.0	46.6	29.4	19.8	0.43

Table 3.4: Saturation Extracts Analysis of Sodic Haplusterts at Village Sadthala, District-Bharuch (Gujarat)

Depth (m)	pH	ECe (dS/m)	Extractable Cations (meq/l)				Extractable Anions (meq/l)			SAR (meq/l)$^{1/2}$
			Ca	Mg	Na	K	Cl	CO$_3$+HCO$_3$	SO$_4$	
0.00-0.21	9.1	1.6	4.0	2.7	8.6	0.1	10.0	2.0	3.2	4.7
0.21-0.53	9.1	1.9	4.0	2.3	8.7	0.1	13.0	2.0	1.0	4.9
0.53-0.96	8.5	4.1	3.0	3.3	30.4	0.1	31.0	2.0	3.0	17.1
0.96-1.32	8.4	10.9	7.0	6.5	76.1	0.4	75.0	2.5	13.0	26.0
1.32-1.70	8.5	11.6	10.0	7.1	152.0	0.2	156.0	1.5	11.5	58.6

between carbonate clay and SAR was not significant, the HCO_3/Ca ratio of the saturation extract has a significant positive correlation with SAR ($r = 0.57$ at the 1 per cent level) (Balpande *et al.*, 1996). This suggests that if this ratio increases, then SAR will also increase as will the ESP.

3.4 Salt Affected Vertisols in the Bara Tract of Gujarat

The prevalence of low to moderate exchangeable sodium is observed in association of with soil salinity in some part of the Bara tract. The exchangeable sodium percentage (ESP) of the surface soils of the Bara tract (Figure 3.3) varies from 0.24 to 22.9 with a mean of 4.5 (SD = 3.9). The value of ESP to distinguish sodic soils

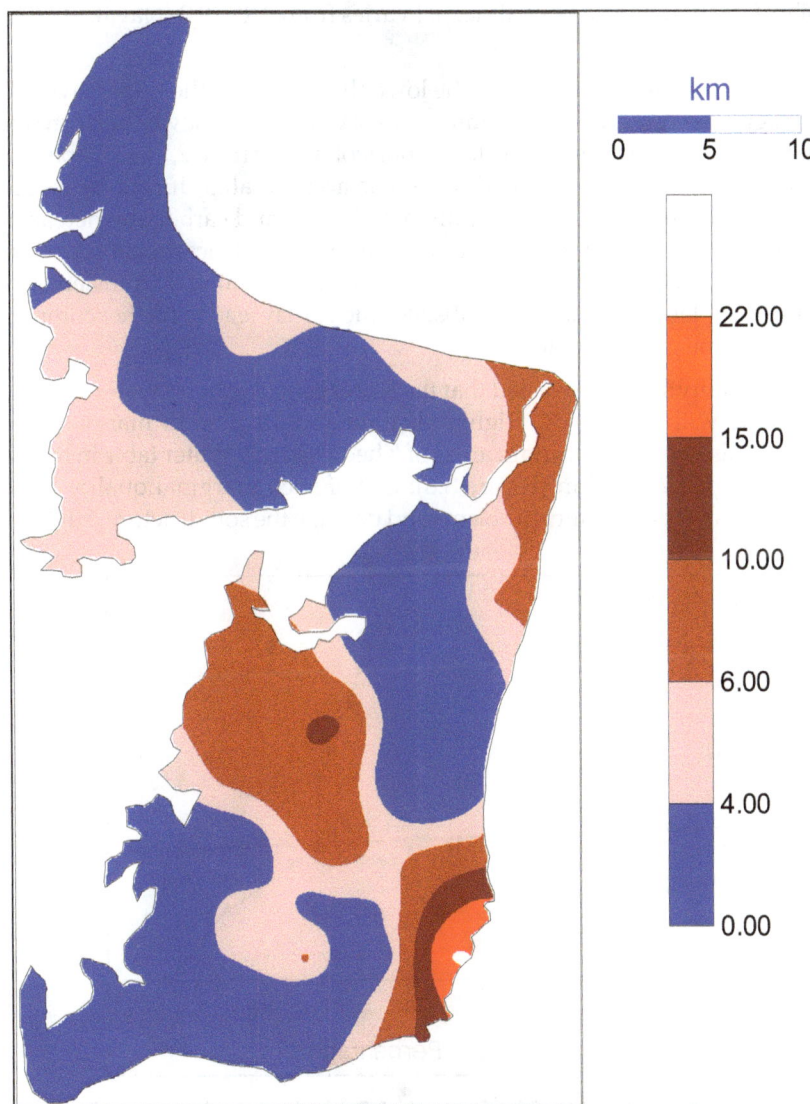

Figure 3.3: ESP of Surface Soil of the Bara Tract.

from non-sodic soil is found to be much less in case of Vertisols as compared to light textured soils. Accordingly these soils are classified into different sodic classes. The soils with electrolyte concentration (EC) < 2 dS m^{-1} and ESP > 6 are classified as sodic soils whereas the soils with EC > 4 dS m^{-1} and the ESP >13 are also termed as sodic soils. Accordingly most of the surface soils (90 per cent) of the Bara tract are free from sodicity (Figure 3.3). Howev, the problem of subsoil sodicity is encountered in a fairly large area of the Bara tract. Salinity of these soils shows a greater degree of temporal variation, the marginal case of 8-10 ESP with soil salinity of 2 dS m^{-1} having temporal variation of 0.5 dS m^{-1}, need no gypsum application, as the increase electrolyte concentration in some part of the year may lead to the flocculation of soil colloids. The soils of the Bara tract are alkaline; pH varies from 7.2 to 9.3 (Nayak *et al.*, 1999a, 2003).

These soils exhibit higher pH at the lower horizon than the upper layer. This is due to progressively increasing accumulation of CaCO$_3$ in the lower horizon (Figures 3.4 and 3.5). The lime content of the lower horizon varies from 2.7 to 27.4 per cent. In the lower layer where calcium carbonate has accumulated during pedo-genesis, sodium accumulation generates sodium bicarbonate and carbonate increasing the soil pH above 9 in addition to the toxicity of carbonate and bicarbonate species. This may lead to Fe, N, Cu, Zn and P deficiency. The subsoil salinity (transient salinity) occurring in dry land dominated by subsoil sodicity may lead to further complication with rising saline ground water.

Water quality studies revealed that the 80 per cent of ground water of the area is mostly saline with high SAR. Highly alkaline and alkaline to marginal alkaline waters were also noticed in certain areas of Kheda district. Water table in this region remains as high 1.2 to 1.5 m (Figure 3.6). Use of such marginal quality water for irrigation leads to permeability problems and damage the soil structure. High ground

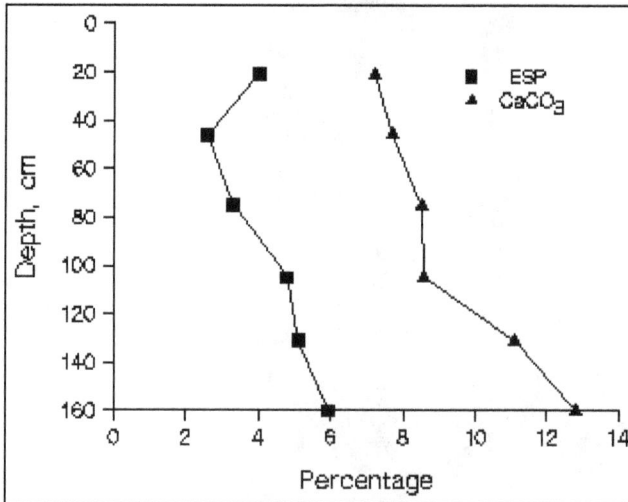

Figure 3.4: ESP and CaCO$_3$ Content of Typical Vertisols of Bara Tract (Vagra Village).

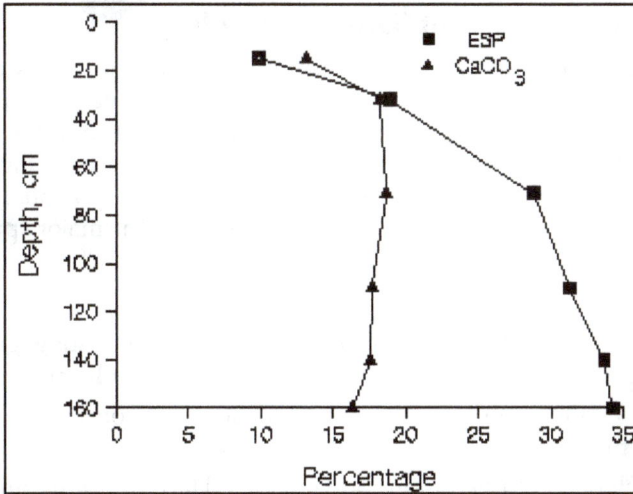

Figure 3.5: ESP and CaCO₃ Content of Sodic Vertisols of Bara Tract (Sadathala Village).

Figure 3.6: High Groundwater Table in Bhal Area of Gujarat State.

water table with highly saline water is quite prevalent in Bhal area. Over-exploitation of ground water resulting in depletion of ground water table makesthe present cropping systems unsustainable. Excessive withdrawal of ground water in coastal region has led to the intrusion of seawater leading to soil salinisation, pollution of drinking water supplies and large-scale migration of people from the affected areas. Since degraded soil and water and low soil fertility are the major constraints to increase and stabilize agricultural production, efforts must be made to conserve the vital resources. Rainwater management and the implementation of soil conservation programme hold the key to an ecologically balanced improvement in the quality of rain fed land (Gururaja Rao, 2012).

4. Management and Use of Saline Vertisols

Vertisols that are widespread in India, by and large remain fallow during rainy season which subjects them to soil erosion. Vertisols have great potential for productive cropping because their stored water can sustain crops through drought/stress periods. Although the water holding capacity is high, permeability is slow and drainability is poor. Vertisols form a considerable agricultural potential but adapted management is a precondition for sustained production. The major approaches for reclamation and management of vertisols are:

☆ Salinity control through ameliorative measures

☆ Conservation of resources through rain water harvesting, ground water recharging, nutrient and water budgeting and recycling of residues

☆ To use saline ground water in conjunction with conserved rainwater for crop production

☆ Developing multi-cropping and integrated farming with components like crops/fruit species/vegetables/biomass species and aquaculture

4.1 Physical Land Management

The physical properties and the soil moisture regime of Vertisols represent serious management constraints. The heavy soil texture and domination of expanding clay minerals result in a narrow soil moisture range between moisture stress and water excess. Tillage is hindered by stickiness when the soil is wet and hardness when it is dry. The susceptibility of Vertisols to water logging is the single most important factor that reduces the actual growing period. Excess water during rainy season must be stored for post-rainy use (water harvesting) on Vertisols.

Surface drainage by using alternate broad beds and furrows, protects crops from water logging of the root zone. The drained water may be stored in small ponds and used for watering vegetables etc. This technology solves problems on individual farmer's fields but the soil erosion caused by furrow systems needs to be tackled by bringing the runoff water safely along the grassed waterways to the lowest part of the field. A participatory approach involving all stakeholders is needed to solve this problem on a watershed scale.

Farm pond technology for storing excess water would serve irrigation needs of the crops. The swell-shrink behaviour of Vertisols poses serious problems and enhances percolation losses. On farm water management can be attained by having small ponds that provide irrigation needs of vegetables and water requirement of livestock.

4.2 Agro-technological Interventions

Since the saline black soils particularly in the costal belt due to climatic vagaries are difficult to manage and need suitable technological interventions those mainly comprising alternate cropping systems with biomass species, economic halophytes, forages and arid fruit species and using saline ground water for irrigation.

4.2.1. Cultivation of Halophyte, *Salvadora persica* on Highly Saline Black Soils (ECe > 30 dS m⁻¹)

Salt affected black soils, which constitute a major portion of saline soils present in the Gujarat state, pose serious threat to the economy of the state. While soils with low and moderate salinity have been put under cultivation, highly saline black soils by and large remain either barren or possess some native hardy species. Thus, for the management of moderate to highly saline black soils, agro-technology for the cultivation of economically important and salt tolerant halophyte has been evolved. *Salvadora persica* L. (Meswak), a facultative halophyte which is a potential source for seed oil has been identified as a predominant species in highly saline habitats of coastal and inland black soils. This species is a medicinal plant of great value and its bark contains resins and an alkaloid called Salvadoricine. The seeds are good source of non-edible oil rich in C-12 and C-14 fatty acids having immense applications in soap and detergent industry (Gururaja Rao, 1995, 2004).

Table 3.5: Cost of Cultivation of *Salvadora persica* on Highly Saline Black Soil (Cost taken per hectare of plantation)

Field Operations (Input costs)	*Cost (Rs.)*
Field preparation (by tractor)	500.00
Pitting (625 pits of 1' x 1' x 1')	625.00
Cost of Saplings @ Rs. 0.90 per plant	565.00
Planting	50.00
Irrigation during first year (saline water)	150.00
Digging of pit of 2.5 x 2.0 x 1m (for saline water)	300.00
Fertilizer (@ 50 g DAP/plant) and FYM	300.00
Plant basin making @ Rs. 0.35/plant	220.00
Miscellaneous (gap filling at 5 per cent)	50.00
Total	**2760.00**

Source: Gururaja Rao *et al.*, 2003.

Through different field experiments undertaken, the Regional Research Station of CSSRI has evolved agro-technology for raising of saplings using saline water; field planting and crop harvest and also worked out cost of cultivation. The studies indicated that the saplings could be raised using saline water of 15 dS m⁻¹, which is an advantageous feature under limited fresh water available situations. Cost of cultivation indicated that total cost for raising 500 saplings works out to be Rs. 455. The cost of cultivation under field conditions including raising of nursery comes to Rs. 2760 per hectare in the first year (Table 3.5). In subsequent years recurring costs would be mainly the labour for fertilizer application and harvesting. By fifth year, the plants would yield about 1800 kg ha⁻¹, thus giving net returns to a tune of Rs. 8400 per hectare. Thus, this species, while giving economic returns for the highly saline black soils with salinity values up to 50 dS m⁻¹, also provides eco-restoration through environmental greening and thus forms a niche for highly saline black soils (Table 3.6 and Figure 3.7; Gururaja Rao *et al.*, 2003, 2004).

Table 3.6: Seed Production and Economic Returns of *Salvadora* Plantation on Highly Saline Black Soils (ECe > 55 dS/m)

Year	Seed Yield tha⁻¹	Returns (Rs. ha⁻¹)		Cost/Benefit Ratio
		Gross	Net	
I Year	Nil	Nil	Nil	Nil
II Year	0.725	3625.00	365.00	10.03
III Year	0.978	4890.00	4340.00	0.13
IV Year	1.58	7900.00	7250.00	0.09
V Year	1.838	9190.00	8440.00	0.09

Source: Gururaja Rao *et al.*, 2003.

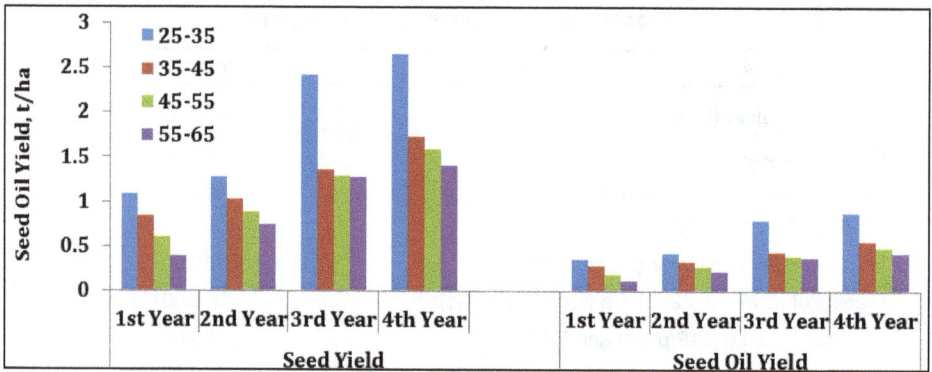

Figure 3.7: Seed and Oil Yield in *Salvadora persica* Grown at different Salinities.

This species was found to grow well on saline black soils having salinity up to 65 dS m⁻¹ and found to yield well. A spacing of 4 m x 4 m has been found ideal for planting on saline black soils (Gururaja Rao *et al.*, 2004). Based on the studies conducted, the National bank for Agriculture and Rural Development (NABARD), Mumbai in association with the Station has developed a bankable model scheme for cultivation of *Salvadora persica* on salt affected black soils through the project sponsored by NABARD (Figure 3.8). Regreening of highly saline black soils that cannot be put under arable farming; reduction in salinity by 4th year onwards that enable to take up intercropping with less tolerant crops/forages.Planting of *Salvadora persica* would fetch about Rs. 7000 per hectare. Apart from this, the species provides a dwelling place for birds and enhances the environmental greening (Gururaja Rao *et al.*, 2012, 2013a).

The targeted areas are saline black soil region in Bhal area and Bara tract regions in Gujarat (Parts of Ahmedabad and Anand districts) and Amod, Vagra and Jambusar talukas in Bharuch district covering about 500 ha and coastal areas of Kachhchh.

NATIONAL BANK FOR AGRICULTURE AND RURAL DEVELOPMENT

Model Bankable Agricultural Projects &
Area Specific State Projects

Home   Vol 1,2,3   IMPROVEMENT OF SALT AFFECTED BLACK SOILS BY USING
SALVADORA SPECIES

LandDevelopment

Biopesticide Unit
Biofertilizers
Agriclinic for
Vermicompost

Reclamation of Saline
Soils
Salvadora
speciesTesting Services

NADEP Compost

**TECHNICAL ASPECTS AND
AGRONOMIC PRACTICES OF
*SALVADORA***

With R&D grant assistance from NABARD Central Soil Salinity Research Institute, Regional Research Station, Anand (now in Baruch) have conducted a field experiments and standardized package of practices for growing Salvadora. Based on a detailed study conducted by NABARD in Gujarat and also discussions had with Scientists of CSSRI, Regional Research Station, Bharuch, the techno economic aspects of *Salvadora* has been finalized.

References:
1. Technical Bulletin No.1/2003 "Salvadora persica: A life support species for salt affected black soils" by G Gururaja Rao, A K Nayak and Anil Chinchmalatpure, published by Central Soil Salinity Research Institute (ICAR), Regional Research Station, Bharuch, Gujarat
2. R&D Project Report "Management of salt affected black soils using *Salvadora* – forage grass based land use system" submitted to NABARD by CSSRI, Regional Research Station, Bharuch in 2002.

**Figure 3.8: Bankable Model Scheme Developed by NABARD Mumbai in
Association with CSSRI, RRS, Bharuch for Cultivating *Salvadora persica*
on Highly Saline Black Soils.**

Output capacity: Seed yield: 1846 kg/ha (by fifth year)

Unit cost: Rs. 2760/- per ha in the first year

Total realization: Rs. 9230/- per ha (net income)

User Agencies: Gujarat State Land Development Corporation, State Agricultural Development, NGOs like Coastal Salinity Prevention cell, Ahmedabad, Saline Area Vitalisation Enterprise and farmers of coastal areas.

4.2.2. Physiological Basis of Salt Tolerance in *Salvadora persica*

4.2.2.1. *Growth*

Salinity in the root medium manifests its effects on growth and physiology of plants. Halophytes like *Salvadora* generally do not get affected by salinity and maintains its growth. Growth measured in terms of plant height and canopy coverage showed that the plants attained a height of 1.95 m by 4th year at 25-35 dS/m and 1.48 m at 55-65 dS/m salinity levels. The canopy coverage of 1.72 m and 1.12 m at 25-35 dS/m and 55-65 dS/m salinity, respectively indicate that this species is capable of withstanding high salinity of 65 dS/m (Figure 3.9). The higher canopy coverage facilitates better light interception. Higher growth is also ascribed to higher synthesis of osmotic substances such as proline, amino acids, sugars which facilitate osmotic adjustment (Gururaja Rao *et al.*, 2001c, 2009a) which are also in conformity with Maggio *et al.* (2000).

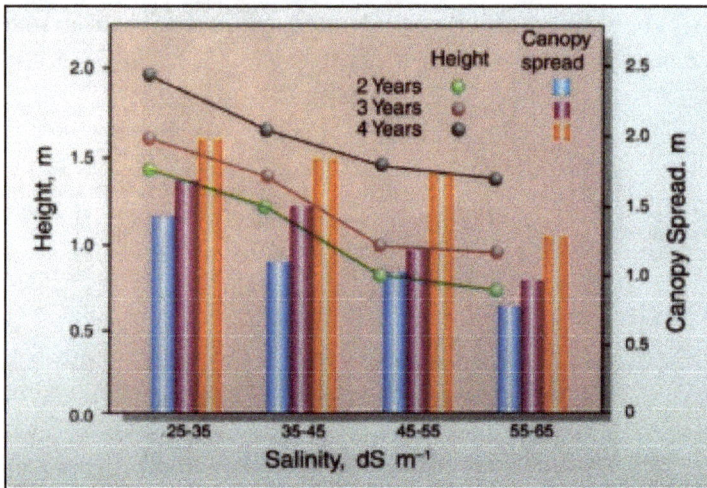

Figure 3.9: Growth of *Salvadora persica* on Highly Saline Black Soil.

4.2.2.2. Salt Compartmentation

Distribution of sodium and chloride ions (Figure 3.10) studied in different plant parts of *S. persica* growing at different in situ salinities indicated bark and senescing leaves as the potential sinks for such toxic ions, there by sparing other plant parts like immature leaves, partially mature and physiologically mature leaves to perform their normal physiological activity and help in normal growth and development, which enable the plants to remain lush green even at high salinity. Further, senescing leaves act as potential sinks for toxic ions thereby reduce the load on other photosynthesizing tissues, which remain by and large salt free (Gururaja Rao *et al.*, 1999, 2003).

4.2.2.3. Na⁺ and Cl⁻ Concentration and Flux

The rate and ion transport (flux) from root to shoot and to whole plant was calculated using the formula, $Js = (Ms2-Ms1) \ln (WR2/WR10/(t2-t1)(9WR2-WR1)$, where Js is the rate of transport (flux), Ms1 and Ms2 are the amounts of ion in the shoot/whole plant and WR1 and WR2 are the fresh weights of the roots at the harvest times t2 and t1 (Pitman, 1975). Concentration of sodium and chloride in plant parts increased with increase in salinity of the soil. Maximum amount of Na⁺ and Cl⁻ ions were retained in the bark, root and senescing leaves sparing immature (expanding) and mature (fully expanded) leaves over the years (Table 3.7). These tissues act as

Table 3.7: Concentration of Na$^+$ and Cl$^-$ (per cent) Ions in different Plant Parts of *Salvadora persica* Grown on Saline Black Soils

Plant Part	Salinity Range, dS m^{-1}											
	25–35			35–45			45–55			55–65		
	2nd Year	3rd Year	4th Year	2nd Year	3rd Year	4th Year	2nd Year	3rd Year	4th Year	2nd Year	3rd Year	4th Year
Na$^+$												
Root	1.13	1.62	1.86	1.61	2.11	2.30	1.91	2.37	2.57	1.93	2.39	2.60
Wood	0.02	0.03	0.03	0.03	0.04	0.04	0.03	0.04	0.05	0.04	0.06	0.06
Bark	1.59	1.72	1.95	2.12	2.29	2.60	2.33	2.52	2.86	2.73	2.95	3.34
Im. Leaf	0.02	0.02	0.02	0.02	0.02	0.03	0.02	0.02	0.03	0.03	0.03	0.03
M. leaf	0.18	0.18	0.21	0.19	0.20	0.23	0.21	0.22	0.25	0.22	0.24	0.27
S. leaf	1.66	1.81	2.06	2.11	2.30	2.61	2.30	2.38	2.71	2.39	2.51	2.83
CD 0.05	0.36	0.13	0.18	0.46	0.21	0.22	0.89	0.16	0.30	0.69	0.46	0.71
Cl$^-$												
Root	2.13	2.65	2.70	2.65	3.36	3.60	2.88	3.00	3.90	2.94	4.00	4.01
Wood	0.04	0.05	0.06	0.05	0.06	0.06	0.05	0.07	0.07	0.07	1.00	0.10
Bark	2.49	2.69	3.05	3.43	3.70	4.01	3.74	4.04	4.41	4.26	4.62	5.14
Im. Leaf	0.03	0.04	0.04	0.04	0.04	0.05	0.04	0.04	0.04	0.5	0.05	0.06
M. leaf	0.28	0.29	0.32	0.31	0.31	0.38	0.33	0.34	0.39	0.35	0.40	0.42
S. leaf	2.63	2.82	3.19	3.17	3.67	2.71	3.58	3.76	4.23	3.58	4.04	4.59
CD $_{0.05}$	0.40	0.16	0.39	1.05	0.28	0.30	0.58	0.22	0.40	0.95	0.98	0.83

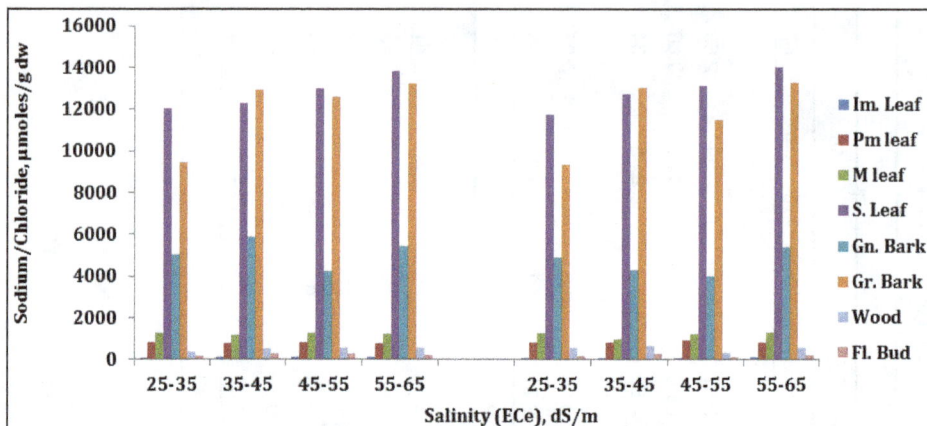

Figure 3.10: Compartmentation of Sodium and Chloride in *Salvadora persica* Grown at different Salinities on Highly Saline Black Soil.

potential sink for excess Na^+ and Cl^- ions. The capacity of the sink increased with age of the plant as well as increase in salinity which indicates that *S. persica* has very well developed salt compartmentation mechanism (Gururaja Rao *et al.*, 2000c). Though Na^+ concentration increases with increase in salinity, the total Na uptake showed a decreasing trend which may be obviously due to decrease in the biomass yield with increase in salinity. Similarly, chloride uptake in root is much higher than that of the shoot. The rate of flux of Na^+ and Cl^- ions to the whole plant while increase with increase in salinity showed a decreasing trend with age (Table 3.7). The flux of these ions from root to the shoot was a fraction of that to the whole plant indicating that roots accumulate more ions than shoots. In this species, roots act as both Na^+ and Cl^- accumulator (Table 3.8).

4.2.2.4 Seasonal Variation in Ionic Concentration

Data presented in Figure 3.11 indicated that immature, partially mature and mature leaves possess 5-8 per cent (of total ion uptake) during dry and wet months, while wood had very low amounts (0.5-1.5 per cent). Contrary to this, senescing leaves accounted 40-48 per cent during dry months and 25-26 per cent in wet season. However, bark accounted 46-53 per cent in dry season and 67-68 per cent in wet season. These studies clearly indicated that bark tissues retained more sodium and chloride during wet season, thus lowering the salt injury to the leaves, and hence low degree of senescence when the soils had adequate moisture coupled with low salinity. Leaves of *S. persica* also showed some degree of succulence which may also facilitate the dilution of salt within the tissues (Tiku, 1976, Gururaja Rao and Rajeswara Rao, 1982, Gururaja Rao *et al.*, 2000c).

4.2.2.5 Tissue Tolerance: Relation between Sodium and Chlorophyll

Data on tissue tolerance as studied by combined chlorophyll and sodium estimations indicates that about 2250 μmoles of sodium is needed to reduce chlorophyll content of leaves by 50 per cent, when compared to leaves of plants growing in non-

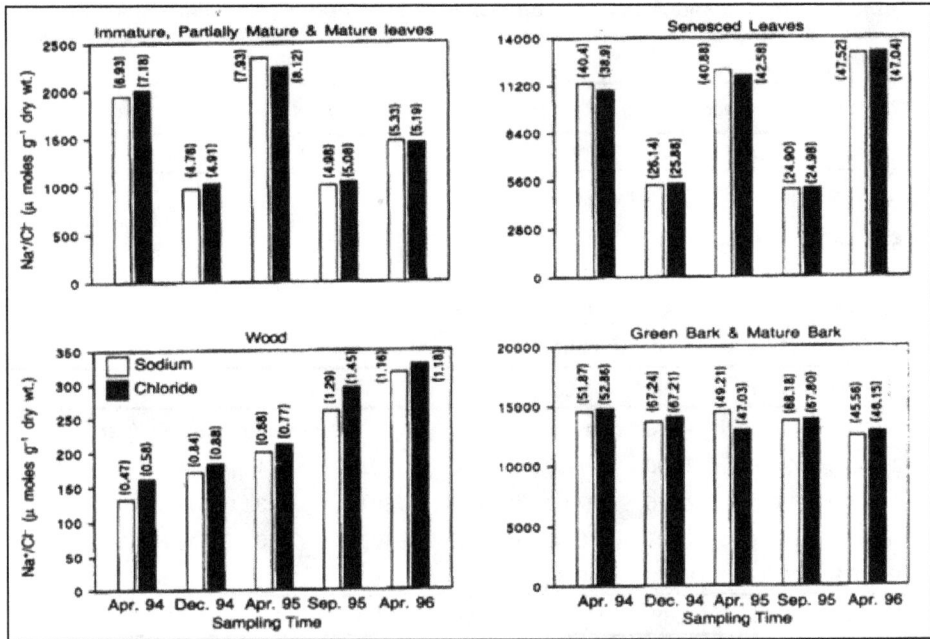

**Figure 3.11: Seasonal Variation in Ion Uptake in *S. persica*
Grown on Highly Saline Black Soils.**

saline environment. The scatter diagram (Figure 3.12) indicates that the high degree of scatter could be due to both variations in individual plants and from leaf-to-leaf. A fitten linear regressions account of 58 per cent of the variability although the subjective

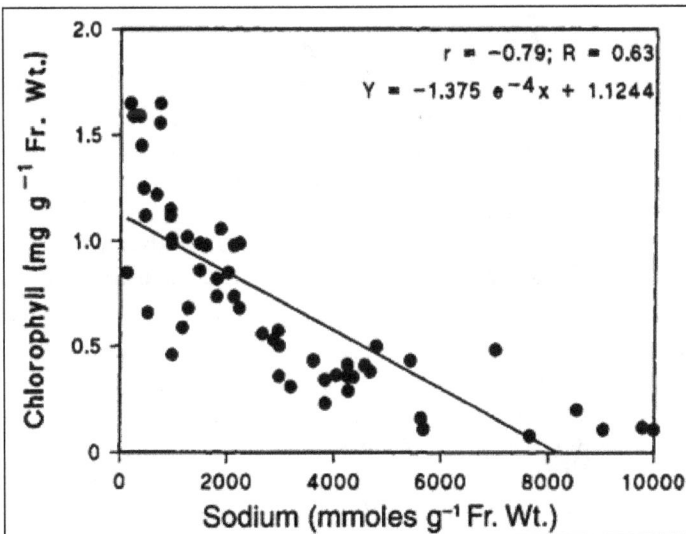

**Figure 3.12: Relation between Leaf Sodium and Chlorophyll in *S. persica*
(Pooled data for all leaves).**

Table 3.8: Uptake and Flux of Na⁺ and Cl⁻ Ions in *S. persica* on Saline Black Soils

Salinity Class, $dS\,m^{-1}$	Uptake (g)				Flux ($\mu g\,g^{-1}\,day^{-1}$)			
	Shoot		Root		Shoot		Root	
	Na^+	Cl^-	Na^+	Cl^-	Na^+	Cl^-	Na^+	Cl^-
	2nd Year							
25 – 35	6.44	10.18	8.40	15.86				
35 – 45	5.12	8.53	9.31	15.29				
45 – 55	4.10	6.57	6.58	9.91				
55 - 65	3.68	5.64	4.97	7.56				
CD 0.05	1.21	1.88	1.93	2.12				
	3rd Year				**Between 3rd and 2nd Year**			
25 – 35	16.01	25.90	27.36	44.93	29.9	46.1	9.8	16.2
35 – 45	14.21	22.95	27.69	44.08	39.0	61.3	12.9	20.4
45 – 55	10.13	16.21	18.56	29.43	50.2	81.3	16.8	26.9
55 - 65	9.82	15.59	13.62	22.84	78.8	131.4	19.5	52.6
CD 0.05	2.11	2.88	3.58	5.35	10.5	13.8	4.3	5.8
	4th Year				**Between 4th and 3rd Year**			
25 – 35	22.31	34.71	38.33	56.66	10.8	12.9	3.9	5.5
35 – 45	18.42	28.69	37.64	58.73	12.3	17.8	3.7	5.0
45 – 55	14.43	22.30	37.23	37.23	17.8	23.5	7.3	9.7
55 - 65	13.51	20.84	29.35	29.35	29.7	410.2	11.9	16.9
CD 0.05	3.95	4.23	0.53	1.88	1.88	3.50	1.20	1.70

appearance of the fit is poor at higher sodium concentration. The salinity induced decrease in chlorophyll can be attributed to a weakening of pigment-protein-lipid complex (Strogonov, 1964). The scatter diagrams showed that as Na concentration increased, chlorophyll content decreased both in immature and mature leaves. The LC_{50} (lethal concentration) value for Na analogous to LD_{50} (lethal dose) of toxicology is defined here as the mean Na concentration in the individual leaves having more than 50 per cent of chlorophyll on non-salinity healthy leaves.

The LC_{50} value for the immature leaves is much less than that of the senescing leaves. Such a high value of LC_{50} in senescing leaves indicated that the leaves of S. *persica* possess a fairly degree of tissue tolerance to Na. The linear regression equation for immature and senescing accounts for 82.1 and 99.0 per cent variability, respectively (Figure 3.13), while the earlier studies in S. *persica* indicated 58 per cent variability both in immature and mature leaves when data of all leaves were pooled in the scattered diagram (Gururaja Rao *et al.*, 1999a, b).

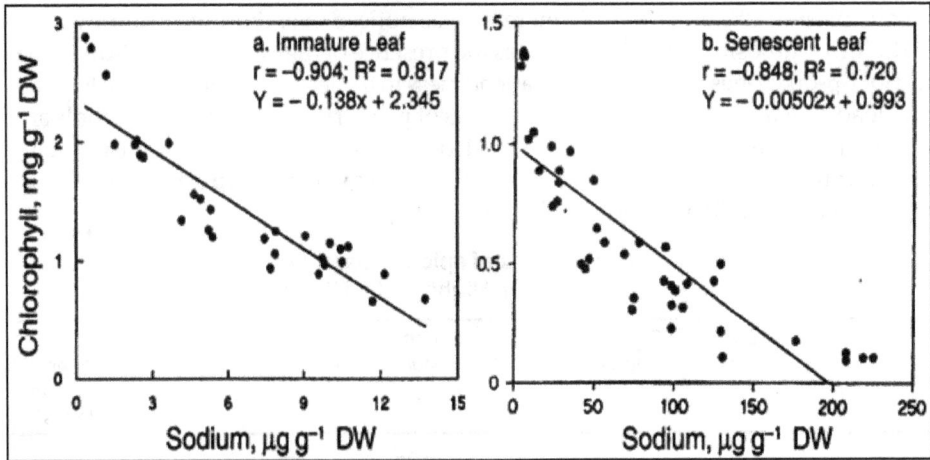

Figure 3.13: Relation between Leaf Sodium and Chlorophyll in
S. persica (Immature and senescing leaves).

4.2.2.6. Free Proline

Proline was found to increase with salinity and the increase was three-fold at 55-65 dS/m when compared to low salinity of 25-35 dS/m (Figure 3.14). Very high amount of free proline was noticed in dry season when compared to wet season, which helps in osmoregualtion as reported earlier (Weretelinyk *et al.*, 1982).

4.2.2.7. Epicuticular Waxes

Another adaptive feature exhibited by *S. persica* is its ability to synthesise very high amounts of leaf epicuticular waxes which increase with salinity and at highest

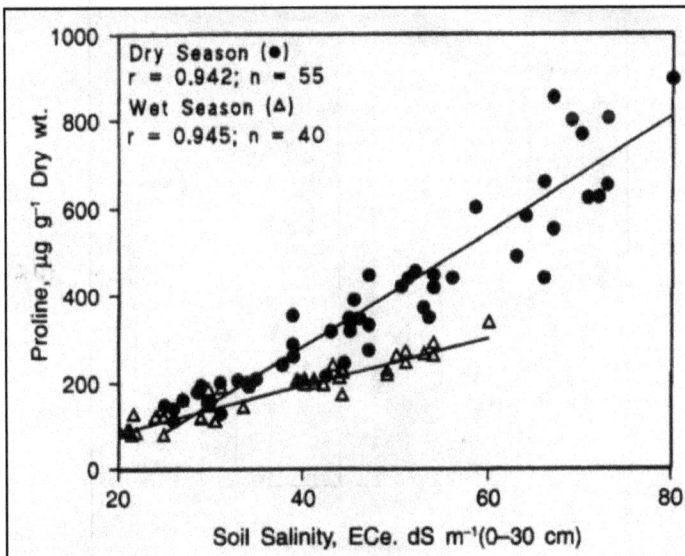

Figure 3.14: Relation between Leaf Sodium and Proline in *S. perisca* under Salinity.

salinity of 55-65 dS/m, a four-fold increase was noticed over the lowest salinity of 25-35 dS/m. Composition of LEW indicates the presence of high amounts of β-diketones and OH-β-diketones which impart glaucousness to the leaves resulting in the lowered heat load on leaves. LEW with high amounts of primary and secondary alcohols and aldehydes (Figure 3.15 and Table 3.9) also plays an important role in reducing cuticular transpiration, a water conservation strategy to cope up stress induced by salinity (Gururaja Rao and Ravindra Babu, 1997).

**Table 3.9: Composition of leaf epicuticular wax (mg dm^{-2}) of
S. persica Grown on Highly Saline Black Soil**

Salinity Range. dS/m	Fatty Acids	OH-β-diketones	Primary Alcohols	Secondary Alcohols	β-diketones	Aldehydes	Hydro-carbons, Esters and Ketones
25-35	0.84	2.54	2.61	2.47	1.90	0.82	0.82
35-45	2.16	3.14	3.24	2.16	2.16	1.06	1.08
45-55	2.75	3.17	3.20	1.76	1.76	0.80	0.90
55-65	3.12	7.70	7.72	7.68	5.16	4.68	1.90
CD $_{0.05}$	0.62	0.53	0.61	0.58	0.48	0.50	0.23

4.2.3 Soil Salinity

The soils grouped a deep, clay loam, hyperthermic, montmorillonitic family of Vertic Haplustepts showed high degree of spatial and temporal variation in soil

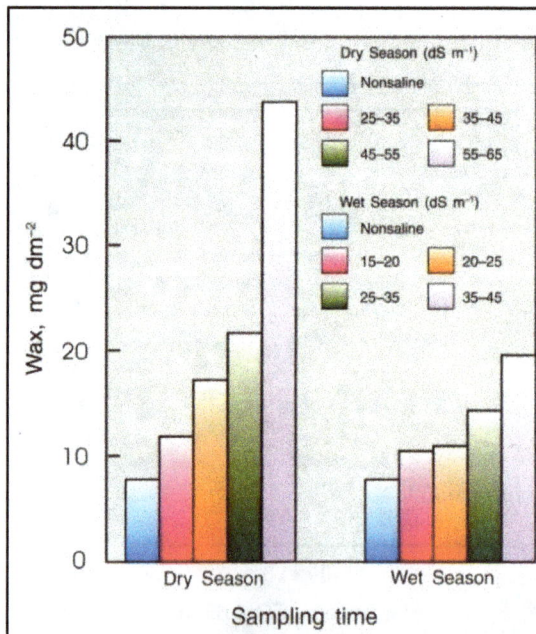

Figure 3.15: Leaf Epicuticular Waxes in *S. persica* during Wet and Dry.

salinity ranging from 65 to 70 dS m^{-1}. Salinity of the soil decreased with depth *i.e.*, from surface to 90 cm depth (Figure 3.16). Cultivation of *S. persica* up to 5 years resulted in slight decline in soil salinity when compared to the pre-planting salinity. Changes in surface salinity are partly attributed to the ability of plants to extract the salt and partly due to root activity which improves the physical properties of the soil. However, the magnitude of fluctuation in salinity was not much at lower layers. The ground water table might be contributing to such small changes at lower depths. The spatial variability of surface salinity under 5 years old plantation (Figure 3.17) showed significant difference from the initial salinity prior to planting.

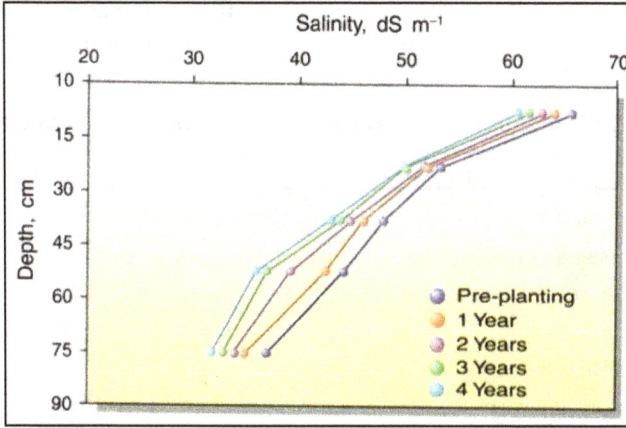

Figure 3.16: Soil Salinity Variations Over the Years under *S. persica* Grown on Highly Saline Black Soil (55-65 dS/m range).

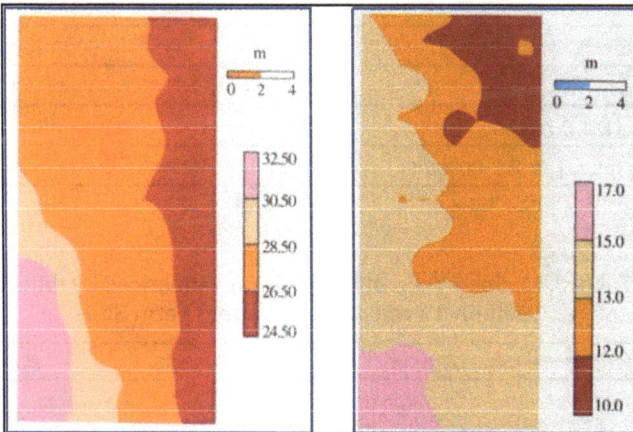

Figure 3.17: Spatial Variability in Soil Salinity Prior and after Planting of *S. persica*.

4.2.4 Cultivation of Dill (*Anethum graveolens*) on Saline Black Soils

Dill, *Anethum graveolens*, a non-conventional seed spice crop has been identified as potential crop for cultivation on saline black soils having salinity up to 6 dS m^{-1} in *rabi* season with the residual soil moisture. It has multiple uses *viz.*, pot herb, leafy vegetable, seeds used as condiments and seed oil for aromatic and medicinal purposes.

The herb contains Vitamin-C as high as 121.4 mg/100g. The oil of dill seeds and its emulsion in water (Dill water) are considered to be aromatic, carminative and effective in colic pains and possesses anti-pyretic and anti-helmenthic properties. The crop responds well to saline water irrigation. Three critical stages for saline water irrigation have been noticed *i.e.*, vegetative, flowering and seed formation stage. A substantial increase in yield can be obtained by using saline ground water in conjunction with best available surface water. Cost of cultivation and economics have been worked for this species. The cost of cultivation comes to Rs. 6000/- ha^{-1} and the crop would yield net returns of Rs. 16500/- ha^{-1}. The benefit: cost ratio works out to be 2.75. This crop thus would help farmers of the region to go for the second crop in the *rabi* season on lands, which hitherto remain fallow due to water and salinity constraints (Tables 3.10–3.12; Gururaja Rao *et al.*, 2000a, 2001a).

Table 3.10: Cost of Cultivation of Dill on Saline Black Soils and Economic Returns

Items of Expenditure	Cost (Rs.)
Field preparation	1200.00
Seed material	150.00
Seed treatment and sowing	300.00
Fertilizers	800.00
Application of fertilizer	200.00
Interculture and weeding	350.00
Irrigation*	1500.00
Harvesting and threshing	1000.00
Miscellaneous	500.00
Total	6000.00
Returns	
Yield of Dill, 0.75 t ha^{-1}	
Gross returns @ Rs. 30000 t^{-1}	22500.00

Source: Gururaja Rao *et al.*, 2000a.

Table 3.11: Seed Yield of Dill (q/ha) as Influenced by different Salinity Levels Under different Farm Sites

Salinity, dS m^{-1}	Khanpur	Warsada	Bamangam
2-4	3.74	4.09	3.16
4-6	3.03	2.25	2.37
6-8	2.34	1.76	1.64
8-10	1.95	1.36	1.56
CD $_{0.05}$	NS		
Farm Salinity x Farm Salinity	0.13 NS		

Source: Gururaja Rao *et al.*, 2000a.

Table 3.12: Effect of Quality and Number of Irrigations on Yield of Dill (t ha⁻¹) on Saline Black Soils

Salinity (dS m⁻¹)	One Irrigation		Two Irrigations		Three Irrigation	
	Seed	Stover	Seed	Stover	Seed	Stover
BAW	0.784	2.352	0.834	2.500	0.914	2.651
4	0.650	1.958	0.815	2.526	0.906	2.808
8	0.354	1.200	0.417	1.334	0.567	1.814
12	0.209	0.689	0.292	0.992	0.367	1.212
$CD_{0.05}$						
No. of irrigations (I)	0.030					
Quality of Water (Q)	0.033					
I X Q	0.081					

Source: Gururaja Rao *et al.*, 2000a.

Non-conventional crop like Dill can be grown using residual moisture resulting in 2.6 q/ha seed yield with net returns of Rs. 8000/-. This crop forms an ideal option for the state in general and the region in particular, which by and large faces water scarcity problems (Gururaja Rao *et al.*, 2000). Under saline water irrigation, crop would yield net returns of Rs. 16500/- ha⁻¹ with Rs. 6000/- per hectare as cost of cultivation. The benefit: cost ratio works out to be 2.75. This crop thus would help farmers of the region to go for the second crop in the *rabi* season on lands, which hitherto remain fallow due to water and salinity constraints. Thus dill crop can be taken up using residual moisture and/or with saline ground water. The green can be used as leafy vegetable, an additional source of income.

The technology evolved in cultivation of this species on saline soils in *rabi* season (after paddy) with residual moisture has been widely adopted by farmers of Bhal region an area of about 1200 ha. Apart from this, through the network of Gujarat State Land Development Corporation, Gandhinagar, the technology has been widely adapted by the farmers in other parts of the Bhal region in Dhandhuka taluka of Ahmedabad district covering about 300 ha. The seed materials have also been provided to Coastal Salinity Prevention Cell, Ahmedabad to its cultivation in Coastal Saurashtra.

Output capacity: Seed yield: 750 kg/ha

Unit cost: Rs. 6000/- per ha in the first year

Total realization: Rs. 16500/-per ha (net income)

4.2.5. Cultivation of Forage Grasses on Saline Black Soils

Gujarat state has one of the largest dairy industries in the country. As the fodder produced on arable lands and grasslands is not sufficient to meet the demands of the cattle population, cultivation of forage grasses, *Dichanthium annulatum* and *Leptochloa fusca* in a ridge-furrow planting system with 50 cm high ridge and 1 m between midpoints of two successive ridges was found ideal in saline black soils having salinity up to 8-10 dS m⁻¹ (Table 3.6). For maximizing forage production on saline black soils, *Dichanthium* on ridges and *Leptochloa* in furrows form ideal proposition. Nitrogen given at the rate of 45 kg ha⁻¹ (in the form of urea) at the time of rooted slip planting, boosts forage production and improves forage quality traits. *Dichanthium annulatum,* has been found most suitable for saline black soils, as it possessed well-defined salt compartmentation, wherein the roots act as potential sinks for toxic ions like sodium and chloride, making the shoot portions relatively salt free (Table 3.7, Gururaja Rao *et al.,* 1998, 2001b).

Table 3.13: Growth and Yield of Forage Grasses Under Ridge and Furrow Planting System (Salinity of the saturation extract (0-30 cm) : 15.4 dS m⁻¹)

Grass Species	Height (m)		Tillers Plant⁻¹		Green Forage Yield (t ha⁻¹)	
	Ridge	Furrow	Ridge	Furrow	Ridge	Furrow
Leptochloa fusca	1.18	1.02	10.62	9351	3.17	3.73
Dichanthium annulatum	0.91	0.74	6.41	5.32	1.85	1.76
CD ₀.₀₅						
Planting method	0.12	0.91	NS			
Grass species	0.16	1.53	0.82			
Planting method x Grass species	NS	2.24	NS			

Source: Gururaja Rao *et al.*, 2001a.

Experiments on forage grasses indicated that *Dichanthium annulatumand Leptochloa fusca* as the ideal ones for saline black soils. *Dichanthium* is having both well-defined salt exclusion mechanism and osmotic adjustment which makes it salt tolerant. *Leptochloa fusca* also gave maximum forage yield. Furrow method of planting is suitable for cultivation of this grass (Figure 3.18 and Table 3.13). Application of nitrogen at 46 kg/ha as urea increased the forage yield by about 70 per cent in *Dichanthium annulatum* (Table 3.14, Gururaja Rao *et al.,* 2001). The cattle and camel populace form the important livestock of the region. Cultivation of salt tolerant grasses like *Dichanthium annulatum* and *Leptochloa fusca* on moderate saline soils results in 1.9 t/ha and 3.2 t/ha, respectively.

4.2.5.1 Growth

Data on plant height and tiller production at the time of harvesting (Table 3.15) indicated that *Aeluropus lagopoides* produced higher number of tillers and showed better growth at all salinity treatments when compared to *Eragrostis* indicating its higher salt tolerance. While plant height decreased with increase in salinity in *Eragrostis, Aeluropus* showed increased growth with salinity. The total tiller production was very high in *Aeluropus.* Though there is an increase in tillers with salinity in both the species, their further growth got slowed down with increase in salinity, particularly in *Eragrostis.*

A decrease was recorded in growth with an increase in the level of saline water over a period of 36 days (flower initiation stage) particularly in *Eragrostis.* However,

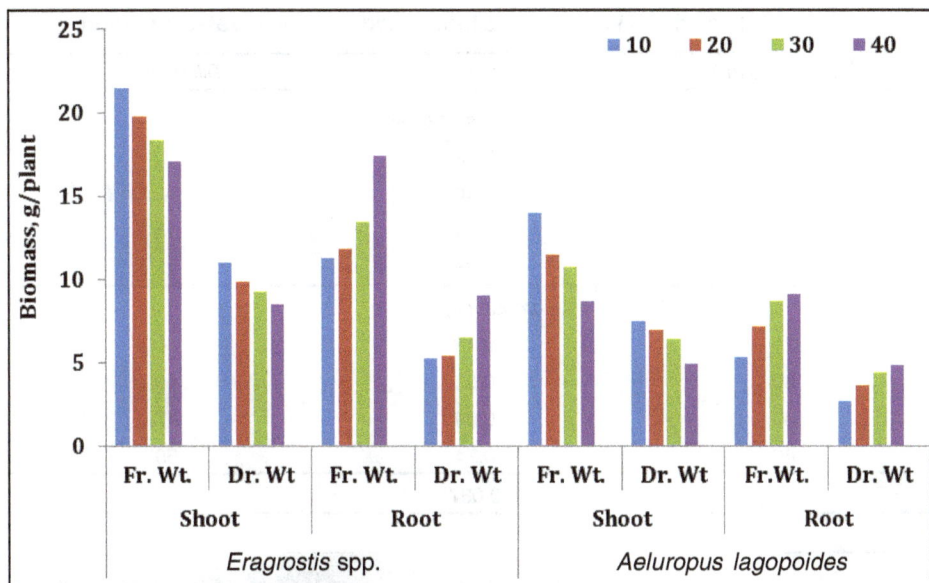

Figure 3.18: Biomass of Halophytica Forage Grasses under Saline Water Irrigation.

the difference between 10 and 20 dS m^{-1} saline water treatments over a period of time is almost similar in *Aeluropus lagopoides* while *Eragrostis* showed only little variation (Figure 3.19). At highest salinity, it was clearly seen that though the buds producing new tillers were emerged, the high salt content of the shoot system affected their further growth. However, these grasses responded well to the saline irrigation even at 30 dS m^{-1} of irrigation water. Tissue ion content of the leaf and stem samples indicated that higher amount of sodium and chloride accumulated in the tissues at highest salinity indicating their ability to absorb the ions (Figure 3.20; Gururaja Rao *et al.*, 2005, 2011).

Table 3.14: Effect of Nitrogen on Growth and Forage Yield of Forage Grasses

Grass Species	Height (m)		Tillers Plant^{-1}		Green Forage Yield (t ha^{-1})	
	+ N	– N	+ N	– N	+ N	– N
Leptochloa fusca	1.39	0.99	12.54	4.46	3.21	2.13
Dichanthium annulatum	1.01	0.87	10.24	7.38	2.24	1.32
CD $_{0.05}$						
Planting method	0.13		3.11			
Grass species	0.22		2.32			
Planting method x Grass species	NS		NS			

Source: Gururaja Rao *et al.*, 2001a.

Table 3.15: Growth of Halophytic as Influenced by Saline Water Irrigation

Salinity, dS m⁻¹	Height, m	Tillers Plant⁻¹
Eragrostis species		
10	0.49	8
20	0.40	9
30	0.34	15
40	0.28	15
Aeluropus lagopoides		
10	0.45	40
20	0.51	46
30	0.69	47
40	0.74	55
CD $_{(0.05)}$	0.087	3.1

Figure 3.20: Grasses Irrigated with Saline Water.

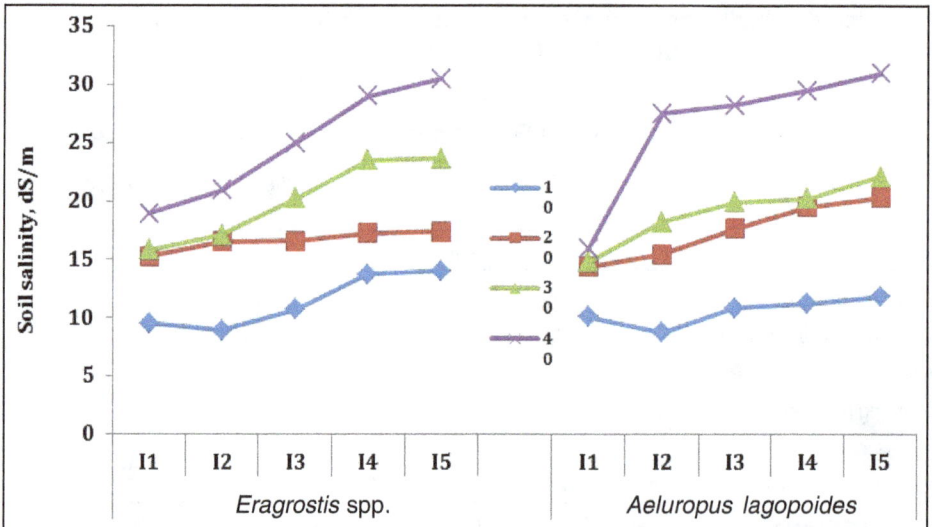

Figure 3.20: Soil Salinity Build-up under Forage Grasses Irrigated with Saline Water.

Irrigation with saline water over a period of time increased the soil salinity under saline water irrigations of 10, 20, 30 and 40 dS m^{-1}. The saline water irrigation was given till the grasses showed flower initiation (about 5 - 6 weeks). The salinity build up (Figure 3.21) indicated highest salinity of 30.5 dS m^{-1} under *Eragrostis* spp. and 31 dS m^{-1} under *Aeluropus lagopoides* when irrigated with 40 dS m^{-1} saline water. There was a significant difference in salinity build up under different treatments indicating the salt build up in the pots (Gururaja Rao *et al.*, 2011).

4.2.5.2. Biochemical Constituents

4.2.5.2.1. Ion Content

The leaf and stem ions *i.e.*, Na$^+$ and Cl$^-$ increased with the increase in salinity. The shoot (leaf and stem) sodium content after two irrigations increased from 2000 µmoles to 5900 µmoles in leaf of *Eragrostis* spp. and 3500 µmoles to 5100 µmoles in *Aeluropus lagopoides*. In stem the Na$^+$ and Cl$^-$ content were higher when compared to the leaves indicating stem as a potential sink. Among the grasses, Na$^+$ and Cl$^-$ contents were found to be more in *Aeluropus lagopoides* than *Eragrostis* spp. Ion partitioning (Na$^+$ and Cl$^-$) in shoots and roots of two grasses indicated that roots do act as sinks for these toxic ions (Figure 3.21). Uptake and flux of Na$^+$ and Cl$^-$ and the total Na$^+$ uptake showed a decreasing trend with increase in salinity of irrigation water in both the grasses. Among the grasses, *Aeluropus lagopoides* showed higher uptake than that of *Eragrostis spp.* though the increase was only marginal (Tables 3.16 and 3.17). The total Na$^+$ content is less in shoot than in the root in both the grasses irrespective of salinity and age of the plant. Chloride uptake, however, is relatively more in root than in shoot. The rate of flux of Na$^+$ and Cl$^-$ to the whole plant though increased with salinity and age of the plant (Gururaja Rao *et al.*, 2011).

4.2.5.2.2 Chlorophyll

The chlorophyll and its components, chlorophyll *a* and chlorophyll *b* (Table 3.18) indicated that there was no significant variation among the chlorophyll *a* among 20, 30 and 40 dS m^{-1} irrigations in *Eragrostis* spp. However, the chlorophyll *b* and

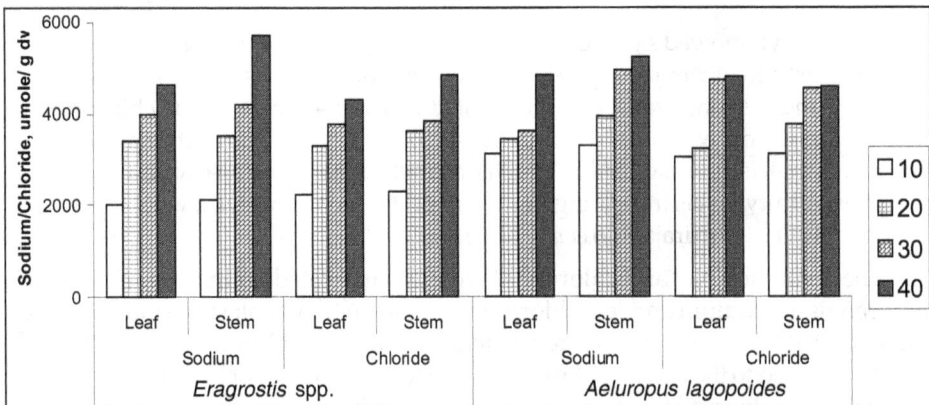

Figure 3.21: Tissue Sodium and Chloride in Forage Grasses under Saline Water Irrigation.

Table 3.16: Uptake and Flux of Na⁺ and Cl⁻ Ions in *Aeluropus lagopoides* Under Saline Water Irrigation

Salinity Class, $dS\,m^{-1}$	Uptake (g)				Flux ($\mu g\,g^{-1}\,day^{-1}$)			
	Shoot		Root		Whole Plant		Shoot	
	Na^+	Cl^-	Na^+	Cl^-	Na^+	Cl^-	Na^+	Cl^-
1st Week								
10	3.91	4.37	5.42	5.14				
20	3.64	4.12	4.86	4.58				
30	3.42	3.90	4.39	4.32				
40	3.08	3.76	4.14	4.24				
2nd Week					**Between 1st and 2nd week**			
10	4.78	4.36	5.96	4.35	7.90	9.92	2.18	3.42
20	4.32	4.04	5.78	4.14	8.62	10.42	3.62	3.75
30	4.02	3.92	4.14	3.62	9.36	12.62	3.92	4.62
40	3.64	3.51	4.04	3.44	10.42	18.80	4.14	5.96
3rd Week					**Between 2nd and 3rd Week**			
10	4.92	4.88	5.84	4.98	10.39	12.82	3.14	4.36
20	4.81	4.64	5.72	4.86	13.86	16.01	3.36	5.62
30	4.32	4.38	4.92	4.64	15.14	22.41	3.92	6.98
40	4.02	4.14	4.44	4.32	19.52	24.62	4.79	9.39
4th Week					**Between 3rd and 4th Week**			
10	5.64	5.84	6.10	5.14	15.76	19.72	3.79	4.72
20	4.92	4.92	5.72	5.02	16.80	25.6	5.16	6.72
30	4.84	4.63	5.32	4.84	18.44	26.8	6.13	8.42
40	4.12	4.24	4.79	4.36	26.12	28.42	7.14	8.92

total chlorophyll showed significant variation among different treatments in *Eragrostis* spp. Similarly in *Aeluropus lagopoides* also the chlorophyll *a* did not show much variation under saline water irrigations of 20, 30 and 40 dS m⁻¹. The chlorophyll *b* and total chlorophyll content showed significant variation with saline water treatments in *Aeluropus lagopoides*. The chlorophyll *a/b* ratio was also found to increase at higher salinity levels indicating higher stability of chlorophyll *a* when compared to chlorophyll *b* (Gururaja Rao *et al.*, 2011).

The combined Na⁺ and chlorophyll analysis indicated an inverse relationship between tissue sodium and leaf chlorophyll. Data on tissue tolerance indicated that about 8950-9000 µ moles of Na⁺ are needed to reduce the chlorophyll content of leaf when compared to the leaves of the plants. The scatter diagram of Na⁺ and chlorophyll showed a high degree of scatter could be due to both variation in induced plant and leaf-to-leaf variations. A significant negative correlation was noticed between tissue Na⁺ and chlorophyll in both the grasses.

Table 3.17: Uptake and Flux of Na+ and Cl− Ions in Eragrostis under Saline Water Irrigation

Salinity Class, dS m⁻¹	Uptake (g)				Flux (µg g⁻¹ day⁻¹)			
	Shoot		Root		Whole Plant		Shoot	
	Na⁺	Cl⁻	Na⁺	Cl⁻	Na⁺	Cl⁻	Na⁺	Cl⁻
	Iˢᵗ Week							
10	4.20	4.85	5.82	5.13				
20	3.92	4.60	5.20	4.93				
30	3.61	4.05	4.85	4.14				
40	3.40	3.70	3.85	3.96				
	2ⁿᵈ Week				**Between 1ˢᵗ and 2ⁿᵈ week**			
10	5.15	4.74	6.40	4.68	8.78	1.88	2.42	3.62
20	4.95	4.44	6.00	4.53	9.92	13.84	3.03	4.81
30	4.45	4.10	5.28	3.96	12.62	14.09	3.84	4.99
40	3.62	3.78	4.62	3.65	13.12	26.32	4.12	8.92
	3ʳᵈWeek				**Between 2ⁿᵈand 3ʳᵈ Week**			
10	5.60	5.66	6.78	5.25	13.42	16.24	3.62	5.14
20	5.05	5.05	6.18	4.94	17.32	20.33	5.14	6.24
30	4.62	4.60	5.85	4.60	18.92	28.75	5.92	8.36
40	4.28	4.20	4.90	4.28	24.32	32.48	7.66	10.62
	4ᵗʰWeek				**Between 3ʳᵈand 4ᵗʰ Week**			
10	6.05	6.20	6.90	6.18	19.70	26.78	6.32	8.32
20	5.25	5.00	6.23	5.54	26.30	32.14	8.44	10.64
30	5.05	5.05	5.63	4.84	28.15	39.36	9.63	14.20
40	4.60	4.60	5.00	4.12	30.10	42.74	10.40	16.32

Table 3.18: Effect of Saline Irrigation on Chlorophyll Content (mg g⁻¹ fr. wt.)

ECe, dS m⁻¹	Eragrostis spp.				Aeluropus lagopoides			
	Chla	Chlb	Total Chl	Chl a/b Ratio	Chla	Chlb	Total Chl	Chl a/b Ratio
10	0.198	0.296	0.494	0.6689	0.209	0.329	0.538	0.6352
20	0.190	0.274	0.464	0.6934	0.200	0.320	0.520	0.625
30	0.190	0.264	0.454	0.7197	0.200	0.292	0.492	0.6849
40	0.190	0.242	0.432	0.7851	0.196	0.268	0.464	0.7313
C.D ₀.₀₅	0.004	0.008	0.0071		0.004	0.008	0.007	

4.2.5.2.3 Forage Quality

Leaf protein content was found to be more in *Aeluropus lagopoides* than *Eragrostis spp.* at all salinity treatments (Figure 3.22). With increase in salinity, a decrease in protein content was noticed in both the grass species. There is a significant variation among the treatments in protein contents. There was an increase in protein with increase in salinity indicating the efficiency of nitrogen metabolism in these grasses.

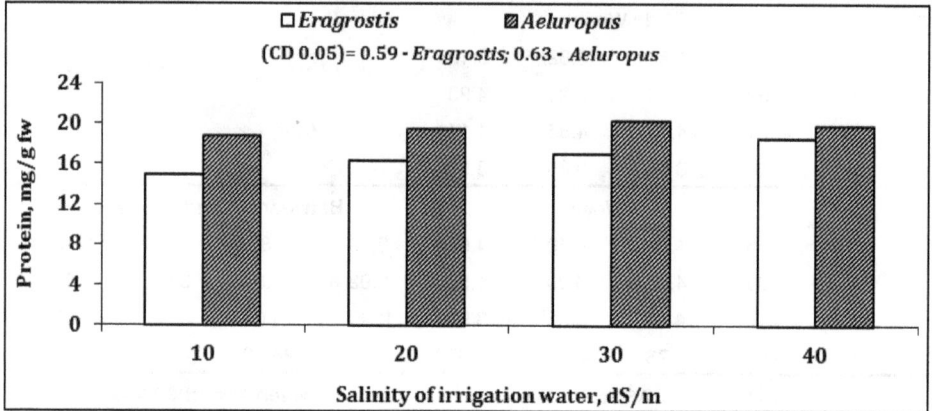

Figure 3.22: Protein Content of Forage Grasses under Saline Water Irrigation.

Sugar content of the shoot showed higher sugar content in *Aeluropus lagopoides* then *Eragrostis spp.* (Figure 3.23). The sugar content was found to increase with salinity in shoots of plants irrigated with 30 dS m^{-1} saline water. However, under 40 dS m^{-1} saline water irrigation, both the grasses showed lesser sugars as compared to 30 dS m^{-1}. Significant variations in sugars were noticed between 10 dS m^{-1} and other treatments in both the grasses.

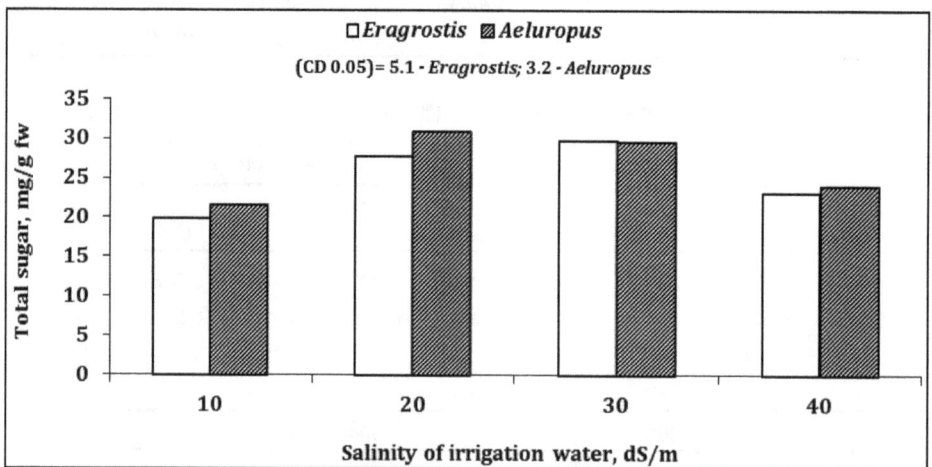

Figure 3.23: Total Sugar Content of Forage Grasses under Saline Water Irrigation.

Proline content also followed the trend of the protein, it increased from 10 dS m⁻¹ to 40 dS m⁻¹ in both grasses (Figure 3.24). Higher Proline as noticed at higher salinities coupled with higher tissue Na⁺ constitutes the osmoregulating substances which favor water uptake from saline medium, thereby enabling the plants to maintain its physiological activity. Higher amounts of proline noticed in these grasses help in turgor regulation and thus physiological activity of these grasses.

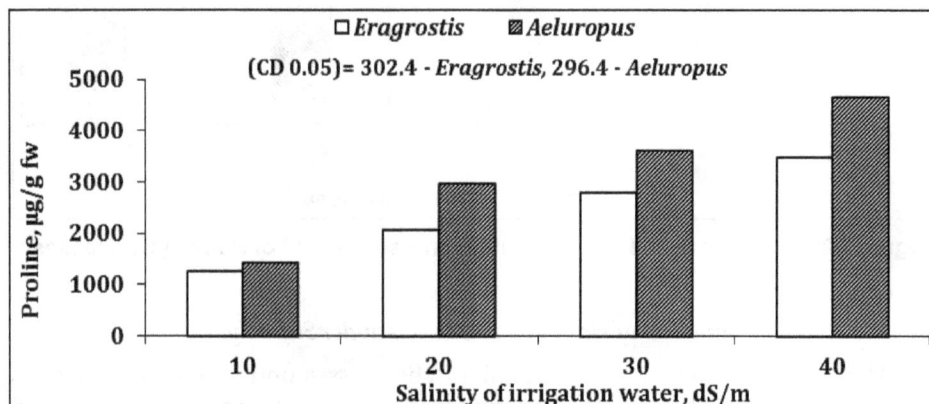

Figure 3.24: Proline Content of Forage Grasses under Saline Water Irrigation.

Crude fiber and ash contents (Figures 3.25 and 3.26) of the halophytic grasses (shoots) indicated that *Aeluropus lagopoides* showed higher fiber and ash content when compared to *Eragrostis*. However, the differences between fiber content were not so more as that of the ash content between the two species.

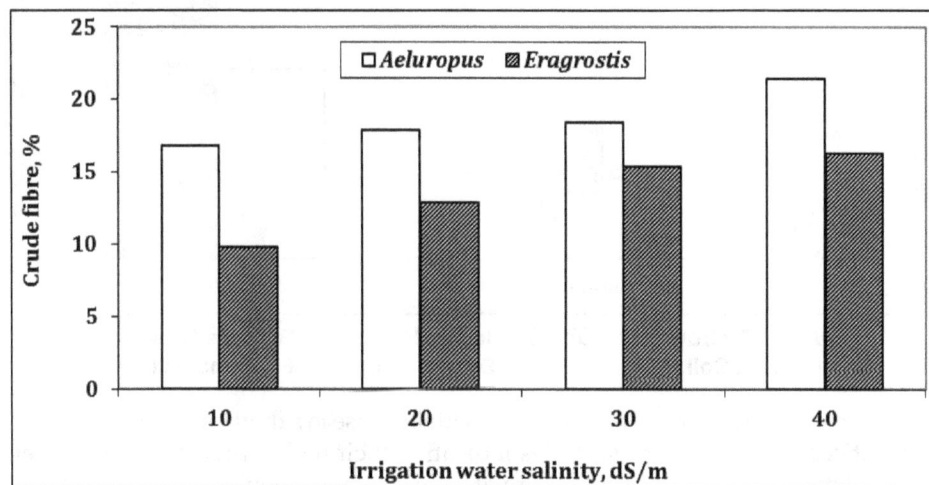

Figure 3.25: Effect of Saline Water Irrigation on Crude Fiber Content (Per cent) of Halophytic Grasses.

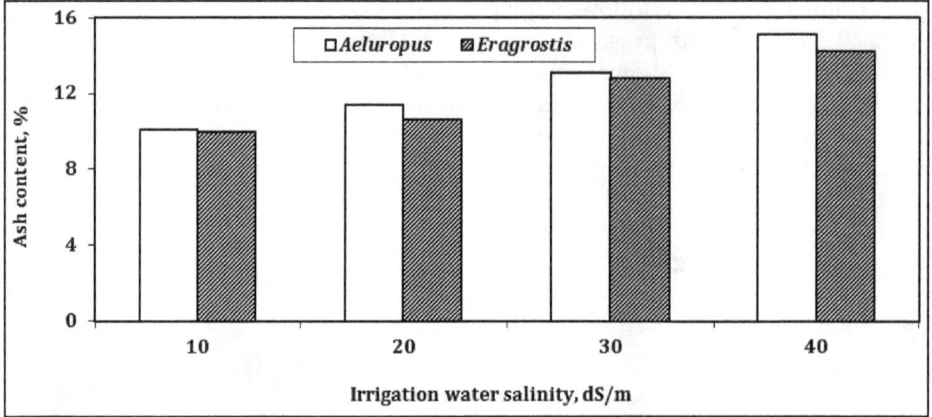

Figure 3.26: Effect of Saline Water rrigation on Ash Content of Halophytic Grasses.

4.2.5.2.4 Forage Production of Halophytic Grasses under Salinity

The green forage yield of these halophytic grasses under field conditions is given in Figure 3.27. The grasses developed well on saline black soils when irrigated with saline water. The data indicated that *Eragrostis* spp. showed higher forage yield under field conditions when compared to *Aeluropus lagopoides* even at salinity of 14.6 dS m^{-1}.

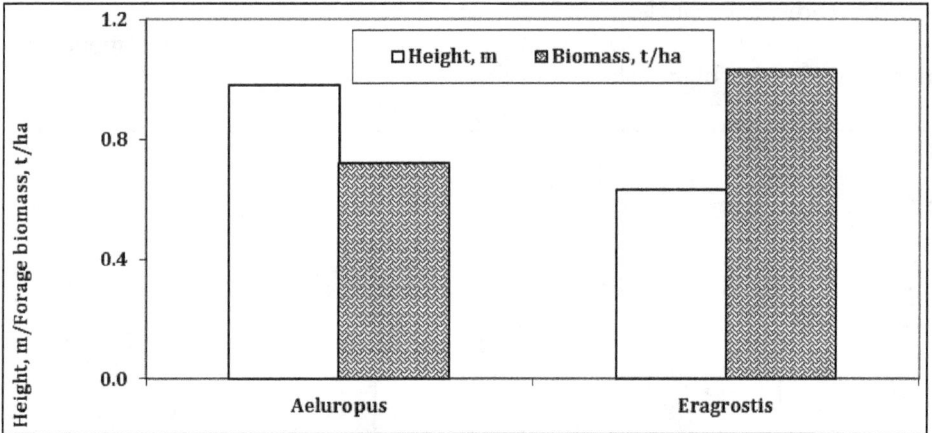

Figure 3.27: Growth and Gorage Yield of Halophytic Grasses Grown on Saline Black Soil (EC 12.8 dS/m; CD 0.05: Height: 0.26; Biomass: 0.418).

Increase in these forage quality traits with increase in salinity of irrigation water was noticed in both the grass species indicating their higher production at higher salinity levels. Crude fiber is a mixture of cellulose, hemicellulose and lignin gives strength and its higher content indicates higher photosynthate production and further its conversion. Higher ash content of *Aeluropus lagopoides* can be ascribed to higher mineral uptake as reported in other grasses.

4.2.5.2.5. Effect of Nitrogenous Fertilizer on Halophytic Grasses

Experiments conducted with nitrogenous fertilizer 0 (F0),30 (F1) and 60 (F2) kg N ha^{-1}) given in association with saline water irrigations (T) given at 15 (T1), 30 (T2) and 45 (T3) days interval indicated significant increase in growth and forage yield of both the grasses. Of the two grasses, *Eragrostis* showed better growth, higher number of tillers, forage yield and forage quality parameters at 60 kg ha^{-1}. Saline water (15 days interval) given with 60 kg N proved beneficial under highly saline conditions. The grasses have been found very effective in salt removal from the soil layers. *Aeluropus* was found to remove more salt than *Eragrostis* (Figure 3.28).

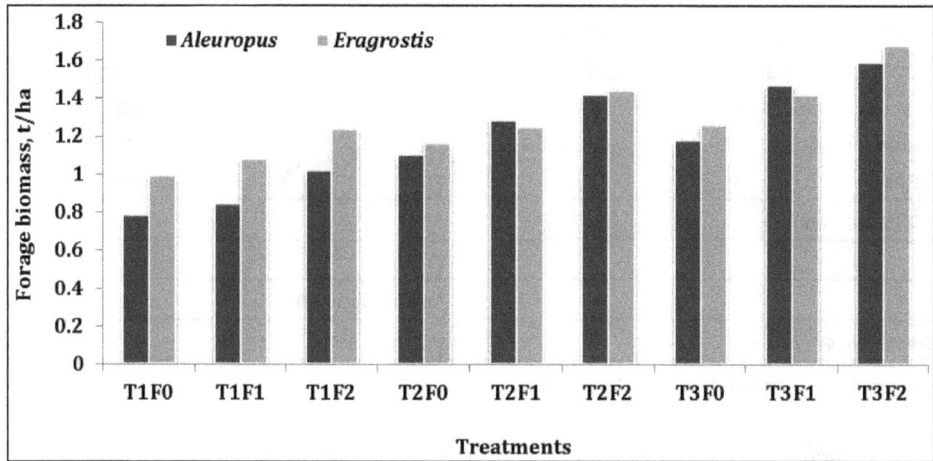

Figure 3.28: Forage Yield of Halophytic Grasses as Influenced by Nitrogen.

The technology has been widely adapted in the Bhal areas covering Tarapur, Dholka, Dhandhuka talukas and also through the NGOs in Coastal saline villages in Cambay talukas. The planting of grasses was also taken up by the National Tree Growers Federation, Anand in Cambay taluka. These grasses being perennial in nature, while providing fodder also bring up the environmental stability in the area which is extremely fragile in nature.

Unit cost: Rs. 3000/-per ha in the first year for planting, fertilizers and labour input. The grasses used to give 3-4 cuts and together gave economic returns of about 10000/- per hectare.

4.2.6. Cotton-pulse Intercropping Beneficial on Moderately Saline Black Soils

Farmers of the Bara tract in Amod, Vagra and Jambusar talukas and other parts of the state who take cotton as rain fed mono-crop, do face crop losses due to salinity development at later stages of crop growth. Under such situations, intercropping with pulses provides some remuneration to the farmer in the event of failure of cotton crop. On-farm trials have indicated cotton - cluster bean proved to be beneficial on moderately saline black soils having salinity of 4-6 dS m^{-1}. Cotton intercropped with cluster bean produced cotton seed yield at par with that of sole cotton. Increase in nitrogen up to 80 kg N ha^{-1} significantly increased the seed cotton yield under saline conditions. Cluster bean while improving the fertility of the soil provides an insurance against the failure of cotton crop (Table 3.19).

Table 3.19: Performance of Cotton Intercropped with Pulses under different Levels of Fertilizers on Moderately Saline Vertisols of Bara Tract

Treatments	Seed Cotton Yield, kg/ha	Treatments	Seed Cotton Yield, kg/ha
Main-plot Treatments		Sub-plot Treatments	
Inter crops (pulses)		Fertilizer levels (kg/ha) N + 40 kgP$_2$O$_5$/ha	
Sole cotton	572.0	Control	355.1
Cotton + Black gram	532.7	20	440.3
Cotton + Cluster bean	559.3	40	537.6
Cotton + Soybean	556.9	60	626.1
S Em (±)	9.3	80	676.3
CD (5 per cent)	22.7	100	685.8
		S Em (±)	15.2
		CD (5 per cent)	30.7
Coefficient of variation		6.7 per cent	

Source: CSSRI Annual Report (2003-2004).

Farmers who take cotton as rain fed mono crop in the Bara tract in Amod, Vagra and Jambusar talukas and other parts of the state do face crop losses due to salinity and due to other climatic vagaries. Under such situations, intercropping with pulses provides some remuneration to the farmer in the event of failure of cotton crop. About 165 ha of land of the Bara tract particularly in Vagra and Amod talukas, the farmers have been adopting the cotton intercropped with pulse technology for maximising the production.

The system would fetch about 16000/-per hectare from cotton and further the pulses due to their nitrogen fixing ability enrich the soils with nitrogen. Cotton as well as pulses can be taken as rain fed crops, providing saline water irrigation, if available further boosts the crop yields. Use of saline water in cotton has been proved beneficial on saline black soils (Gururaja Rao *et al.*, unpublished).Targeted Areas are Bara tract covering Vagra, Amod and Jambusar talukas of Bharuch.

4.3. Conjunctive Use of Saline Water with Surface Water for Crop Production

As the real potential of land and water resources in the Bara tract was not assessed, the proper utilization of the land and groundwater resources have not been done to its fullest extent for the agricultural production. The non-use of saline groundwater is not only making the crop production stagnant but also contributing to the increase in the ground water table and salinity.

The salinity of groundwater is too high in the saline tracks of Bhal area in Gujarat that cannot be used for irrigation purposes as such and thus it needs to be mixed with limited surface water available. In the absence of inadequate irrigation water supplies in the region, technologies evolved for conjunctive use of saline groundwater in mixing and cyclic modes for growing *rabi* season crops like dill, mustard, safflower and wheat proved to be remunerative due to its long-term potential impacts on the economic development, employment generation and environmental improvement (Gururaja Rao, 2004).

Dill, a potential seed spice crop can be grown during *rabi* season on saline black soils having salinity up to 4-5 dS m^{-1} with a seed yield of 3 t ha^{-1} without any irrigation, which otherwise remained fallow during the *rabi* season. However, the conjunctive use of saline groundwater with surface water can improve the productivity manifold. In dill, if surface water is available for one irrigation, it should be applied at the seed formation stage and saline water at the vegetative and flowering stages. If surface water is available for two irrigations, it should be applied at the time of flowering and seed formation stage and saline water at the vegetative stage. In areas with high groundwater table and lack of sufficient surface water, surface water up to 66 per cent can be saved by application of saline groundwater (4 dS m^{-1}) at branching and flowering stage and surface water at seed formation stage without further increase in soil salinity (Gururaja Rao *et al.*, 2001b; Nayak *et al.*, 1999b, 2001a). This method can increase seed yield by 150 per cent over the yield obtained under unirrigated condition.

In safflower, branching and flowering stages were found to be sensitive stages for saline water irrigation. If surface water is available for one irrigation, it should be applied at branching stage and saline water at vegetative and flowering stages. If surface water is available for two irrigations it should be applied at branching and flowering stages and saline water at vegetative stage. In safflower by applying saline groundwater (4 dS m^{-1}) at flowering and grain filling stages and surface water at branching stage, 86 per cent increase in yields over the yield obtained under unirrigated conditions (3.7 q ha^{-1}) can be obtained (Singh *et al.*, 1996; Nayak *et al.*, 2001a; Gururaja Rao *et al.*, 2001a).

Indian mustard can be grown on saline black soils with saline ground water having EC of 4 dS m^{-1} in conjunction with the limited surface water. Flowering and pod formation stages are relatively more sensitive to saline water irrigation. In mustard, application of two saline water irrigations (4 dS m^{-1}) at branching and pod formation stages and surface water at flowering stage resulted in yield of 5.59 q ha^{-1} (Nayak *et al.*, 2001b). Flowering and pod formation stages are relatively sensitive to saline water irrigations. This method while saving 66 per cent surface irrigation water increases the yield by 123 per cent over the yield obtained under unirrigated conditions (2.2 q ha^{-1}). In wheat, good quality water, if available, should be applied at crown initiation stage and saline water at maximum tillering and flower initiation stages. If Best Available Water (BAW) for two irrigations is available, it should be applied at crown root initiation and flower initiation stages. Under high saline water table prevailing in saline black soils, in wheat, when saline water is applied, flowering and maximum tillering stages are equally sensitive as that of crown root initiation stage for salinity because of their exposure to increase in groundwater salinity

(Gururaja Rao *et al.,* 2000; 2001). Application of saline water (SW, 4 dS m^{-1}) at these stages and good quality water at crown root initiation stage resulted in an increase of 180 per cent seed yield when compared to the yield obtained under unirrigated condition (Table 3.20).

Table 3.20: Effect of Conjunctive Use of Saline Water and Surface Water in Cyclic Mode on the Yield of Arable Crops on Saline Black Soils

Treatment	Dill	Safflower	Mustard	Wheat
T1 - All BAW	8.32	8.25	7.35	29.1
T2 - SW at Branching stage/crown root initiation stage in wheat + rest BAW	7.93	7.45	6.65	27.2
T3- SW at Flowering stage/maximum tillering stage in wheat + rest BAW	7.84	7.65	6.29	28.1
T4 - SW at Seed formation stage/flower initiation stage in wheat + rest BAW	7.68	7.75	6.45	27.4
T5 - (SW– Branching/tillering and flowering) rest BAW	7.52	6.35	5.59	22.2
T6 - (SW – Branching/tillering and Seed) + rest BAW	7.44	6.85	5.19	23.0
T7 - (SW – Flowering and Seed) + rest BAW	7.36	6.88	5.35	22.3
T8 - (All saline)	6.82	4.99	3.26	16.8
CD 0.05	0.06	0.21	0.75	0.50

Source: Gururaja Rao *et al.*, 2001a.

The salinity of ground water is too high in the saline tracks of Bhal area in Gujarat that cannot be used for irrigation purposes as such and thus it needs to be mixed with limited surface water available. Even the assured supply of canal water in the *rabi* season is not possible. For such conditions, the following technologies have been evolved. The strategies developed for conjunctive use of saline water and ground water for four important crops of the region would improve the dill yield by 4.46 q/ha, mustard by 2.37 q/ha, safflower by 5.18 q/ha and wheat by 12 q/ha over the yield under unirrigated condition. The surface water saved per hectare would create 2 acre additional command under irrigation. This technology as well can be extrapolated to other canal command areas where the salinity problems are prevailing. This would pave for reclamation of waterlogged soils by 52.1 per cent in Ukai-Kakrapar Command and by 65 per cent in Mahi Irrigation Command.

Surface water up to 66 per cent can be saved by application of saline ground water (4 dS m^{-1}) at branching and flowering stage and surface water at seed formation stage without further increase in soil salinity in dill. In Safflower, irrigating with saline ground water (4 dS m^{-1}) at flowering and grain filling stages and surface water at branching stage, 86 per cent increase in yield over the yield obtained under un-irrigated conditions.

In mustard, conjunctive use of saline water while saving 66 per cent surface irrigation water increases the yield by 123 per cent over the yield obtained under unirrigated conditions. In wheat, application of saline water (4 dS m^{-1}) at vegetative and tillering stages and good quality water at crown root initiation stage resulted in an increase of 180 per cent seed yield over the yields under unirrigated conditions.

The technologies hitherto evolved were disseminated and demonstrated at the farmers' sites through field days, workshops and farmers trainings. The targeted area includes saline black soil region in Bhal and Bara tract regions in Gujarat (Parts of Ahmedabad and Anand districts) and Amod, Vagra and Jambusar talukas in Bharuch district.

4.4. Development of Cotton (*Gossypium* spp.) for Salt Affected Soils

Cotton has been one of important commercial fiber crops in India since ancient times. Cotton has become an important commercial crop instead of a component in the mixed cropping system, primarily due to Green Revolution which made this crop dependant on chemical fertilizers. Gujarat has emerged as the largest producer of cotton, producing 116 lakh bales from 26.9 lakh ha area with productivity of 733 kg/ha (Cotton Corporation of India, 2013). In Gujarat cotton is cultivated mainly in Bharuch, Surat, Vadodara, Anand, Ahmedabad, Bhavnagar, Rajkot, Junagadh, Amreli, Porbandar and Jamnagar districts.Experiments were conducted with different species of cotton under saline water irrigation and the studies have indicated that herbaceums and arboreums as salt tolerant and superior to hirsutums and Bt hybrids. To corroborate this finding, apart from seed cotton yield, tissue ion content, K/Na ratio and water productivity were studied both in pot and field conditions (Gururaja Rao, 2013; Gururaja Rao *et al.*, 2012, 2013a). The physiological basis of salt tolerance is described below.

4.4.1 Leaf Sodium and Potassium

Tissue sodium was found to be more in herbaceums when compared to hirsutums and Bt line. Concomitantly, potassium was also found to be more in herbaceum, G. Cot DH 7 and other herbaceums and arboreums, which makes these species maintain higher K/Na ratio. The results indicated that accession G. Cot DH 7 showed better salt tolerance and maintained higher K/Na ratio followed by GBav 105, GBav 374 and GShv 368/05 (Figures 3.29 and 3.30).

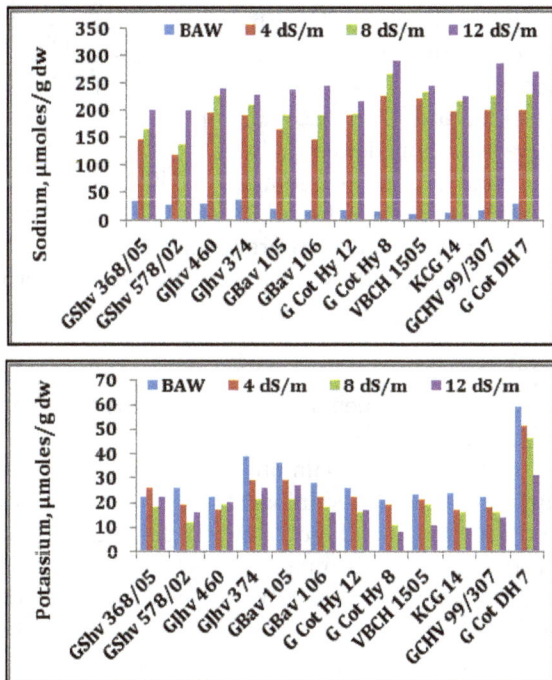

Figure 3.29: Leaf Tissue Sodium and Potassium of Cotton Species under Saline Water Irrigation.

Among 12 accessions tried, G. Cot DH 7 was found superior in seed cotton yield (Figure 3.31). This accession yielded higher seed cotton yield of 119 g/plant at 4 dS/m irrigation and this yield is at par with the control. Even with the increase in sanity, G. Cot DH 7 showed only 0.6, 6.1 and 15.1 per cent decrease in yield at 4, 8 and 12 dS/m salinity respectively, over the control. Among the other cultivars, GJhv 374, followed by GBav 106, GBav 105 along with G. Cot. Hy 8 gave moderated yield at 4 dS/m with further decline at 8 and 12 dS/m salinity levels. The studies are also in conformity with earlier observations which indicated the superiority of G. Cot DH 7. The accessions *viz.* VBCH 11505, KCH 14 and GSHV 99/307 were the poor yielders. Lint yield and ginning percentage of herbaceums was found to be more than that of arboreums.Based on the seed cotton yield at 12 dS/m saline water irrigation, the cultivars are placed in the order G. Cot DH 7 > GJhv 374>GBav 106> GBav 105 > G.Cot. Hy 8.

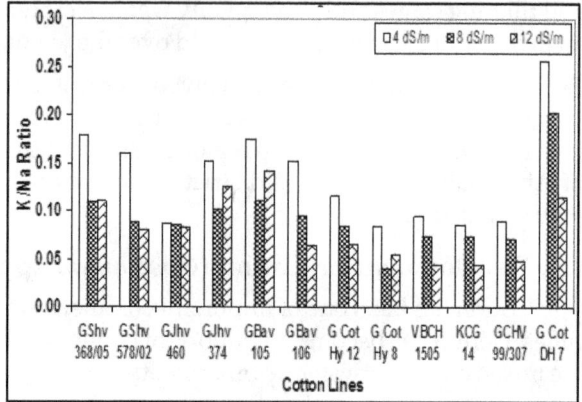

Figure 3.30: K/Na Ratio in Leaves of Cotton Species Under Saline Water Irrigation.

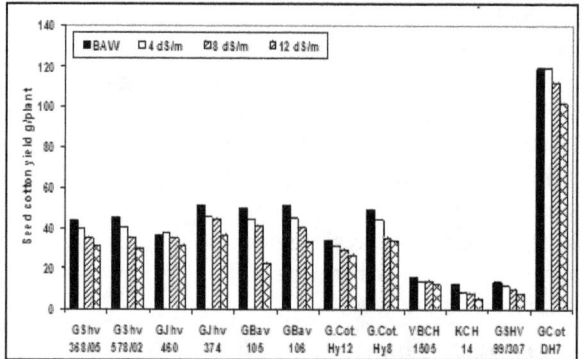

Figure 3.31: Seed Cotton Yield of Cotton Species under Saline Water Irrigation.

4.4.2. Ions

Field experiments with 36 accessions of cotton covering herbaceums, hirsutums, arboreums, Bt hybrids and two checks irrigated with saline water having salinity of 4.2 to 11.8 dS m^{-1}. Data on leaf sodium and potassium indicated high potassium uptake was observed in herbaceums and arboreums when compared to hirsutums and Bt cotton lines (Figure 3.32) suggesting better K/Na ratio that enables the plant to overcome the stress impact. This is further seen in better seed yields of these accessions (Figure 3.32). Relation between seed cotton yield and soil salinity and chlorophyll and soil salinity are depicted in Figures 3.34 and 3.35 which indicated the superiority of herbaceums and arboreums over hirsutums and Bt hybrids.

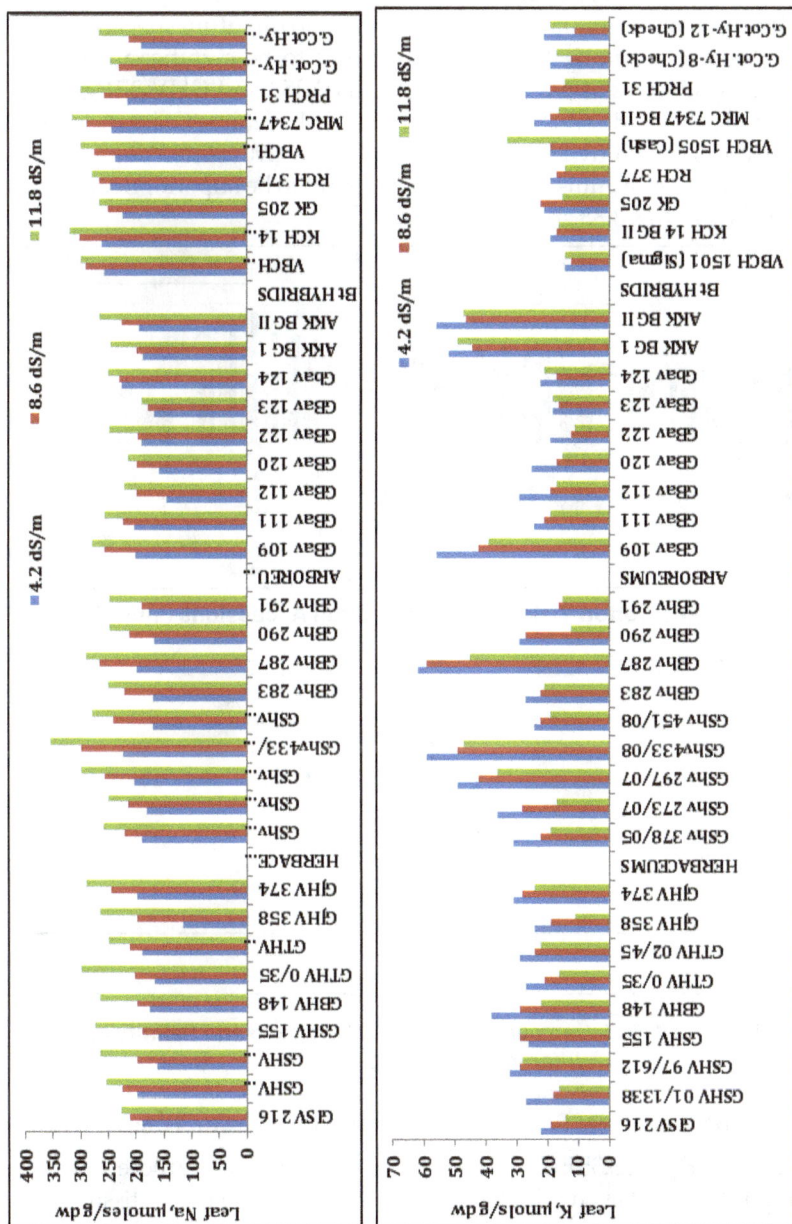

Figure 3.32: Leaf Sodium and Potassium in Cotton Species under Saline Water Irrigation (Field condition).

4.4.3 Seed Cotton Yield

Seed cotton yield (productivity), was found to be maximum in arboreum G Bav 124 followed by herbaceum, GShv 297/07 which in fact yield more than the checks G. Cot Hy 12 and G. Cot Hy 8. Both herbaceums and arboreums gave higher seed cotton yields over hirsutums and *Bt* hybrids except on hirsutum species, GSHV 97/ 612Seed cotton yield among the lines was found to be significant. Among the hirsutums, the yield was at par in GISV 216,GJHV 358 and GHV 374 gave at par yields at salinity of 11.8dS/m. Similarly, among herbaceums, GShv 378/05, GShv 433/08, GShv 451/08, GBhv 283, GBhv 287 and GBhv 290yielded at par. Based on seed cotton yield, herbaceums and arboreums found to be superior over hirsutums and Bt hybrids (Figures 3.33 and 3.34).

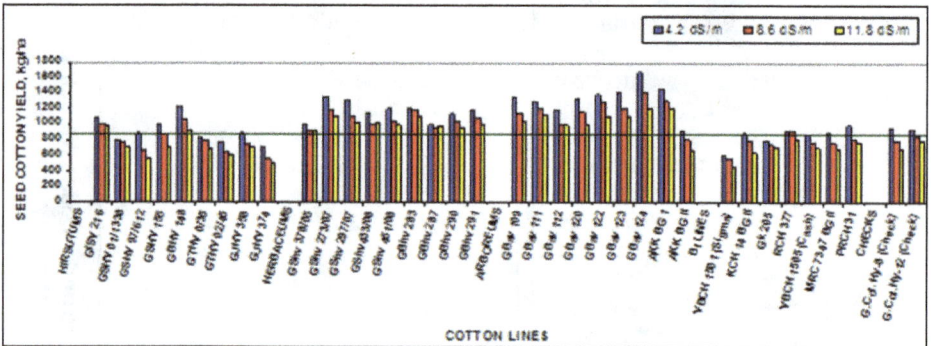

Figure 3.33: Seed Cotton Yield of different Cotton Accessions under Salinity.

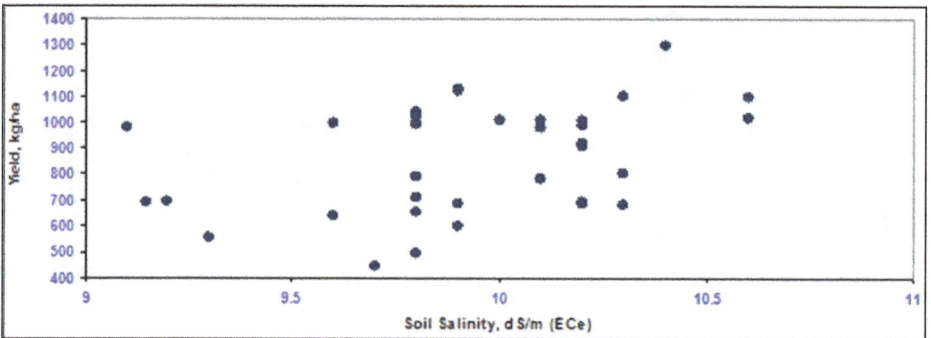

Figure 3.34: Relation between Seed Cotton Yield and Salinity in different Cotton Accessions.

4.4.4. Relation between Leaf Sodium and Chlorophyll (Tissue Tolerance)

Data given in Figure 3.35 indicates that both herbaceums and arboreums maintain high chlorophyll content even at high tissue sodium indicating higher tissue tolerance. Contrary to this, both hirsutums and Bt hybrids showed a lowered chlorophyll at higher tissue sodium indicating their susceptibility to salinity (Figure 3.35). It is because of this higher salt tolerance, both arboreums and herbaceums showed higher water productivity (Figure 3.36).

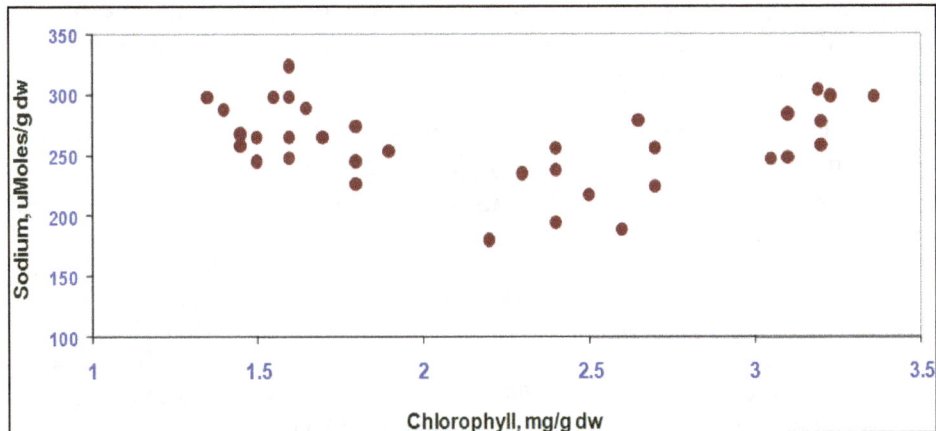

Figure 3.35: Relation between Leaf Sodium and Chlorophyll in Cotton Species.

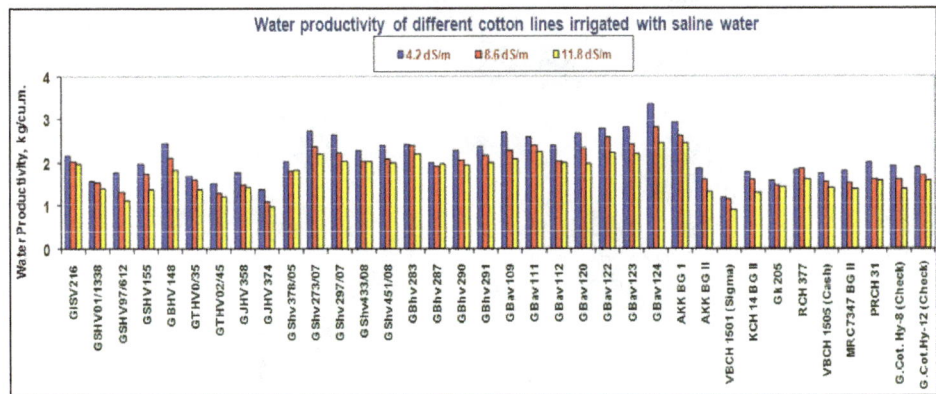

Figure 3.36: Water Productivity of different Cotton Species under Salinity.

4.4.5. Lint Yield

Lint yield, like seed cotton yield was found to be more in Bt lines and checks when compared to the other three groups of cotton accessions. Among all the lines, lowest lint yield was noticed in arboreums. Significant differences were noticed in lint yield among all the cotton lines. Bt lines and hirsutums were found to be superior in lint yield also. Quality parameters like ginning percentage though found significant along the lines, the variation within the groups seem to be minimum. Ginning percentage has been found to be more in hirsutums followed by Bt lines, herbaceums and some arboreums. Among the cotton lines, highest ginning percentage was noticed in GBav 111, followed by GISV 216 and G Cot Hy 8 and the minimum in VBCH 1505 (Cash). Seed index was recorded highest in Bt lines and hirsutums and the least in arboreums. The higher water productivity of herbaceums and arboreums, thus clearly indicates their suitability to water scarce regions, unlike hirsutums and Bt lines, which, however, need more irrigations.

Biomass: Biomass was found to be maximum in herbaceums, GBhv 290 (8.65 t/ha) and GBhv 291 (8.11 t/ha) followed by GBhv 283 (6.19 t/ha). In the remaining accessions, it was in the range of 1.9 to 4.8 t/ha.

4.5. Ideal Farming System Model for Arid Saline Soils of Bara Tract

The Bara tract comprising Amod, Vagra and Jambusar tehsils of Bharuch district of Gujarat state is characterised by various soil and ground water related constraints such as sub-surface salinity, sodicity, highly saline ground water and accumulation of calcium carbonate. Farmers in this region are resource poor with low and insecure farm income. Thus, evolving appropriate technological interventions with special emphasis on enhancing the water productivity of crops to provide them a staggered income throughout the year have immense importance. The components of the system comprise a rain water harvesting structure (farm pond), fruit, vegetable, biomass species and field crops (spices) covering an area of 1.12 ha (Figure 3.37).

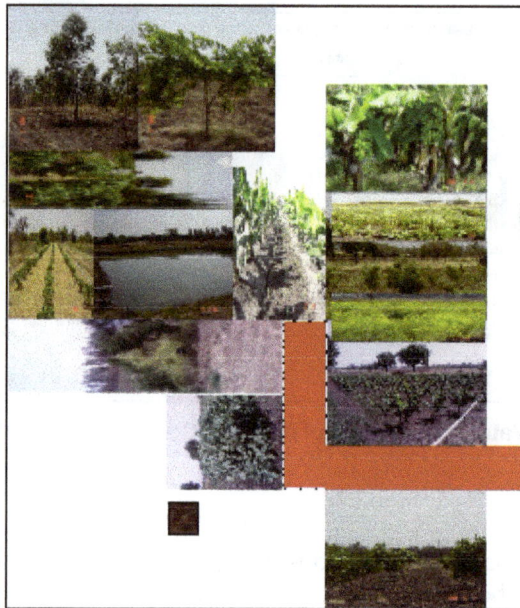

Component	Area, m^2	Crop
Pond	2000	
Dykes	900	Banana and papaya, Vegetables
Fruit species	4000	Banana, Jamun and Aonla
Biomass species	3500	Eucalyptus and Pongamia
Compost pit (m^3)	48	
Total	11200	

Figure 3.37: Components of Farming System.

Table 3.21: Water Productivity and Benefit/Cost Ratios of Components in the System

Component	Banana	Papaya	Dill	Coriander	Brinjal	Tomato	Bottle Gourd	Cabbage
Plot Area, m²	190	225	1600	1600	150	160	150	250
Water applied(mm)	953	180	120	122	94	104	112	122
Economic yield,(kg/plot)	887	652	122	116	208	220	324	364
Water productivity,(kg m⁻³)	0.931	3.62	1.017	0.951	2.21	2.12	2.89	2.98
Water productivity (Rs m⁻³)	3.58	19.25	19.85	33.28	15.21	10.00	13.88	17.90
Cost of cultivation, (Rs.)	1590/-	1450/-	960/-	1120/-	400	280/-	490/-	440/-
Gross Income (Rs.)	5002/-	4916/-	3342/-	5180/-	1830/-	1320/-	1944/-	2184/-
Net Income (Rs.)	3412/-	3466/-	2382/-	4060/-	1430/-	1040/-	1554/-	1744/-
B/C ratio	2.15	2.39	2.48	3.63	3.8	3.71	3.17	3.96

Total Net Income from the system (under yielding stage, Rs/ha): 43840=00.

Data on crop components, crop production, water productivity and benefit/cost ratio are given in table 5 indicated that spices had higher water productivity (in terms of monetary gains) when compared to vegetables. Vegetables, however, had higher water productivity when yield is taken into account. High water requiring crops like banana had the lowest water productivity (both in terms of economic yield and monetary gains) and also the Benefit/Cost ratio when compared to low and moderate water requiring crops like species and vegetables. Among the fruit species, papaya was found beneficial both in terms of water productivity and benefit/cost ratio and found suitable for Bara tract areas. The vegetable crops like tomato, brinjal, bottle gourd and cabbage, are moderate in water consumption and however, better money earners due to their continued yielding and thus provide subsistence income to the farmers. The spices, however, because of very low water requirement and high water productivity have been found ideal for water scare areas like Bara tract (Gururaja Rao *et al.*, 2009b; Tables 3.21 and 3.22).

Water productivity over the years in spice and vegetable components did not vary much when expressed in terms of economic yield did not show much variation. However, water productivity in terms of monetary gains of spices was found much higher than that of vegetables (Figure 3.38).

Table 3.22: Water Productivity and Benefit/Cost Ratio of different Components of the Farming Systems on Saline Vertisols

Component	Water Productivity, kg/m³	Water Productivity, Rs/m³	B/C Ratio
Fruit Species	Papaya > Banana	Papaya > Banana	Papaya > Banana
Vegetables	Bottle gourd > Brinjal > Tomato	Brinjal > Bottle Gourd > Tomato	Brinjal > Tomato > Bottle Gourd
Seed Spices	Dill > Coriander	Coriander > Dill	Coriander > Dill

Woody biomass plants such as *Eucalyptus* and *Pongamia* had yielded significant amounts of litter and their corresponding growth in terms of plant height and girth were also good. The litter fall was found to be more in *Syzygium* when compared to

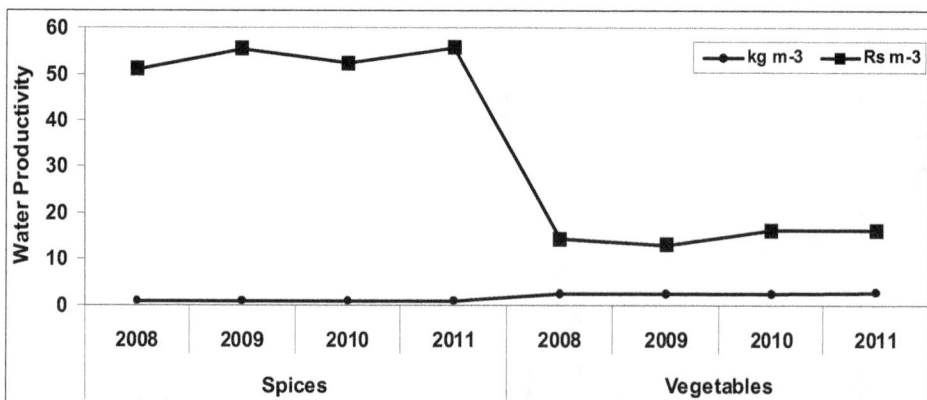

Figure 3.38: Water Productivity of Spices and Vegetables under the Farming System Study.

Eucalyptus and *Pongamia*.The compost generated from the litter of different components was analyzed and the data indicated (Figure 3.39) that it had dry matter of 54.9 per cent and 15 percent organic matter. The composition of the compost indicated that it is rich in nitrogen and calcium and thus forms a good nutrient source and provides a scope for minimizing the use of inorganic fertilizers (Gururaja Rao *et al.*, 2009a,b).

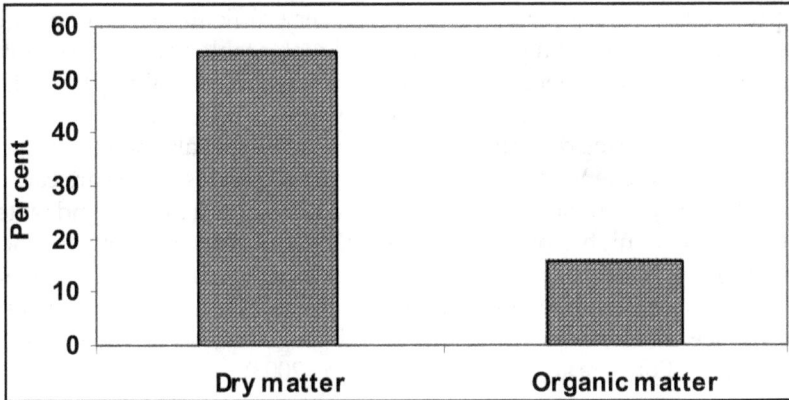

Figure 3.39: Dry Matter and Organic Matter Content of Biomass Species.

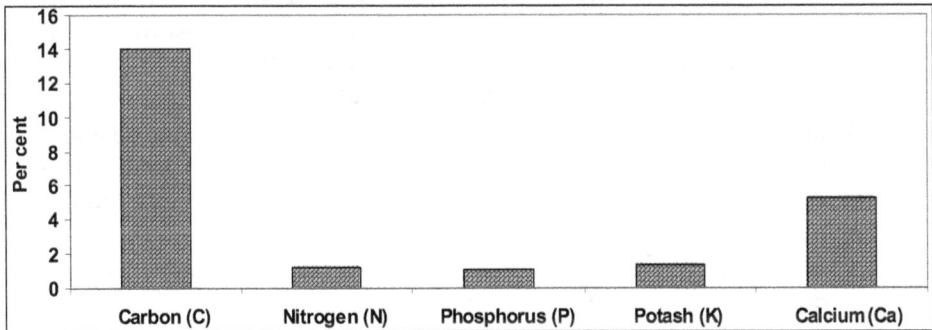

Figure 3.40: Nutrient Composition of the Compost.

The studies revealed that low water requiring crops like papaya, dill and coriander had higher water productivity than banana, which is high water requiring species. The B/C ratio of papaya, dill and coriander were higher than banana. Vegetables, brinjal, tomato and bottle gourd had high water productivity and also higher B/C ratio than banana. These crops with low crop duration and low water needs, form ideal components that provide regular income to the farming community. The crops like papaya, dill and coriander along with vegetables because of their higher water productivity and B/C ratio are found to be suitable for the saline Vertisols of Bara tract.

4.6. Remedial Measures for Saline Vertisol of AER Region V of Bara Tract of Gujarat

The soils of Bara tract have significant concentration of soluble salts in sub-soils, although the concentration in surface layer is low. This pattern of salinity build up may be because of previous continuous contact with saline or brackish water due to the proximity of sea. The sub-soil salts are very difficult to leach down further because of the presence of high saline ground water table and very low saturated hydraulic conductivity. After the introduction of irrigation the salinity hazards would be much greater involving too much use of water, which will not only bring the salinity to surface during dry season by capillary action but also accelerate the rise in groundwater table.The Vertisols and associated soils of this region are categorized into (1) potentially saline soils with sub-surface salinity and low ground water table (2) saline soils with high ground water table (3) saline soils with low ground water table and (4) highly saline lands. As the problems of soil and ground water salinity are dynamic in nature, the management options and land use patterns vary from place to place and season to season. Based on research efforts, the following management options are suggested (Nayak *et al.*, 2003).

☆ In the area with low ground water during *rabi* season, the land should be ploughed upto a shallow depth of 4-5 cm where the moisture controlled section strata. Seeds of *rabi* season crops should be sown in this moisture controlled section to attain perfect germination. The moisture controlled section would provide the moisture to the crop for the rest of the period and the upper ploughed soil will act as soil mulch.

☆ In the area with high saline ground water during *rabi* season and with fine textured soils there should be a deep summer ploughing in order to break the capillaries so that the ground water flux to the surface is minimized.

☆ In the area with high saline ground water and where water stagnation occurs for a period of over 10-15 days during *kharif*/monsoon season, short duration paddy or ragi (finger millet) are ideal options followed by less water requiring crops like seed spices and oil seed crops.

☆ Canal irrigation should be restricted to *rabi* season crops only with a provision of single distribution system within the water distribution network for applying both the surface water and ground water either by cyclic or mixing modes. The main canal should be deeper and below the surface, so that the farmers can lift the wearer from the canal for irrigating crops and the same canal will be used as surface drain during *kharif* season.

☆ Non-conventional crops like dill and fennel can be grown with the residual moisture (during *rabi* season) and ragi can be taken up in the *kharif* season with limited irrigation. While dill and fennel form cash crops, ragi provides the stable food.

5. Summary and Future Strategy

Soil salinity is an important environmental constraint in improving agricultural productivity. Salts in the root medium generally cause a wide array of metabolic changes culminating in reduced plant growth and finally yield. Salt affected Vertisols, due to their inherent physical and chemical constraints pose severe threats for crop production even at low salinity. The high clay content and associated low infiltration rates, further hampers drainage options and thus identifying ideal technologies for overcoming the salinity is of paramount importance. Understanding the physiological mechanisms of salt tolerance further helps in identifying ideal crops/varieties suitable to soils of different salinity regimes. This approach has resulted in evolving different agrotechnological interventions comprising economic halophytes for highly saline black soils, halophytic forages ideal for saline water irrigation, cotton pulse intercropping for moderately saline soils, herbaceums and arboreums as ideal cotton species for rainfed coastal saline tracts and also provides a scope for a farming system model with crops of high water productivity. While taking up these interventions on saline black soils, further research is need to be intensified giving primacy to crops of high economic value and high water productivity, identifying alternate nutrient sources to meet the plant nutrient requirements and a special emphasis on horticultural crops. Further efforts are needed to disseminate the technical know-how to the end users who are instrumental in improving the economy.

References

Balpande, S.S., Deshpande, S.B. and Pal, D.K. 1996. Factors and processes of soil degradation in Vertisols of the Purna valley, Maharashtra, India. *Land Degradation and Development* **7**: 313-324.

Chaudhary, S. K. 2001. Dependence of saturated hydraulic conductivity on dispersion, swelling and exchangeable sodium under different quality waters. *Journal of the Indian Society of Soil Science* **49**: 12-20.

Gupta, Ram K. and Verma, S.K. 1984. Influence of soil ESP and electrolyte concentration of permeating water on hydraulic properties of a Vertisol. *Z. Pflanzenernaehr Bodenk.* **147**, 77-84.

Gururaja Rao, G. 1995. India discovers a salt species. *Salt Force News*, Victoria (Australia). **45**, October (1995): 4.

Gururaja Rao, G. 2004. Diversified cropping systems and socio-economic prospects in coastal saline soils of Gujart state. *J. Indian Coastal Agric. Res.* **22** (1 and 2): 178-184.

Gururaja Rao, G. 2012. Crop Management for Saline Vertisols. Lead Paper presented in the National Seminar on Management of Salt Affected Soil and Waters: Challenges of the 21st Century during 16-17 March, 2012, Lucknow (India).

Gururaja Rao, G. 2013. Crop production technologies for saline Vertisols. Lead Paper presented at National Seminar on "Technology for Development and Production of Rainfed Cotton", October 24-25, 2013, Regional Cotton Research Station, Navsari Agricultural University, Bharuch, Gujarat, India.

Gururaja Ra,G., Chinchmalatpure, A.R., Khandelwal, M.K., Arora, Sanjay and Gurbachan Singh. 2009a. Management of salt affected black soils – impact of technological interventions. *J. Soil Sal. Water Qual.*, **1**(1 and 2): 55-62.

Gururaja Rao,G., Chinchmalatpure, A.R., Khandelwal, M.K. and Gurbachan Singh, 2009b.Studies on enhancing water productivity of farming system components on saline Vertisols of Bara tract of Gujarat State, India. Paper presented at the 4th World Congress on Conservation Agriculture, New Delhi, 4 – 7 February, 2009.

Gururaja Rao, G. Chinchmalatpure, Anil. R., Meena, R.L. and Khandelwal, M.K. 2011. Saline Agriculture in Saline Vertisols with Halophytic Forage Grasses. *J. Soil Sal. Wat. Qual.* **3**(1): 41-48.

Gururaja Rao, G., Chinchmaltpure, Anil R., Sanjay Arora, Khandelwal, M.K. and Sharma, D.K.2013.*Coastal Saline Soils of Gujarat: Problems and their Management*. Technical Bulletin 1/2013, CSSRI, RRS, Bharuch.

Gururaja Rao, G., Nayak, A.K. and Chinchmalatpure, Anil. R, 2000a. *Dill (Anethum graveolens) - A Potential Species for Salt Affected Black Soils*. Technical MonographNo.1, CSSRI, RRS, Anand, p. 10.

Gururaja Rao, G., Nayak, A.K. and Anil R. Chinchmalatpure, 2000b. Yield and yield attributes of wheat as influenced by conjunctive use of saline water with surface water on saline black soil. *J. Indian Soc. Soil Sci.* 51 (1): 86-88.

Gururaja Rao, G., Nayak, A.K and Chinchmalatpure, Anil. R, 2001a. Conjunctive use of saline water with surface water for cultivation of some arable crops on salt affected black soils with high saline ground water table. *J. Indian Soc. Coastal Agric. Res.*, 19(1): 103-109.

Gururaja Rao, G., Nayak, A.K and Chinchmalatpure, Anil. R, 2001b.Growth and yield of dill (*Anethum graveolens* L.) on saline black soils of different unirrigated farm sites of Bhal area in Gujarat state.*Indian J. Agric. Sci.*,71 (11): 712-714.

Gururaja Rao, G., Chinchmalatpure, SA.R.,Meena, R.L., Indivar Prasad, Shrvan Kumar, Kumar, V. and Sharma, D.K. 2013. *Herbaceum* cotton: An ideal option for saline Vertisols. In National Seminar on "Technology for Development and Production of Rainfed Cotton", October 24-25, 2013, Regional Cotton Research Station, Navsari Agricultural University, Bharuch, Gujarat, India.

Gururaja Rao, G., Nayak, A.K and Chinchmalatpure, Anil. R, Abhay Nath and Ravindra Babu, V. 2004. Growth and yield of *Salvadora persica*, a facultative halophyte grown on highly saline black soil.*Arid Land Res. and Manage.*,18(1): 51-61.

Gururaja Rao, G., Nayak A.K., Chinchmalatpure Anil R. and Ravindra, V. 2001b. Growth and yield of some forage grasses grown on salt affectedblack soils. *J. Mah. Agric. Univ.***26** (2): 195-197.

Gururaja Rao, G, Nayak A.K, Chinchmalatpure Anil R and Babu, V.R. 2004a. Growth and yield of Salvadora persica : a facultative halophyte grown o highly saline black soil. *Arid Land Res. and Manage.***18**(1): 51-61.

Gururaja Rao, G., Nayak, A.K, Chinchmalatpure, Anil. R, Sumana Mandal and Tyagi, N.K. 2004. Salt tolerance mechanism of *Salvadora persica* grown on highly saline black soil. *J.Plant Biol.* **31**(1); 59-65.

Gururaja Rao, G., Nayak, A.K, Chinchmalatpure, Anil. Rand Ravindra Babu, V. 2000. Influence of planting density on growth and development of *Salvadora persica* grown on highly saline black soil. *Indian J.Soil Conser.***28** (2): 167-174).

Gururaja Rao, G., Polra, V,N. andRavindra Babu, V. 1999a. Salt tolerance of *Salvadora persica* - A facultative halophyte.*Indian J.Soil Conser.***27** (1): 55-63.

Gururaja Rao, G., Polra, V,N., Ravindra Babu, V. and Girdhar, I.K. 1999b. Growth and development of *Salvadora persica* on highly saline blacks soils: Salt tolerance during immature phase. *Indian J. Plant Physiol.***4** (3): 152-156.

Gururaja Rao, G., Ravindra Babu, V., Abhay Nath and Rajkumar, 2000c. Salt tolerance in *Salvadora persica* : Osmotic constituents and growth during immature phase. *Indian J. Plant Physiol.***6** (2): 131-135.

Gururaja Rao, G., Prakash R. Patel, Bagdi, D.L., Chinchmalatpure, Anil. R, M.K. Khandelwal and Meena, R.L. 2005. Effect of saline water irrigation on growth, ion content and forage yield of halophytic grasses grown on saline black soil. *Indian J. Plant Physiol.*, 10(4): 315-321.

Gururaja Rao, G., Ravindra Babu, V. Nayak, A.K and Girdhar, I.K. 1998. Growth and yield of some forage grasses on saline black soils (Vertic Ustochrepts). Paper presented at the National Seminar on *"Salinity Management in Agriculture"*, CSSRI, Karnal 2-5 December, 1998.

Gururaja Rao, G., Nayak, A.K., Chinchmalatpure, Anil. R, Ravender Sigh and. Tyagi, N.K. 2001c. Resource characterisation and management options for salt affected black soils of Agro-ecological Region V of Gujarat State. Technical Bulletin1/ 2001, Central Soil Salinity Research Institute, Regional Research Station, Anand, Gujarat, p. 83.

Gururaja Rao, G., Nayak, A.K and Chinchmalatpure, Anil. R, 2003. *Salvadora persica*: A Life Support Species for Salt Affected Black soils. Technical Bulletin 1/2003, Central Soil Salinity Research Institute, Regional Research Station, Bharuch, Gujarat, p. 54.

Gururaja Rao, G. and Ravindra Babu, V. 1997. Epicuticular wax of *Salvadora persica* grown on saline black soils. *Indian J. Plant Physiol.*,**2** (4): 290-292.

Kadu, P.R., Pal, D.K. and Deshpande, S.B. (1993).Effect of low exchangeable sodium on hydraulic conductivity and drainage in shrink-swell soils of Purna Valley, Maharashtra.*Clay Res.*, **12**, 65-70.

Loveday, J. and Pyle, J. (1973).The Emerson dispersion test and its relationship with hydraulic conductivity.Division of soils, Technical paper, 15, 7 (CSIRO, Melbourne).

Maggio, A., Reddy, M.P. and Jolly, R.J. 2000.Leaf ga exchange and solute accumulation in the halophyte, *Salvadora persica* grown under moderate salinity. *Environ. Exp. Bot.* **44** (1): 31-38.

Murthy, R.S., Bhattacharjee, J.C., Lande, R.J. and Pofali, R.M.(1982).Distribution, characterization and classification of Vertisols.*Transaction 12th International Congress Soil Science*, New Delhi, India. **2**: 3-22.

Nayak, A.K., Chinchmalatpure, Anil. R, M.K. Khandelwal, Gururaja Rao, G. and N.K. Tyagi, 2003. Soil and Water Resources and Management Options for the Bara Tract under Sardar Sarovar Canal Command: A Critical Appraisal. Status Paper No. 1, Central Soil Salinity Research Institute, Regional Research Station, Bharuch, Gujarat, p. 14.

Nayak A.K., Chinchmalatpure, Anil R, Rao, G. Gururaja and Tyagi, N.K. 2004.Effect of electrolyte concentration, ESP and soil texture on the flocculation, hydraulic conductivity and bulk density of Vertisols.Extendeed Summaries, International Conference on sustainable management of sodic lands, Lucknow, India.203-205.

Nayak, A.K., Girdhar, I.K. and Zaman, M.S. 1999. Characterization of benchmark soils of Bara tract of Bharuch district of Gujarat State. *Journal of Indian Society of Coastal Agricultural Research*, **17**(1 and 2): 66-77.

Nayak, A.K., Gururaja Rao, G. and Chinchmalatpure, Anil. R, 1999.Conjunctive use of Saline water for production of dill (*Anethum graveolens* L.) on saline black soil.*Indian J. Agric. Sci.***70**(12): 860-862.

Nayak, A.K., Gururaja Rao, G. and Anil R. Chinchmalatpure, 2001. Conjunctive use of saline water and surface water in safflower on salt affected black soil. *J. Oil Seeds Res.* **18**(2): 247-249.

Nayak, A.K., Gururaja Rao, G. and Chinchmalatpure, Anil. R. 2001. Conjunctive use of saline ground water in Indian mustard on salt affected black soil. *J. Indian So. Soil Sci.*, **49**(2): 328-331.

Northcote, K.H and Skene, J.K.M. 1972. Australian soils with saline and sodic properties. Soils Publ.27, CSIRO, Canberra, Australia.

NRSA and Associates, 1996. Mapping of salt affected soils of India, 1: 250000 map sheets, Legend, NRSA Hyderabad.

Patel, A.M., Patil, R.G., Raman, S., Singh, T.P. and Shukala, P. 2000. Extent of salt affected and waterlogged soils in Surat Branch of Ukai Kakrapar Command, SWM Tech. Bull. 13, Soil and Water Management Research Unit, GAU, Navsari, Gujarat.

Pitman, M.G. 1975. Ion transport in plant cells and tissues. In: *Whole Plant*, Baker, D.A. and Hall, J.L (Eds.), pp. 267-308, Noth-Holland, Amsterdam-Oxford.

Rana, C.H. and Raman, S. 1999. Issues related to water management in Agriculture-An over view. In : SWMP pub. 10 (Ed. Raman, S.).Soil and Water Management Research Unit, GAU, Navsari, Gujarat, pp. 1-10.

Ravender Singh, Gururaja Rao, G. and Bhargava, G.P. 1996. Conjunctive use of saline water for raising fennel and safflower.*Indian J. Agric. Sci.***65** (10): 727-732.

Rengasamy, P. and Olsson, K. A. 1991.Sodicity and soil structure. *Australian Journal of Soil Research***29**: 935-952.

Robinson, G.H. 1971. Exchangeable sodium and yields of cotton on certain clay soils of Sudan.*Journal of Soil Science* **22**, 328-335.

Shainberg, I. 1984. Response of soils to sodic and saline condition. *Hilgardia.* **52**(2), 1-57.

Strogonov, B.P. 1964. *Structure and function of plant cells in saline habitats. New trends in the study of salt tolerance.* Israel Progrmme for Scientifc Translations, Sci. Transl. jarusalem.

Tiku, B.L. 1976. Effect of salinity on photosysthesis of the halophyte, *Salicornia rubra* and *Distichlis stricta. Physiol. Plant.* **37**: 25-28.

Weretilnyk, E.A., Bodnanek, S., McCue, K.F. and Hanson, A.D. 1982. Compartive biochemical and immunological studies of the glycine betaine synthesis pathways in diverse families of dicotyledons. *Planta,* **178**: 342-352.

Recent Advances in Crop Physiology Vol. 2 (2016)   *Pages* **107–172**
Editor: **Dr. Amrit Lal Singh**
Published by: **DAYA PUBLISHING HOUSE, NEW DELHI**

Chapter 4

Physiological Basis for Maximizing Yield Potentials in Coffee

Chandra Gupt Anand* and P. Prathima

Central Coffee Research Institute,
Research Station, Chikmagalur – 577 117, Karanataka

1. Introduction

Coffee is perennial beverage crop plays very important role in commodity trade worldwide. In India, the coffee cultivation is confined mostly to hilly tracts of Western and Eastern ghats of southern part of country having well distributed annual rainfall preferred for coffee and major coffee growing areas experience South-West monsoon (85 per cent) in the states of Karnataka, Kerala and whole of the North-East and Sikkim. It is also grown in areas which receive predominant North-East monsoon as in Tamil Nadu, Andhra Pradesh and Odisha.

The yield potential of any crop depends on environmental and genetic interactions and coffee is not an exception. If, we have to understand the physiological characteristic of the coffee tree more clearly, we have to know its habitat at the center of origin of commercially important Arabica and Robusta coffee. Arabica coffee is native of Ethiopian tropical forests at little seasonal fluctuation of air temperature on an annual average of 20°C in well distributed rainfall of 1600 to more than 2000 mm with a dry season lasting for three to four months with coolest period. Robusta coffee is native to the lowland forests of the Congo River basin, which extends up to Lake Victoria in Uganda with average temperature between 24-26°C without large oscillations, abundant rainfall over 2000 mm distributed for 9 to 10 months with

---
* *Corresponding Author:* E-mail: cganandccri@gmail.com; hdpphysioccri@gmail.com

atmospheric humidity almost to saturation (Wrigley, 1988). Therefore, for better crop prospects preferably, we need to have such conditions. However, all these above conditions do not exactly exist in Indian condition, so most of the coffee is grown on hilly slopes and the intense tropical rains of south Indian conditions under indigenous naturally established shade trees. The climate change and erratic weather events ranging from droughts to floods under Indian conditions and elsewhere generates substantial losses on crop production especially drought and erratic rainfall at crucial crop growth stages and outbreaks of pests and diseases may further result in heavy losses.

As far drought is concern, a number of methods exist for improving the efficiency of water use. The methods classified in three broad categories: (i) increasing the efficiency of water delivery and the timing of water application (ii) increasing the efficiency of water use by the plants and (iii) increasing the drought tolerance of the plants. Water use efficiency (WUE) is defined as "the total dry matter produced by plants per unit of water used." The fraction of the crop that is economically valuable, termed the "harvest index," is part of the total dry mass and thus part of WUE. The most accurate means of measuring water use efficiency is to monitor the evapotranspiration and harvest the crop for biomass measurements at the end of the season. With the increased ability to transform plants genetically in recent years, it is desirable to apply the tools of molecular biology to the improvement of WUE. Plants showing improved growth with limited water are considered to tolerate drought regardless of how the improvement occurs or whether the water use efficiency is affected. Of the three forms of drought tolerance, dehydration tolerance is most intriguing because it often requires only slight repartitioning of dry mass (Boyer, 1996). Growth exhibited by the plants is an important physiological phenomenon. Healthy and vigorous growth is an external manifestation of the normal metabolic functions which are continuously taking place. Often it is difficult to demarcate it from a number of other physiological processes with which it is intimately linked. Various factors, both external and internal influence or affect growth and development. Some of the most important climatic and edaphic factors concerned with growth process are light, temperature, rainfall, humidity, soil nutrients, soil moisture and soil temperature (Boss, 1951; Sylvain, 1954; Whyte, 1946; Curtis and Clark, 1950). Carbohydrates too have a significant role in growth and developments of plants (Priestly, 1962; Janardhan *et al.*, 1971). The blossom, fruit set and fruit ripening have a profound influence on the vegetative growth. Thus, physiology of coffee plant is affected and crop loss encountered depending on the critical crop growth stages at which such abiotic stresses are experienced. Under these conditions, to realize yield potential of coffee, need based physiological interventions are required at the critical crop growth stages.

2. Cultivation Areas

Presently, India is sixth in production among major coffee producers of world. The area under coffee is around 4, 15,341 ha of which arabica and robusta account for 2, 05,775 and 2, 09,566 ha respectively each (Table 4.1). The annual average production is around 3, 00,000 MT. and above 2/3rd of the produce is exported annually (Table 4.2) to Europe, Russia, USA and other Asian countries (Anonymous

Table 4.1: Coffee Area, Production and Productivity – India (Historic data)

Season	Bearing Area (ha.)			Production (MT)			Productivity (kg/ha)		
	Arabica	Robusta	Total	Arabica	Robusta	Total	Arabica	Robusta	Overall
1950-51	67613	24910	92523	15511	3382	18893	229	136	204
1960-61	70649	49672	120321	39526	28643	68169	559	577	567
1970-71	80433	55030	135463	58348	51883	110231	725	943	814
1980-81	98005	92071	190076	61262	57384	118646	625	623	624
1985-86	108476	106000	214476	72311	50139	122450	667	473	571
1986-87	108500	107000	215500	88975	103119	192094	820	964	891
1987-88	108500	109500	218000	64556	58157	122713	595	531	563
1988-89	108500	112500	221000	94781	119934	214715	874	1066	972
1989-90	108500	112500	221000	62572	55481	118053	577	493	534
1990-91	108500	115000	223500	78311	91415	169726	722	795	759
1991-92	108500	115000	223500	88320	91680	180000	814	797	805
1992-93	108500	115000	223500	73120	96275	169395	674	837	758
1993-94	108500	118000	226500	98300	113700	212000	906	964	936
1994-95	108500	120000	228500	79000	101100	180100	728	843	788
1995-96	120100	121900	242000	103250	119750	223000	860	982	921
1996-97	125017	126267	251284	90450	114550	205000	724	907	816
1997-98	130664	154988	285652	99300	129000	228300	760	832	799
1998-99	143007	159227	302234	97000	168000	265000	678	1055	877
99-2000	146052	162381	308433	119000	173000	292000	815	1065	947
2000-01	146502	167432	313934	104400	196800	301200	713	1175	959
2001-02	149056	171681	320737	121050	179550	300600	812	1046	937
2002-03	146780	173835	320615	102125	173150	275275	696	996	859
2003-04	148389	176735	325124	101950	168550	270500	687	954	832
2004-05	153280	180058	333338	103400	172100	275500	675	956	826
2005-06	151547	189804	341351	94000	180000	274000	620	948	803
2006-07	151861	191179	343040	99700	188300	288000	657	985	840
2007-08	151013	193495	344508	92500	169500	262000	613	876	761
2008-09	156421	194079	350500	79500	182800	262300	508	942	748
2009-10	159828	195674	355502	94600	195000	289600	592	997	815
2010-11	163737	196748	360485	94140	207860	302000	575	1,056	838
2011-12	169906	198781	368687	101500	212500	314000	597	1,069	852
2012-13	176131	200174	376305	102200	202300	304500	580	1,011	809
2013-14*	205775	209566	415341	102200	202300	304500	497	965	733
2014-15**	NA	NA	NA	105500	239250	344750	NA	NA	NA

*: Final Estimate.

Source: Coffee Board of India, 2014.

2014a). Being an export oriented commodity, it contributes nearly Rs 4,700 crores of foreign exchange annually. The coffee area is distributed in a total of 2, 80,241 holdings, out of which more than 98.4 per cent belongs to small growers and only 4,595 holdings (1.6 per cent) belongs to large grower sector (above 10 ha). The coffee consumption is increasing globally at steady rate and in domestic consumption consistently at 5-6 per cent annually (Table 4.3). Hence, to meet consumption, the production and productivity has to be improved but, in India it has reached at plateau and more serious efforts are needed. Since, coffee is sub tropical and tropical crop and thus most of Latin American countries, African and Asian countries, coffee is cultivated under open countries, but in Indian conditions because of large variation in climate and weather conditions it is cultivated under two tier shade canopy, thereby additional efforts are needed now.

Table 4.2: Production and Exports: India Vs World * (In '000 bags of 60 kilo each)

Year	Production		India'sShare (per cent)	Exports		India'sShare (per cent)
	World	India @		World#	India @	
1993-94	90366	3533	3.91	73911	2907	3.93
1994-95	95154	3002	3.15	65718	2070	3.15
1995-96	85250	3717	4.36	74014	3572	4.83
1996-97	101865	3417	3.35	81745	2476	3.03
1997-98	95872	3805	3.97	77806	3685	4.74
1998-99	106163	4417	4.16	82554	3442	4.17
1999-00	115117	4867	4.23	92282	4214	4.57
2000-01	116619	5020	4.30	89248	4229	4.74
2001-02	108451	5010	4.62	90564	3730	4.12
2002-03	123723	4588	3.71	90007	3567	3.96
2003-04	103982	4508	4.34	87527	3826	4.37
2004-05	116062	4592	3.96	91095	2790	3.06
2005-06	111247	4567	4.11	85648	3359	3.92
2006-07	128209	4800	3.74	98143	4150	4.23
2007-08	116455	4367	3.75	95271	3569	3.75
2008/09	128636	4372	3.40	98843	3547	3.59
2009/10	122953	4827	3.93	92455	3005	3.25
2010/11	132983	5033	3.78	105557	4647	4.40
2011/12	132304	5233	3.96	104042	5414	5.20
2012/13	145116	5303	3.65	111523	5043	4.52
2013/14**	145775	5192	3.56	NA	5029	NA

*: Production and exports of ICO members; * *: Provisional, subject to revision.

Source: # ICO Exports -ICO Coffee Statistics 12/ICA 2007 - January 2014 and Coffee Board.

@: Board's Estimate.

Source: Coffee Board of India, 2014.

Table 4.3: Estimated Domestic and World Consumption (From 2000)

Year	Quantity (in MT)	
	Domestic	World
2000	60000	1,626,480
2001	64020	1,700,460
2002	67980	1,726,080
2003	70020	1,794,540
2004	75000	1,897,140
2005	80220	2,003,700
2006	85020	2,100,600
2007	90000	2,212,440
2008	94,380	2,311,620
2009	102,000	2,419,500
2010	108,000	2,512,380
2011	115,020	2,564,700
2012	115,020	2,621,640

Source: ICO, 2014.

3. Cultivars, Botanical Types and their Adaptability

Botanically, coffee belongs to the genus *Coffea* of the family Rubiaceae. There are approximately 70 species under the genus *Coffea*, most of which are native of Africa including the two species *viz.*, *Coffea arabica* and *Coffea canephora* which are commercially cultivated in India. Another species, *C. liberica* is grown to a small extent. Apart from these, some with related genera namely *Psylanthus bengalensis*, *Psylanthus wightiana*, *Psylanthus travencorensis* and *Nostolachma khasiana* are native to India occurring in forests of Kerala, Tamil Nadu, Meghalaya and Assam (Anonymous, 2014).

There are many species of coffee but arabica and robusta coffee species are commercially cultivated in almost 80 developing and undeveloped Central American, Asian and African countries. Arabica (*Coffea arabica*) is a tetraploid species with 2n=44 chromosomes and is popularly known as "arabica coffee". Under natural conditions arabica grows like a small tree but when plant growth is regulated through training, it looks like a bush. Robusta (*Coffea canephora*) is a diploid species (2n=22) and is a bigger bush with robust growth than arabica coffee and is popularly known as robusta coffee. *Coffea liberica* is also a diploid species with 2n=22 chromosome number. In view of its appearance it is a large tree, it is popularly known as 'tree coffee'. It is generally grown as tree coffee along the fence and boundary line of estate as it is not economical to train into bush for commercial exploitation and it is used as wind belt in most of the coffee estates.

The success of breeding programme in any crop depends on the proper evaluation and exploitation of the available gene pool. Hence, a massive gene bank over 250 collections from indigenous sources between 1925 and 1940 formed the base for developing early Indian selections. After 1940, exotic material from all over coffee growing countries including Ethiopia, the home land of arabica there are about 360 surviving collections of arabica, 15 types of robusta and 17 other different species have thoroughly been assessed and exploited in evolving superior coffee types. In India, the emphasis for varietal improvement of arabica coffee has been on evolution of superior types through breeding for rust tolerance, high productivity, wide adaptability and improved quality. So far, India has released 13 improved elite Arabica varieties (both tall and semi dwarf) and three high yielding robusta varieties for commercial cultivation.

Arabica coffee (*Coffea arabica* L) is a small tree in its original habitat but grows like a shrub with bushy growth when trained. Arabica produces superior quality coffee but is susceptible to diseases like leaf rust and pests like white stem borer. This necessitated the systematic breeding efforts for superior genotypes with respect to leaf rust resistance, high yielding potential, wide adaptability and superior quality. So far, CCRI has developed 13 arabica selections for commercial cultivation of which seven selections are grown on large scale (Anonymous, 2014b; Sreenath and Prakash, 2006). A brief description of different arabica and robusta selections developed by CCRI and their adaptability is shown as follows.

3.1. Arabica Cultivars

The first variety Sln.1 (S.288) was developed by pure line selection from S.26, a putative natural hybrid between *C. arabica* and *C. liberica* released in India during 1936-37. It is moderate yielder (800-1000 kg/ha) with general adaptability as its liquor quality rated was Fair Average Quality (FAQ). It is a tall and resistance to leaf rust races I and II superiority other varieties like 'Chiks' and 'Kents' under cultivation then, but produces high percentage of defective beans (20-30 per cent). Subsequently a cross of S.31 and S.22 was released as Sln.2 because of its hybrid vigour and resistance to leaf rust recommended for cultivation initially, but later due to high production of bean abnormalities, the seed distribution has been taken back.

Sln.3 (S.795) selection was developed from the cross between S.288 and 'Kents' coffee and achieved in minimizing bean abnormalities in S.288. It is a vigorous and wide spreading with profuse growth and a yield potential of 1500-2000 kg/ha as well as bold beans of 60-65 per cent 'A' grade and having balanced quality with good body, good acidity and fairly good flavour. This showed resistance to leaf rust races I and II, prevalent at the time of release, but later fell susceptible to new virulent races VII, VIII, XII, XIV and XVI. It is having broader adaptability even now.

Sln.4 was released as a strategy to check the devastation of race VIII on S.795 plantations on outburst of leaf during 1960s and given for commercial cultivation during late 1960s. It is a composite selection of three Ethiopian arabica collections *viz*. Cioccie, Agaro and Tafarikela.

Sln.5 (a and b) includes two separate families of cross bred lines namely Devamachy and S.881 (a wild Arabica collection from Rume in Sudan) and Devamachy and S.333. Devamachy is a spontaneous Robusta x Arabica hybrid characterized as intermediate plant phenotype to arabica and robusta with high vegetative vigour but low fruit set. The plants of Devamachy x S.881 shows vigorous vegetative growth with erect branching habit exhibits field resistance to leaf rust with a consistent yield potential of 1200 to 1500 kg/ha but beans are medium in size with high percentage of B grade beans (~ 40 per cent) adaptable to hot and humid climate. Devamachy x S.333 line (S.2931) is characterized by uniform, medium sized and semi drooping plants exhibits higher resistance to leaf rust with a yield potential of 1500 kg/ha. Fruits are round, bold with bold beans 60-65 per cent 'A' grade and liquor in both Sln.5A and Sln.5B is FAQ to FAQ+ in quality adaptable to different arabica growing regions and performs well in medium to high altitudes.

Sln.6 (S.2828) is a cross between robusta cv.S.274 and 'Kents Arabica' and the progenies from the 2nd backcross to Kents exhibiting vigorous with Arabica phenotype and robusta type tight fruit clusters having yield range between 1200 and 1500 kg/ha and bold beans with 60- 65 per cent `A` grade having liquor quality is FAQ to FAQ+. The Sln.6 is well adapted to mid altitudes and shows a mixed type of rust reaction with about 80 per cent of plants manifesting high tolerance to rust.

Sln.7.3 is multiple cross hybrid involving 'San Ramon', a dwarf mutant and tall arabicas like S.795, Agaro and Hibrido de Timor (HDT). The 'San Ramon' mutants were very compact in branching habit with closer internodes and deep roots with 70 per cent dwarf and 30 per cent tall types progeny but highly susceptible to rust. In order to improve resistance to leaf rust and to exploit the dwarf phenotype for high density planting, San Ramon was crossed with S.795 and developed Sln.7.1. The dwarf type Sln 7.1 was crossed with Agaro and derived Sln. 7.2. Dwarf type Sln 7.2 was further crossed with HDT and Sln.7.3 was developed. The dwarfs in Sln.7.3 are suitable for high density planting (4ft x 4ft, 5ft x 4ft.) with a yield potential of over 1500 kg/ha but with annual variations for production. It is a late ripener, drought tolerant and suitable for marginal areas, however, susceptible to virulent races of leaf rust. The variety produces FAQ to above FAQ cup quality.

Sln.8 is naturally-occurring Robusta x Arabica hybrid identified in Timor island (in South East Asia) and a pure line selection of 'Hibrido de Timor' (HDT) and was introduced in India during 1961. The plant phenotype closely resembles arabica and possesses high vertical resistance to leaf rust a moderate yielder (1000-1200 kg/ha) and produces medium to bold beans. Cup quality is similar to Arabica (FAQ to FAQ+). Sln.8 is suitable for cultivation in mid altitudes.

Sln.9 is an exploitation of HDT as a source of rust resistance, several crosses were effected between HDT and other tall arabicas such as Tafarikela, S.795, Geisha, S.12 Kaffa and Bourbon has given HDT x Tafarikela among others as highest resistance to leaf rust under field conditions. It has an average yield of 1700 kg/ha (23 years) at CCRI and thus it has been released as Sln.9 for commercial cultivation. The Sln.9 manifested fairly good tolerance to leaf rust, drought and widely adaptable across all the coffee regions. Fruits are bold with 60 to 65 per cent 'A' grade beans and excellent liquor quality and possessing distinct flavour in cup.

Sln.10 is a double cross hybrid evolved by 'Caturra' a semi dwarf with high yielding potential but highly susceptible to rust crossed with Cioccie and S.795 and again crossing respective progenies. The plants are compact and semi dwarf with profuse bearing habit exhibiting resistance to common races of rust but susceptible to virulent races. Fruits are bold and 65 per cent 'A' grade beans with cup quality similar to S.795.

Sln.11 is an interspecific hybrid, a cross between *C.arabica* and *C.eugenioides* obtained by amphiploidy. It is late ripening, resistant to leaf rust and moderately tolerant to drought with moderate yield potential of 1000 to 1200 kg/ha and thus exhibited better adaptability to hot and humid climate. It produces smaller bean size with similar cup quality to arabica.

Sln.12 (Cauvery/Catimor) is a semi dwarf genotype derived from a cross between Caturra and HDT and was released in 1985. The variety is an early bearer and exhibits high resistance to leaf rust in the initial years. The yield potential is 2000 kg/ha and under different management conditions ranges from 1000-3000 kg/ha and thus, requires intensive management practices for realizing consistent yields with bold bean and high liquor quality of FAQ to FAQ+ with pleasant aroma and taste. It is widely adaptable across all the coffee regions, but best suitable for cultivation in high elevation areas (above 3300 feet MSL).

Sln.13 (Chandragiri) is a latest variety released in 2007, which has derived from the cross Villasarchi x HDT. It manifests high field tolerance to leaf rust with yield potential of 1500 – 1800 kg/ha and better bean characteristics having FAQ and above liquor quality. It is best suited for cultivation at higher elevations (3300 ft and above).

3.2. Robusta Cultivars

Another most important commercially cultivated 'Robusta' (*Coffea canephora*) played very important role in development of Coffee Industry of India. The devastating influence of leaf rust and white stem borer on arabica coffee as well as robust nature of robusta coffee, it encouraged the Indian coffee growers towards robusta coffee in 1903-1906 onwards. Robusta is a native of Central Africa and is predominantly a low land coffee that adapted readily to the Indian soils. Robusta coffee possesses several useful characters like tolerance to leaf rust, white stem borer, nematode invasion as well as the potential to give consistent yields and thus, relatively lower cost of cultivation compared to arabica coffee. On the other hand due to inability to cope with long droughts, late cropping, late stabilization of yields and inferior quality compared to arabica it has some of the negative aspects too. The robusta coffee is self incompatibility and has limited gene pool, which are bottlenecks for robusta improvement in our country. However, to meet the industry need and keeping all above aspects in mind, improvement to robusta was made and three superior robusta cultivars were developed for commercial cultivation.

Selection 1R (S.270 and S.274) is the seedling progenies of two individual high yielding mother plants. Both these type progenies are vigorous and grow into moderately large trees. The fruits are medium bold with tight clusters of 30-50 fruits each. Ripening is usually late compared to arabica and outturn ratio is 5:1. Beans are medium to round and green to greyish in colour when wet processed and they produce nearly 43 per cent 'A' grade beans. The cup is neutral with a quality rating FAQ to good. These two genotypes have recorded yields of nearly 1000 kg/ha on an average over 35 years under rain fed conditions. Among these two cultivars, the S.274 produces more bold beans compared to S.270 and thus, S.274 has become more popular and also due to its wide adaptability.

Selection 2R (BR Series 9, 10 and 11) is identified from seventeen superior plants which have yielded twice the family mean yield were identified from the initial twelve single seedling progenies of S.267 to S.278 and named as 'Balehonnur Robustas' (BR), named for the eponymous town in Karnataka's Chikmagalur district. Based on the progeny performance of these plants BR 9, 10 and 11 were found to be superior in production and as such seed mixture of these clones was issued for commercial cultivation as Sln.2R. Clones were also issued for the establishment of bi/polyclonal gardens. This selection resembles S.274 in growth habit, yield potential and bean/ cup characteristics. However, in these clones, high degree of stability for 'A' grade beans and other prominent characters are seen. Only seed mixture of all the three clones BR 9, 10, 11 is advocated for planting, since, planting separately may result in poor fruit set due to self incompatibility.

Selection 3R (C x R) is an interspecific hybrid developed through involving *C. congensis* and *C. canephora* (Robusta) in its hybridization. *C.congensis* is a species closely related to both Arabica and Robusta with compact bush, drooping branches and better bean quality than robusta. The F_1 was backcrossed to both the parents (robusta and congensis) and then progeny to robusta, through selection and sib-mating a commercial CxR hybrid was evolved. Its plants show intermediate to both its parents in size with compact secondary growth of drooping nature and each branch consists of 9-12 bearing nodes. This hybrid is most suited to closer spacing of 2.7x2.7m or 2.4x2.4m (9x9ft or 8x8ft) compared to pure robusta S.274 which requires a wider spacing of 3x3m (10x10ft). Fruits are in tight clusters with 30-40 per node, orange to crimson red with prominent and projected naval. Early, uniform ripener and easy to harvest with an out turn ratio of 4.5:1.0. Beans are medium to bold, golden brown and about 54 per cent are A grade. Liquor is soft, neutral with light to fair acidity and good cup quality.

3.3. Genetic Improvement through Molecular and Biotechnological Approaches

There always has been curiosity on development desired plant material and it has been achieved through success of breeding programme in any crop, which depends on the proper evaluation and exploitation of the available gene pool. However, it is time consuming and all desirable characters are not translated into the new plant material. Now the applications of molecular biology and biotechnology have proved to make progress in same direction within short span of time and overcome the difficulties of breeding programme (Zamarripa and Petiard, 2004; Sreenath and Praksh, 2004). Accordingly it gained movement for coffee and in recent year's significant progress is made. While there are known resistance genes to some fungal diseases of coffee, the long-term breeding cycle and the need for durable resistance due to the perennial nature of coffee, make biotechnology an important tool for sustainable coffee production. The ability of biotechnology to move natural coffee resistance genes into established cultivars and to pyramid resistance genes will save considerable time and provide sustainable and high quality coffee production. There appears to be naturally occurring resistance gene for certain important insect pests like coffee berry borer (*Hypothenemus hampei*). Although biological control mechanisms are being developed, coffee berry borer is still considered the most economically important insect pathogen of coffee and effective resistance introduced through biotechnology would have major effect on the lives of coffee farmers in many parts of the world. Abiotic stresses, such as freezing and drought also represent significant problems in the coffee industry. For example, periodic freezes experienced in certain production areas of Brazil disrupt commodity markets affecting importers, roasters and techniques would be of great importance in alleviating the disruption that a frost can cause to the coffee industry. Finally cup quality is becoming increasingly important as a result of the growth of the specialty coffee industry. Biotechnology will undoubtedly play a progressively more important role in assuring cup quality by reducing defects resulting from contaminating by microorganisms and insects. However, it may also play direct role by modifying the cup quality-related and nutraceutical components in the green bean. Various biotechnology focused approaches on stem, leaf integument, apical bud and node cultures for micropropagation, anther culture, embryo culture, endosperm culture, protoplast culture, genetic transformation and molecular markers for genetic improvement and *in vitro* preservation techniques including cryopreservation for conserving the biodiversity have tremendous potential for improvement of coffee.

The use of the tissue culture technology for genetic improvement resulted in plant regeneration through embryo culture by direct germination as well as somatic embryogenesis and these plants have been established in the field. Three interspecific and five intervarietal crosses through embryo rescue method also established in the field and are being evaluated. Plantlets regenerated from the anther culture of CxR cultivar is the first achievement in a diploid species of coffee. Callus cultures established from endosperm tissues and plantlets regenerated through embryogenesis from the endosperm calli of S.2803 also have been achieved in coffee. DNA markers are becoming powerful tools in the hands of the coffee breeder. Protocols have been developed for isolation of DNA and producing RAPD and AFLP markers. The RAPD markers are being used for evaluating leaf rust differentials and genetic fingerprinting. Genetic transformation protocols are being developed with the objective of genetic engineering of coffee. Transient expression of gus genes was achieved in leaf tissue of CxR cultivar after co-cultivation with Agrobacterium tumfaciens. *In vitro* zygotic embryos preservation up to 2 years under slow growth condition in *C. Arabica* and successful zygotic embryos cryopreservation has been achieved in three species (Anonymous, 2003). Genetic transformation with Agrobacterium mediated transformation; Construction of cDNA library has been constructed during early stage of infestation white stem borer in coffee. It has revealed strong presence of transcripts putatively coding for a gene involved in the biosynthetic pathway of a plant defence system and confirmed the existence of defence system against herbivory. Similarly, in identification of Bt isolates active against WSB DBT sponsored research project, out of 3000 native Bt isolates, 14 *Cry* genes tested against WSB one Bt isolates containing a *Cry* gene was active against WSB (Anonymous, 2014).

4. Establishment of Young Coffee Plants

Plants are the biological entity capable of mining the soil and releasing the inorganic elements in an assimilating form. The micronutrients play an important role in growth and development of plants and insufficient quantity can reduce growth and yields. Furthermore, if the mineral elements are not available to plants adequately, the vital process like photosynthesis will be adversely affected. The carbon fixation depends on adequate supply of macro and micronutrients to the plants. Different nutrients in combination or even individually can alter the photosynthesis and associated processes. Therefore, balanced supply of nutrients to the plants is essential for maintenance of growth and carbon fixation during sustained loss due to leaching, evaporation and crop removal. Hence early and better establishment of young coffee with required canopy structure is very much important to achieve sustainable crop production in coffee. Different sources of fertilizers were recommended to coffee based on the requirement at different ages of the plants (Anonymous, 2000). Normally, coffee seedlings raised in poly bags were planted to the field during the period of cessation of South West monsoon. For such young coffee lower dosage of macronutrients are being suggested to maintain sustained growth. In general, straight fertilizers like Urea (N), Rock phosphate (P) and Muriate of Potash (K) are preferred for young coffee. However, information on the impact of different sources of fertilizers on growth and carbon fixation of young coffee is sparse. Keeping this in view, to generate information on the effects of different sources of fertilizers on growth characteristics and carbon fixation rates in young coffee, the sources of fertilizers application of Factomphos, Urea+Single Super Phosphate (SSP) + Muriate of Potash (MOP) and Ammonium sulphate+ Di ammonium phosphate (DAP) combinations

**Figure 4.1: Arabica in Extreme Conditions Showing Wilting and
Dadap Defoliate Completely.**

were tested during pre and post monsoon periods @ 25 and 50 grams per plant per application respectively during first and second year. A significantly (p=0.05) increase in growth characteristics such as plant height, number of primary branches and stem girth coupled with improved bush spread of young coffee was noticed in both the cultivars applied with Factomphos than other sources of fertilizers. Higher increase in net photosynthesis and carboxylation efficiency was also observed after twenty and forty days of Factomphos application as source of fertilizer compared to rest of the fertilizer sources. It was concluded that Factomphos would be very much useful as a source of fertilizer to improve the growth and development of young coffee for better establishment (D'souza *et al.*, 2002).

4.1. Boosting of Seedling Growth

In case of delayed sowing and transplanting, the seedlings may not attain desired growth in time for field planting. Under such conditions the growth of seedlings can be boosted by spraying of the growth regulator options such as Planofix 50ml + Urea 1 kg or Agronaa 50ml + Urea 1 kg or Cytozyme crop plus 60ml +Urea 1kg in 200 litres of water at least one month before field planting or when the seedlings are at 4-5 pairs leaf stage. It helps to produce healthy seedlings of 5-6 pair of leaves with just emergence of first pair of primaries (Anonymous, 2014b).

4.2. Seasonal Influence on Clone Preparation

Robusta (*Coffea canephora* Pierre ex Froehner) is an important commercially cultivated coffee species. Much importance is given to this species when arabica coffee found susceptible to leaf rust caused by *Hemeleia vastatrix*. As robusta is tolerant to this disease it could occupy arabica areas also. To find the suitable season for planting and to investigate the favorable factors for successful rooting of cuttings revealed that cuttings planted in different periods of the years exert notable influence on rooting. Higher percentage of rooting was recorded in June cuttings of S.274. Even the cuttings in control rooted to cent percent while cuttings planted in December rooted to very low percent (35.5). High percentage of rooting cuttings coincided with the rainy season. It was observed by earlier workers also that rainy season was most suitable for planting coffee cuttings (Evans, 1958; Anonymous, 1960). Regarding root system, an increase in the number and length of primary roots were obtained in June. Thus planting of cuttings in June and August appears to be congenial of cuttings for good success in rooting and for better sprout number, length and leaf production were improved in cuttings planted during February, but rooting was comparatively less. Considering both root and shoot characters, it can be clearly understood that June and August plantings recorded good percent of success. When a relation was tried to find between rooting and carbohydrate status of cuttings, it was observed that maximum content of total sugars obtained in August cuttings enhanced development of good root system. It is pertinent to mention here that the root system and total sugar were reported to have positive correlation between them in various other crops (Stoltz, 1968; Adarsha Bala *et al.*, 1969; Basu *et al.*, 1972; Reuveni and Adato, 1974). Starch content was high in August and June and this corroborates well with the better rooting. There seems to be positive correlation between the starch content and rooting in robusta (Raju, 1972).

Highest auxin content in April cuttings showed low rooting percentage as this period usually remained dry and hot, the production of auxin increased with the increase in atmospheric temperature. Nanda and Anand (1970) also found increase in auxin content reaching supra optimal level with exogenous application of auxin adversely affecting rooting in *Populus nigra*. In October, the auxin content increased in S.274 but the rooting was low as the carbohydrate status was at very low level. Earlier it was reported by several investigators that though there was an increase in auxin content, if carbohydrates are not present in sufficient quantities, cuttings of different crop plants failed to root (Adarsha Bala *et al.*, 1969; Gauthert, 1969; Oliemann *et al.*, 1971; Haissig, 1971). However, it was not possible to draw any definite relationship between rooting and auxin content in robusta coffee. As a further step, total phenol content was estimated and found that the highest production was in February and October when rooting response was the lowest. Lowest phenol content was estimated in June cuttings which rooted to the highest percent. Auxin in combination with phenol content showed some relation with respect to rooting. It seems that a proper balance between total auxin and total phenol contents is necessary for satisfactory rooting. Increase or decrease in any of the two contents reduced rooting. A synergistic effect of phenols with the endogenous auxins in stimulating rooting has been suggested by Mahadevan (1964), Tomaszowski and Thimann (1966) and Bose *et al.* (1972). Rooting co-factors were identified in S.274 to find whether there has any significant relation to rooting. February material had co-factor 2-in low quantity and co-factor-4 was absent. Hence rooting of this material was also poor. In April all the co-factors were present but at very low levels. It is in June and August all the co-factors were at high levels and highest rooting was obtained with the increase of inhibitors in October and December, there was a decrease in rooting of these cuttings. Though the seasonal differences were not reported, the changes in rooting capacity of cuttings were proved by many investigators (Lanphear and Meahl, 1963; Fadl and Hartmann, 1967). However, with the present studies it was found possible to draw relation between rooting of cuttings to the presence/absence of rooting co-factors and their levels in addition to the ratio between total phenol and total auxin contents while in the other factors no definite conclusion could be drawn.

4.3. Rejuvenation of Unproductive Field Robusta

The process of uprooting 'off type/passenger plants' by conventional methods and filling those gaps with young seedlings is a cumbersome and time consuming process. Moreover, such replanting usually starts yielding only after 4 or 5 years. To avoid these difficulties rejuvenation of off-type plants through top working is practised. In this technique, the off-type plants are collar pruned at a height of about 30-45 cm in March-April, soon after summer showers. The stumps start sprouting and produce many vertical shoots (suckers) in about 30-45 days. However, there are chances of 20 to 30 per cent failure of sucker production in old robusta plants. To induce sucker production, a spray of IAA 200 ppm in combination with 1per cent DMSO (200 mg of Indole Acetic Acid and 10 ml Dimethyl Sulphoxide in 1 litre of water) could be applied on the collar-pruned stumps (Anand *et al.*, 2000). Two or three healthy suckers that arise from just below the cut end are retained and the remaining suckers are removed at the time of grafting. During the monsoon season, suckers of the same age and thickness are collected from elite desirable plants. Single node scions are prepared and grafted by cleft method on to the suckers arising from the collar pruned 'off type' plants. Union takes place within 1 or 2 months depending upon weather conditions. Around 65 to 70 per cent success can be achieved using compatible scions. Hence it is advisable to graft 2-3 suckers in each stump. The scion starts growing vigorously by drawing nutrients from the already well established root system of the stock (off-type plant) and produces a crop within 2 or 3 years of grafting. Thus there is a saving of about 2-3 years time towards the first crop when compared to the conventional methods. Besides, yields from top-worked plants are assured as the scion material is collected from superior mother plants.

5. Nutrition Management

The balanced fertilization is key component for sustainable crop growth and yield. For this the macro nutrients nitrogen, phosphorus and potassium are of importance along with calcium, magnesium and sulphur which are secondary nutrients, as well as iron, manganese, copper, zinc, molybdenum, boron, sodium and chlorine which are considered micro/minor elements. Their availability and uptake depends on type of soil texture and structure, thus management is most important in coffee prospects.

5.1. Soil Acidity

Coffee in south India is cultivated in high rainfall areas. Leaching out or loss of calcium and magnesium present in the soil through runoff during heavy rains and continuous use of acid forming fertilizers like ammonium sulphate and ammonium chloride are few of the major causes for soil acidity. The soil pH indicates whether the soil is acidic, neutral or alkaline and can be measure easily. The optimum plant growth of coffee takes place in the pH of 6.2 at which all the essential nutrients are in available form and their uptake is maximum. In order to grow healthy coffee plants and to realize a better crop, it is very important to sustain soil health. The various plant nutrients in the soil need to be balanced so that soils remain productive for a longer time. The acidity (pH) determines the productivity of the soil and nutrient assimilation/mineralization in the soil. To ensure maximum fertilizer use efficacy, the soil pH should be maintained at around in the range of 6.0 -6.5. Hence the soil pH has to be tested once in two to three years and corrective action is taken if the pH is not within the optimum range of 6.0 to 6.5. Highly acidic soils are harmful to useful soil micro-organisms and in such soils, coffee growth tends to be stunted. At the same time, if the soils are alkaline (above pH 7.0) most of the essential plant nutrients are not available to the plants.

5.2. Leaf Analysis

The analysis of coffee leaf is another approach in determining the nutrient requirement of the plants. Coffee being a perennial crop, the plant growth, development and yield has to be balanced with energy allocation between vegetative and reproductive simultaneously throughout the year. It is well established that a tonne of clean arabica coffee removes approximately 40, 7 and 45 kg of nitrogen, P_2O_5 and K_2O from the soil while a tonne of clean robusta coffee removes 45, 9 and 58 kg of nitrogen, P_2O_5 and K_2O, respectively (Ananda Alwar, 1985). Taking into consideration the crop removal of nutrients, biomass production, nutrient loss through leaching and fixation and fertilizer use efficiency (FUE) in coffee soils, fertilizer doses for different quantities of yield per unit area vary.

The leaf analysis successfully helps in distinguishing nutritional disorders, identification of sufficiency or deficiency of nutrients in plants and corrective steps can be taken. Thus, analytical results may be used for adjusting the balanced application of fertilizers.

The leaf analysis values established to adjust the fertilizer levels are as follows:

Nutrients	Per cent Concentration in Leaf	Classification	Fertilizer Recommendation
Nitrogen	< 2.5	Low	Increase the N level
	2.5-3.5	Optimum	Continue present N level
	> 3.5	High	Reduce N Level
Phosphorus	< 0.1	Low	Increase the P level
	0.1-0.15	Optimum	Continue present P level
	> 0.15	High	Reduce P Level
Potassium	< 1.5	Low	Increase the K level
	1.5-3.5	Optimum	Continue present K level
	> 3.5	High	Reduce K Level

5.3. Soil Testing

Soil testing is one of the most important tools which provides precise information on the nutrient status of the soil. Such information aids for making fertilizer recommendations and suggesting suitable fertilizer dosages. The table showing the upper and lower limits of available NPK in soils is given below for reference.

Rating	O.C. per cent	Available N (kg/ha)	Available P (kg/ha)	Available K (kg/ha)
Low	< 0.5	< 250	< 9	< 125
Medium	0.5-2.5	250-500	10-22	126-250
High	> 2.5	> 500	> 22	> 250
Nutrient Index	<1.67		1.67-2.45	>2.45

5.4. Deficiency Symptoms

Non application of balanced fertilizers and over exploitation without adequate nutrient supply generally result in deficiency of nutrients in plants. This can be noticed in the decolouration or malformation of leaves in the plant. Deficiency of nutrients not only affects the plant growth but also the yield. Deficiency symptoms of major and micro nutrient elements and corrective measures are given below.

Nutrient	Symptoms	Corrective Methods
Nitrogen	Yellowing of older leave, defoliation and die-back in severe cases. Poor, stunted growth	Foliar spray of 0.5-2.0 per cent urea alone or with Bordeaux mixture
Phosphorus	Irregular patches of yellow colour and mottled appearance in older leaves. Leaf turns red or violet and drops under severe condition.	Foliar spray of 0.5 per cent SSP or 1 per cent DAP not compatible with B.M.
Potassium	Necrosis or scorching of leaf tip and margins of older leaves	Foliar spray of 0.1 per cent MOP alone or with Bordeaux mixture
Magnesium	Yellowing of older leaves between the lateral veins on either side of mid rib. Yellow area surrounded usually by green lines	Foliar spray of 0.1 per cent Magnesium sulphate or soil application of dolomite
Calcium	Death of growing tip. Brittle leaves with yellowish appearance.	Soil liming
Iron	Chlorosis of leaves	Good drainage
Zinc	Short internodes, stunted and chlorotic small elongated leaves. Rosette shaped leaf bunch	Foliar spray of 0.25 per cent Zinc sulphate
Boron	Terminals shoot death and leathery leaves with malformation of small leaves	Foliar spray of 0.05 per cent boric acid

5.5. Fertilizers and Quantity

Achieving yield potential of coffee, application of right quantities of appropriate fertilizers in right time is the key factor. The most of the nitrogen, phosphorus and potassium required quantities need to be applied to a given block depends on the estimated yield, performance of plants for the last three or five seasons and the soil test values of the nutrients (Coffee Guide, 2014; Ananda Alwar, 1985; Krishnamurthy Rao, 1991).

A dose of 20:20:20 kg of NPK must be given compulsorily for mature arabica coffee, where coffee yields do not exceed one M.T. per acre. Further, for every 100 kg of bearing coffee a dose of 10:7:10 kg of NPK up to 1000 kg crop of 120:90:120Kg NPK must be applied. Where the production levels range between 1 and 1.5 M.T., the mandatory/sustenance dose is 30:30:30 kg may be added accordingly with 10:7:10 kg NPK for each 100 kg bearing coffee dose. Thus, it will be at the dose of 130:100:130 and 170:125:170 kg NPK for bearing coffee 1100 and 1500 kg per unit area. It is advisable to apply the annual dose of fertilizers in as many splits as possible. Certainly, it should be restricted to a maximum of 40:30:40 kg of N: P_2O_5:K_2O per acre for every round of application.

The nitrogen and potassium requirement of coffee plants grown without shade trees is generally higher than that of plants grown under normal shade. This is because shade trees contribute organic matter to the soil, increasing the efficiency of these nutrient elements. Similarly coffee under irrigated conditions requires more nutrients as these plants tend to produce higher crops. The annual dose of nutrients required to be applied will have to be worked out based on the anticipated yield and soil fertility status. Nutrients applied without considering the soil test values will be inaccurate, uneconomical and incomplete.

Subsistence agriculture is being transformed to intensive agricultural practices through the use of improved varieties, greater use of fertilizers and intensive cropping patterns resulting in the creation of a large gap between the sulphur supply and demand in soil-crop cycle. Even though sulphur is recognised as secondary nutrients for plants (Ananda Alwar and Krishnamurthy Rao, 1992a), the soil survey on sulphur status in coffee plantations revealed deficiency of its availability in the coffee soils. Thus, its requirement for coffee is as much as phosphorus and considered as a major nutrient. It has been found that sulphur is not only an essential nutrient for optimum yield of coffee but is also essential to maintain and improve the quality of coffee. Sulphur requirement is 15-20 kg per 1 MT of production of clean coffee for arabica and 20 to 25 kg per 1 MT of production of clean coffee for robusta coffee. Sulphur can be applied in the form of ammonium sulphate, calcium or magnesium sulphate, elemental sulphur, potassium sulphate or Single Super Phosphate preferably during the post monsoon period (Anonymous, 2014).

5.6. On Farm Waste Recycling

Plant nutrient input is one of the major constraints of coffee cultivation in India. Harvesting the nutrient energy of biodegradable on-farm wastes is of prime significance for maximising production. Recycling on-farm waste can minimize the pollution of water. The potential for exploitation of manure values of biodegradable coffee plantation wastes is large in Indian plantations in the form of shade tree leaf litter, coffee leaves and prunings, weeded materials, fruit skin/pulp and cherry or parchment husks.

It is estimated that shade trees contribute roughly 10×10^3 kg of leaf litter/ha/ year in India resulting in the addition of 40-60 kg N, 30-33 kg P_2O_5 and 40-60 kg K_2O to the soil on total decomposition. In estates where dadaps are used as temporary or permanent shades, the annual return of nutrients is around 96 kg N, 8 kg P_2O_5 and 67 kg K_2O in the form of degradable stakes, tender branches and leaves. Similarly, coffee leaves (fallen) and annual prunings return significant amount of plant nutrients. If, all the weed materials are returned directly to the field after weeding or after composting, significant amou, 1992b). The fruit skin/pulp obtained after pulping 6000 kg of fruits to get a tonne of coffee returns 14-15 kg N, 3-3.7 kg P_2O_5 and 29-37 kg K_2O to the coffee soil if properly decomposed and recycled. These wastes are found to be superior to cattle manure. Similarly, if all cherry husk is recycled after composting, they contribute 1.66-2.0 per cent N, 0.4-0.5 per cent P_2O_5 and 2.4-2.6 per cent K_2O. It is estimated that nearly 84-95 kg N, 40-42 kg P_2O_5 and 108-123 kg K_2O are available for recycling in a hectare of coffee field per year besides dadap wastes. Thus, recycling of all farm wastes in the coffee plantations is strongly recommended for improving the production and quality of coffee. It may help reduce the fertilizer input costs and increase efficiency of applied fertilizers on account of enriched organic matter content of the soil (Anonymous, 2014b).

6. Plant Growth and Development

6.1. Shoot and Leaf Growth

In coffee, two types of branches formation can be distinguished. The branches which grow vertically are orthotropic shoot commonly called suckers and the branches placed 45 to 90° angle in relation to main axis are plagiotropic and commonly called primaries. The leaf axils of plagiotropic shoots have buds which may develop into vegetative shoot or flower buds depends upon the environmental conditions. Growth of lateral branches exhibits a seasonal fluctuation depending upon the climatic conditions. The pattern of growth in coffee under south Indian conditions is typically sigmoidal nature. Vasudeva and Ramaiah, (1979) observed resumption of shoot growth in March after the dry period on receipt of blossom showers in Arabica S.288 and S.795 plants. In the initial period the branches started increasing in length gradually and there was a proportionate increase in the number of nodes also. The period of slow growth continued during monsoon up to July may be due to leaching of soil nitrates (Rayner, 1946) when rainfall is heavy (Sylain, 1954). After the peak period of South-west monsoon rains, a period of rapid growth was recorded from August to October when the mean maximum temperature ranged from 25.8 to 27.1°C and particularly during September when the mean maximum temperature was 26.3°C after October, the growth slowed down which was generally associated with the low moisture content of the soil (Gopal and Vasudeva, 1973) and ripening of fruits. Mayne (1944) is of the opinion that crop development has an antagonistic effect on the vegetative growth as during maximum fruit growth period the vegetative growth is observed minimum. The minimum vegetative growth was from December to February which represented the resting period or dormant period of coffee. During this period the mean maximum temperature ranged from 27.4 to 32.4 °C.

The understanding of coffee growth characteristics of the 4 month old seedlings of S.274, Devamachy, San Ramon and S.2828 revealed faster root growth in the earlier stages of seedling's growth than stem growth in all the five coffee cultivars. The increase in the dry weight production of seedlings was in proportion to its age. Quantitative analysis during early phase of growth in S.795, San Ramon and S.274 coffee cultivars grown under diffused and filtered light showed markedly high Net Assimilation Rate (NAR) and Relative Growth Rate (RGR) in the plants grown under diffused light in all the plant genotypes. Leaf area ratio (LAR) was lesser in plants grown under diffused light in all the varieties except in San Ramon than those grown under filtered light. Significant variations in RGR and its component NAR was noticed between seven arabica, one robusta and four hybrids during the two periods of growth (30 and 38 weeks). There was high NAR and RGR and NAR in S.288 compared to other plant types but not much variation exhibited between HDT, robusta, San Ramon hybrid (Venkataramanan *et al.*, 1983). The vegetative growth slows down with advent of dry period from November. The minimum growth is observed from December to February (dormant period) due to depletion of soil moisture and is resumed in March after the drought period. Thus, the role of environmental and edaphic factors like light, temperature, rainfall and soil moisture have significant influence on growth and development of coffee.

Leaf growth of coffee shows a periodicity as observed in shoot growth. Vasudeva *et al.* (1973) observed the initiation of new leaves continuously from March-April (after receipt of blossom showers in summer months) to January in Arabica coffee. The numbers of leaves initiated were more during September and had the maximum leaf area but those initiated during summer months had the minimum leaf area. The variation is influenced by many factors other than water, mineral nutrients, endogenous hormones (Humphries and Wheeler, 1963), such as atmospheric temperature, relative humidity, duration of sunlight and rainfall. Leaves initiated in September were found significant association with maximum temperature of 25.8 °C and minimum of 18 °C, sunshine for 3 hours, atmospheric relative humidity of 92 per cent (at 8.00 hours) and rainfall of 230.4 mm. Similar results were reported by Venkataramanan (1985) and the leaves initiated in August and September had higher dry weight of leaves. Apparent photosynthesis and nitrate reductase activity were also higher in these leaves. The higher photosynthetic rate was due to the rapid increase in chlorophyll content. Vasudeva (1969) observed variations in the leaf area even between the leaves of a pair. The leaves on right side (with reference to the position of observer) of the branch showed more area compared to the area of the leaves of the left side. In an attempt to evaluate the relative importance of leaf breadth and length in determining the leaf area, a study conducted on 10 varieties of Arabia showed correlation co-efficient between both length and leaf area and breadth and leaf area in eight varieties out of 10 studied which clearly indicate that breadth influences more in the determination of leaf area than leaf length. The relative importance of these two factors in determining the leaf area was further confirmed by the high regression co-efficient between breadth and leaf area (Vasudeva *et al.*, 1971). The life span of leaves of arabica coffee ranged from 40-510 days. Maximum numbers of leaves were defoliated between 3 to 9 months after their initiation (Vasudeva and Gopal, 1975). The main periods of severe defoliation were well correlated to the dry hot months of the year and also to the maturity and ripening of the crop on the branches. During maturity and ripening of fruits, it was observed that the leaves become yellow and defoliate due to exhaustion of carbohydrates (Janardhan *et al.*, 1971; Gopal,1974). A positive relationship of SLW has been observed with Pn ($r=0.514$), E ($r=0.563$) and CE ($r=0.771$) indicating the role of SLW as a prominent growth trait governing the carbon fixation in coffee (D'souza *et al.*, 2013).

6.2. Flowering and Pollination

Coffee is a short day plant *i.e.*, floral initiation takes place during short day conditions of 8-11 hrs of day light which is prevalent between September to December in South India. Flower buds are produced at the axils of mature green wood on short stalks which are known as peduncles. The group of flowers, technically called 'inflorescence' is a condensed cymose type subtended by bracts. In robusta, bracts are leafy and expanded whereas they are small and scaly in arabica. In arabica, 4-5 inflorescences of 1-4 flowers each are produced per axil while in robusta more number of flowers per inflorescence (5 to 6) is commonly produced. The axillary buds in arabica coffee are indeterminate *i.e.*, they may produce either vegetative shoots or flower buds depending upon the seasonal factors like temperature, moisture and photoperiod. In robusta floral differentiation is faster than in arabica and also appears to be determinate. The flower buds grow to a length of 4.0 mm (Frederico and Maestri, 1970) to 7-8 mm (Gopal and Vishveshwara, 1971) after initiation and then remain quiescent until stimulated into flowering. Rain or irrigation to an extent of 25 to 40 mm after a dry period induces further growth in flower buds which open into flowers within 8-10 days in arabica and 7 to 8 in robusta coffee. This kind of flowering, which occurs simultaneously called gregarious form of flowering. Only buds attaining these critical sizes are able to perceive the several stimuli which break dormancy (Schuch *et al.*, 1990; Crisosto *et al.*, 1992). From growth until anthesis, a large and fast importation of assimilates from leaves and branches to the buds occurs. Thus the number of opening flowers on branch is highly correlated to number of leaves and starch reserves (Gopal *et al.*, 1975a). It is also known that leaf area of about 470 mm^2 is required to sustain the complete expansion of a flower bud and that the buds are able to drain assimilates from leaves and internodes near to or away from them (Barros *et al.*, 1992). The nature of the imported materials is not clear, however the contribution of non-structural carbohydrates from alternative sources for flower bud expansion in *Coffea arabica* revealed that current photosynthesis in leaves alone provided all assimilates required for bud expansion, irrespective of buds being connected to leaves or both leaves and branch segments. Under these conditions, open flowers were also bigger than when they drained assimilates from other sources. Starch and reducing sugars to a less extent contributed to the expansion of the flower buds on leafless branch section. The role of pre-existing carbohydrate has been demonstrated from darkened leaves, it contributed little to bud expansion; firstly, non-reducing sugars were mobilized, and lately reducing sugars assisted in the process. In this experimentation the starch levels were kept identical in both exporting and non-feeding darkened leaves (Eunice Melotto *et al.*, 1993).

Under different adverse conditions, particularly at high temperatures, abnormal flowers called "star-flowers" may occur in some plants which are under genetic control also. High temperatures, radiation, prolonged dry weather and inadequate blossom showers are the reasons for production of 'star-flowers'. The petals of these flowers are small, fleshy and often green in colour. The male and female parts are rudimentary. They may remain in this condition till the receipt of water and grow into partially successful blossom but the fruit set would be poor (Anonymous, 2014b).

Description of pollination is available and Coffee Guide (Anonymous, 2014), the arabica coffee is autogamous with different degrees of natural cross-pollination in contrast to robusta coffee which is strictly allogamous with an in built gametophytic system of self-incompatibility. Robusta is having adaptive advantage in having longer styles compared to arabica which may also facilitate cross pollination. On the other hand, in some cases of arabica, self pollination occurs before the opening of flower buds within the bud itself. But in interspecific hybrids between robusta and arabica, the tendency of cross pollination is observed to be high. The pollination in coffee takes place within 6 hours after flower opening under bright light and warm windy conditions. Rain during morning hours before or after flower opening affects pollination and thereby lowers fruit set. Wind gravity and bees movements play very important role in pollination.

6.3. Fertilization and Fruit Set

After pollination, the fertilization occurs when pollen grain germinates and produces a pollen tube on the stigmatic surface. The pollen tube reaches the embryo sac by growing through micropyle and subsequently bursts open releasing the two male nuclei of which one unites with the egg to form zygote and the other fuses with the secondary nucleus to form the primary endosperm nucleus (double fertilization). The process of fertilization is completed within 24 to 48 hours after pollination. The zygote and endosperm nucleus formed as a result of fertilization, undergo a rest period for nearly 45 days in arabica and 60 days in robusta. Meanwhile, the integument (protective coat) of the ovule begins to increase in size to perform nutritive function for zygote. The endocarp or the parchment cover is laid down after the integument (perisperm) grows to its maximum size in 100-120 days after blossom. This determines the size of the future bean. Finally the endosperm grows into the space that has been previously occupied by the integument and restricts it to a thin layer surrounding the endosperm. This forms the "silver skin" or the seed coat of the bean. Commensurating with the growth of endosperm, the zygote grows into an embryo with a hypocotyl and two (sometimes 3-4) cotyledons. Embryo is situated at one end of the bean on its convex surface. The normal duration of a flower to develop into a fruit is about 6 to 8 months in arabica and 9 to 11 months in robusta. Ripe fruits have a thick fleshy pericarp with a mucilagenous layer surrounding the parchment which is made up of stone cells or scleroides (Anonymous, 2014b).

The fruit is a "drupe" and normally contains two seeds. Abortion of one ovule due to non-fertilization leads to the formation of a single seeded fruit, called the 'pea berry'. Occasionally 3 or more seeds may be present due to trilocular ovaries or false poly-embryony (more than one ovule per locule).

6.4. Fruit Growth and Development

The growth of Arabica coffee fruits under South Indian conditions showed a bisigmoidal curves (Ramaiah and Vasudeva, 1969) and similar trend in arabica coffee were observed by Wormer (1964). The berry growth remained minimal up to 42 days after flower opening under South Indian conditions. This phase followed by a rapid increase in the fresh weight up to 102 days and thereafter between 102 and 117 days a drop in the fresh weight has been noticed, further followed by a steep rise in fresh weight up to about 152 days after which again a fall in weight has been observed. Finally a rise in fresh weight occurs extending up to 212 days followed by a recession growth of the berry. The overall pattern of growth of endosperm development was almost the same as berry growth as a whole. No appreciable fall in fresh weight of fruit wall has been noticed whereas the dry matter accumulation shows a definite pattern with well defined periods of growth of the berry as a whole and endosperm separately. Although there is dry matter accumulation in the berry from 42 days after flower opening but it was observed gradual and slow compared to the rate of increase in fresh weight. Thus, during this period, the main growth of berry may not be on the basis of any significant dry matter accumulation rather it was mainly composed of mellow tissue with full of moisture. This phase continues up to about 117 days. During this period about 50 percent of the total final weight of the fruit wall and 15 per cent of the final weight of endosperm is accomplished. After 117 days, a distinct period of dry matter accumulation is seen extending up to 152 days. This can be regarded as the first peak period of growth of the berry and about 45 per cent of final total dry matter of the endosperm is formed in this period. This stage is followed by a recession in the growth. The second grand period of growth is observed after 182 days which represents final filling of the endosperm and lasts for about a month. This generally coincides with the month of October. During this stage the increase in dry weight in the fruit wall is seen to be mainly contributed by filling up and hardening of parchment. Thus three clear periods of dry weight increase of the coffee berry were observed with two grand periods of growth (Ramaiah and Vasudeva, 1969). The berry growth of robusta coffee differs from Arabica showing only two periods of fresh weight increase and a steady and gradual increase in dry weight as a function of time giving an appearance of linear growth. The endosperm filling in the case of robusta was also observed to be typically sigmoidal in nature. The berry growth in case of robusta is slower and maturity and ripening generally take 6-8 weeks more than Arabica.

6.5. Beans/Seeds

Seeds are elliptical or egg shaped, planoconvex possessing longitudinal furrow on the plane surface. Seed coat is represented by the "silver skin" which is also made up of scleroides. The size, thickness and number of pits in the walls of scleroides are considered as important taxonomic characters in evaluating differences between species. Bulk of the seed is formed by endosperm which is hard consisting of polyhedral (many sided) cells. Seeds do not exhibit any dormancy. Viability is also short and germination in coffee takes place in about 45 days (Anonymous, 2014b).

Figure 4.2: Changed Canopy due to Dieback.

'Pea berry' formation in a normal coffee plant (balanced constitution) is due to abortion of one of the ovules in a fruit at or subsequent to fertilization. The fertilized ovule in the other locule grows into a round seed called 'pea berry'. At times, three locules having single ovule in each locule are formed in the ovary, which give rise to triangular seeds. Endosperm sometimes shows partial development with or without any embryo. These occur commonly in floats (Jollu). Formation of more than one ovule per locule is seen occasionally in arabica, but quite frequently in S.288 and S. 795. Seeds resulting from this are called "elephant beans". There are different types of elephant beans, the beans inside the parchment cover inter-locked (hollow and bit) and second the seeds occurring side by side (bits). When one of the ovule gets aborted, the other functional ovule usually assumes irregular shape (defective). Black or spotted bean is a physiological abnormality where the endosperm is completely or partially blackened (Gopal and Venkataramanan, 1974; Anonymous, 2014b). The black bean disorder in north eastern region of India has been found negatively correlated with elevation in arabica coffee (Anand *et al.,* 2000).

6.6. Die-back

Die-back refers to death of younger tertiary branches which is a physiological disorder occurring due to exposure during dry period. The various aspects of the problem studied for more than a decade indicated that it occurs due to adverse environmental factors such as high temperature, high light intensity, low soil moisture and low relative humidity (Gopal and Ramaiah, 1971). Under controlled conditions of the glass house, typical new malady die-back symptoms were induced successfully and confirmed that these factors are responsible. Depletion of reserve carbohydrate was also found to be associated with the disorder.

Younger tertiary branches start drying from the third and fourth node progressing downwards as well as forward and also from apex and progressing downwards. Premature yellowing of leaves is the initial symptom. Interveinal chlorosis andnarrow crinkled small leaf formation usually occurs with abnormal branching after the onset of first summer showers.

Table 4.4: Micro-climatic and Soil Moisture of Healthy and Die Back Affected Plants (Monthly mean)

Months	Light Intensity		Temperature (ºC)		Per cent Relative Humidity		Soil Moisture	
	Healthy	Affected	Healthy	Affected	Healthy	Affected	Healthy	Affected
November	707	1469	26.0	26.8	57.00	55.00	16.10	13.60
December	729	1539	26.5	26.6	59.00	55.00	13.10	9.80
January	739	1544	26.8	29.6	56.00	38.00	10.70	8.60
February	1026	1938	29.2	33.8	53.00	29.00	8.50	6.20
March	1003	1938	29.8	34.9	49.00	27.00	8.00	4.80
April	1000	1938	29.6	33.4	50.00	31.00	16.80	13.20
May	997	1938	28.2	31.8	53.00	36.00	13.70	11.80
June	745	1935	26.4	26.8	57.00	52.00	19.80	15.60

The remedial/control measures to overcome the problem require removal of dead and whippy wood; providing judicious shade of 50-60 per cent; pre blossom and pre monsoon spray of 0.5 per cent Bordeaux mixture; conserving soil moisture by thick mulch; adequate and timely fertilizer application; foliar application of nutrients; lime application to correct the soil pH.

7. Photosynthesis

The quantity and quality of coffee produce are controlled by genetic makeup and the environmental interaction, which in turn cause the changes in physiological and biochemical processes. Vegetative growth of coffee is regulated genetically and interaction with environmental conditions or by various cultural operations. Production of new hybrids by breeding is a continuous and long term process to achieve ideotypes with desirable characters. However, there is a need to assess the newly developed hybrids for physiological efficiency.

In this direction, the knowledge of habitat of coffee in humid and subtropical for arabica and humid and tropical for robusta coffee plays important role. Hence, photosynthetic rates of coffee plants grown under the full sun and shade were compared (Kumar *et al.*, 1979). The saturating irradiance of 300 μE (μmoles $m^{-2} s^{-1}$) of shade plants shifted to near 600 μE in sun grown plants, but shade plants possessed substantially higher photosynthetic rates. Photosynthetic rates decreased above 25°C, which was apparently due to a decline in the mesophyll conductance, as stomatal conductance remained more or less unchanged between 25° and 35°C. Most of these measurements were made on seedlings in plastic bags. It is concluded that coffee is more suited to high density plantings where mutual shading, resulting in low light intensities and lower leaf temperatures, are likely to produce a favourable environment for efficient photosynthesis and growth. The studies on physiological parameters such as carbon exchange rate (CER) and instantaneous water use efficiency (IWUE) and critical assessment as they reflect the overall physiological efficiency of the genotype under Indian conditions was felt very much essential to identify the plant materials which are physiologically more efficient on account of higher carbon exchange rate (CER) and instantaneous water use efficiency (IWUE) among new hybrids. The seven new arabica hybrid were studied for physiological efficiency in comparison with popular station released cultivars by assessing parameters like net photosynthesis (Pn), Stomatal conductance (gs), transpiration loss (E), sub stomatal CO_2 concentration (Ci), carboxylation efficiency (CE) and leaf growth trait like specific leaf weight (SLW) for two seasons. Among the new arabica coffee genotypes, S.4712 and S.4710 showed significantly (p>0.01) higher Pn, gs, E, Ci, CE and SLW compared to rest of genotypes. The SLW showed a positive relationship with Pn (r=0.514), E(r=0.563) and CE (r=0.771) indicating the role of SLW as a prominent growth trait governing the carbon fixation in coffee. The genotypes S.4712 and S.4710, exhibited higher maintenance of Pn, E, gs and CE at water stress condition compared to rest of the hybrids and station released cultivars like Sln.7.3 and HDT. Hence, genotypes such as S.4712 and S.4710 are physiologically more efficient genotypes on account of higher carbon exchange rate and drought adaptive mechanisms. (D'souza *et al.*, 2013).

Prolonged drought decreases soil water potential and soil hydraulic conductivity, which ultimately affects physiological and biochemical functions of the coffee plants (D'souza *et al.*, 1995). Inadequate and uneven distribution of rainfall causes drought conditions in coffee which affects the vegetative growth and induces floral abnormalities resulting in poor fruit set and less yield (Venkataramanan *et al.*, 1996). Drastic reduction in gas exchnage parameters due to water stress has also been reported in coffee (Mallikarjun *et al.*, 2000; D'souza *et al.*, 2002). However, tolerant genotypes show higher maintenance of net photosynthesis and associated processes during water stress conditions by curtailing transpiration loss and also show better recovery rates compared to susceptible cultivars (D'souza *et al.*, 2002). Studies conducted for two seasons using thirteen coffee genotypes indicated significantly (p=0.01) higher net photosynthesis (Pn) in Sln.9, Sln1(R) and Sln.12 compared to rest of the cultivars. The physiological water use efficiency (PWUE) and carboxylation efficiency (CE) were significantly (p=0.01) higher in the cultivars such as Sln.9, 1(R), 12, 7.3 and 10. Assessment of physiological parameters during soil moisture stress indicated reduction in Pn and associated parameters in all the genotypes. However, maintenance of higher Pn, CE and PWUE was observed in Sln.9, Sln.7.3, Sln.10, and Sln.4 (T) at soil water deficit of 30.12 mm which curtailment in transpiration loss. The drought adaptive materials like Sln.9, Sln7.3 and Sln.10 with high carbon exchange rates (CER) could be utilized for drought prone areas and also to evolve a suitable idiotype with desirable characters for crop improvement. (D'souza *et al.*, 2002).

The export of nitrogen (N) from senescent plant parts is important for the efficient use of this macronutrient. The study established correlations among the photosynthetic pigment content, total N, and the photosynthetic variables with the SPAD-502 readings in *Coffea arabica* leaves at different months shown positive correlation increased linearly with N doses and with total chlorophyll presented a direct linear correlation with readings of the portable chlorophyll meter (Anonymous, 2007). It is in conformity with results of Taylor and francis, (2009). The SPAD readings have shown to be a good tool to diagnose the integrity of the photosynthetic system in coffee leaves. Thus, the portable chlorophyll SPAD-502 instrument can be used to evaluate the N status and can also help to evaluate the photosynthetic process in coffee plants.

8. Light and Temperature Requirements

8.1 LUX Requirement

The Coffee is shade loving plant and under Indian conditions it is cultivated under mixed shade canopy. The shade essentially helps in preventing large variations in soil temperature and moisture levels and reduces high intensity of sun light during dry period as well as protects the plants from low temperature hail stone damage, high wind velocities and prevents over bearing of the plants. Adequate shade improves soil fertility by way of returning large amounts of leaf litter to the soil. The fallen tree leaf litter improves the organic matter in the soil thereby increasing the useful micro-organism status in the soil. However, during rainy season, excessive shade adversely affects plant growth development and crop yields as it causes heavy dripping effects on coffee plants, severe wet feet conditions and cooling effect, less light intensity and

Figure 4.2: Shade Plants for Coffee Plantation

congenial conditions to pest and diseases. Hence, a proper maintenance of shade is equally important and helps in increasing photosynthesis of coffee bushes and also limiting the incidence of coffee pest and diseases.

Based on this regression curve of μmol photon $m^{-2}s^{-1}$ in combination with LUX, the coffee growers can measure light intensity with the help of LUX meter for optimum shade for arabica and robusta coffee plantations is as follows:

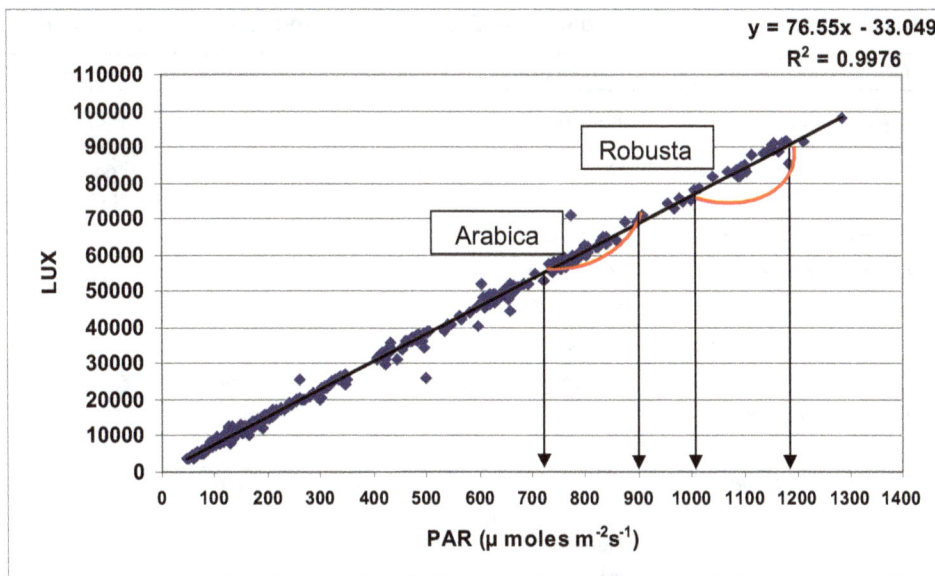

Figure 4.3: Light Intensity Regression Analysis of LUX *vs.* μmol Photon m⁻²s⁻¹ (Photosynthetic Active Radiation - PAR) during February 2009.

The studies on photosynthesis light saturation curve revealed the optimum light intensity in the range of 700 to 900 and 1000 to 1200 μmol photon m-²s-¹ for arabica and robusta coffee, respectively (Anonymous, 2014). To achieve the optimum light, shade of 50 to 60 per cent for arabica and 40 to 50 per cent in robusta coffee is to be maintained. However, there is no easy method to measure light intensity in μmol photon m-²s-¹ or in percentage of shade. Hence, correlation studies of the LUX with light intensity (μmol photon m⁻²s⁻¹) were under taken and standardized (Figure 4.3).

The recommended range of LUX for optimum growth is 55,000 -70,000 LUX for Arabica and 75,000-90,000 for LUX Robusta. The LUX measurements need to be between 12.00 to 2.00 PM above the canopy of coffee bushes. A minimum of 50 readings per acre have to be taken during peak dry period at randomly moving in the field for assessing the shade requirement.

8.2. Shade Trees Commonly Grown in South Indian Coffee Estates

The maintenance of two tier shade canopy in coffee plantations need some scientific inputs apart from managerial skill. The extensive studies revealed that the different shade species for suitability to coffee plantation are *Albizzia sp* (Albizzia. Durazz) and *Ficus* spp. (Achyuta Rao, 1962, 1963); *Dalbergia latifolia* Roxb. (Balaram Menon, 1968); *Bischofia javanica* Blume (Rama Murthy, 1965). Ananth *et al.*, 1960) emphasized mixed shade trees in coffee plantation for providing optimum condition suitable for coffee. The importance of shade for coffee is well addressed for right type of shade species to be maintained in the Plantation of location specific (Anonymous, 2014; Venkataramanan, 2004; Venkataramanan and Govindappa, 1987). Based on all these studies, a list of shade trees commonly grown in South Indian coffee plantations are as follow:

Tale 4.5: List of Shade Trees Commonly Grown in South Indian Coffee Plantations

Sl.No.	Botanical Name	Kannada	Tamil	Malayalam	Remarks
1.	Albizzia lebbeck	Kalbage	Vagai	Vella vaga	Permanent shade trees, generally preferred
2.	Albizzia moluccana	Bage mara	Karuvagai/ Chittilavaga	Pulivaga	
3.	Albizzia odoratissima	Bili wara	Chilavagai	Pottavaga	
4.	Albizzia stipulate	Pottabage	Pilivagai	–	
5.	Albizzia sumatrans	–	–	–	
6.	Atrocarpus integrifolia	Halasu	Pilavu; Pila	Pilavu; Pila	
7.	Chlorozylon swietenia	Biligandha garige	–	Veetti	
8.	Dalbergia latifolia	Beete	Eravadi/ Thothagathi	Cherla	
9.	Ficus infectoria	Basari	Malaiichchi	Ecchamaram	
10.	Ficus nervosa	Mallegaragathi/ Neerabasari	Kal Athi/Nir-al	Nir-al	
11.	Ficus retusa var. nitida	Elibasari	Kal-ichchi	Karigal	
12.		Ficus tsiela	Bilibasari	–	–
13.	Maesopsis eminii	Honne	Vengai	Venga	
14.	Pterocarpus marsupiam	Nerale	Nagal/Nova	Naga	
15.	Syzygium jambolana	Tare			
16.	Terminalia bellarica				
17.	Sapondias mangifiera	Gaddamatte	Katmaa/	Ambalam Mampulichchi	Not very much in favour
18.	Stereospermum tetragonum	Ulimara	Pambathiri	Kaingura	
19.	Cedrella toona	Gadhagarige/ Thandu/ Mundurike/Noge	Santhana vembu/ Thunmarm vembu	Alareveppu/ Mathagiri	Defoliating during rainy season
20.	Ficus glomerata	Atthi	Atthi	Atthi	
21.	Ficus retusa	Mitli/Pillala/Chegoli	Ponichchi	Ittiyal	
22.	Ficus tjakela	Karibasari	Kra-al	Kra-al	
23.	Acrocarpus fraxinifolians	Havolige	–	–	
24.	Erythrina lithosperma	Dadap/Palwana	Dadap	Marukku/ Velapadari	Temporary and generally preferred
25.	Grevillea robusta	Silver oak	Silvar	Silvar	
26.	Gliricidia	Gobbrada mara			

8.3. Temperature Requirements

Coffee plants were sensitive to low and wide diurnal variations in temperature. However, a minimum of 15°C and maximum of 24°C was found to be ideal temperature for growing coffee (Willson, 1985). In the present study unshaded coffee showed reduction in leaf size, formation of distorted leaves and arrest of the extension growth and etiolating of young leaves with leathery/brittle texture during winter. This was found to be associated with 0.5 to 15 °C of minimum and 26 to 36.5°C of maximum in the first season and a range of 3.5 to 16 °C of minimum and 24 to 35.5°C of maximum temperature in the following season. These symptoms corroborate with the earlier report on the effect of chilling treatment under laboratory conditions (Franco, 1956), which was termed as 'hot and cold' disease of coffee (Haares, 1962; Willson, 1985). Slow and stunted growth of coffee was reported by McDonald (1930) and Haarres (1962) at freezing temperatures. Franco (1956) observed reduction in leaf size and discolouration of leaves when coffee plants were treated at 3°C for 6 hours. When night temperature dropped to 3 to 4°C or even less, Gindel (1962) pointed out hindered leafing and shortening of internodes in coffee. This was later supported by the report that shoot elongation was sensitive to even slight temperature changes (Lopez *et al.*, 1972) and also cool short days of winter caused slow shoot growth in Kenya (Cannell, 1975). These reports indicated that the small leaf formation, reduction in internodal length and chlorophyll content observed may be due to the low temperature of winter season and wide variations in diurnal temperatures. The decrease in the leaf dry matter content may be due to the low temperature of winter season and wide variations in diurnal temperatures. The decrease in the leaf dry matter content may be due to reduction in leaf area and also decrease in net photosynthetic rate during wide diurnal temperature variations (Gomez and Robledo, 1974; Buttler, 1977; Castano and Robledo, 1978). Neverthless, reduction of leaf area and leaf dry matter production in unshaded plants was observed during summer month of both the seasons. This may be due to high atmospheric temperature and light intensity during this period as reported earlier in other coffee growing areas (Alvim, 1958; Gindel, 1962; Huxley, 1967; Nunes *et al.*, 1968; Cannell, 1971; Vasudeva *et al.*, Barros and Maestri, 1974; Venkataramanan, 1985). Plants under shade during both winter and summer seasons have shown less adverse effects on leaf growth, as well as extension growth and higher dry matter production. This may be due to scattered light and least temperature fluctuations over the seasons due to shade (Gindel, 1962). Similar effects of shade in keeping the micro climate conducive for normal growth of coffee plants have been reported earlier (Alvim, 1960; Maestri and Gomez, 1961; Decastro *et al.*, 1962; Huxley, 1967; Venkataramanan, 1985). Discolouration of leaves associated with the reduction of other growth characteristics was more in S.1934 compared to Sln.7 during both the seasons. The severity of the symptoms increased when the low temperature followed the bright clear sunny days. Similar symptoms were observed by Franco (1956) during chilling treatments. Discoloration of young leaves was mainly as a result of loss of chlorophyll due to severe and prolonged winter. However, the effect of winter was found to be greater in exposed plants of S.1934 and during 1986-87 and of Sln.7 during 1987-88 seasons than in shaded plants. Probable causes for these differences could be, the ability of shade in minimizing diurnal variations in temperature within the coffee by keeping it cooler during the day and warmer at night (Kirkpatrick, 1936), severe winter during the first season and less cold tolerance in S.1934. It is worthy to note that in spite of 'chilling injury' due to low temperature, the plants of San Ramon selection (Sln.7) exhibited vigrous and healthy growth with less reduction in growth characteristics and chlorophyll content compared to S.1934 plants. These results corroborate with the higher average yield of 378 kg clean coffee ha^{-1} recorded for five seasons in Sln.7 compared to S.1934 (186 kg/ha) under adequate shade management in Andhra Pradesh (Anonymous, 1988 and D'souza *et al.*, 1992).

9. Drought and its Management

Climate change is universal phenomenon and has threatened coffee production in every major coffee producing region of the world mainly due to aberrations in high temperatures, long droughts, punctuated by intense rainfall, more resilient pests and

Figure 4.4: Multiple Shade Trees Canopy Maintenance in Coffee.

plant diseases. In India, coffee cultivation is taken up under multiple shade trees canopy due to prevalence of very harsh climatic conditions, poor soil fertility and difficult land escapes not like in most of other coffee producing countries in the world. Under South Indian conditions, coffee experiences drought for a period of more than three to four months in a year and in extreme conditions up to 6 months. It is evident during 2002-03 seasons under Silver oak canopy with and dadop shade combination, even dadap plant showed complete defoliation as rain stopped during September and drought prolonged.

Coffee being a perennial and economically important plantation crop is predominantly cultivated as a rain fed crop under shade in the hilly tracts of south India and thus experiences various abiotic stresses and among them drought is the major factor affecting various phases of growth and development of the plant (Venkataramanan, 1985). Though the coffee tracts usually experience heavy rainfall (1500-3000 mm per annum) during rainy season (June to October), but it also encounters a minimum of 3 to 4 months of drought (January to April) under South Indian conditions some up to 6 month in a year. The prolonged drought affects the sphere of soil-plant water relationship, which is vital for normal functioning of assimilative processes and reproduction. Due to prolonged drought the soil water potential and soil hydraulic conductivity decreases making it difficult for the plants to utilize water. As a consequence, the leaf water potential tends to decrease and other physiological and biochemical functions of the coffee plants are affected (Venkataramanan, 1985; D'souza *et al.*, 1995).

In India, young coffee seedlings of 6 to 7 months age are normally planted to the field during the months of August/September. The young coffee plants in the field subsequently suffer from moisture stress during the drought period, which starts from January onwards. Generally, the young seedlings growing in open sun have poor development during drought period due to soil moisture stress and high light intensity (Venkataramanan, 1985; Barros, 1999 and Anand *et al.*, 2013). Artificial shading (hutting) is a common practice to protect young coffee plants from high light intensity. Neverthless it may not completely protect the seedlings from the ill effects of soil moisture stress during prolonged drought. Sub soil irrigation has been suggested for the better establishment of young coffee in the field (Mallikarjun *et al.*, 2000). However, irrigation is an expensive operation and water resources/irrigation facilities are limited in many of the coffee plantations. Further, continous irrigation during drought period may encourage development of shallow root system. Under such circumstances, feasibility of using nutrient mixtures as drought ameliorative measures on young coffee is found relevant. Hence, based on the earlier information on effectiveness of nutrient mixtures as drought ameliorative measures on grown up coffee a study was carried out using 11 month old arabica Sln.12 seedlings planted in the field under 70 per cent (1356-1450 $\mu m^{-2}s^{-2}$) open day light conditions (Venkataramanan *et al.*, 1996). Coffee plants planted at a spacing of 6'X 6' with a plant in the middle were used for the study. The design of the trial was completely randomized with three replications each consisting of four plants. The treatments such as NPK+ZnSO$_4$ (Urea 0.25 per cent +Single super phosphate 0.25 per cent + Muriate of Potash 0.175 per cent +Zinc sulphate 0.125 per cent), Urea alone (0.5 per cent) and cattle urine (20 per cent) were imposed as foliar application at monthly intervals during the second week of January and February (dry season). Water spray treatment was maintained as control. During the trial period, the plants were not subjected to artificial shading (hutting). No summer showers were received during the trial period. The study indicated that, the percent increase of growth characteristics was significantly (p=0.05) higher in the seedlings treated with NPK+ZnSO$_4$ and cattle urine compared to urea alone and control (water spray). No significant differences were observed in respect of leaf area, but maintenance of significantly (p=0.01) high leaf turgidity was observed in all the treatments at a soil water deficit (SWD) increased from 22.86 mm (initial) to 29.61mm (final) during the trial period. Among the treatments, maintenance of higher leaf turgidity (84.96 per cent RWC) was observed in the seedlings treated with cattle urine followed by NPK+ZnSO$_4$ (74.23 per cent RWC) at the maximum SWD of 29.61mm. Thus, based on these results it was inferred that treatments like NPK+ ZnSO$_4$ or cattle urine could be useful as drought ameliorative measures for young coffee. Earlier studies indicated that foliar application of nutrient mixture could induce osmotic adjustments in robusta coffee on account of accumulation of solutes such as free proline, nitrogen, calcium and potassium (Anonymous, 1990; Venkataramanan *et al.*, 1996). These osmolites were found to increase bound water capacity (Paleg *et al.*, 1981), regulate stomatal movement (Kumar, 1979), trigger water uptake (Munns *et al.*, 1979) and also maintain cell membrane integrity (Stadelmann and Lee, 1974) in the plants during water stress conditions. Maintenance of higher net photosynthesis in the plants treated with nutrient mixture as foliar application during drought was also reported in coffee (Anonymous, 2001). Hence, increased growth and maintenance of higher leaf turgidity observed during drought period in the seedlings treated with nutrient mixture and cattle urine might have been due to occurrence of osmotic adjustments due to accumulation of osmolites. Thus under non-irrigated conditions the treatments like NPK+ ZnSO$_4$ at the concentration of Urea 0.5kg +Single Super Phosphate 0.5 kg+ Muriate of Potash 375 grams and Zinc sulphate 250grams per 200 liters of water or spraying of cattle urine diluted five times (20 per cent) twice at monthly intervals during drought period (2nd week of January and February) could be helpful for better establishment of young coffee in the field in both shaded or open conditions. Cattle urine, which is well known for its value as a biofertilizers could be used as a drought ameliorative measure under organic coffee cultivation (D'souza *et al.*, 2002).

In the breeding programme of coffee, improving productivity and resistance to disease and pest are the underlying factors; several new arabica hybrids have been developed in India. The vegetative and reproductive growth of coffee plant is controlled or modified by genetic constitution as well as environmental interactions, which cause changes in physiological parameters in newly evolved promising genotypes is very much essential to categorize the superiority of these materials over the existing cultivars. Keeping this in view two promising new hybrids were assessed for two seasons to find out the variability existing in carbon exchange rates and changes in physiological traits in S.4595 (Sln.11 cv S.2464) X HDT) and S.4369 (Sln.6 cv. S.1156 X HDT cv. S.2769) in comparison with two important hybrids such as Sln.9 and Sln.3 were assessed for some physiological parameters for two seasons.

Based on the results of the study, it could be concluded that among the two hybrids S.4595 was a better CER (Carbon Exchange Rate) type on account of high net photosynthesis, carboxylation efficiency, physiological water use efficiency, leaf area index; specific leaf weight and leaf area duration with drought tolerant traits similar to Sln.9. Hence, S.4595 genotype could be suggested for drought prone areas for cultivation and may be useful material for further breeding programmes (D'souza *et al.*, 2005). Another study undertaken at different soil moisture regime in newly released Chandragiri in comparison with the known drought tolerant cultivar Sln.9 revealed significantly ($p < 0.05$) increase of soluble proteins, proline and epicuticular wax in both the cultivars during moisture stress at incipient wilting. However, the relative water content, nitrate reductase enzyme activity and all the gas exchange parameters such as net photosynthesis, stomatal conductance, transpiration rate, carboxylation efficiency and instantaneous water use efficiency reduced significantly ($p<0.05$) under moisture stress and after alleviation of moisture stress the contents returned to normal levels in both the cultivars. The reduction at stress and recovery after the alleviation of stress of the parameters indicated Chandragiri is a moderately tolerant compared to Sln.9 (D'souza *et al.*, 2009).

Evaluation of eight robusta accessions *viz.*, S.879, S.880, S.1979, S.3656, S.1902, S.1932, S.1509 and S.1481 for drought tolerance capacity in comparison with the local robusta cultivars, *Coffea canephora* Pierre ex Froehner cv. S.274 and an improved series of diallelic crosses of robusta (S.4202) by studying the pattern of solute accumulation such as free proline, nitrogen, phosphorus, potassium, calcium and total carbohydrates before inducing stress, at wilting and after alleviation of moisture stress showed S.1932 and S.1979 have higher accumulation of these solutes except total carbohydrates during moisture stress and decrease of the contents after alleviation of stress compared to other accessions. Thus these two accessions are found to be more tolerant to drought among the exotic robusta (Saraswathy *et al.*, 1992).

Thus, these researches have shown higher "Osmotic adjustment" in drought tolerant coffee cultivars due to higher accumulation of solutes such as proline, nitrogen, phosphorus, potassium, calcium etc. and lowering the osmotic potential associated with drought tolerance mechanism. Hence, based on osmotic adjustment arabica coffee is found more tolerant to drought than robusta coffee, but it has lot of variability among the cultivars. These studies have revealed the importance of drought tolerant genotypes and several other studies based on above parameters and solute accumulation, all the station released coffee have placed in order of tolerance as follows. Among the order of drought tolerance in arabica coffee, the most tolerant is Sln.7.3: (San Ramon Hybrid X HDT) followed by Sln.9: (Tafarikela X HDT); Sln.10: (Catura Hybrid X Cioccie and S.795) ; Sln.4: Tafarikela ; Sln.5(B): Devamachy X S.333 (S.2931); Sln.1: (S.288); Sln.11: (Liberica X Eugenioides); Sln.3: S.795 (Kents X S.288); Sln.12: (Catimore/Cauvery); Sln.5(A): (Devamachy X S.881); Sln.6: (Arabica X Robusta) and least is Sln.8: (HDT). The first five selections are more tolerant and can be cultivated in drought prone areas. In robusta cultivars S.4040 (BR 9 X BR 10) is most drought tolerant followed by C X R (Congensis X Robusta); S.274 (Sln.1(R)) and S.4042 (BR 9 X BR 11) (Anand *et al.*, 2003; Saraswathy *et al.*, 1992).

9.1. Impact of Drought

Inadequate and uneven distribution of rainfall causes drought in the coffee, which affects vegetative growth of both arabica and robusta coffee and wilting symptoms are exhibited, induces floral abnormalities and results in poor fruit set and less crop yield. The drought affects coffee plants in the sphere of soil-plant-water relationships, which are vital for normal functioning of assimilative processes and reproductive functions. Hence, coffee in India is grown under shade. The light

Figure 4.5: Arabica (a) and Robusta (b) Coffee showing Wilting.

a. Normal flower buds

b. Paddying

c. Pinking

c. Snake mouthed flower

d. Star flowers

Figure 4.6: Floral Abnormalities.

illumination intensity varies considerably within and between coffee estates. It is very much necessary to determine the optimum light requirement for better growth and development in coffee, since the high light intensities and high temperature during dry period of the year cause yellowing of leaves, defoliation and die-back. Venkataramanan and Govindappa (1987), conducted two identical experiments for 2 years on the seedlings of selections 1(R), 3 and 9 giving 25, 70 and 100 per cent daylight using growth analysis technique and showed more height, leaves, primaries and total dry matter production in the seedlings kept under shade due to the increase in net assimilation rate (NAR) as this parameter shows a significant positive correlation with the total dry matter (r=0.8894). The relative leaf growth rate showed a positive significant correlation with RGR and NAR indicating that RLGR was responsible for the changes in these parameters support the cultivation of coffee under shade in South India.

Thus, under exposure conditions whenever, water deficit is encountered, it affects normal physiological and biochemical processes of coffee plants, results in poor vegetative growth and exhibits floral abnormalities such as pinking, paddying, star flowers and snake mouthed flowers. It results in poor fruit set and reduction in crop yield. If the moisture stress continued even after the receipt of blossom showers, the development of berry is affected, which results not only in the production of more floats and blacks, but produces smaller size beans and delays ripening.

The vagaries of weather in recent years continued to affect the coffee production in India. Correlation studies on 30 years yield and rainfall data revealed that delay in blossom shower every week after March, resulted in a reduction of crop yield of 61.14 and 43.51 kg/clean coffee per hectare in Balehonnur and Kodagu zone, respectively (Figures 4.7 and 4.8).

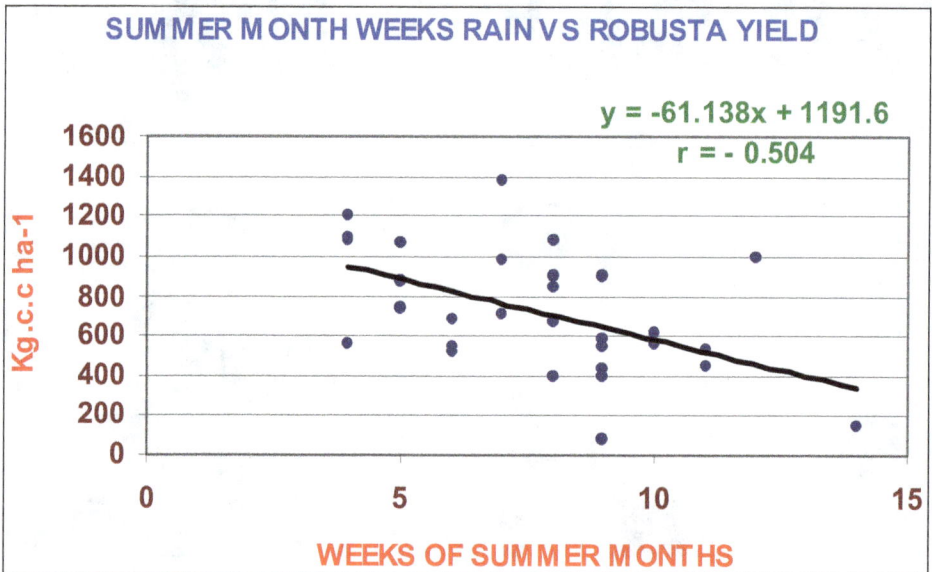

Figure 4.7: Crop Loss in Robusta Coffee due to Delayed Blossom Showers in Balehonnur Zone.

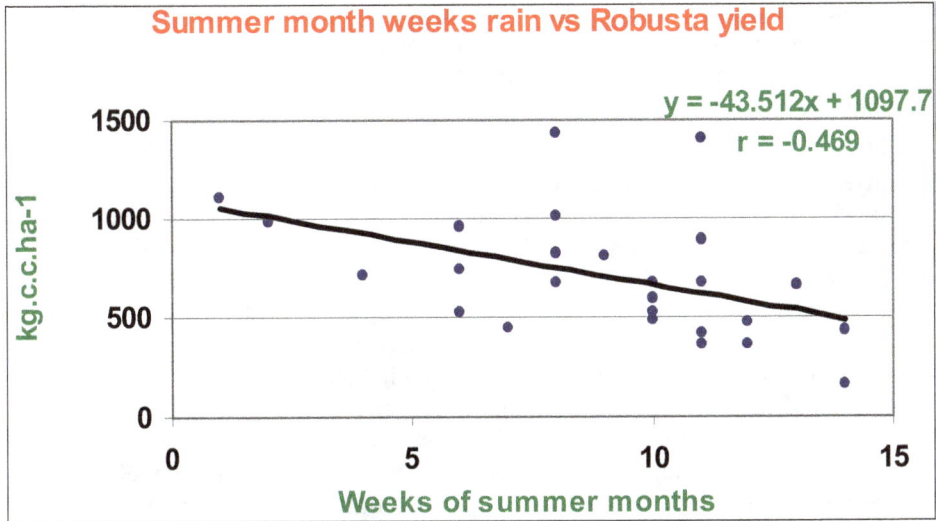

Summer month weeks rain vs Robusta yield

$$y = -43.512x + 1097.7$$
$$r = -0.469$$

(y-axis: kg.c.c.ha-1; x-axis: Weeks of summer months)

Figure 4.8: Crop Loss in Robusta Coffee due to Delayed Blossom Showers in Kodagu Zone.

9.2. Drought Amelioration

Since, coffee is shade loving plant and the maintenance of adequate shade canopy is most important component in coffee production system in India. It essentially helps in preventing large variations in soil temperature and moisture levels and reduces high intensity of sun light during dry period as well as protects the plants from low temperature hail stone damage, high wind velocities and prevents over bearing of the plants. Adequate shade improves soil fertility by way of returning large amounts of leaf litter to the soil. The fallen tree leaf litter improves the organic matter in the soil thereby increasing the useful micro-organism status in the soil and conservation of soil moisture.

The arabica and robusta coffee have different habitat and have to be dealt accordingly. Arabica coffee under adequate shade of 50 to 60 per cent with proper management practices does not need much attention due to deeper root system and adoptability. But robusta coffee is more sensitive to drought conditions; hence, drought management is an important aspect in coffee productivity. Plants vary in their capacity to tolerate drought depending on the genetic constitution. High 'osmotic adjustments' occur in the drought tolerant cultivars through accumulation of solutes such as free proline, nitrogen, phosphorus, potassium, calcium etc. These compounds increase the bound water capacity, regulate stomatal movements, trigger water uptake, induce heat stability and maintain the integrity of the membrane, ultimately photosynthetic activity and crop growth and development.

9.3. Inducing Drought Tolerance Robusta Coffee

Since, Robusta coffee is susceptible to drought, which occupies about 50 per cent of total area under coffee cultivation and it is understood the role of nutrients in osmotic adjustment, the attempts were made to overcome the adverse effects of drought in robusta coffee by inducing osmotic adjustment. Anand *et al.* (2014) reviewed all the work and based on the various studies the drought ameliorative measures are summarized as follows:

Multilocational field trials for seven seasons on drought ameliorative measures of nutrients mixture containing Urea (0.5 per cent), SSP (0.5 per cent), MOP (0.375 per cent) and Zinc sulphate (0.5 per cent) could increase crop yield to an extent of 63 kg ha^{-1} (22.03 per cent). The increment due to spray of drought ameliorative measures was mainly on account of improving leaf turgidity, reducing floral abnormalities and enhancing fruit set. The Nutrient mixture spray was found more effective under prolonged drought period. The suggested concentration and schedule of spray is Urea, SSP and ZnSO$_4$ each 1 kg + MOP 750 gm in 200 L. of water should be given foliar spray 45 days after the last rainfall @ 1 L/plant (usually during 2nd fortnight of January) and if rain is not received, the 2nd spray should be repeated after 30-45 days of 1st spray at same rate. Its economics was worked and the cost benefit ratio of nutrients spray was found to be 1:3, when the drought exceed for 110 days.

Lantana camara is one of the common weeds in coffee plantation areas and it is found abundantly everywhere. It has insecticidal properties and also found growing very well in abandoned dry areas to have the drought adoptability characteristics. Hence, attempts have been made to evaluate *Lantana camara* hot water leaf extract as eco-friendly and cost effective drought ameliorative measures in robusta coffee under un-irrigated conditions, where drought prolongs for more than 110 days. The results revealed maintenance of high photosynthetic activity, high relative water content and better physiological water use efficiency and better crop yield of 252 kg/ha in robusta coffee un-irrigated conditions (Anonymous, 2014b).

Collect @ 2 kg fresh leaves of *Lantana camara,* chop them into small pieces and immerse in 10 liters of hot/warm water for 24 hours for complete extraction. Filter it by using cloth and mix entire filtrate in 200 liters (in one Barrel) of water and spray on the plants concentrating on lower surface of the coffee leaves.

☆ 1st spray: 45 days after the last rainfall

☆ 2nd spray: 30 days after the first spray (If rain for more than 15 mm is not received)

As the spray does not require any input cost except the deployment of workers for collection of naturally available leaves material and spraying, the increment in net income is substantial as the improvement in crop yield was 252 kg ha^{-1}.

Generally, under exposed conditions robusta coffee exhibits scorching and has reduced photosynthetic activities. In absence of proper shade, foliar application of lime (CaO) along with 0.4 per cent starch as contact shade has been found to be good in mitigating drought by protecting robusta coffee plants from photo induced damage to the leaves. It has been noticed that in the robusta coffee leaves coated with lime had high retention of chlorophyll and maintaining high net photosynthetic rates. In a replicated field trials with 10 per cent lime application during second fortnight of January for four years from 1996-97 to 1999-2000 seasons in S.274 robusta coffee increased crop yield of 292 kg ha^{-1} (27.6 per cent). The spray timings is once during second fortnight of January every year.

9.4. Establishment of Young Coffee and Protection from Drought

The young coffee establishment in the field needs special attention and apart from maintenance of adequate shade, hutting, mulching and regular use of agro inputs, the following few vital operations need to be taken up.

Figure 4.9: Young Coffee Plants showing Wilting.

The use of sub soil injection has been found effective in inducing deeper root system by sub soil injection of water @ of 2 litre at 15 days interval and increasing plant growth and development thereby better establishment of young coffee in the field (Malliakarjun *et al.*, 2000; D'souza *et al.*, 2012).

Foliar spray of drought ameliorative measures such as half strength of nutrient mixture or same dose of *Lantana camara* leaf extract and or lime application certainly help to protect plant from adverse effect of drought and better establishment in the field.

Therefore, depending on the availability of the materials, the foliar spray of drought ameliorative measures need to be taken to overcome the adverse effect of drought and improve crop yield. Also, it is necessary that the recommended agronomical practices such as maintenance of adequate shade, timely application of fertilizers, mulching after cessation of monsoon, opening of trenches or cradle pits in time across the slope during post monsoon period have to be taken up regularly for getting better benefits of drought ameliorative measures and improving crop production.

10. Physiological Constraints during Monsoon

Coffee growing regions in India, experience different harsh climatic conditions in a year such as three to four months of dry period from January to May and heavy and continuous rainfall during monsoon, which results in premature fruit drop and infestation of black rot and stalk rot. Erratic climatic conditions have influenced and changed pattern of monsoon rainfall and also summer showers during last two decades. The onset of monsoon is getting delayed to the second fortnight of June from the last week of May. High rainfall also shifted from June second fortnight to July second fortnight and August first fortnight. More ever the average monsoon rainfall keeps on showing a lot of

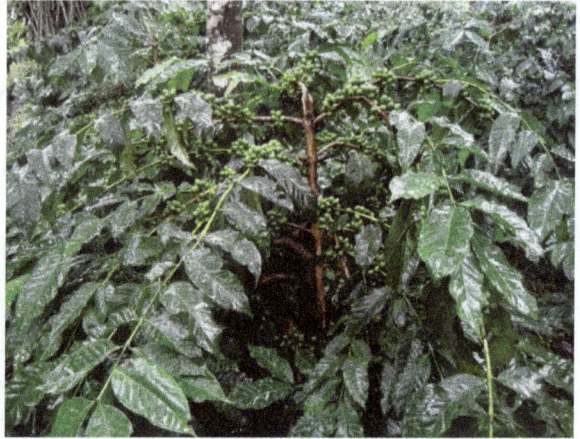

Figure 4.10: Good Fruit Bearing in Coffee.

variation in different seasons. This results in differences in extension growth besides berry development. All these climate changes triggering increased pre mature berry

Figure 4.11: Coffee with Berries in Cluster.

drop and defoliation in coffee plants resulting in substantial crop loss (Gopal 1971; Anand *et al.*, 2013; Anonymous, 2012).

Pre mature fruit drop is a major constraint in coffee plantations and occurs during early stage of berrry growth and development due to various reasons, while fruit drop is absent at maturity (Anand *et al.*, 2013; Anonymous, 2014). Barros *et al.* (1999) described three well defined periods of fruit drop in arabica coffee, one occurring in the first four weeks, during the pinhead stage (pinhead drop), second from the 5th to 11th week during the endosperm filling stage (early drop) and the last after the 11th week of fruit growth. Fruit drop takes place mostly in the first three months after blossom. The first wave of fruit drop results from fertilization failure and seems to be unavoidable. Pinhead drop seems to be unrelated to the level of fruit set and therefore to insufficient assimilate supply. Severe defoliation can cause some shedding, especially of younger fruits. The second wave appears to be linked to the beginning of endosperm formation, which has been associated at least partially with low carbohydrate supply and water deficit. The third period of fruit loss seems less important and more erratic and probably results from competition, since it is affected by leaf and shoots diseases.

The extent of pre mature fruit drop during the developmental stage, fruit growth is subjected to variation (10 to 40 per cent) depending upon seasonal factors. Apart from the adverse climatic factors, the fruit drop occurs due to carbohydrate deficiency, auxin/carbohydrate imbalance and nutritional factors. The maximum fruit drop generally occurs between 90 to 120 days after set in the month of June, which coincides with the peak period of South-west monsoon. Under normal conditions pre-harvest fruit drop is absent in coffee (Anonymous 2014; Anand *et al.*, 2013).

The coffee production and productivity in general depend upon the plant growth, development and balance between vegetative and reproductive growth. The pattern of shoot growth in coffee under south west monsoon conditions is typically of sigmoidal in nature

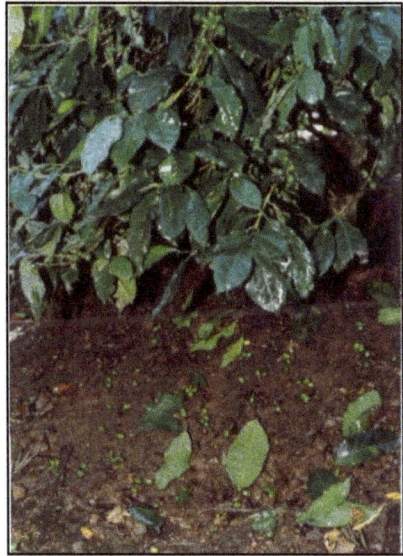

Figure 4.12: Premature Fruit Drop.

with a slow growth from March/April to July and a rapid growth period from August to October. The vegetative growth slows down with onset of dry period from November onward and remains minimal during December-February (in winter period) till receipt of blossom and backing showers and once again resumes growth after receipt of March rain. The cropping wood, which develops during September - October plays very important role in achieving maximum crop yield as these branches gets maximum time for production of a minimum of 8–10 cropping nodes at the time of floral bud initiation during August/September and provides better opportunity to achieve maximum crop yield.

The growth of arabica coffee fruits under south Indian conditions shows a bi-sigmoidal curve with two grand periods of growth during 120 to 150 and 180 to 210 days after blossom. While the berry growth in robusta coffee is linear and takes 6-8 weeks more than arabica coffee for fruit maturity and ripening. The normal growth and development of berry results maximum crop yield. However, pre-mature fruit drop during early stage of berry growth and development in June/July is a common phenomenon due to various reasons irrespective of the crop seasons and ranges from 10 to 15 per cent, but in extreme condition due to secondary infection of stalk rot and black rot, the decrease more than 20 to 50 per cent in different years. However, fruit drop does not take place at the maturity/ripening in coffee until and unless winter rain is received at the time of ripening which results in swelling of berries due to absorption of water by mucilage and splitting and dropping.

The extent of fruit drop varies based on environmental conditions (water logging - 'wet feet') as well as internal factors such as reduction in cytokinins coupled with high content of abscissic acid (ABA), reduction in carbohydrate content and nutrient deficiency. The 'Wet feet' conditions in plantations is observed due to continuous heavy rainfall with cool ambient temperature, high relative humidity, cloudy weather and improper drainage lead to soil saturation and water logging. Multiple blossoms (running blossom) due to early receipt of blossom showers in instalments result in formation of different size berries and lead competition among them for carbohydrate reserves, which increase premature fruit drop. Over bearing plants also show high percentage of fruit drop due to competition among the developing berries. Therefore, it is very important to take care of physiological remedial measures and adopt agronomical practices for proving congenial conditions for proper berry growth and development and maximum crop yield.

10.1. Pre-mature Fruit Drop

In coffee, the blossom is water dependent and adequate rainfall of 25 to 40 mm is very much necessary. Failure of blossom showers or inadequate rainfall and under adverse conditions, particularly high temperature leads to floral abnormality. During normal blossom condition, pollination takes place within 6 to 8 hours after floral opening. The fertilization completes within 24 to 48 hours and there after zygote and endosperm nuclei undergo a resting period of 45 days in arabica and 60 days in robusta coffee. The failure of backing showers, high temperature and exposure further results in poor fruit set. Prolonged drought after blossom and soil saturation during monsoon adversely affect on carbon exchange rates, extension growth besides altering the biosynthesis of plant growth regulators and inducing pre mature berry drop. The physiological effects of water logging on coffee roots have been well explained by Gopal and Ramaiah (1990) that the deficiency of O_2 (anoxia) due to water logging interferes with Gibberellin and Cytokinin production in the roots and also causes a loss of nitrates, decrease of soil pH, reduction of sulfate into H_2S and fixation of phosphorus. Mallikarjun *et al.* (2002) based on periodical assessment of hormonal status, have proved that ABA level increases five folds in coffee plants under soil saturation and it was found to be associated with reduction in Cytokinin: ABA ratio and photosynthetic activity. This corroborate with other studies as cytokinins and other plant growth regulators play important role in growth and development including photosynthetic processes and its components have been well demonstrated in different crops (Hall, 1973). Similarly, the review on biochemistry and physiology of abscissic acid (ABA) revealed fast change of ABA during abiotic stresses (Daniel Watson, 1980). Hence, the hormonal imbalance occurs due to the changes in increase of ABA and reduction in cytokinins and other growth promoting hormones under such conditions results in heavy berry drop.

Normally, the occurrence of pre-mature berry drop is common phenomenon in coffee and ranges to an extent of 8 to 10 per cent. However, soil moisture stress with high temperature during April and May and prolonged soil saturation effects during monsoon induces heavy pre- mature berry drop to an extent of 20 to 30 per cent due to hormonal imbalance, low level of starch content (Janaradhan *et al.*, 1970; Wormer and Egabole, 1965) and 10 to 20 per cent due to infestations of diseases like black rot and stalk rot (Anonymous, 2003; Muthappa, 1970). Hence, occurrence of pre-mature berry drop and defoliation depends upon more than one reason, causing significant crop loss in coffee (Anand *et al.*, 2013). D'souza *et al.* (2013) demonstrated variations of pre mature berry drop of both arabica and robusta coffee and mostly remain dependent on variations in climatic conditions and rainfall pattern during 2011, 2012 and 2013 seasons under Kodagu coffee growing regions to an extent of 25.8, 30.2 and 39.8 per cent in arabica and 29.21, 35.25 and 44.81 per cent in robusta coffee, respectively.

Ultimately, based on these studies conducted almost for four decades, it could be inferred that there would be crop loss to an extent of 5-10 per cent due to blackening and browning at pin head stage berries, 10-15 per cent physiological fruit drop due to soil saturation effects and around 10-15 per cent due to stalk and black rot during monsoon period in coffee growing regions of India. It could further increase under severe conditions. The studies also shown that adequate timely receival of blossom and backing summer showers, timely onset and well distributed monsoon rains with frequent break have beneficial effect of crop prospects it results in reduction of pre-mature berry drop. The erratic climatic conditions and changes in monsoon rainfall pattern in India from last decade, an alarming problems of pre-mature berry drop and improper berry development is noticed and leading to reduce crop yield and production of quality coffee.

10.2. Remedial Measures for Premature Fruit Drop

The premature fruit drop occurs due to many factors and its management is looked in totality. The physiological and agronomical as well as plant protection measures are equally important in management of coffee plantations.

10.2.1. Physiological Approaches

The major physiological constraints of premature fruit drop, defoliation and less development of next year cropping wood during rainy season are mainly due to hormonal imbalance and depletion of carbohydrate reserves. To overcome these physiological constraints following measures need to be under taken in time.

The foliar application of plant growth regulators (PGR) can bring the balance between the endogenous level of growth promoters and growth retarding substances and thereby helps in controlling the premature fruit drop and increasing the crop yield. Many growth regulators have been tried for controlling the premature fruit drop. Foliar sprays of Planofix (Vasudeva and Venkataraanan, 1981), Atonik (Vasudeva *et al.*, 1981), Ascorbic acid (Vasudeva and Ratageri, 1981), Cytozyme crop plus (Vasudeva, 1983), Hormonol, Miraculan and Agronaa (Vasudeva, 1984) etc., were found to be useful in controlling the fruit drop, which increased the crop yield. The first round of these PGR application to be taken up after 15 to 20 days of blossom and second application after one month of first application during pre-monsoon period. The commercially available plant growth regulator formulations, which have been found useful for coffee plantations are Planofix, Agronaa, Hormonol, Ascorbic Acid, Miraculan, Protozyme, Cytozyme crop plus, Atonik (Green Magic) and Vrudhi. These have significantly increased mean crop yield of 130 to 212 kg clean coffee per hectare (*i.e.* 11.9 to 19.4 per cent over control) of arabica coffee for four seasons.

Table 4.7: The Concentrations of the above Formulations for Increasing Crop Yield

Proprietary Formulations	Quantity/ 200 Litres	Quantity per Acre	Per cent Yield Increase
Hormonol	50 ml	150 ml	19.42
Planofix	50 ml	150 ml	19.32
Vrudhi	50 ml	150 ml	15.98
Protozyme	60 ml	180 ml	15.83
Agronaa	50 ml	150 ml	14.25
Miraculan	50 ml	150 ml	13.64
Ascorbic acid	20 g	60 g	13.66
Cytozyme crop plus	60 ml	180 ml	12.30
Atonik (Green Magic)	50 ml	150 ml	11.94

Source: Coffee Guide, 2014.

The foliar application of plant growth regulators (PGR) helps in increase of fruit set of 3 to 9 per cent, control of premature fruit drop of 8 to 12 per cent, control of defoliation to the extent of 7 per cent and promotion of vegetative growth in terms of cropping wood for the following year. The PGRs treated plants show less variation in yield for the consecutive years and did not affect the cup quality.

The second time foliar application of PGRs coincides with per-monsoon Bordeaux mixture application and it becomes difficult for the growers to take up PGRs and Bordeaux mixture application separately. Keeping this in view, the compatibility of PGR formulations with Bordeaux mixture was tested. Among various PGRs formulations, the Planofix was found compatible and 50 ml of Planofix could be mixed in 200 litres of Bordeaux mixture solution during pre-monsoon application. Hence, first spray for aforesaid PGR alone after 15 to 20 days of blossom and second application of Planofix with Bordeaux mixture will give equal benefit.

The foliar application of PGRs involves extra cost of cultivation and recommended dose of fertilizer application schedule is commonly taken up during pre and post monsoon period. Hence to minimize cost of cultivation, the soil application of PGR formulations along with fertilizers was taken up. The "Biozyme" granules @ 10g per plant along with NPK fertilizer applied during pre and post monsoon period improved crop yield of 15.31 and 17.99 per cent in arabica and robusta coffee, respectively by control of premature fruit drop and defoliation (D'souza *et al.*, 2004).

In general, the floral bud initiation begins during September and continued till January/February and if there is excessive vegetative growth, the floral bud induction is less. To achieve desired level of floral bud initiation, it is important that vegetative growth has to be checked and balance between reproductive and vegetative growth is brought. To improve the flower bud induction and the cropping nodes under such situation, foliar application of Mepiquat chloride at 1000 ppm concentration (4 liters of Chmatkar in 200 liters of water) or 0.75 per cent Single super phosphate (SSP) @ 1.5 Kg per 200 liters of water twice at monthly intervals, during August and September is found useful (George, 2009; Mallikarjuna *et al.*, 2007; Anand *et al.*, 2013). Similarly, Bayleton 160 g/barrel during August followed by 1.5 kg of SSP in September is equally useful. These treatments under excessive vegetative conditions will improve the crop yield significantly by enhancing flower bud production and if Bayleton is used it will control leaf rust also (Anand *et al.*, 2012, 2013).

The bulk of berry growth and endosperm development occur during monsoon period, it is inevitable to provide adequate nutrients during the grand period of bean filling. The additional dose of fertilizers application during break in monsoon helped in increase of "A" grade bean and reduction of "B" grade beans (Anand *et al.*, 2001). These findings are in confirmatory to the report of Bruno *et al.* (2011) in a study of ^{15}N dynamics, using labeled Nitrogen (N) that demonstrated high requirement of N at specific period coinciding with maximum N consumption by leaf and fruit. In fact, the maximum demand occurs on the final stage of bean filling, corresponding to the natural leaf senescence and an intense N remobilization from leaf to fruit, so that the major quantity of N should be applied prior to this stage. Therefore, to achieve maximum yield, the additional dose of fertilizer application during break in monsoon is very important. This will help reduction in fruit drop, defoliation, increase of bean size and proper development of endosperm.

10.2.2. Adoption of Agronomical Practices

In addition to the above, it is equally important to undertake following agronomical operations timely for better response of physiological remedial measures.

10.2.2.1. *Removal of Mulch*

The mulching is carried out to cover the soil under the coffee bushes with dry leaves, weed slashing, cherry husk or any other organic wastes etc. after digging/scuffling operations. This helps in maintenance of optimum soil temperature, conservation of soil moisture and suppression of weed growth during dry period. However, during onset of rainy days, all the much below the bush canopy need to be removed and kept in centre of four plants. This helps in avoiding stagnation of water and early evaporation from below the bush canopy.

Figure 4.13 Mulching with Weed and Dry Leaves.

10.2.2.2. Drainage Measures

In coffee fields, conservation of soil moisture is important for protecting from adverse effect of prolonged dry period. However, in low laying flat lands, soils become saturated due to water stagnation during monsoon period and will not be able to support the activities requiring oxygen in the root zone. In prolonged water logging conditions, coffee roots suffer badly and develop wet feet symptoms. In such areas, drainage channels of 45cm depth and 30cm width should be opened at suitable intervals between the rows to drain out the excess water from the plots. All these channels in a plot are connected to a main channel which is connected to a catch pit at the end of the plot. Proper drainage is essential for better soil structure, root respiration and plant growth development.

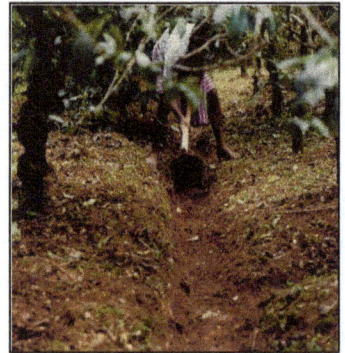

Figure 4.14: Drainage Channel.

10.2.2.3. Trenches/Pits Opening

In sloppy areas, trenches or cradle pits should be opened across the slope during post monsoon period. These are dug in a staggered manner in between the rows of coffee all along the contour. Trenches are dug 30 cm wide, 45 cm deep and of any convenient length. The cradle pits are short trenches measuring about 1 to 1.5 m long. Trenches and cradle pits act as mini compost pits for fallen leaves, weed bio-mass and dadap lopping besides conserving of soil moisture and preventing soil erosion. The trenches/pits are to be renovated once in 2- to 3 years by removal of the deposits and spreading them near by the surrounding plants.

Figure 4.15: Pits for Organic Waste.

10.2.2.4. Handling, Centring and De-suckering

Apart from main pruning after one month of harvesting and regular light pruning the two rounds of handling is most important. During the first handling operation at early monsoon (June-July) the new flush arising after the main pruning is thinned out to a desired numbers of well spaced branches. Later during August-September, the excessive branches have to be thinned out mercilessly to bring the balance between current year crop and next year cropping wood to avoid alternate year bearing of crop. The centring is done by removal of new shoots arising within 15 cm radius of the main stem. De-suckering has to be undertaken by removing the suckers growing from the main stem thereby conserving food reserve considerably, which is utilized berry growth and development. Gourmandizers and all criss-cross branches have to be removed. These will help more aeration and provide more

Figure 4.16: Light Pruning.

opportunity to penetration of light to the lower canopy for better photosynthetic activity during rainy season. This activity could be sanitation and will minimize the black rot disease (*Koleroga noxia* Donk).

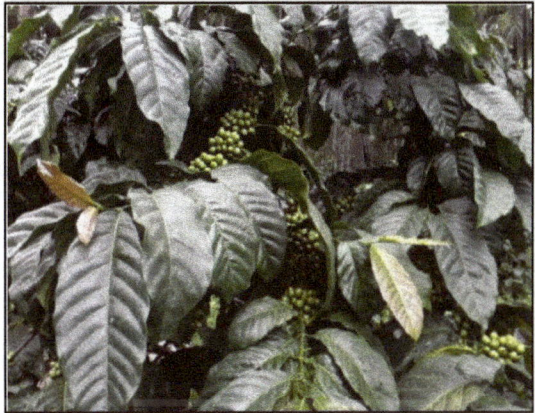

The adoption all the above physiological remedies and agronomical package of practices in time, helps in overcoming of the physiological constraints during monsoon and helps in achieving maximum crop yield.

11. Yield and Yield Attributes

Increase in crop yields has been largely attributable to the improved partitioning of biomass to fruits *i.e.* harvest index. Studying the pattern of partitioning of biomass in different varieties/genotypes is essential to understand their yield potential. Hence, a study was undertaken on partitioning of the above-ground biomass *i.e.* fruits, branches, stem, leaf and architecture traits such as harvest index, leaf area index, number of primary/secondary/tertiary branches, internodal length and gas exchange parameters in two arabica cultivars, Sln.9 (tall variety) and Sln.12 (semi-dwarf) for three seasons. A significant (p=0.05) differences were noticed among the cultivars for leaf, branch, stem, fruit biomass, leaf area, leaf area ratio (LAR), leaf weight ratio (LWR), branch weight ratio (BWR), Stem weight ratio (SWR), harvest index (HI) and leaf area index (LAI). The maximum partitioning of biomass of 36 per cent towards the fruits was recorded in Sln.12 with better harvest index of 0.353, whereas, in Sln.9 the maximum partitioning and accumulation of biomass in main stem was to an extent of 32 per cent followed by primary/secondary/tertiary branches (28 per cent) and showed high leaf area index (2.291) with relatively low harvest index of 0.221. however, total biomass production and fruit biomass was high in Sln.9. The correlation studies indicated a significant and strong positive correlation (p=0.05) between harvest index (HI) with fruit biomass and total dry matter production. Significant relationship of fruit biomass with leaf area index (LAI) was also observed in Sln.9 but a significant relationship of LAI with total biomass was recorded in this variety. Higher harvest index and partitioning efficiency of photo-assimilates towards fruits in Sln.12 due to high photosynthetic rate, shorter internodal length and significant relationship of fruit to LAI resulted in increased translocation of reserves towards fruits compared to Sln.9 (Mallikarjun *et al.*, 2013).

The photosynthetic component and current photosynthesis of leaf play very important role in berry growth and development. It has been demonstrated that the maintenance of leaf to fruit ratio of 1:3 in Arabica and 1:6 in robusta coffee is important and also in other wards 17, 72 cm^2 leaf area for Arabica and 27 cm^2 for robusta coffee is needed (Vasudeva and Ratageri, 1981). The great deal of variation in coffee yields is due to environmental conditions. This variation may be classified in two types, the variation in the yield of a population over years and the variation in the yield of individual plants within the population. Some reports on yield variation in coffee are available (Anstead, 1921; Gilbert, 1938; Ferwerda, 1948; Carvalho *et al.*, 1959; Manaco and Carvalho, 1964; Srinivasan and Vishveshwara, 1978a, b; Srinivasan *et al.*, 1979). Most of these reports make use of progenies in replicated trials for evaluating the environmental effect. When a large number of progenies are to be evaluated, replicating becomes a tedious job especially in a perennial crop like coffee, needing considerable space between plants. Further, genetic heterogeneity is likely to exist within most of the progenies, contributing to variation between replications which cannot be distinguished. A relatively simple method developed and used by the authors for differentiating progenies in unreplicated trials showed that, the coefficient of variation for yield over years as well as between plants, the percentage of plants giving an average fruit yield of 3 kg and above and their percentage contribution to total yield of the progeny were used to differentiate 25 high yielding *Coffea arabica* and 34 high yielding *C. canephora* (robusta) progenies at a single location (Central Coffee Research Institute, India). Arabica progenies had in general lower variation for yield and lower mean yield than robusta. The clonal progenies in robusta obtained by rooted cuttings had lower plant to plant variation of mean yield and higher mean yield than seedling progenies of the same mother plants. Seven progenies each in arabica and robusta indicated as superior for all the parameters were suggested as suitable for further testing in multi locations (Srinivasan and Subbalakshmi, 1981).

12. Pest and Diseases

The carefully cultivated and trained canopy of shade trees, under which coffee is grown in India, greatly influences the microclimate in the coffee ecosystem. This is of great significance in the context of pest and diseases management. The management strategy for tackling coffee pests and diseases is based on the factors specially anticipation of pest and diseases outbreaks; judicious management of ecological factors by shade regulation; phyto-sanitation, handling and pruning; maintenance of optimal overhead shade; continuous monitoring of diseases and pest populations; conservation and manipulation of indigenous natural enemies of pest; introduction, mass multiplication and field release of exotic natural enemies; timely and need-based use of bio-pesticides and fungicides and integrating the available interventions for effective management.

Nearly a hundred species of insects, other invertebrates and a few mammals have been recorded on coffee in India, but only a few of them are of economic importance. As a perennial plant, coffee is subject to attack not only by passing insects but also sedentary insects, with several generations colonizing the plant. Almost all parts of the coffee plant are susceptible to one pest or the other.

12.1. Diseases

Arabica and robusta both coffees are susceptible to fungal diseases, but in India bacterial and viral diseases have not been noticed so far. The Arabica coffee is more susceptible to diseases than robusta coffee. There are approximately ten different types of diseases affecting coffee and can be classified as parasitic and non-parasitic. The parasitic diseases are Leaf rust *Hemileia vastatrix* B. and Br.; Black rot *Koleroga noxia* Donk.; Pink disease *Corticium salmonicolor B.* and Br.; Anthracnose *Colletotrichum gloeosporioides* Penz.; Root diseases *Fomes noxius* Corner *Poria hypolateritia* Berk. *Rosellinia bunodes* (B. and Br.) Sacc. *Fusarium oxysporum* f. sp. *Coffeae*; Berry blotch *Cercospora coffeicola* B. and Cke.; Brown-eye-spot *Cercospora coffeicola* B. and Cke.; Collar rot *Rhizoctonia solani* Kuhn.; Diseases of minor importance; Flowering parasites and *Cuscuta reflexa* - shoot parasite *Balanophora indica*-root parasite *Santalum album* - root parasite as well as non-parasitic disorder Kondli (copper toxicity). The disease monitoring and control measures play very important role in the management of a coffee estate. The most of the diseases affecting coffee are classified as parasitic except one is non-parasitic (Anonymous, 2014; Sudhakar *et al.*, 2014). Therefore, better response of physiological intervention on achieving yield potential, timely control measures of disease is key factor in crop management of coffee industry.

12.2. Pests in India

Coffee being a perennial plant is vulnerable to attack not only by passing insects but also sedentary insects, with several generations colonizing the plant. Although, there have been nearly a hundred species of insects, other invertebrates and a few mammals in coffee plantations of India, but only a few of them are of economic importance. Almost all parts of the coffee plant are prone to one pest or another. In India, the carefully cultivated and trained canopy of shade trees, under which coffee is grown, greatly influences the microclimate in the coffee ecosystem. This is of great significance in the context of pest management. The management strategy for tackling coffee pests is based on the anticipation of pest outbreaks; judicious management of ecological factors by shade regulation, phyto-sanitation, handling and pruning; maintenance of optimal overhead shade; continuous monitoring of pest populations; conservation and manipulation of indigenous natural enemies; introduction, mass multiplication and field release of exotic natural enemies; timely and need-based use of bio-pesticides and integrating the available interventions for effective management. Detailed accounts of perfect management of coffee pests are provided in Coffee Guide (Anonymous, 2014; Vinod Kumar *et al.*, 2009; Naidu, 1997; Bhat *et al.*, 1992) and reviewed by Venkatesha *et al.* (1998) and biological coffee pest (Prakasan *et al.*, 1986); field evaluation of synthetic male Phermone (Vinod Kumar *et al.*, 2000).

12.2.1. Major Insects

The most important coffee pests are White Stem Borer (*Xylotrechus quadripes* Chevrolat; Coleoptera: Cerambycidae). It attacks on arabica coffee stem and thick primary branches and alternate hosts include robusta and tree coffee, teak, *Olea dioica* etc. (Anonymous, 2014b; Venkatesha *et al.*, 2012). Coffee berry borer (*Hypothenemus hampei* Ferrari; Coleoptera: Scolytidae) is another most serious pest of coffee world over. Its host plants are all cultivated coffee varieties are attacked and female beetles may take shelter in the seeds of a variety of plants like *Crotalaria, Lantana, Maesopsis,* tamarind, tea etc., without feeding and breeding. The pest damages young as well as ripe berries. Shot-hole borer or Black twig borer (*Xylosandrus compactus* Eichhoff; Coleoptera:Scolytidae) attacks young branches a major pest of robusta coffee with a wide range of host plants occasionally infests arabica and also *Annato,* Avocado, *Clerodendrum,* cocoa, *Crotalaria,* croton, dadap, *Dendrobium,* mahogany, mango, neem etc. Mealy bug (*Planococcus* spp.; Homoptera: Pseudococcidae). The most important sucking pests of coffee common Mealy bugs species are *P. citri* (Risso) and *P. lilacinus* (Cockerell) infesting coffee. The host plants of Mealy bugs are highly polyphagous, infesting a wide range of cultivated and uncultivated plants like citrus, guava, cacao, mango etc. Green Scale *Coccus viridis* (Green); Homoptera:Coccidae) is a serious sucking pest of coffee, particularly arabica. Green scale is polyphagous, infesting host plants like citrus, mango, guava, sapota and a number of weeds. Brown scale (*Saissetia coffeae* Walker; Homoptera: Coccidae has a wide range of host plants including tea, citrus, guava, etc. Cockchafers or white grubs (*Holotrichia* spp.; Coleoptera: Melolonthidae). Cockchafers are sometimes serious pests in new clearings and replanted areas. The grubs feed on roots of many plants. Hairy caterpillars (*Eupterote* spp.; Lepidoptera: Eupterotidae). Host plants are Cardamom, trees like *Bischofia javanica* (neeli), *Syzygium jambolana* (nerale), *Eugenia jambolana* and *Erythrina lithosperma* (dadap). Coffee bean beetle (*Araecerus fasciculatus,* De Geer; Coleoptera: Anthribidae) is a pest of coffee berries in the field and of beans in storage.

12.2.2. Nematode

There are several species of nematodes attacking coffee. Of these, the root-lesion nematode, Zimmermann, phylum Nematoda; Tylenchida: Pratylenchidae). *Pratylenchus coffeae* has been found to be highly destructive to arabica coffee. Robusta is tolerant to the attack of this nematode. Host plants are as many as 200 hosts including weeds, shade trees and plantation crops grown in coffee estates are known to harbour *P. coffeae.*

12.2.3. Other Pests

Snail (*Ariophanta solata* Benson; Phylum Mollusca, Gastropoda: Ariophantidae). *A. solata* is a non insect pest with a wide range of host plants including coffee and dadap. The snail feeds on the leaves of arabica coffee and dadap, and occasionally on the bark of tender branches and the skin of fruits.

Red borer (*Zeuzera coffeae* Nietner; Lepidoptera: Cossidae) is a minor pest of arabica and robusta coffee which bores into the young stem, primaries and secondaries. The other hosts including tea are cinnamon, sandal, cotton, orange, teak and many more forest trees.

Thrips (*Heliothrips haemorrhoidalis* Bouche, *Retithrips syriacus* Mayet, *Scirtothrips bispinosus* Bagnall, *S. sweetmanni* Bianchi and *Thrips nilgiriensis* Ayyar; Thysanoptera: Thripidae) are minor pests of coffee. Though several species feed and reproduce on coffee, none is capable of causing economic damage when the plants are grown under adequate shade. Thrips lacerate the leaf tissues with their mouth parts and suck in the sap which oozes out. Damage is more pronounced in exposed areas during dry weather. Some species of thrips feed on leaf rust and disperse the uredospores.

Termites attack dadap stakes planted in certain areas which may result in poor establishment. This, coupled with exposure, results in high casualty of dadap stakes. Termites are also found to attack silver oak trees during dry conditions.

Nursery pests are grasshoppers, crickets, wireworms, aphids, cutworms, loopers and other leaf eating caterpillars are some of the common nursery pests.

12.2.4. Minor Pests

The tea totrix *Homona coffearia* Nietner, the cutworms *Condica illecta* Walker, *Spodoptera litura* Fabricius, the arctiids *Pericallia ricini* Fabricius and *Lemyra* sp. occasionally feed on coffee leaves under sporadic outbreaks. Larvae of the swift moth *Sahyadrassus malabaricus* (Moore) sometimes bore into robusta suckers. The star scale *Cerococcus ornatus* and the thread scale *Ischnaspis longirostris* (Signoret), the guava mealybug *Chloropulvinaria psidii* and the stink bug *Udonga montana* (Distant) pests of minor importance. *Tropicomyia* sp. and *Acrocercops caerulea* Meyrick are leaf miners. The scolytid beetles *Xyleborus globus* Blandford, *X. interjenctus* Blandford, *X. nitidus* Hagedorn, *X. noxius* Sampson and *Xylosandrus crassiusculus* Motschulsky bore into the main stem and thick branches of arabica and robusta. Among the mammals, squirrels and bats occasionally feed on the succulent branches of coffee. Monkeys feed on fruits and sometimes destroy young plants. Most of the minor pests are usually under natural control and seldom build up to cause economic damage.

13. Harvest and Post-harvest Practices

Coffee cultivation is labour intensive crop and more than 40 per cent its labour is used annually for harvesting and processing, thereby it is both an art and a science to mange coffee plantation and workforce to meet social obligations. It is a matter of pride and joy to be a planter to make his contribution to the production of good quality coffee which has always been our greatest asset in the international market and which in the past has enabled India to maintain a second position in the world market. This is because of our natural coffee habitat adopted and harvesting and processing technology used. As far as quality concerned it is a summative index of many characteristics of coffee such as its appearance in the raw, roast and liquor qualities comprising factors like aroma, body and acidity. Quality is also defined as the conformance of the product to the requirement of the consumer. Quality of coffee depends on the variety, environmental factors (soil and altitude), insect, disease or fungal attack, nutritional factors, harvest quality, method of processing, drying, hulling, grading, storage and transport. While it is possible to overcome the influence of these by adopting improved cultural practices, correct processing techniques are necessary to prevent deterioration in quality. Faulty processing can bring about deterioration of even the best quality coffee. Proper processing on the estates can go a long way to preserve and enhance the inherent qualities of good coffee.

13.1. Harvesting

As India is known for quality coffee production and for the preparation of both parchment and cherry coffee, picking of the right type of fruits forms an essential part of processing. Coffee fruits should be picked as and when they just ripe for wet processing. On gently squeezing the ripe fruit, the bean inside pops out easily. Under-ripe and over-ripe fruits cause deterioration in quality, the former tending to produce "immature beans", and the latter "foxy" tasting coffee. If, for any reason, it is not possible to pick coffee as and when it ripens, the over-and under-ripe fruits are scrupulously sorted out before using the fruits for pulping. However, for dry processing also fruits should be picked in similar manner for better quality but for the sake of economy fruits are harvested when about 90 per cent of fruits are ripe. The green and under-ripe fruits should be sorted out and dried separately for quality maintenance. It is advisable to wash and dry frequently the bags used for collecting the harvested fruits. The bags in which fertilizers, pesticides and fungicides are stored should never be used for this purpose. No any cherries should be left on the trees after the end of the harvest otherwise it leads to high infestation of coffee berry borer in coming year and neighbour estates. An indefinite delay in pulping and keeping heaped fruits for more than 8-10 hours causes fruity/winey taste. Provisional storage of fruits before pulping in water will avoid overheating/fermentation of fruits.

13.2. Processing of Harvested Fruits

Coffee processing is under taken by two ways; one is wet processing by which plantation or parchment coffee is prepared and second one is by dry method in which cherry coffee is prepared directly drying the fruits under sun. The parchment coffee, prepared by the wet method, is generally favoured by the market. Cherry coffee, because of its very nature of preparation and due to its longer contact with the mucilage and fruit skin, is usually associated with a characteristic "fruity" flavour. Hence, it is desirable to process the largest quantity possible by the wet method.

13.3. Drying

The next stage in coffee processing is drying the parchment in the sun until the moisture content is sufficiently reduced to permit storage of beans till they are despatched to curing works. It is necessary to emphasize that proper drying contributes to the healthy colour of the bean and other quality factors. Both over drying and under drying will lead to poor quality. Under drying leads to rapid deterioration of beans and thus, turns to "mouldy" and gets "bleached" during storage and subsequent curing operations. The wet parchment coffee has a water content of around 50-55 per cent which has to be brought down to 11 per cent. While preparing arabica or robusta cherry, the moisture content of cherries to be reduced to 11-11.5 per cent, respectively for safe storage. At this moisture level enzymatic activity and mould growth will be minimal. The proper drying methodologies have been well described in Coffee Guide (Anonymous, 2014) and for north east region (Anand *et al.*, 2010).

13.4. Hygiene

Improvement of coffee quality depends on maintenance of hygienic conditions, because coffee beans absorb foreign taints and odours easily. Pulper yard, pulper vats, siphon and channels as well as sieves and processing machineries should be checked daily and kept clean. No fruit, fruit skin or beans from the previous day's harvest should be allowed to remain and mix with fresh coffee fruits or pulped mass. Fermented beans of the previous day's lots when mixed with fresh and clean parchment will result in deterioration of quality of the entire lot. Clean water should be used for pulping and washing and all extraneous matters such as leaves, twigs, etc. should be excluded. Thus, it becomes inevitable to follow Good Agricultural Practices (GAP) and Good Manufacturing Practices (GMP).

13.5. Storage and Despatch

Stores should be kept well ventilated and dry without letting in moisture or rain water. The bags containing dried parchment or cherry should be stored on a raised wooden platform to ensure circulation of air underneath the bags. Parchment and cherry coffees should not be stacked together. It is desirable that they are stacked in separate compartments in the store. Other materials, especially fertilizers, pesticides, etc., should not be stored in the same room.

The bagged coffee should be despatched to the curing works at the earliest opportunity. The bags must bear labels as to their grade, lot number and other details, such as parchment, cherry, estate pounded coffee etc., with instructions to cure them separately. Estate hulled coffee should be transported to the curing works in clean gunnies for further cleaning and grading of coffee to the presented standards. All gleaning, floats (Jollu) should be packed and sent separately for curing.

14. Conclusions and Future Research Strategies

The increasing consumption in domestic and international scenario, limited coffee cultivable area, reaching productivity at its peak and remains at plateau as well as climate change are putting difficult challenges ahead to Coffee industry. In

India coffee is grown against various geological and climatic adversities. In recent years extremely unpredictable and adverse climatic changes have contributed to frequent flare up of pest and diseases especially the Coffee White Stem Borer and significantly crop losses due to premature fruit drop. The cultural operations which hitherto have been linked to months and weeks during a year are increasingly getting linked to the prevalence of specific climate conditions such as heat, unseasonal rains, drought etc. due weather uncertainties, calling for a grater acquaintance of plant requirement in different conditions. To meet the increasing coffee demand, it is inevitable to increase vertical production through increase of productivity in a unit area. This can be achieved only through achieving yield potential of both arabica and robusta coffee. The physiological interventions certainly play a vital role such as minimizing effects of drought and maintaining growth and development of plants during gestation period of drought, minimizing floral atrophy and increase fruit set and berry growth; curtailing premature fruit drop, defoliation and improving cropping wood for next year crop; increasing floral bud induction under excessive vegetative growth and shade in high elevations; improve berry growth and perfect bean filling without bean disorder due to understanding of carbohydrate status in the plants during bean filling stages; providing handy tools to the coffee growers for measuring shade in coffee plantations for maintenance of adequate shade with suitable shade trees and providing optimum light; help in scheduling of irrigation for uniform blossom, rejuvenate unproductive robusta coffee, boost seedling growth etc. However, it is equally important that other variable such timely adoption of standard agronomical practices, fulfilling adequate nutritional requirement and taking care of plant protection measures in time are not the limiting factors in coffee production system.

The climatic change is playing wide range of effects on crop production including shifting of growing areas. The fluctuation in rainfall pattern, prolonged drought, high temperatures and growing of coffee with low shade or no shade has tremendous pressure on evolution of suitable technologies to meet growing demand of coffee consumption. Hence, it is imperative to look plant type having better light use efficiency, water use effectively, withstanding high temperature, more nutritional responsive, having less biennial bearing habit, resistance to drought, pest and diseases and adoptive to less shade, thereby increasing production and productivity of coffee without compromising quality. However, all these characters getting together may be a dream. The application of biotechnology for genetic transformation could be a great potential tool in improvement of coffee cultivars and growing coffee in an eco-friendly manner.

As part of strategies to give thrust to develop coffee industry in India, the focus areas on to increase coffee production perfect could be utilization of natural resources, high yielding cultivars, drip irrigation along with nutrition with smart technologies and precision agriculture as well as emphasis on mechanization of most of the vital operation may be helpful in achieving desired results. Diversification of crops which are compatible with coffee is need of the hour and if one fails the other can have sustainability. Further, the selection of right type of plant material to suitable areas, maintaining complete plant population in unit area, adopting proper mixed type shade trees to provide adequate shade, timely and perfect dose of fertilizers application in critical phases of the crop with emphasis on integrated nutrient management, timely adoption of agronomical practices and integrated plant protection measures would be added inputs in crop production system.

References

Achyuta Rao, Y.R. (1962). Shade trees for coffee. III *Albizzia* sp. (Albizzia. Durazz). *Indian Coffee*. 26(7): 195-202.

Achyuta Rao, Y.R. (1963). Shade trees for coffee. IV. *Ficus* spp. *Indian Coffee*. 27(5): 133-136.

Anand, C.G., Pradeep Kumar, D'Souza, G.F., M. Shalini, Badru Lamani, R Nagarathanamma and N Sadananda (2014). Drought management for minimizing crop loss and increasing coffee production under changing climatic scenario. *Indian Coffee*. January-2014 pp.4-7.

Anand, C. G., D'Souza, G. F., Pradeep Kumar, and Jayaram (2013). Physiology of premature fruit drop and remedial measures for increasing coffee production in India. *Proceedings Book of National Conference of Plant Physiology-2013 on "Current Trends in Plant Biology Research"* pp. 149-159.

Anand, C.G., Mallikarjun, G. Awati, D'souza, G.F, Pradeep Kumar, Prakash Koler, Nagarattnamma and N. Sadanand. 2013. Physiological constraints in coffee production during monsoon and remedial measures for achieving maximum crop yield. *Indian Coffee*. Vol. LXXVII (4): 4-10.

Anand, C.G., B. Barman, A. Saikia, D. Venkataramanan and C. S. Srinivasan (2000), Black bean disorder: A major limiting factor to crop productivity in arabica coffee in North-East region. In "Plant Physiology for sustainable Forestry Agri.-Horticulture and Industry" *Vol.-III Production and Developmental Plant Physiology"* pp. 186-194.

Anand, C.G., C. B. Prakasan and Anil Kumar 2010. Coffee drying Practices in North East Region. *Indian Coffee*. LXXIV (2): 4- 6.

Anand, C.G., D'Souza, G.F., Mallikarjuna, G. Awati and Venkataramanan, D., 2000. Inducing sucker production in robusta coffee: Use of DMSO and PGR. In: *International Conference on Plantation Crops, Placrosym – XIV*, Dec, 12-15, Hyderabad, pp. 392-395.

Anand, C.G., Venkataramanan, D., D'Souza, G.F., Mallikarjuna, G. Awati and Naidu, R., 2003. Influence of seasonal variation on physiological parameters of four different cultivars of robusta coffee. In: *2nd International Congress of Plant Physiology*, Jan. 8-12, 2003, New Delhi.

Anand, C.G., D'Souza, G.F., Saraswathy, V.M., Venkataramanan, D. and Mallikarjuna, G. Awati, 2001. Effects of mid-monsoon application of NPK fertilizers on physiological and biochemical parameters of berry development in coffee. In: *National Seminar on role of plant physiology for sustaining quality and quantity of food production in relation to environment*. Dec. 5-78, 2001, Dharwad. (Abst. No. 198 page 97).

Anand, C.G., Awati, M.G., Pradeep Kumar, D'Souza, G. F., Badru Lamani, Prakash Koler, Nagarathnamma, R and Sadananda, N. (2013). Physiological constraints in coffee production during monsoon and remedial measures for achieving maximum crop yield. *Indian Coffee*. Vol LXXVII NO.4: 4-10.

Anand, C.G. Mallikarjun G. Awati, Badru Lamani and D'Souza, G. F. 2012. Physiological response of coffee in Tamil Nadu under north-east monsoon conditions. In: *Silver Jubilee Souvenir*, Dec. 08, 2012, pp. 127-132.

Anand, C.G., Venkataramanan, D., D'Souza, G.F., Mallikarjuna, G. Awati, and Naidu, R. (2003). Influence of seasonal variation on physiological parameters of four different cultivars of robusta coffee. 2nd International Congress of Plant Physiology: Abstract: P-39. page no: -517.

Ananda Alwar, R.P.(1985). Nutrients extracted by crop – A basis for fertilizer recommendation. *Journal of Coffee Research* Vol.15 : 15-16.

Ananda Alwar, R.P., and Ramaiah, P.K. (1991). By products of coffee berries and their possible utilization. *Indian Coffee*. 55(5): 3-7.

Ananda Alwar, R.P., and Rao. W. Krishnamurthy. (1992a). Sulphur requirements of arabica coffee cultivars. *J. Coffee Res.* 22(2): 123.

Ananda Alwar, R.P., and Rao. W. Krishnamurthy. (1992b). Secondary and micronutrients composition of a weeds of coffee plantation. *J. Coffee Res.* 22(2): 143.

Ananth, B.R., Iyengar, B.R.V., and Chokkanna, N.G. (1960). Studies on seasonal variations of plant foods under different shade trees. *Indian Coffee*. August 1960: 347-351.

Anil Kumar, and Srinivasan, C.S. (1993). Response of different growth regulators on rooting of mallet cuttings in coffee. *Indian Coffee*. 57(8): 3-6.

Anonymous (2011). Assessment of Photosynthetic contribution by developing green berries in arabica coffee. In 64th Annual Report (2010-11), Central Coffee Research Institute, Published by Director of Research, CRS (Post), Chikmagalur, Karnataka. pp. 98-99.

Anonymous (2014 a). Database on Coffee. Published by Market Research and Intelligence Unit, Coffee Board, India. http: //www.indiacoffee.org/ Database/ DATABASE_Sep14_I.pdf.

Anonymous (2014 b). Shade Management. Coffee Guide "A manual of coffee cultivation" Published by Director of Research, CCRI, Chikmagalur, Karnataka India. Revised Eighth edition. pp.-262.

Awatramani, N.A., and Subramanya, H.,(1973). Measurement of growth in single stem arabica coffee. *J. Coffee Res.* 3(2): 37-41.

Balaram Menon, P. (1968). Shade trees for coffee. VII. *Dalbergia latifolia* Roxb. *Indian Coffee*. 32(6): 170-173.

Bhaskara Reddy, K., Venkatesha, M.G., and Seetharama, H.G. (1996). S Bioefficacy of kemisal against green scale *Coccus viridis* (green) in coffee. *J. Coffee Res.* 26(2): 7378.

Bhat, P. K. Ramprasad, A. B. and Gowda, D.K.S. (1992).Effective control of white stem borer by debarking the main stem and thick primariesof *Arabica coffee*. *Canadian Journal of Plant Science*. Vol. 72: 25-28.

Boopathy, R. (1988). Metabolism of protien, carbohydrates and lipid durin anaerobis fermentation of coffee pulp. *J. Coffee Res.* 18(1): 1-22.

Bopaiah, B.M., Vittal Rai, P., and Khan, N.A. (1977). The influence of leaf surface organisms on the vegetative growth of coffee plants. *J. Coffee Res.* 7(3): 59-64.

D'Souza, G.F., Venkataramanan, D., Gopal, N.H., and Vasudeva Rao. (1992). Seasonal effect on coffee varieties in Andhra pradesh. *J. Coffee Res.* 22(2): 87-102.

D'Souza, G.F., Venkataramanan, D., Saraswathy, V.M., and Mallikarjuna, G. Awati, (1995). Effect of moisture stress on nature reductase activity to drought tolerance. *J. Coffee Res.* 25(2): 80-87.

D'Souza, G.F., Mallikarjuna, G. Awati, Anand, C.G., and Devaraj Achar, A.M. (2002). Variation in photosynthesis parameters amongst station released coffee genotypes and their importance in drought tolerance. *J. Coffee Res.* 30(1); 14-23.

D'Souza, G.F., Mallikarjuna, G. Awati, Venkataramanan, D., Anand, C.G., and Naidu, R. (2002). Impact of fertilizer sources on growth and carbon fixation in young arabica coffee (*Coffea arabica* L. CV. SLN.12 and SLN.9). *J. Coffee Res.* 30(2): 120-127.

D'Souza, G.F., Venkataramanan, D., George Daniel, Sadananda, N., Ramamurthy, N., and Mallikarjuna, G. Awati, (2004). Biozyme granules as biostimulants for crop improvement in coffee. *Journal of Plantation Crops*, 32(suppl.). 61-63.

D'Souza, G.F., Mallikarjuna, G. Awati, and Venkataramanan, D. (2005). Field performance of two new arabica (*Coffea arabica* L.) hybrids in relation to carbon uptake and drought tolerance. *J. Coffee Res.* 33(1 and 2): 34-42.

D'Souza, G.F., Pradeep Kumar, Mallikarjuna, G. Awati, Nagarathnamma, R., and Anand, C.G. (2013). Assessment of new arabica coffee hybrids based on carbon exchange parameters. *J. Coffee Res.* 41(1 and 2): 75-86.

D'Souza, G.F., Rajeshwari, N., Ramesh Babu, H.N., Renukaswamy, N.S., Mallikarjuna, G. Awati, Lamani, B., and Anand, C.G. (2008). Physiological changes in arabica coffee due to root-stock scion interaction.

D'Souza, G.F., Rajeshwari, N., Ramesh Babu, H.N. (2010). Influence of root stock-scion combinations on biochemical composition and nutrient uptake in coffee (*Coffea arabica* L.). *J. Coffee Res.* 38(1 and 2): 29-47.

D'Souza, G.F., Renukaswamy, N.S., Nagaraj, J.S., Mallikarjuna, G. Awati, Anand, C.G., and Jayarama, (2010). Diagnosis and recommendation integrated system (DRIS) using leaf nutrient norms of robusta coffee (*Coffea canephora* Pierre ex. Froenher). *J. Coffee Res.* 38 (1 and 2): 1-10.

D'Souza, G.F., Venkataramanan, D., and Mallikarjuna, G. Awati, (1999). Effect of triacontanol on physiological attributies of coffee seedlings. *J. Coffee Res.* 27(2): 103-109.

Erwin J. Benne. (1972). The essential elements for growth and how they function. *Indian Coffee.* 36(1): 21-26.

Eunice Melotto, Raimundo S. Barros and Moacyr Maestri, (1993). Carbohydrate accumulation and mobilization by expanding flower buds of *Coffea arabica.* *J. Coffee Res.* 23(2): 63-74.

George Daniel, M.G. Awati, C.G Anand, G.F. D'Souza, N.S Renukaswamy, N.Sadananda and D.venkataramanan, 2009. Effect of foliar application of mepiquat chloride and ethephon on floral bud induction and crop yield in robusta coffee. *J. Coffee Res.* 36 (1 and 2): 60-63.

German Valencia, A. (1970). Physiological study of defoliation caused by *Cercospora coffeicola* in coffee plant. *Journal of Coffee Research.* 21: 105-114.

Gopal, N.H. (1971). Preliminary studies on the control of fruit drop in arabica coffee. *Indian Coffee.* 35(10): 413-419.

Gindel (1962).

Gopal, N.H. (1972). Root exudates. *Indian Coffee.* 36(1): 21-26.

Gopal, N.H. (1973). Defoliation and fruit drop after bordeaux mixture spray in arabica coffee. *J. Coffee Res.* 3(1): 14-23.

Gopal, N.H. (1974). Some aspects of hormonal balance in coffee. *Indian Coffee.* 38(7). 168-175.

Gopal, N.H. (1974). Some physiological factors to be considered for stabilization of arabica coffee production in South India. *Indian Coffee.* 38(8): 217-221.

Gopal, N.H. (1981). Coffee production in Andhra pradesh. *Indian Coffee.* 45(6): 171-173.

Gopal, N.H., and Balasubramanian, A. (1975). Studies on phosphorus nutrition using ^{32}P in coffee plants. *Indian Coffee.* 39(2): 58-63.

Gopal, N.H., and D'Souza, G.I. (1975). Radio isotopes- A new food for coffee research in India. *Indian Coffee.* 39(4/5): 139.

Gopal, N.H., and D'Souza. G.I. (1977). Some aspects of quality of *Indian Coffee.* *Indian Coffee.* 41(1): 14-16.

Gopal, N.H., and Raju. K.I. (1978). Physiological studies on flowering in coffee under South Indian conditions. VIII. Number of flower buds in relation to wood starch of cropping branches. *Turrialba.* 25(4): 311-31.

Gopal, N.H., Raju, K.I., and Janardhan, K.V. (1976). Carbohydrates of different components of ripe fruits in coffee. *Indian Coffee.* 40 (4 and 5): 114-117.

Gopal, N.H., and Ramaiah, P.K. (1971). Flowering of coffee under South- Indian conditions. *Indian Coffee.* 35(4): 142-143 and 154.

Gopal, N.H., and Ramaiah, P.K. (1971). Studies on wilting and die-back of Arabica coffee plants. *Indian Coffee.* 35(11): 459-464.

Gopal, N.H., and Ramaiah, P.K. (1972). Studies on the physiology of germination of coffee seed. I. observations on sprouting. *J. Coffee Res.* 2(1): 14-19.

Gopal, N.H., and vasudeva, N. (1973). Physiological studies on flowering in arabica coffee under south Indian conditions. I. Growth of flower buds and flowering. *Turrialba* 23(2): 146-153.

Gopal, N.H., and Vasudeva, N. (1974). Studies on ascorbic acid in coffee plants. III. Distribution in pulp and pulp water of ripe fruits. *J. Coffee Res.* 4(4): 121-122.

Gopal, N.H., Vasudeva, N., and Balasubramanian, A. (1976). Studies on absorption and translocation of phosphorus using radioactive sulphur phosphate (32 P) in coffee plants. *J. Coffee Res.* 6(3/4): 69-75.

Gopal, N.H., Vasudeva, N., Venkataramanan, D., and Janardhana, K.V. (1975). Physiological studies on flowering in coffee under South Indian conditions. III. Flowering in relation to foliage and wood starch. *Turrialba.* 25: 239-242.

Gopal, N.H., Vasudeva, N., Venkataramanan, D., and Raju, K.I. (1975). Some physiological differences in arabica and robusta coffee plants grown in normal soil. *Indian Coffee.* 39(7/8): 220-225.

Gopal, N.H., and Venkataramanan, D. (1974). Studies on black bean disorder in coffee. *Indian Coffee.* 38(9/10): 259-267.

Gopal, N.H., Venkataramanan, D., and Raju, K. (1975). Physiological studies on flowering in coffee under South Indian conditions. II. Changes in water content, growth rate, respiration and carbohydrate metbolism of flower buds during bud enlargement and anthesis. *Turrialba.* 25(1): 29-36.

Gopal, N.H., Venkataramanan, D., and Ramaiah, P.K. (1993). Bearing nature of arabica coffee. *Indian Coffee.* 57(3): 11-13.

Gopal, N.H., Venkataramanan, D., Mrs. Ratna, N.G.N. (1975). Physiological studies on flowering in coffee under South Indian conditions. IV. Some physiological properties and chromatographic assay of a gum-like substance exuded by flower buds. *Turrialba.* 25: 4 (410).

Gopal, N.H., Venkataramanan, D., and Ratna, N.G.N. (1976). A quick biochemical test for assessment of coffee quality. *Indian Coffee.* 40(1): 29-32.

Gopalakrishnan, R., and Ananth, K. C. (1960). Transpirational studies on coffee. *Indian Coffee.* 24(4): 20-28.

Gopal Ram, Reddy, A.G.S., and Ramaiah, P.K. (1992). Effect of drip irrigation on flowering, fruit set retention and yield of *Coffea canephora* sub var robusta cv.s. 274. A preliminary study. *Indian Coffee.* 56(11): 9-13.

Glory Swarupa, S., Basavraj Naik, T., Reddy, A.G.S., and Raju, T. (1997). Effect of biofertilizers on germination of coffee seed. *Indian Coffee.* 61(3): 21-28.

ICO,(2014). Domestic consumption. Historical Data. http://dev.ico.org/historical/2000-09/PDF/DOMCONSUMPTION.pdf.

Jamshed Ahmed and Sreenivasan, M.S. (1990). Studies on crop turn relationship with yield and bean size in exotic robusta coffee. *Indian Coffee*. 54(6): 9-12.

Jamshed Ahmed, Radhakrishna, S., and Sreenivasan, M.S. (1992). Blossom and backing showers an analysis of its impact on coffee yield. *Indian Coffee*. 56(2-3): 3-7.

Janardhan, K.V., Gopal, N.H., and Ramaiah, P.K. (1971). Carbohydrates reserves in relation to vegetative growth, flower bud formation and crop levels in Arabica coffee. *Indian Coffee*. 35(4): 145-148.

Jnardhan, K.V., Gopal, N.H., and Ramaiah, PK. (1971). Starch scoring by visual observation in fresh wood of coffee plants. *Indian Coffee*. 35(6): 219-221.

Janardhana, K.V., Raju, K.I., and Gopal, N.H. (1973). Studies on cuticle of coffee plant. I. isolation of cuticular membranes from leaves. *J. Coffee Res*. 3(3): 54-57.

Janardana, K.V., Raju, K.I., and Gopal, N.H. (1977). Physiological studies on flowering in coffee under South Indian conditions. VI. Changes in growth rate, indole acetic acid and carbohydrate metabolism during flower bud development and anthesis. Turrialba. 27(1): 29-38.

Kamala Bai, S., Nanjappa, H.V., Raghuramulu, Y., Jayarama, Bhaskar, S., Babou, C., Venkatesha, M.M.M Dinesh Kumar, S.P., and Roopa, T.K. (2013). Effects of levels and sources of phosphorous and phosphorous utilization efficiency on growth of young coffee (*Coffea arabica* cv. Chandrgiri). *J. Coffee Res*. 41(1 and 2): 1-13.

Krishnamurthy Rao, W. (1991). Rationalization of application fertilizer inputs in coffee. Planters Chronicle 86(2): 649-652.

Krishnappa, C.S. Naik, Deb, D.L., and Ramaiah, P.K. (1989). Studies on phosphorus fixation and release in coffee growing oxisols. *J. Coffee Res*. 19(1): 46-50.

Mallikarjuna, G. Awati, D'Souza, G.F., Venkataramanan, D., and Naidu, R. (2000). Irrigation schedule for establishment for young coffee. *J. Coffee Res*. 28(1 and 2): 1-8.

Madhav Naidu, M., and Srinivasan, C.S. (1999). Effect of growth regulators in coffee (*Coffea canephora* Pierre ex. Froenher). *J. Coffee Res*. 27(1): 121-126.

Mathur, P.K., and Chokkana, N.G. (1991). Studies on intake of water and nutrients during development of flower buds to blossoms in coffee. *Indian Coffee*. 55(4): 3-10. Reprinted from *Indian Coffee*. 25(9): 272. 1961.

Marimuthu, R., and Iruthayaraj, M.R. (1994). Studies on the impact of irrigation at different leaf water potential on physiological and growth parameters in the two coffee cultivars. *J. Coffee Res*. 24(1): 23-32.

Mallikarjuna, G. Awati, D'Souza, G.F., and Venkataramanan, D. (2004). Enhancing flower bud initiation and crop yield in coffee. Journal of plantation crops, 32(suppl.). 58-60.

Mallikarjuna, G. Awati, D'Souza, G.F., Renukaswamy, N.S., Venkataramanan, D., and Anand, C.G. (2007). Influence of foliar application of mepiquat chloride and ethephon on flowering and crop yield in arabica coffee. *J. Coffee Res*. 35(1 and 2): 10-21.

Mallikarjuna, G. Awati, D'Souza, G.F., Renukaswamy, N.S., Venkataramanan, D., Anand, C.G., and Jayarama, (2011). Effect of root pruning on growth and development of coffee seedlings. *J. Coffee Res.* 39(1 and 2): 10-21.

Mallikarjuna, G. Awati, Renukaswamy, N.S., D'Souza, G.F., Anand, C.G., Venkataramanan, D., and Jayarama, (2013). Distribution efficiency of biomass in arabica coffee genotypes. *J. Coffee Res.* 41(1 and 2): 47-60.

Mallikarjuna, G. Awati, D'Souza, G.F., Seetharama, H.G., Anand, C.G., Vinod Kumar, K., and Jayarama, (2013). Effect of white stem borer infestation on photosynthetic efficiency in coffee. *J. Coffee Res.* 41(1 and 2): 108-111.

Mallikarjuna, G. Awati, D'Souza, G.F., Venkataramanan, D., Anand, C.G., and Naidu, R. (2003). Diurnal seasonal fluctuation of photosynthetic efficiency of coffee in open and shade conditions. 2nd ICPP; Abstracts: 2nd *International congress of plant physiology.* p. 517.

Mathew, P.K., and Chokkanna, N.G. (1961). Studies on the intake of water and nutrients during development of flower buds to blossoms in coffee. *Indian Coffee.* 25(9): 264-272.

Naidu, R. (1997). White stem borer in Coffee, current management and future strategies. *Planters Chronicle* Vol. 8: 627.

Nataraj, T., and Subramanian, S. (1975). Effect of shade and exposure on the incidence of brown eye spot of coffee. *Indian Coffee.* 39(6): 179-180.

Paulo De T. Alvim. (1977). Factors affecting flowering of coffee. *J. Coffee Res.* 7(1): 15-25.

Prakasan, C. B.(1987). Biological control of coffee Pests. *J. Coffee Res.* 17: 114-117.

Peter A.Huxley. (1968). Crop thinning as a possible method for improving quality and reducing storage problems. *Indian Coffee.* 32(2): 69-71.

Raghuramulu, Y. (1999). Training and pruning in coffee. *Indian Coffee.* 63(3): 17-18, 26.

Raghuramulu, Y., Padmajyothi, D., and Pharis, R.P. (1996). Influence of gibberellin A$_4$ on flowering and fruiting in coffee (*Coffea arabica* L.). *J. Coffee Res.* 26(1): 17-22.

Raju, K, Suryakantha, Srinivasan, C.S., and Vishweswara, S. (1975). Vegetative- floral balance in coffee. III. Effect of thinning of blossom on set and bean size. *Indian Coffee.* 39(7/8): 217-219.

Raju, K.I., Ratageri, M.C., Venkataramanan, D., and Gopal, N.H. (1981). Progressive quantitative changes of caffeine and total nitrogen in reproductive parts of arabica and robusta coffee. *J. Coffee Res.* 11(3): 70-75.

Raju, K.I., and Vasudeva, N. (1982). The distribution pattern of caffeine from seed coffee to seedling in two commercial species of coffee. *Indian Coffee.* 46(7): 165-170.

Raju, K.I., Venkataramanan, D., and Vasudeva, N. (1984). Variations of caffeine, total nitrogen and dry matter in the leaves of arabica and robusta coffee. *J. Coffee Res.* 14(1): 38-42.

Raju, K.I., and Vasudeva, N. (1984). Caffeine content in different arabica and robusta selections grown in India. *J. Coffee Res.* 14(1): 1-6.

Ramaiah, P.K. (1984). Coffee research in India. *Indian Coffee.* 48(8): 9-14.

Raimundo S. Barros, Moacyr Maestri, and Alemar B. Rena, (1994). Purine alkaloids in coffea. *J. Coffee Res.* 24(2): 55-86.

Raimundo S. Barros, Moacyr Maestri, and Alemar B. Rena, (1999). Physiology of growth and production of the coffee tree- a review. *J. Coffee Res.* 27(1): 1-54.

Raju, K.I., and Gopal, N.H. (1979). Distribution of caffeine in arabica and robusta coffee plants. *J. Coffee Res.* 9(4): 83-90.

Ramaiah, P.K., and Chokkanna, N.G. (1964). Observations on die-back of coffee. I. percentage incidence of 'Die-back' in different liaison zones. *Indian Coffee.* 28 (9): 195-204.

Rama Murthy, A. (1965). Shade trees for coffee. VI. *Bischofia javanica* Blume. *Indian Coffee.* 29(8): 7-11.

Rao. W. Krishnamurthy. (1978). Trace element nutrition of coffee. *Indian Coffee.* 42(11): 315-316.

Rao. W. Krishnamurthy., Raju, L., and Iyengar, B.R.V. (1979). Effect of liming on growth and chemical composition of four cultuvars of *Arabica Coffee* seedlings. *J. Coffee Res.* 9(4): 95-101.

Reddeppa Raju. V., Radhakrishnan, S., Venkataramanan, D., and Rao. W. Krishnamurthy. (1991). Leaf area determination in *Coffea arabica* L. *J. Coffee Res.* 21(2): 109-117.

Reddy, Sambamurthy, A.G. (1979). Quinescence of coffee flower buds and observations on the influence of temperature and humidity on its release. *J. Coffee Res.* 9(1): 1-13.

Robinson, J.D.B., and Bull, R.A. (1962). Debility growth symptoms in *Arabica coffee. Indian Coffee.* 26(8): 242-245.

Santa Rama, Sreenath, H.L., and Srinivasan, C.S. (1999). Morphogenetic responses of cultured root and hypocotyl tissues of *Coffea arabica* L. *J. Coffee Res.*27(2): 87-92.

Sisay Asrat, Shivashankar, K., Venkataramanan, D., Thimma Raju, K., and Jaganath, M.K. (1992). Effect of growth regulators on fruit setting of coffee. *J. Coffee Res.* 22 (1): 45-56.

Srinivasa, C.S. (1972). Estimation of sample size for stem girth, leaf, fruit and bean measurements in coffee. *J. Coffee Res.* 2(3): 21-25.

Srinivasan, C.S., Suryakantha Raju. K., and Visveswara, S. (1973). Vegetative-floral balance in coffee. I. Effect of distributing the floral phase on vegetative growth. *Indian Coffee.* 37(12): 390-393.

Srinivasan, C.S., and Vishveshwara, S. (1980). Selection in coffee: some criteria adopted and result obtained in India. *J. Coffee Res.* 10(3): 53-62.

Srinivasan, C.S. (1980). Association of some vegetative characters with initial fruit yield in coffee (*Coffea arabica* L.). *J. Coffee Res.* 16(2): 21-27.

Srinivasan, C.S., and Vishveshwara, S. (1980). A study of robusta propagation clone vs seedling. *Indian Coffee.* 44(6): 105-111.

Subbalakshmi. V. (1991). A note on abnormality of fruits in *coffea arabica* var cauvery (catimor). *J. Coffee Res.* 21(1): 63-65.

Sudhakar, S. Bhat, K. C. Kiran Kumar, **S. D. Mani**, B. T. Hanumantha, N. S. Prakash and Jayarama. **2014**. Chapter 2. Diseases of coffee. **In:** Diseases of Plantation Crops, p. 55 - 109. Ed.: P. Chowdappa, Pratibha Sharma, M. Anandaraj and R. K. Khetarpal. Indian Phytopathological Society, IARI, New Delhi, Today and Tomorrow's Printers and Pulishers, New Delhi, India. 263 pp. ISBN 81-7019-492-X (India) ISBN 1-55528-349-7 (USA).

Sudhakara Rao, K. (1978). Stimulation of bean growth in coffee by exogenous application of ethylene. *Turrialba.* 28(2): 157-158.

Sundar, K.R. (1981). Methods for estimating empty fruit locules in *Coffea arabica.* *J. Coffee Res.* 11(3).

Suryakantha Raju. K., Srinivasan, C.S., and Visveswara, S. (1974). Vegetative floral balance in coffee. II. Effect of crop thinning on the growth of axillary vegetative shoot, extension of nodes, defoliation and fruit drop. *J. Coffee Res.* 4(2): 21-24.

Suresh Kumar, V.B., Keshavaiah, K.V., Ramachandran, M., George Daniel and Naidu, R. (1996). Influence of growth regulators on graft union success in coffee. *J. Coffee Res.* 26(2): 67-72.

Tapley, R.G. (1962). The life span of coffee leaves in Tanganyika. *Indian Coffee,* 26(10): 365-366.

Thimma Reddy, G.S., and Srinivasan, C.S. (1979). Variability for flower production fruit set and fruit drop in some varieties of *Coffea arabica* L. *J. Coffee Res.* 9(2): 27-34.

Srinivasan, C.S. (1984). Performance of arabica varieties of three locations in coorg. *Indian Coffee.* 48(4); 11-13Vasudeva, N. (1983). Effect of cytozyme crop plus on growth and yield of arabica coffee. *Indian Coffee.* 47(5): 9-11.

Srinivasan, C.S., and Subbalakshmi, V. (1981). A biometrical study of yield variation in some coffee progenies. *J. Coffee Res.* 11(2): 26-34.

Purushotam, K., Sulladmath, U.V., and Ramaiah, P.K. (1984). Seasonal changes in biochemical constituents and their relation to rooting of coffee (*Coffea canephora* Pierre) sucker cuttings. *J. Coffee Res.* 14(3): 117-130.

Raju, K.I., and Vasudeva, N. (1984). Stomatal studies in coffee diurnal variation robusta and san Ramon. *J. Coffee Res.* 14(4): 149-155.

Radhakrishnan, S., Ramaiah, P.K., and Rao. W. Krishnamurthy. (1985). An evaluation of economic optimal level of fertilizer to coffee. *J. Coffee Res.* 15(1/2): 14-20.

Purushotam, K., and Sulladmath, U.V. (1985). Studies on the rooting of invigorated sucker cuttings of coffee. *J. Coffee Res.* 15(1/2): 21-28.

Raghuramulu, Y., Gopal Ram and Siddaramappa, S.N. (1990). Influence of growth regulator sprays on fruit retention in robusta coffee under delayed backing shower conditions. *J. Coffee Res.* 20(2): 157-160.

Raghuramulu, Y., and Purushotam, K. (1987). Flower induction in juvenile coffee (*Coffea canephora* Pierre) seedlings. *J. Coffee Res.* 17(1): 31-39.

Raghuramulu, Y., and Purushotam, K. (1987). Root stock trials in coffee. I. Studies in grafting and stionic influence on growth plants in some graft combinations. *J. Coffee Res.* 17(2): 8-15.

Raghuramulu, Y., and Purushotham, K. (1986). Studies on rootability of different exotic robusta collections. *J. Coffee Res.* 16(1/2): 28-30.

Raju, T. (1986). Investigation of zinc defeciency in coffee (*Coffea arabica* L.). *Indian Coffee.* 50(3): 7-11.

Raju, T., and Deshpande, P.B. (1986). Phosphorous and zinc interactions in coffee. (*Coffea arabica* L.) Seedlings. *J. Coffee Res.* 16(1/2): 1-13.

Raju, T., and Deshpande, P.B. (1986). Lime interactions in coffee. (*Coffea arabica* L.) Seedlings. *J. Coffee Res.* 16(3/4). 79-85.

Raju, T., and Ramaiah, P.K. (1986). How important is phosphate for coffee. *Indian Coffee.* 50(4): 13-20.

Reddy, A.G.S., Sreenivasan, M.S., and Ramaiah, P.K. (1986). Adaptation of coffee selection in non-traditional areas. *Indian Coffee.* 50(10): 11-16.

Reddy, L. Shivaram., Ramachandran, M., and Govindrajan, T.S. (1986). Bethel (2-chloroethyl phosphonic acid) spray to overcome dormancy in coffee seedlings. *J. Coffee Res.* (1/2): 31-34.

Ramaiah, P.K., and Venkataramanan, D. (1985). Growth and development of coffee in south Indian conditions. A review. *J. Coffee Res.* 15(1/2): 1-13.

Ramaiah, P.K., and Venkataramanan, D. (1988). Studies on the effect of gibberlic acid on arabica coffee in India. *J. Coffee Res.* 18(1): 47-51.

Sreenath, H.L. and Praksh, N. S. (2006). Coffee. In Book on Plantation crops. Edited by V. A. Parasarathy, P. K. Chattopadhyay and T. K. Bose. Published by Naya Udyog, Kolkata, India. Vol.1: 149-264.

Srinivasan, C.S. (1988). A comparitive study of juvenile growth characters of coffee varieties in North-east India. *J. Coffee Res.* 18(1): 36-46.

Srinivasan, C.S. (1988). Yield response of some new arabica coffee selections in Bordeaux spray. *Indian Coffee.* 52(3): 29-32.

Srinivasan, M.S. (1989). Maximising coffee yield through cauvery (catimor). Selection. *Indian Coffee.* 53(4). 5-9.

Vasudeva, N. (1967). Preliminary studies on the growth of coffee leaves. *Indian Coffee.* 31(8): 5-6.

Vasudeva, N. (1979). The roles of carbohydrates in growth and development of coffee. *Indian Coffee.* 43(5): 127-128, 136.

Vasudeva, N. (1983). Preliminary studies on the effect of experimental growth regulator (AC/99,524) of coffee. *J. Coffee Res.* 13(2): 44-47.

Vasudeva, N. (1983). Vegetative propagation through seedlings in coffee. *J. Coffee Res.* 13(3): 78-80.

Vasudeva, N.H., and Gopal, N.H. (1972). Studies on ascorbic acid in coffee plants. I. distribution in green fruits. *J. Coffee Res.* 2(2): 23-26.

Vasudeva, N., and Gopal, N.H. (1977). Distribution of nitrogen in ripe fruits of coffee plants. *Indian Coffee.* 41(4): 121-123.

Vasudeva, N., Gopal. N.H., and Ramaiah, P.K. (1973). Studies on leaf growth IV. Influence of environmental factors on leaf growth. *J. Coffee Res.* 3(4): 80-88.

Vasudeva, N., and Gopal, N.H. (1974). Studies on ascorbic acid in coffee plants II. Distribution in ripe fruits and its relationship with coffee quality. *J. Coffee Res.* 4: 25-28.

Vasudeva, N., and Gopal, N.H. (1977). Physiological studies on flowering in coffee under South Indian conditions. VII. Changes in iron and copper enzymes and ascorbic acid during flower bud development and anthesis. *Turrialba.* 27(4): 355-360.

Vasudeva, N., and Gopal, N.H. (1979). Physiological studies on flowering in coffee under South Indian conditions. IX. Effect of different irrigation treatment on blossom and yield of arabica coffee. *J. Coffee Res.* 9(3): 74-79.

Vasudeva, N. (1980). Effect of ascorbic acid on the growth of arabica coffee seedlings. *J. Coffee Res.* 10(1): 1-3.

Vasudeva, N. (1980). Upward translocation and deposition of copper sulphate in arabica coffee seedlings. *J. Coffee Res.* 16(2): 36-38.

Vasudeva, N., Raju, K.I., Venkataramanan, D., and Ratageri, M.C. (1981). Studies on the effect of atonik on yield of arabica coffee. *J. Coffee Res.* 11(2): 39-43.

Vasudeva, N., Raju, K.I., and Gopal, N.H. (1984). The distribution pattern of boron in seedlings and mature plants of arabica and robusta coffee. *J. Coffee Res.* 14(3): 133-135.

Vasudeva, N., and Ramaiah, P.K. (1979). The growth and development of arabica coffee under South Indian conditions. *J. Coffee Res.* 9(2): 35-45.

Vasudeva, N., and Ratageri, M.C. (1979). Effect of certain chemicals and growth regulators on pollen germination in robusta coffee. *J. Coffee Res.* 9(2): 46-48.

Vasudeva, N., and Ratageri, M.C. (1981). Exogenous application of ascorbic acid and burelex on yield of arabica coffee. *Indian Coffee.* 45(10): 295-296.

Vasudeva, N., Ratageri, M.C., Venkataramanan, D., and Raju, K.I. (1981). Caffeine, carbohydrates and protein in spent coffee from commercial blends. *J. Coffee Res.* 11(2): 49-51.

Vasudeva, N., and Ratageri, M.C. (1981). Studies on leaf to crop ratio in two commercial species of coffee grown in India. *J. Coffee Res.* 11(4): 129-136.

Vasudeva, N., Raju, K.I., Ratageri, M.C., and Venkataramanan, D. (1982). Correlation between visual scoring and chemical analysis of starch in robusta coffee. *Turrialba.* 32(3): 336-338.

Vasudeva, N., and Venkataramanan, D. (1981). Increase in yield of arabica coffee. *Indian Coffee.* 45(7): 193-194.

Venkatesh, M. G., H. G. Seetharama and K Sreedharan (1998). Coffee Pest. In "A compendium on pest and Diseases of coffee and their management in India". Published by Dr. R. Naidu, Director of Research, Central Coffee Research Institute, Coffee Research Station (Post) – 577 117, Chikmagalur, Karnataka, India. pp. 1-28.

Venkatesulu. K., Gokuldas kumar and Chako, M.J. (1981). Studies on the effect of sevidal 4: 4G on roots of coffee seedlings. *J. Coffee Res.* 11(4): 137-139.

Venkataramanan, D. (1988). Metabolic changes in relation to growth of coffee leaves. *J. Coffee Res.* 18(2): 90-119.

Venkataramanan, D. and Govindappa, D.A. (1987). Shade requirements and productivity of coffee. *J. Coffee Res.* 17(2): 16-39.

Venkataramanan, D., Ramaiah, P.K., Vasudeva,N., Srinivasan, C.S. (1989). Effect of certain growth regulators on yield on arabica coffee. *Indian Coffee.* 53(11): 3-6.

Venkataramanan, D., Vasudeva, N., Raju, K.I., Ratageri, M.C., and Gopal, N.H. (1983). Growth analysis of coffee the relative growth and net assimilation rate of certain cultivars of coffee. *J. Coffee Res.* 13(1): 6-13.

Venkataramanan, D., Vasudeva, N., Ratageri, M.C., Raju, K. I., and Narasegowda. N.C. (1983). Total chlorophyll content in some coffee types grown in India. *Indian Coffee.* 47(6): 7-9 and 21.

Vinod Kumar, P. K., Sreedharan, K., Jayarama, Seetharama H. G. Samuel, S. D., Raphael, P. K. and Irulandi, S. (2000). Field evaluation of synthetic male pheromone of white stem borer *Xylotrechus* quadripes (Coleopptra: Cerambycidae) pp.467-469. In Plantation crop Research and development in the new Millennium, Plantation Crops Symposium XIV, PLACROSYM XIV. 12-15 December 2000, Hyderabad, India edited by Rethinam, H. Khan, V. M. Reddy, P.K. Mandal and K Suresh). Coconut Development Board, Cochin, ISBN-87561-05X.

Vinod Kumar, *P. K.*, Seetharama H. G., Balakrishna, M.M., Irulandi, S. and Jayarama 2009. The Coffee Whitw Stem Borer – An insight. *Indian Coffee,* 73: 8-11.

Wintgens, J. N. (2004). Coffee: growing, processing, sustainable production. A guidebook for growers, processors, traders and researchers. 976pp.

Wrigley, G. 1988. Coffee Longman scientific and technical Essex, England. 639 p.

Zamarripa, C. A. and Petiard, V. (2004). In coffee: Growing, processing, sustainable production. (Wintgens, J. N. Ed.) Wily- VCH. Weinheim) pp. 137-163.

Recent Advances in Crop Physiology Vol. 2 (2016)   *Pages* **173–202**
Editor: **Dr. Amrit Lal Singh**
Published by: **DAYA PUBLISHING HOUSE, NEW DELHI**

Chapter 5

Bioregulators Improve the Productivity and Quality of Indian Table Grapes

S.D. Ramteke*

ICAR-National Research Centre for Grapes,
Manjri Farm, PB 3, Pune – 412 307, Maharashtra

1. Introduction

Grape, (*Vitis vinifera*) is one of the major important fruit crops of India where the reference of its cultivation is found in ancient text. In modern history, commercial *V. vinifera* is believed to be introduced by Persians in 1300 AD and its cultivation spread and flourished during the Mughal empire (1526–1758 AD). Grape cultivation declined after the fall of the Mughal empire. Systematic research started in the early 20th century. Natural habitat for the grape crop is temperate region. Presently, grape cultivation covers the largest area *i.e.* 119.0 thousand ha and the production is around 2400.4 thousand tons in 2013-2014 (Second advance estimate NHB, 2014). Major grape-growing states in India are Maharashtra, Karnataka, Andhra Pradesh, Mizoram, Tamil Nadu, Punjab, Haryana, western Uttar Pradesh, Rajasthan and Madhya Pradesh. In India mainly Maharashtra, Karnataka, Andhra Pradesh, Tamil Nadu and Mizoram jointly contribute to more than 99.3 per cent of the total area and production of grapes (Indian Horticulture Database, 2013).Total export of grape from

* *Author:* E-mail: sdramteke@yahoo.com

India is 173 thousand MT, valuing Rs. 1259 cores during 2012-13 (Indian Horticulture Database, 2013). The grape export from India was started in the year 1991 after economic liberalization, major grape exporting countries are Chile, South Africa, Spain, Italy, France etc. (Thorat *et al.*, 2012). The grapes in the country are grown for various purposes like table purpose, wine, juice and resins. The survey has indicated that among the table grapes Thompson Seedless and its clones Tas-a-Ganesh, Sonaka and others like Sharad Seedless are commercially being cultivated for domestic and international use. To achieve export quality standards, the cultivation practices play important role in any crop. In Grapes, nutritional factors mainly related to proteins and carbohydrates and utilization of these nutrients depends on the endogenous hormone levels. Precise use of growth regulators at particular growth stage assist in development and improve quality and quantity of final product.

The Agricultural Produce (Grading and Marking) Act, 1937 was enacted to maintain the quality of agricultural produce in India. The Act authorizes the central government to frame rules related to the fixing of grade standards and the procedure to be adopted to grade the agricultural commodities included in the schedules. In accordance with this Act, specifications have been drawn up for grapes considering various quality factors. As per AGMARK, Extra, Class I and Class II grades are available based on bench weight. As per the AGMARK standards fixed, the grapes selected for export should fulfil the following requirements: Clean, sound, free of any visible foreign matter; free of pest affecting the general appearance of the produce; free of damage caused by pest and disease; free of abnormal external moisture; free of any foreign smell; free of all visible trace of mould; berries shall be intact, well formed and normally developed. Along with these qualities table grape shall have minimum soluble solids of 16° Brix with minimum sugar/acid ratio of 20:1 and grapes shall comply with the residue level of heavy metals, pesticide and other food safety parameters as laid down by the Codex Alimentarius Commission for export.

Plant growth regulators are the compounds other than nutrients that affect the physiological process of growth and development in plant when applied at low concentration. Bioregulators have been used for many years to alter the behaviour of fruit or fruit trees for the economic benefit of the fruit grower, control of vegetative vigor, stimulation of flowering, regulation of crop load, reduction of fruit drop, and delay or stimulation of fruit maturity and ripening.The different processes in fruit and fruit trees that can be regulated with exogenous applications of bioregulators. Different types of bioregulators and their general functions are listed in Table 5.1.

As grapevine is perennial fruit crop with long generation time, response of plant or plant part to bioregulator may vary with variety, age, environmental conditions, physiological state of development and nutrition.

The present work summarizes the significance of judicious use of different bioregulators for improving productivity and quality of table grapes. It includes role of each bioregulator, its mode of application, phenological stage of application, additive cultural practices and the parameters evaluated to study its efficacy.

Table 5.1: Role of Bioregulators in Horticultural Crops

Type of Bioregulators	Use in Horticulture
Auxin	Propagation of cuttings, prevention of preharvest drop of fruits, increasing parthenocarpy, increasing fruit set, prevention of sprouting by inhibiting buds, inhibition in prolonged dormancy, control of flowering, prevention of leaf fall, thinning of compact fruits and selective weed killer
Cytokinins	Cell division and enlargement, counteract apical dominance, induction of flowering in short day plant
Gibberellins	Cell elongation promoting less tight bunch and bold berries, induces flowering in long day plants and increased fruit set.
ABA	Inhibitory chemical compound that affect bud growth, seed and bud dormancy.
Ethylene	Inducing fruit ripening, abscission, de-greening of fruits.

2. Budbreak

Grape is perennial fruit crop with long generation life cycle. New generation starts every year by bud break after pruning. Bud breaking chemicals were being used for grapevine to enhance uniformity in bud breaking. Timing, extent and uniformity are the key parameters that used to evaluate the efficacy of chemical for inducing bud break.For deciding dose of chemical and effective application time other crucial factor includes the stored nutrients in grapevine, mode of action of bioregulator and microclimatic conditions within vineyard.Among the different chemicals, Hydrogen cyanamide (HC) is the most useful for bud breaking chemical for grapevine.

Hydrogen Cyanamide (HC 50 per cent) a formulation manufactured in India was assessed in the growers' fields in three varieties *viz*. Tas-A-Ganesh, Sharad Seedless and Sonaka. To evaluate the effectiveness of chemical for bud break following parameters were tested: per cent bud break, duration of bud break, mean number of days from pruning to bud break. The cumulative effect on yield parameters based on TSS, Acidity and yield/vine were also evaluated. Results recorded from Tas-A-Ganesh grape variety show highly significant difference in bud break and the time taken for bud break. HC @ 20 ml/l provided maximum bud break of 99.20 per cent and highest yield as compared to control. The duration of bud break was reduced significantly from 12.19 days in control to 2.20 days in treated vines, thus ensuring the uniformity in shoot growth and the phenological stages of cluster development with this treatment. The same treatment reduced the number of days required for bud break from 20.59 (control) to 8.34 days. Berry quality was not adversely affected by HC treatment. Continuous use of HC for increased budbreak was found to reduce the total shoot length in a vine in a long run, but did not depress any of the yield or quality attributes in Tas-A-Ganesh grapes (Ramteke *et al.*, 2003).

In case of 'Sharad' Seedless grapes, per cent bud break was increased from 23.2 in control to 92.0 with HC applied @ 40 ml/l. Duration of bud break was significantly reduced from 7.86 days in control to 2.72 days in HC @ 30 ml/l treatment. Time taken for bud break was reduced from 19.5 days in control to 11 days with the application

@ 30 ml/l treatment. The yield/vine also increase in comparisons to control. HC treatment did not affect the fruit quality adversely.

Treatment of HC in Sonaka grapes resulted in increased bud break from 38.4 per cent in control to 92.0 per cent when applied @ 40 ml/l. Duration of bud break was significantly reduced from 7.71 days in control to 4.52 days in 20 ml/l treatment. Its application in Sonaka was effective in hastening the bud break. The days required for bud break were reduced from 18.61 in control to 12.64 in 30 ml/l treatment. The yield increased to 6.95 kg/vine in 30 ml/l treatment compared to 3.74 kg in control. Berry quality was not affected adversely with its treatments.The above observation on different cultivars of Vitis vinifera provides the effectiveness of HC in inducing uniform bud break which is preliminary requirement for enhancing productivity of grapes.

To understand the reproductive development occurs during bud burst induce due to the effect of rest-breaking treatments with HC and pruning practices, the effect on xylem sap and bud Zeatinriboside (ZR) content was determined in 'Sultanina'during late winter by Lombard et al. (2006). Bud ZR levels of HC treated 'Sultanina' canes showed an earlier and increased cytokinin peak, in particular in the distal buds. Large ZR peaks before bud burst and noticeable differences between distal and proximal buds were absent on non-sprayed canes. The xylary ZR content of long pruned 'Sultanina' canes showed a significantly higher and earlier cytokinin peak than in short pruned spurs. By contrast, the bud ZR content tended to be higher in spurs than canes. Pruning and rest-breaking treatments enhanced budburst and increased endogenous cytokinins in these experiments. Long pruning in 'Sultanina' is necessary for acceptable fruit set and yields. Possibly, in longer canes more cytokinin is available for reproductive development.

Various methods like spraying; dipping, swabbing and dropping the droplets of HC on the bud were used. No significant differences were observed among different methods of application of HC for duration and uniformity of bud break. However, application of it through spraying although consumes more spray solution but requires less time, thus reduces the cost of the labour. Terminal residue of hydrogen cyanamide in berries was also studied to understand the efficacy of treatment (Banerjee et al., 2000).

After analysing efficacy of HC in inducing bud break, the potential of this chemical and its associated biochemical changes were examined in field experiment conducted on four-year old Thompson Seedless vines grafted on Dog-Ridge rootstock (Ramteke et al., 2010). Manual swab application of commercial formulation containing 50 per cent HC at a concentration of 1.5 per cent was applied and rest were untreated. Results show that metabolic activity during sprouting involves utilization of sugar, protein synthesis and high peroxidase activity. The metabolic activities are temperature and time dependent. After pruning, normally 1.5 per cent concentration is sufficient when the day/night temperature is 30/15 °C. If temperature is lower than 30-35 °C then the concentration needs to be increased to 2 per cent for effective results. The concentration of HC also depends on the thickness of the cane. Concluding statements includes the use of HC treatment to induce early and uniform bud break in grafted Thompson Seedless grapes and continuous use of it for the last 3 years did not have any deleterious effect on vine growth, productivity or quality of fruits. The treatments also did not result in any residues in grapes at harvest. While, thiourea @ 6 per cent level resulted in higher filage compared to HC at 2 per cent level.

The mechanism of action of HC to induce the breakage of endodormancy (ED) in grape was explained by Francisco J. Pérez *et al.* (2007). They observed that applications of HC to grapevine buds produce oxidative stress and transient respiratory disturbances which are related to the breakage of ED. Moreover, since the expression and activity of catalase (Cat) is inhibited by HC, enhancements in the levels of H_2O_2 have also been associated to the breakage of ED in grapevine buds. Further, they reported that increases in H_2O_2 level in HC-treated grapevine buds are due to the inhibition of Cat activity and enhancement of the respiratory activity of buds. In addition, exogenous applications of H_2O_2 partially reproduced the inducing effect of HC in the breakage of ED, thus providing further support for the hypothesis that H_2O_2 mediates the effects of HC. On the other hand, mitochondria isolated from both control and HC-treated buds respired equally well when NADH was used as a respiratory substrate, but when succinate was used as an electron donor mitochondria respiration was non-detected, suggesting that the stimulatory effect of HC on bud respiration is related to metabolic alterations leading to increase of the concentration of NADH rather than to changes in mitochondrial functionality.

Other research also examined the effect of HC on budburst and fruit maturity in nine year old vines of cv. Thompson Seedless grafted on 110R rootstock. HC @ either 2 per cent or 4 per cent was sprayed and compared with unsprayed vines. All the treatments of HC induced more uniform budburst compared with the control. The best uniformity was obtained by the earliest spray using the higher concentration of 4 per cent. It had no significant effect on yield, but a positive effect on fruit quality. A higher index of maturity was obtained by the treatment @ 4 per cent HC, 67 d before bud break. The 4 per cent HC treatment enhanced bud break and advanced fruit maturity by 17 d. (Carreno *et al.*, 1999).

Other than HC, alternative treatments also tested for inducing bud break in grapevine. Some of these treatments include wounding, hot water, dinitro-ortho-cresol (DNOC) and thiourea. Although apart from thiourea, all the treatments enhanced respiration. The HC solutions and calcium cyanamide ($CaCN_2$) suspensions at 0.25–1.25 M caused a rapid bud opening when applied to dormant intact buds of grapevine (Shulman *et al.*, 1983).

All the previous systematic research work concludes that the uniform bud break leads to increased and uniform growth in vineyard which makes future cultural practices efficiently manageable. Also higher number of shoots is additive into higher productivity.

3. Shoot Development

Under the Indian condition grapevine is pruned twice in a year (back and fruit pruning). The shoot growth and development is required for effective fruit bud differentiation after back pruning and bunch development after fruit pruning in grape. On actively growing shoots primordial are formed that develop into tendrils where as those in latent buds develop into inflorescences. The tendrils are generally regarded as vegetative appendages which provide support for climbing plant and inflorescence both arise from Anlagen- meristematic protuberances formed by terminal and axillary bud meristem (Shrinivasan and Mullins, 1976). Anlagen may be induced to form either tendril primordia or inflorescence primordia. Gibberellin causes stem elongation and leaf expansion.To promote the shoot growth in grapevine, use of nitrogenous fertilizers is common. General recommended practice for proper shoot development includes the growth restriction of shoots to 15-18 leaves after back pruning by shoottopping after 75 days of back pruning and thinning out the excess shoots to retain only one per two square feet of ground area occupied by the vine, to build strong canes. (Manual on Good Agricultural Marketing Practices for Grapes, 2012). However the application of Gibberellins @ 10-15 ppm at proper stage after fruit pruning helps in better way for shoot development.

Growth retardants to regulate shoot vigour at the time of fruit bud differentiation. The growth regulators like 6-BA and Uracil activate the physiological processes by increasing cytokinin: gibberellin ratio and RNA: DNA ratio in the buds, thus helping the vine for fruit bud formation.

Excess vigour was recorded in grafted vines which has adverse effect on fruitfulness. To enhance the productivity of vine the storage of food material in canes and endogenous levels of hormones play crucial role. To assist at this point of development, controlled use of bioregulators depending on their mode of action, time of application can help in sustainable production under various macroclimatic conditions. The key parameters taken in consideration for evaluating efficacy of bioregulator on growth reduction in grapevines include total shoot length, cane diameter, internodal distance, number of leaves, mean leaf area and pruning weight. In case of Thompson Seedless grafted on Dog-ridge rootstock, initial shoot growth is very slow particularly from bud break (10 days after pruning) to 5 leaf stage (15 days from bud break); the growth rate shoots up and continues for a longer period (about 90 days). During this period growth retardants like Cycocel (CCC), ethephon and propiconazole (tilt) are frequently used to check the excess shoot vigour.

CCC reduces the vegetative growth with reduction in internodal distance which aids in better source-sink relationship. Also increases endogenous cytokinines level which is essential for the formation of inflorescence primordial. Management of vine with pinching/shoot topping at 3, 5 and 7th leaf stage along with CCC application allows the better growth and canopy development. Foliar application of CCC @ 500 ppm for 2-3 times depending on the intensity of vine growth is helpful. The highest weight of bunch, length of berry, diameter of berry and weight of berry were recorded in cane density 30 per vine, leaf density 16 per cane with application of cycocel 500 ppm at five leaf stages after October pruning (Ramteke *et al.*, 2005).

Holzapfel *et al.* (2012) conducted study across four consecutive seasons to understand the extent which carbohydrate reserve accumulation could be altered in Shiraz grapevines using existing, or variations of existing, cultural practices. The results suggested that developmental stage or seasonal climatic factors had a stronger influence on carbohydrate mobilization and storage, and that the manipulation of reserve accumulation through cultural practice change may be more suited to effecting longer-term adjustments in vine productivity than short-term responses to seasonal yield fluctuations.

To assist proper shoot development, cultural practices such as tipping was carried out to decrease apical dominance. Hence, along with CCC application cultural practices such as topping and removal of side shoot at different growth stages were carried to reduce the shoot vigour during the growth season and promote fruitfulness of buds and subsequent yield of Tas-A-Ganesh grafted on Dogridge. Observations were recorded with same key parameters for evaluating growth reduction and observed that shoot length was reduced significantly with CCC applications at 5+5+3+3 leaf stages. On second consecutive year Cycocel sprays did not influence vigour as indicated by the weight of pruning in October. However, mean shoot length was reduced significantly from 198.70 in control to 155.13 when cycocel was applied at 5+5+3+3 leaf stage with topping and side shoot removed. Under the same treatments, quality and yield parameters were evaluated by recording bunch weight, berry diameter, berry length, yield per vine and TSS and Acidity. These parameters tested the bioefficacy of bioregulator in quality and yield of grapes and found significantly effective on improving yield/vine. Concluding results are four sprays of CCC at 5+5+3+3 stage coupled with shoot topping after 18 leaves during the growth season (April-October) were effective in increasing the yield and quality of Tas-A-Ganesh grapes grafted on Dogridge rootstock.

The bioefficacy and phyto-toxicity of Chlormequat chloride (Lihocin 50 per cent SL) in Thompson Seedless grapes showed significant differences in treatments with respect to vegetative parameters like cane diameter, inter-nodal distance, shoot length, leaf area and pruning weight after 90 days of foundation pruning. The parameters studied after harvesting like bunchweight, pedicel thickness, skin thickness, TSS and acidity also showed significant differences with varying concentration of CCC applied. The highest cane diameter, leaf area and reduced inter-nodal length was recorded with treatment of single application of CCC@500g a.i/ha at 3-5 leaf stage after foundation pruning. No symptoms of any abnormality, toxicity were found either on the leaves, canes/shoots or berries when CCC was applied up to 1000g a.i/ha. The residue analysis of leaf, petiole and cane samples was also done at harvesting stage and found below MRL in most of the samples.

Dissipation of CCC residues in grapevine is important issue hence study was conducted in grower's fields. The quantity of CCC used from 2100 to 9600 ml/ha. Samples of different growth stages and plant parts (Stage 1: Cane/Stem, Stage 2: Leaf, Stage 3: Flowering, Stage 4: Veraison, Stage 5: Harvest) were collected and analyzed for CCC residues. The residues of CCC were traced at levels above the EU-MRL of 0.05 mg/kg in all growth stage samples before veraison (Stage 4) in plots where no application of CCC was done after October pruning. At veraison (Stage 4), the residues of CCC dissipated to below the EU-MRL of 0.05 mg/kg. In trial plots where no CCC was applied both after foundation and forward pruning, CCC was traced in all the growth stage samples monitored, although the residues were below the MRL and in the range of 0.005-0.02 mg/kg. In the trial plot where, CCC was applied after both the foundation and forward prunings, residue was detected at above the EU-MRL in all growth stages and in the range of 0.2 mg/kg at veraison stage of grape berries, which is significantly above the EU-MRL.

Alternative formulation for increased fruitfulness in Thompson Seedless was tested. Results from includes, formulation with sea weed extract @ 2 ml/l at 15 days interval from 3 to 5 leaf stage up to 60 days found to be helpful in keeping the endogenous levels of cytokinines high leading to increased fruitfulness. Application of Tilt twice @ 100 ppm at 15 days after bud break (DABB) and 200 ppm at 45 days after shoot pinching (DASP) reduces shoot length, cane diameter and internodal length.

4. Fruit Bud Formation

Fruit bud formation is the transformation of vegetative primordium into reproductive primordium in a bud. This transformation takes place in three stages. The factors like light, cytokinins or RNA play an important role in fruitfulness. On the other hand if gibberellins and other growth promoting substances are in excess and the soil and environmental conditions are conductive for vegetative growth, the anlagen do not proliferate resulting in the formation of either tendril primordia or shoot primordia. Formation of inflorescence primordium would make the anlagen to differentiate into either tendril or shoot. Sometimes, a partly differentiated inflorescence primordium can revert back and convert into tendril primordial, which is referred to as 'Fillage". The orderly events of fruit bud formation in the grapevine involve three main steps: 1. Formation of anlagen or uncommitted primordia 2. Differentiation of anlagen to form inflorescence primordia 3. Differentiation of flowers. The growth regulators play important role in formation of fruit buds in grapes.

The fruitful buds are formed during 40-60 days after April pruning. To increase fruitfulness in buds, the moisture stress is to be imposed as well as N fertigation has to be stopped to reduce the excess vigour of the vine.

Position of fruitful buds in Tas-A-Ganesh in relation to cane thickness was determined through microscopic observations of the cane. It was observed that canes with thickness of 6-8 mm had maximum fruitful buds at 6-8 nodes, 8-10 mm canes at 9-10 nodes and more than 10 mm canes at 11-12 nodes. Irrespective of the cane thickness, subcanes had fruitful buds at first 2-3 nodes above the joint. Pruning for fruiting, therefore, should be done accordingly at the level indicated above to get maximum clusters. Warm weather at budburst favours further inflorescence differentiation, resulting in more clusters per shoot, while cool weather favours differentiation of more flowers per clusters and fewer clusters per shoot. Under the condition of late pruning when the period of fruit bud differentiation coincides with insufficient sunlight required for photosynthesis and clouds in the atmosphere, the spray of 6-BA@10 ppm and Uracil @ 50 ppm helps in increasing the cytokinin in the vine and also the higher proportion of RNA : DNA ratio for effective fruit bud differentiation in grape.

Melo *et al.* (2012) in brazil analysed effect of the growth retardants in the bud fertility rate in Thompson Seedless grafted on Hormony rootstock. Two bioregulators,

Paclobutrazol (PBZ) and Cycocel (CCC) were applied in spraying alone or in combination with two cytokinins: Benziladenina (BAP) and Thidiazuron (TDZ), and a nitrogenous base, the Uracil. Results recorded that separate use of the products had not promoted significant differences in the bud fertility, but, the association of the growth retardants with the TDZ it induces the 'Thompson Seedless' to a bigger productivity.

Fruitfulness is a combination of number of inflorescences per vine, flower number per inflorescence, fruit set and berry development. The fruitfulness in grapes depends mainly on the climatic condition during the fruit bud differentiation stage and also the cultural practices followed during period directly or indirectly via their impact on photosynthesis and nutrient availability.

Study was conducted for evaluating fruitfulness in seven table grape varieties *viz.*; Thompson Seedless, Sharad Seedless, Flame Seedless, Fantasy Seedless, Crimson Seedless, A - 17/3 and 2 A clone (Somkuwar and Ramteke, 2007). Among table grapes, maximum fruitful canes were recorded on Flame Seedless followed by Thompson Seedless, Sharad Seedless, A-17/3 and 2A Clone. Least was on Crimson Seedless and Fantasy Seedless. Significant difference in position of fruitful bud was recorded among varieties and basal bud fruitfulness (3-5th bud) was observed in Flame Seedless, Fantasy Seedless and 2 A clone. In Sharad Seedless and A - 17/3 maximum fruitfulness was recorded on 6th bud position, while onThompson Seedless bud fruitfulness was recorded at 7th and 8th bud. Weight of 50 berries was highest in Fantasy Seedless and A - 17/3 while least in Flame Seedless and Thompson Seedless. Maximum berry length was noted in Crimson Seedless, Fantasy Seedless and A - 17/3 while least was in Flame Seedless and Thompson Seedless. In case of berry diameter, TSS and acidity no significant differences were observed.

5. Rachis Elongation and Panicle Thickness

The main axis of cluster is called rachis, from which branches arise at irregular intervals and divide to form pedicle and flower stalk. The main rachis may divide to form secondary and tertiary branches in some varieties. The length of main rachis, it's branching and the number of flowers per unit length of the rachis is dependent on endogenous levels of gibberellins and shoots vigour. Over- compact clusters results in yield loss due to crushing of inner berries in the clusters at the time of fruit ripening which subsequently leads to berry rot. Compact clusters get pressed in the boxes and are bruised leading to rot during transit and storage. Manual thinning of the berries is cumbersome and expensive as compared to chemical thinning. Cluster clipping and berry thinning in Thompson Seedless help in quality improvement (Somkuwar, 2008).

In order to achieve adequate berry thinning and obtain loose clusters by chemical means, a field trial was conducted with the use of GA$_3$ for rachis elongation in Thompson Seedless grapes. The schedule for pre-bloom application of GA$_3$ for better panicle growth was standardized in Thompson Seedless grapes for export production. It includes 1st spray of GA$_3$ @ 10 ppm at parrot green stage, 3-4 days after 1st spray GA$_3$ @ 15 ppm for rachis elongation; at 50 per cent flowering stage GA$_3$ @ 40 ppm as selective spray if rachis elongation was not enough. GA$_3$ spray solution should be

Sharad Seedless Grapes   **Thompson Seedless Grapes**

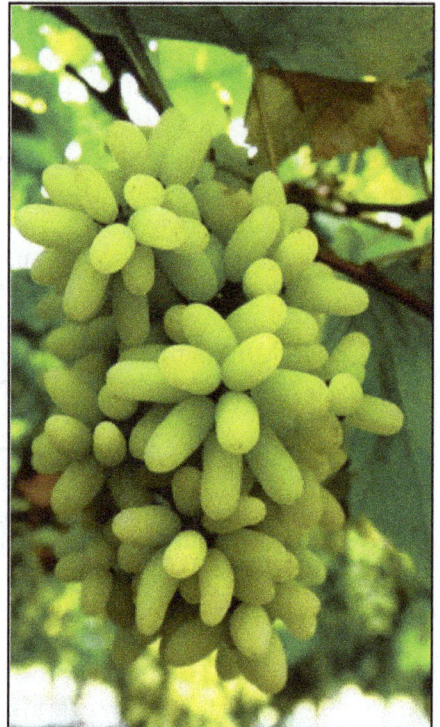

Red Globe Variety of Grapes   **Influence of GA3 on Berry Elongation
                                     of Sonaka Variety of Grapes**

Figure 5.1: Commonly Grown Grapes Varieties.

acidic (pH 5.5 - 6.5). Use citric *or* phosphoricacid *or* urea phosphate as a adjuvant to lower down the pH of spray solution. Another recommended practice for producing loose bunches includes dipping of clusters with 40 ppm GA_3 at 50 per cent flowering if necessary and cut the tips of clusters immediately after set by retaining 8-10 apical branches depending on the number of leaves available for a bunch.

Forward or Fruit Pruning   **Back Pruning**

Pre Bloom Stage

Figure 5.2: Pruning and Blooming in Grapes.

In Tas-A-Ganesh grapevines normal fruit set is high resulting in over- compact clusters. In order to achieve adequate berry thinning and obtain loose clusters by chemical means, a field trial was conducted at different concentrations and stages along with Carbaryl @ 1000 ppm a.i. at 90 per cent bloom to assess their efficacy in berry thinning. The bunch compactness ratio was assessed based on total number of berries, total length of rachis and compactness ratio = where > 4 = compact, 3-4 = well filled, <3 = loose and <2 = straggly. Although, the berry diameter was highest under GA_3 10 ppm at 5 leaf + GA_3 15 ppm at 8 leaf + GA_3 40 ppm at 30-50 per cent bloom +

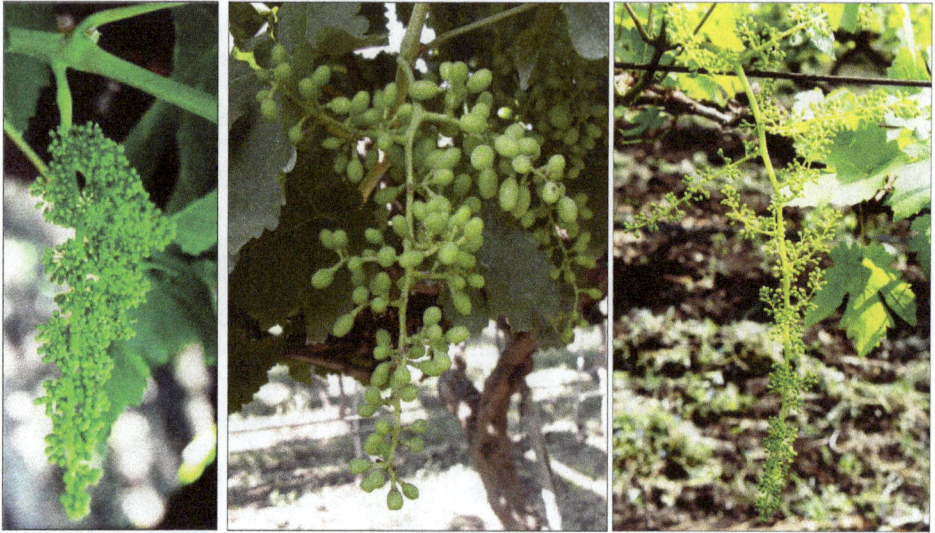

Start of Flowering Stage

Ideal Rachis Elongation from Gibberellic Acid

Figure 5.3: Flowering in Grapes.

Carbaryl 1000 ppm at 90 per cent bloom treatment. Number of berries in a bunch was reduced significantly when the clusters were treated with either GA_3 or Carbaryl at various stages of bloom. Total rachis length was increased significantly with three sprays of GA_3 at different stages on clusters (GA_3 10 ppm at 5 leaf + GA_3 15 ppm at 8 leaf + GA_3 40 ppm at 30-50 per cent bloom + Carbaryl 1000 ppm at 90 per cent bloom) as compared to one spray of GA_3 10 ppm at 5 leaf stage. As a result of increased length of rachis and reduced number of berries per bunch, the compactness ratio was the least in GA_3 10 ppm at 5 leaf + GA_3 15 ppm at 8 leaf + GA_3 40 ppm at 30-50 per cent bloom + Carbaryl 1000 ppm at 90 per cent bloom treatment (Ramteke *et al.*, 2006).

Addition of *Tally plus* to GA_3 has the potential to increase the efficacy of GA_3 in producing loose clusters and improves berry weight and berry size. Tally plus contains Binsol EX which acts as a booster and is used to increase rachis length, berry size and yield in grapes. The *Tally plus* was applied in GA_3 solution at different concentrations at prebloom stages and observations were recorded for evaluating berry characteristics and bunch compactness. Berry diameter was recorded significantly higher in all the GA_3 treatments as compared to control. The 100- berry weight was recorded significantly highest in *Tally plus* treatments as compared to control with no use of GA_3. Similar trend was found for compactness ratio, which is resultant of total number of berries in a bunch and rachis length. However, yield per vine was influenced by the application of *Tally plus* and other treatments and significantly higher yield was recorded in Tally plus treatments. It also increases the pedicel thickness, which is prerequisite for longer shelf life.

6. Berry Set

Pollination and fertilization must take place for ovule development, seed formation and development in normal seeded grapes. Berry set is also regulated by endogenous growth regulators and carbohydrate content at the time of fruit set.

Gibberellins are known for cell enlargement and berry expansion. The application of GA_3 become important to increase the berry set. Therefore the berry number decreases while their weight and volume increase with increase in GA_3 concentration. The effect of GA_3 on fruit set varies with the concentration, stage of development and also with the cultivar. With the seedless cultivar, the GA_3 reduces the number of seeded berries but increases the number of seedless berries.

Studies on Thompson Seedless showed that treatment of flower clusters with GA_3 @ 10 ppm at 5 leaf stage followed by 15 ppm at 8 leaf stage, 40 ppm at 30-50 per cent bloom and Carbaryl spray @ 1000 ppm at 90 per cent bloom produced loose clusters, eliminating manual thinning. Inhibition of natural fruit set by GA_3 application is attributed to its pollenicidial effect. A similar study includes GA_3 @ 30 ppm and 25 per cent berry thinning was found to be the best treatment for improving the fruit quality of Thompson Seedless grape (Singh, 1992).

Effect of CCC on floral bud drop in Thompson Seedless grape was studied in Punjab, India by Kumar *et al.* (1990). The application of CCC at 250 ppm produced the lowest percentage of floral bud drop, highest bunch weight and a yield in treated grapevines in comparison with control vines.

7. Berry Growth and Development

The grape berry is simple fruit consisting of two seed cavities surrounded by ovary wall called as pericarp. In seeded varieties there may be as many as four seeds. In case of stenospermocarpic varieties, the locules contain seed traces resulting from the abortion of the ovules early in their development. The fleshy pericarp consists of an exocarp called as skin and a mesocarp, the pulp. In most seedless varieties, the mesocarp accounts for 85-90 per cent of berry's fresh weight. Increase in berry weight, volume or diameter during development are typically characterized by a double sigmoid curve resulting from two consecutive stages of growth separated by a phase of slow or no growth. The berry development is found in three different stages, they are as follows.

Stage I: A period of rapid berry growth comes immediately after bloom. During this time, berries grow both through cell division and cell enlargement. Berry texture is firm, while its color is green due to the presence of chlorophyll. The sugar content of the berry remains low, while organic acids accumulate. This stage lasts between 3 and 4 weeks in table grapes.

Stage II: Berry growth slows markedly during the second period while the berries organic acid concentration reaches its highest level. The stage is called as lag phase. At this stage, the berries remain 2 and 3 weeks depending on climatic conditions of that region and grapevine variety.

Stage III: In this stage of berry development, the initiation of ripening commences in beginning. The berry starts softening which characterizes the initial stages of color development. Berries soften and lose chlorophyll, while in colored varieties red pigments begin to accumulate in the skin. Sugar also begins to accumulate and the concentration of organic acids decreases. Aroma and flavour components accumulate in the fruit.

To produce export quality grapes, bunch and berry characteristics taken into considerations are as follows: The bunch should weight 400-500 g and uniform in green color. The berries in the bunch should be 3.5 to 5.0 g and diameter more than 18 mm. The berries in a bunch should contain 18°Brix at the time of harvest. The bunch should be well filled, free of scars, pest and diseases and bunch should have maximum shelf life.

Growth regulators play an important role in berry development. To improve the berry size, the basic requirement is to improve source-sink relationship. The carbohydrates synthesized in leaves/stored in canes were mobilized through endogenous levels of growth hormones particularly auxin and help in accumulating the sugar in developing berries. Factors influencing the uneven berry size in bunches include the number of berries in a bunch, leaf area available per bunch, the size of the mother cane and effective coverage of berries on all sides of the cluster by the spray solution. Various chemicals were routinely used to improve the bunch and berry characters which ultimately lead to improved productivity and quality.

Berry size of seedless grapes is generally increased by application of GA_3 sprays at the time of fruit set. For certain cultivars a number of applications are required to obtain a commercially acceptable sized berry and this may have deleterious effects on fruit bud initiation for the following year.

GA_3 sprays are used at bloom to tin berries and increase berry size. GA_3 practices for increase in length between two rachis, increase in length of individual rachis, increase in length of bunch asa whole so as to achieve the cylindrical shaped bunches as preferred in the international market and ultimately increase the berry diameter above 18 mm. The effect of GA_3 on increasing berry weight in grapes is well known. Studies showed that generally seedless varieties responded better than seeded ones and optimal concentration of GA varied with the stage and variety. Endogenous concentration of GA_3 is inadequate in seedless berries hence exogenous application can increase the berry weight 100-150 per cent. To achieve uniform size berries in bunch practices such as the spray application ensuring that all berries in a cluster receive all GA_3 treatments uniformly and adequate leaf/fruit ratio for a developing bunch (6-8 berries/leaf) is necessary.

Bio-efficacy of various formulations of GA_3 in enhancing berry size and quality was assessed by studying berry and bunch characteristics of grapevine. Results from such formulations were cumulatively presented in Table 5.2. Efficacy of GA_3 could be improved by using adjuvant like phosphoric acid and urea phosphate in acidic range of pH 3.0 - 6.0.

GA_3 spraying, during the three different periods of the vine growing: before blooming, after blooming and before veraison influence on the technological

Table 5.2: Different Formulations of GA$_3$ and its Effect in Improving Berry Characteristics

Products	Active Ingradient	Dose and Methods of Applications	Observed Effects
GA$_3$ (China make)	Powder of 90 per cent Gibberellic acid	10-50 ppm at different phenological stages	Increase the berry size and quality of grapes
Earth care GA$_3$ Plus Seasol spray adjuvants	Liquid GA$_3$	Applied at pre-bloom (10 ppm), 40 per cent capfall (20 ppm), 80 per cent capfall (20 ppm) as a foliar spray and at 3 mm and 6mm berry size stage as a dip @ 60 ppm.	Increasing the berry size, yield and shelf life.
Progibb	Powder form of GA$_3$ with 40 per cent active ingredient	Prebloom to 6-7 mm berry size stage in variable doses along with GA$_3$ technical grade, tricontanol 0.05 per cent EC	Increase in bunch weight, berry weight, berry size, pedicel thickness, berry crispness, berry thickness, quality and yield parameters.
GALA	Gibberellic acid 100 g/litre along with ethyl oleate and surfactants	Full dose	Bunch compactness was reduced in Thompson Seedless grapes. Total no. of berries/bunch reduced to the extent of 25.32 per cent.
GA$_3$	Gibberellic acid (20 per cent)	Applied in different concentrations at different phenological stages	Increase in bunch weight, berry size, quality and yield parameters

characteristics of the Thompson and Belgrade berries in all grape growing periods. The addition of GA_3@ 20 mg/L increased the weight of the cluster and berry and increased the transportability of the berries belonging to the two seedless varieties (Dimovska *et al.*, 2011).

Other widely used cytokinin like chemical is forchlorfenuronor CPPU (N (2-Chloro-4-pyridinyl)-N-phenyl urea). Mode of action of CPPU is not well understood but it is known to absorb by leaves, stem, cotyledon and germinated seeds. It promotes cell division, differentiation and development; induces budding of callus, and controls apical dominance; breaks dormancy of lateral buds and promotes germination; delays ageing process and maintains chlorophyll in excised leaves; regulates the transport of nutrients; promotes fruit formation, etc. the bioavailability of CPPU is 10-100 times of that of 6BA. Selection of growth substance in relation to their mode of action and stage of application is very crucial decision in increasing berry characteristics.

The experiment was conducted to study CPPU influence on various characters of Thompson Seedless grape. The highest bunch weight was obtained with the application of 2 ppm CPPU + 30 ppm GA_3 at 3-4 mm and 1 ppm CPPU and 40 ppm GA_3 at 6-7 mm berry size stage when 90-100 berries were retained in a bunch as compared to other treatments. The highest 50 - berry weight was recorded with the application of 3 ppm CPPU + 20 ppm GA_3 at 6-7 mm berry size with retention of 50-60 berries per bunch. The berry size was recorded highest with the application of 3 ppm CPPU + 20 ppm GA_3 at 3-4 mm and 2 ppm CPPU + 30 ppm GA_3 with retaining 70-80 berries per bunch. Irrespective of number of berries retained in the bunch, the pedicel thickness was more in CPPU and GA_3 treatments than GA_3 alone. Similar trend was observed for shelf life also. To achieve berry quality by using CPPU the number of leaves available per bunch (overall leaf area of the vine) should be at least 10 or more above the bunch (Ramteke, 2008). No any abnormalities due to application of CPPU were found in any of the treatments. Application of CPPU on bunch having more than 10 leaves resulted in thick pedicel and higher berry diameter. Similar work with GAs-treatment alone determined a significant increase in berry size and bunch weight as compared with GA_3-treatment. Inclusion of CPPU results in an additional increase in berry growth when added both to GA_3 and GAs, with a reduction in berry drop. Increases in berry growth results in retarded soluble solids accumulation and later harvest (Retamales *et al.*, 1995, 1998).

In Chile, vineyards over 54,600 ha of table grapes (*Vitis vinifera*), mainly cvs. Thompson Seedless, Flame Seedless and Red Globe, are planted. Almost the entire production is exported to the USA, Europe, Asia, or one of several Latin American countries, which typically requires 15–40 d of maritime transportation. During this period, several physical, physiological, and pathological factors cause table grape deterioration. Because berry size is the main quality factor in international markets, farmers often overuse the growth regulators, GA_3 and CPPU in an effort to increase berry size. The effect of preharvest growth regulators on seedless (Thompson Seedless and Ruby Seedless) and seeded (Red Globe) table grape cultivars during cold (0°C) storage plus a shelf life period of 3d at 20°C was examined. The overuse of GA_3, eight instead of two GA_3 applications on Thompson Seedless, and the use of one GA_3 application on Red Globe and Ruby Seedless, increased berry pedicel thickness and lowered cuticle content but induced shatter and predisposed grapes to gray mold caused by *Botrytis cinerea*. In contrast, CPPU increased berry pedicel thickness and cuticle content but did not increase shatter or gray mold incidence. Clusters that were subjected to overuse of combined GA_3 and CPPU were highly sensitive to shatter, had the thickest pedicel, and developed a high gray mold incidence during cold storage. Hairline, a fine cracking developed during cold storage, was induced on 'Thompson Seedless' and 'Ruby Seedless' by growth regulators, but no hairline occurred on 'Red globe' table grapes. Therefore, Zoffoli *et al.* (2009) concluded that berry quality during cold storage is greatly influenced by growth regulator management in the vineyard. Other studies in Chile showed that replacing GA_3 by Promalin has a similar effect on berry growth, but reduces berry drop. Application of GA_4 also resulted in less berry drop than GA_3 but induced less berry growth (Cooper *et al.*, 1993).

Comparable increase in berry size can be obtained by a single application of CPPU to Perlette, Superior and Thompson Seedless cultivars as with GA_3, fruit maturation ripening and were delayed following CPPU application, sometimes even more severely than with GA_3 (BenArie *et al.*, 1998). They also found that other aspects of development of the cluster were affected differently by each growth regulator. Such aspects include the growth of the rachis and pedicels, berry shatter after harvest and susceptibility of the berry to fungal attack and decay. These differences may be attributed to the anatomical effects of the regulators, in that CPPU increased the number and density of cells, whereas GA_3 enhanced cellular expansion and decreased cell density.

While on adverse side recently in 2014 it was noticed that berries treated by CPPU may be more astringent and it was therefore hypothesized that CPPU-treated berries produce more tannins or delay tannin decomposition during ripening. CPPU was applied to Thompson Seedless grapes (*Vitis vinifera* L.) after fruit set in three different vineyards and over three seasons. As previously reported CPPU affected berry size and delayed maturation. Application of CPPU at 6 mm berry diameter (early treatment) had greater influence on increasing berry size, while application at 10 mm berry diameter (late treatment) had greater influence on delaying berry maturation. Berry ripening was also measured by following changes in berry autofluorescence, and it was shown that CPPU caused higher chlorophyll-related autofluorescence throughout the ripening period. The CPPU treatments consistently elevated the level of condensed tannins with up to a four-fold increase as measured by the protein precipitation assay and total phenol content. The level of total tannins in either control or CPPU-treated fruit did not seem to decrease during ripening. These results were in agreement with a sensory assessment in which CPPU-treated berries were more astringent than untreated berries.

Parallel to these observation experiments conducted in Chile to understand the effect of overuse of growth regulators, GA_3 and CPPU, in an effort to increase berry size. (Juan Pablo Zoffoli *et al.*, 2009). Overuse of GA_3, eight instead of two GA_3 applications on Thompson Seedless, increased berry pedicel thickness and lowered cuticle content but induced shatter and predisposed grapes to gray mold caused by *Botrytis cinerea*. In contrast, CPPU increased berry pedicel thickness and cuticle content but did not increase shatter or gray mold incidence. Clusters that were subjected to overuse of combined GA_3 and CPPU were highly sensitive to shatter, had the thickest pedicel, and developed a high gray mold incidence during cold storage. Hence they concluded that, berry quality during cold storage is greatly influenced by growth regulator management in the vineyard.

Equivalent increase in berry size can be obtained by a single application of CPPU to Perlette, Superior and Thompson Seedless cultivars. As with GA_3, fruit ripening and maturation were delayed following CPPU application, sometimes even more severely than with GA_3. However, other aspects of development of the cluster were affected differently by each growth regulator. Such aspects include the growth of the rachis and pedicels, berry shatter after harvest and susceptibility of the berry to fungal attack and decay. These differences may be attributed to the anatomical effects of the regulators, in that CPPU increased the number and density of cells, whereas GA_3 enhanced cellular expansion and decreased cell density.

The quantification of the abscisic acid and its metabolites and a range of cytokinins and gibberellic acid in *Vitis vinifera* L. cv. Shiraz, berry pericarp at developmental series were examined by Böttcher *et al.* (2013). Results showed some relatively minor differences in concentrations for some abscisic acid metabolites and gibberellic acids. The biologically active gibberellic acids 4 and 7 (along with precursor and inactive forms) were present before veraison but at very low concentrations after veraison. Abscisic acid levels peaked shortly after veraison, but high levels of dihydrophaseic acid before veraison and high levels of abscisic acid glucose ester after veraison suggest that two different catabolic pathways are developmentally controlled. The analysis of the concentrations of certain cytokinins provided some previously unreported and surprising results, as the concentration of isopentenyladenine increased rapidly at veraison and remained at elevated levels throughout ripening. The reason for this phenomenon is unknown, but a similar increase in the concentrations of some cytokinins has been observed during kiwifruit development and may be involved in the control of ripening and/or senescence.

The spray or dip of GA_3 along with other bioregulators was also practiced in vineyards (Table 5.3). GA_3 with 6-BA and CPPU is generally followed to achieve bunch elongation at pre-bloom stage and also for increasing the berry size.

Table 5.3: Combination of Bioregulators Used to Improve Quality of Table Grapes

Combination of Bioregulators	Observed Effects
2 ppm CPPU + 35 ppm GA_3 at 2-3 mm berry size stage and 1 ppm combine (Homobrassinolide) + 50 ppm GA_3 at 5-6 mm berry size stage (Ramteke *et al.*, 2005)	Export quality grapes in Thompson Seedless grapes
2 ppm CPPU + 30 ppm GA_3 at 3-4 mm and 1 ppm CPPU and 40 ppm GA_3 at 6-7 mm berry size stage when 90-100 berries were retained in a bunch	Highest mean bunch weight in Thompson seedless grapes

Considering all the factors and research studies resulted in recommended schedule for commercial practices in India to spray GA_3 on Thompson Seedless to increase the berry growth. For export quality grapes, recommended 1st spray at 3-4 mm berry size with 30 ppm GA_3 + 2 ppm CPPU for berry elongation and 2nd spray at 6-7 mm berry size with 40 ppm GA_3 + 1 ppm CPPU. For local market another schedule can also be followed: 1st spray at 3-4 mm berry size with 40-50 ppm GA_3 + 10 ppm 6-BA which induces berry length and 2nd spray at 40 ppm GA_3 at 6-7 mm berry size. Spray given after the shatter stage nourishes the clusters and the bold berries formed.

Now a day, Sharad Seedless is gaining importance as an export variety particularly in Arab countries. The quality of Sharad Seedless in terms of berry size, weight and diameter could be improved by the application of GA_3 @ 40 and 30 ppm along with 10 ppm 6BA at 3-4 mm and 6-7 mm berry size stage, respectively (Ramteke *et al.*, 2008).

Use of bioregulators is not limited by our studies, it was extensively studied by different researchers worldwide. Some relevant supported research work is summarised in the following section: Effect of cytokinins and brassinosteroid with GA_3 for yield and quality attributes and to standardize the application of these plant growth regulators in Thompson Seedless vineyards at Rahuri, indicated that the application of GA_3 in combination with brassinosteroid and 6-BA was found effective for cell elongation and cell division, which leads to increase berry size, yield and quality of Thompson Seedless grapes (Warusavitharana *et al.*, 2008). Results was found better for application of GA_3 at 10 ppm at pre-bloom stage, 15 ppm at initiation of flowering, 25 ppm at 3 to 4 mm berry diameter stage in combination with 1 ppm brassinosteroid, benzyladenine at 10 ppm twice at 4 leaf and berry setting stage, respectively and last application of GA_3 at 25 ppm at 7 to 8 mm berry diameter stage was found better for improving berry and yield characteristics. Analogous results were recorded in Ahmednagar vineyards, with the pruning for fruiting done in third week of October with application of CPPU at 2 ppm at 3–4 mm berry diameter along with application of recommended plant growth regulators (Salunkhe *et al.*, 2008).

To achieve uniform size and colour of berries in bunch, general recommended practices from Manual on good agricultural marketing practices for grapes include:

1. Orientation the rows in North-South direction when trained to flat roof gable system
2. Irrespective of the size of the leaf, retain minimum 10 leaves above the cluster
3. Orient the bearing shoots horizontally or diagonally on a slanting curtain
4. Angle of the curtain should be 40°-45° to the vertical plane
5. Position the side shoots to cover the bunches borne on short shoots

The use of CPPU at 3-4 mm and 6-7 mm berry size stages may also be helpful to retain the green colour of berries at harvest. Shade net are used and also the individual bunches are covered with paper bags to retain the uniform colour of the berries. The excess use of these bioregulators some time may lead to abnormalities in berries. To standardize the dose for export of grapes, certain quality standards have to be followed, a minimum 18 mm berry diameter is one of the essential requirements in many markets. To achieve this, grape growers are using higher doses of GA_3 and CPPU bioregulators. To standardize the dose based on the actual requirements of bioregulators, and its dissipation studies were carried out. Gibberellic acid (GA_3), Auxins (IAA) and Cytokinin (6-BA) were analysed from the grape varieties *viz.*; Red Globe, Fantasy

Seedless, Rizamat, Perelette, A-17/3 and Thompson Seedless at berry set, berry development and maturity stages. Using rapid method for determination of 67 pesticides residues in foods by LC-TOF/MS Agilent technology (WP06113 - ASMS 2006). Samples were collected at pre-bloom stage, flowering, berry setting, 3-4 mm size, 6-7 mm size, veraison and harvesting. The results showed a variation in endogenous levels of hormones in different varieties. Ugare *et al.* (2013) supported the use of recommended concentrations of bioregulators with residue dissipation studies. PGR application with recommended and double dose was analysed and results showed non-linear two-compartment first + first-order kinetics for all the PGRs. CPPU, 6-BA and ethephon residues dissipated with preharvest intervals (PHIs) of 33.5, 12 and 32 days at recommended dose with no PHI applicable for GA_3. The PHIs successfully minimised residue problems as observed from survey results of traceable field samples (Oulkar *et al.*, 2011).

To increase the sugar content in berries the following practices are recommended:

1. Train the shoot in such a way that all leaves are exposed to sunlight and the vine canopy is well illuminated and ventilated

2. Girdle the vines at 3-4 mm size of berries

3. Remove the weak canes of < 6 mm diameter at forward pruning

4. Ensure adequate leaf area per bunch

5. Encourage the shoot growth up to berry setting to achieve proper leaf fruit ratio

With recommended practices, formulations other than PGR used as bio stimulants in vineyards also enhance the quality of grapes. The various formulations tested in NRCG vineyards and results are summarized in Table 5.4.

8. Shelf Life

By considering perishable nature of grapes, along with its production, keeping quality of grapes is also one of the important parameters which affects the market cost. Generally every chemical used as pre-harvest growth regulators, their concentration and time of application, coverage was examined for their effect on keeping quality during cold (0°C) storage plus a shelf life. Berry quality during cold storage is greatly influenced by growth regulator management in the vineyard. Judicious uses of bioregulators are important in enhancing cold storage and hence export of grapes. Detail studies include the examples from every study conducted worldwide.

Shelf life depends on berry attachment (Pedicel thickness), skin thickness, berry crispness, berry freshness and firmness which are assessed by physiological loss of weight (PLW). Increase in the shelf life of grapes is possible by minimizing PLW, and reduction in fallen berries. Observations were also made every day in shelf on PLW per cent fallen berries, per cent rotten berries and pedicel browning after 30 days of cold storage.

Pre-harvest foliar application of putrescine (Put), GA_3, ascorbic acid (AA), ethephon (Eth), salicylic acid (SA), cytofex (CPPU) and calcium chloride ($CaCl_2$) at

Table 5.4: Effect of Biostimulant on Grape Quality Parameters

Products	Active Ingredients	Dose and Methods of Application	Observed Effects
Plantozyme	Biological derivative from sea algae and herbs	Foliar spray @ 2.5 ml/l and soil application for granules @ 25 g/vine	Increased berry length and diameter, average bunch weight, TSS and yield of Thompson Seedless grapes.
Milagro L	Biostimulant-extract from sugarcane flowers	Foliar spray @ 30 mL/100 L	Higher bunch weight, the berry weight and diameter.
Biopower and Bioforce	Mixture of amino acids, plant growth promoting triterpenoid, siderophore and attenuated bacteria fortified with BGA	Granules @ 25 kg/ha along with recommended dose of fertilizers, 50 per cent bioforce and 50 per cent regular bio-regulators at different phenological stages (as per package of practices)	Yield vine was significantly higher in Thompson Seedless grapes.
Silixol	Formulation of stabilized, highly concentrated silicic acid	Spray @ 1.0 L/ha in four splits at 15, 30, 45 and 60 days after October pruning.	Increase in berry weight and berry crispness.
Quantum (N-Acetyl Thiozolidine-4-Carboxylic Acid)	Bio-stimulant	Foliar spray @ 0.5 ml/l at 45 and 75 days after pruning and dipping once at 102 days	Increasing the yield and berry characters in Tas-A-Ganesh grafted on 1613 rootstock.
Aura XL	Algal extract protein hydrolysate and chelated micronutrients	Spray @ 2-3 g/L	Increased leaf chlorophyll, yield quality of Thompson Seedless grapes.
Cabrio Top 60 WG	Metiram 55 per cent and 5 per cent pyraclostrobin	1000-3000 g/ha as a spray	Increased leaf thickness, yield and quality of Thompson Seedless grapes.
Cynoboost	Tetraethyl rhodamine, natural β hydroxytricarboxylic acid mixed with surfactants	0.5 ml/l as a swabbing and pruning	Uniform sprouting in Thompson Seedless grapes.
Elanta Super	Derivative of L-cysteine, folic acid with adjuvants	0.5 ml to 2.0 ml/l as a spray	Increased berry length, diameter, average bunch weight, TSS and yield/vine in Sonaka grapes.

two stages of berry development; pea stage (4-5 mm fruitlet diameter, <"30-35 days after fruit set) and veraison stage (when approximately 20 per cent of the berries on 50 per cent of the clusters had softened) in order to investigate their influence on yield and postharvest fruit quality characters at commercial harvest day as well as the berry keeping quality in Thompson Seedless grapevine was studied by Marzouk and Kassem (2011). Cluster and berry quality characters as well as vine yield were improved by all sprayed chemicals, especially Put, GA_3, SA, CPPU and $CaCl_2$. GA_3 increased cluster and berry width, and resulted in better clusters shape. A positive increase in berry firmness was obtained by Put, GA_3, CPPU, SA and $CaCl_2$ sprays. Berry adherence strength increased and the percentage of unmarketable berries decreased by all sprayed compounds except Eth. Shelf life (keeping quality) was increased by spraying Put, GA_3, SA, CPPU and $CaCl_2$, as they increased berries firmness and decreased the percentage of unmarketable berries after keeping at ambient temperature for seven days after harvest.

Grapes detachment of berries is generally due to weak and poor attachment to the pedicel. Detachment of their pedicel could be of two types viz; i. Wet drop and ii. Dry drop. In wet drop brush being pulled out from the berry. In case of dry drop the abscission forms at attachment point of berry to the pedicel. Hence to reduce the post-harvest berry drop one has to control and implement the good agricultural practices. The factor responsible for dry drop is mainly attributed to the synthesis of high level of endogenous ABA and other growth inhibitors like ethrel. Soil moisture stress in hot dry weather just before harvest increases the dry drop by increasing higher levels of endogenous ABA and on the contrary, lower level of auxins or growth promoters are responsible for dry drop. NAA application at 8-10 days prior to harvest reduced the post harvest berry drop appreciably.

If the pedicels are thick, it will take some time to dry, which is also responsible for dry drop. The use of growth regulators like GA_3 and CPPU at 3-4 and 6-7 mm berry size stage helps in increasing the pedicel thickness and increases shelf life.

Field experiment was conducted to study the effect of growth regulator to increase the pedicel thickness of Tas-A-Ganesh grapes grafted on 1613 C rootstock for two seasons. CPPU @ 2 ppm applied only at veraison was found to be effective in reducing fallen berries. The same treatment or 6 BA @ 10 ppm either at veraison only or at 6 mm berry size and veraison showed promise in reducing the percentage of fallen berries in shelf. Pedicel browning reduced significantly in all the treatments (Ramteke et al., 2002).

In general, the application of either 4-CPA @ 10 ppm or 6 BA @ 10 ppm or 2 ppm CPPU twice, found to be effective in reducing the pedicel browning up to 5th day in shelf. Application of CPPU @ 2 ppm at 6 mm berry size and also at veraison stage was found more effective in reducing the PLW in the shelf.

Pre-harvest application of CPA, CPPU and their effect on shelf life of Thompson Seedless grapes and result described that the dipping the clusters in 10 ppm 4-chlorophenoxy acetic acid (4-CPA) or 2 ppm of CPPU at 6 mm size of berries and at berry softening (veraison) in addition to the normal practice of treating with GA_3 @ 25 ppm at 3 mm and 6 mm size of berries increased the pedicel thickness and retained the freshness of grapes and reduced berry drop after 30 days of cold storage. Similarly usefulness of pre-harvest treatment with 1 per cent calcium nitrate solution on 75+ 90+ 105th days after pruning appears to be beneficial in retaining the freshness of Thompson Seedless and reducing their rotting after 30 days of cold storage (Somkuwar *et al.*, 2008).

Skin thicknesses and berry crispness are next important parameters affecting consumer acceptability. Thick skin of the berries reduces the loss of the water and keeps the berries fresh for longer time. The thickness of skin can be increased by using GA_3 and CPPU in combination at 3-4 mm and again at 6-7 mm berry size stage. Berry crispness is depends on pulp content in the berries. If pulp content is more the shelf life will be more.

Flesh firmness of grape berries determines its eating quality. It is important as cracked berries lead to fungal decay. The cell wall structure appears to play a role in the firmness of table grapes. Calcium determines the structure of cell wall. Calcium nitrate and calcium chloride increase the shelf life of the berries.

Calcium application to bunches and calcium fertilization as well as GA_3 and CPPU application is used to enhance berry firmness. When high concentration of calcium is maintained in the fruit tissue, during its development, the process of fruit ripening slows down. The softening of berries will also slow down. Application of GA_3 enhances the division and expansion of the parenchyma cell in the pericarp of berry. With the application of CPPU in grape berries, cell division takes place and also increase in shelf life is achieved. Berry attachment to the pedicel and its attachment to the rachis accounts for berry adherence. Shrivelling and loss of freshness in grapes is due to loss of water from the berries during storage and transit. Grapes shrivel with 1 to 2 per cent loss of water.

Somkuwar *et al.* (2008) studied the effect of rootstocks and pre-harvest treatments on storage life of Thompson Seedless grapes. A trial was conducted on seven-year-old Thompson Seedless grapes grafted on four rootstocks, *viz.*, Dogridge, Salt Creek, 1613 C and St. George. Bunches were dipped in 1.0 per cent calcium nitrate solution. Among the different rootstocks, Dogridge rootstock was found better for minimizing PLW in Thompson Seedless compared to other rootstocks. On the 3rd day under shelf, minimum PLW of 6.26 per cent was recorded when the berries were treated with calcium nitrate compared to untreated control. On the same day under shelf, fallen berries were minimum on Dogridge rootstock in calcium nitrate treated compared to untreated grapes. Maximum berry fall was recorded in Thompson Seedless grafted on St. George rootstock. The percentage of rotten berries also increased with increased duration of storage at room temperature. The treatment differences were apparent from the first day itself. The percentage of rotten berries exceeded 10 on the fifth day in treatments with calcium chloride. Rotting was more with calcium chloride treatments as compared to the treatments with calcium nitrate or control. The least rotting was in treatment with calcium nitrate on the 75+90+105 days after pruning. The pedicel browning was more on St. George rootstock having shrivelled berries leading to the reduced shelf life.

Flesh firmness is a sensory characteristic proposed by Organisation Internationale de la Vigneet du Vin (OIV) as an ampelographic descriptor for grape varieties and *Vitis* species (OIV code 235). To accurately define the flesh firmness of table grapes, an instrumental texture parameter that makes it possible to classify the five reference table-grape cultivars according to this descriptor was selected. The mechanical properties of the berry flesh were determined by texture profile analysis and cutting tests. The results showed that the berry hardness and gumminess, normalized by berry diameter, are accurate instrumental indicators of flesh firmness because the variations among the cultivars were evident, independently of berry size and maturity grade. The interquartile ranges of berry hardness normalized by berry diameter for each flesh-firmness group established by OIV (soft, slightly firm, very firm) were 0.074 to 0.117, 0.121 to 0.158, and 0.205 to 0.391 N/mm, respectively. This standardized protocol can be applied to better compare the performance of grape varieties (Giacosa *et al.*, 2014).

Peppi *et al.* (2008) examined application of CPPU followed by ABA might increase the size and firmness of 'Flame Seedless' grapes without excessively inhibiting colouring. Grapes were treated with 0 or 20 g·ha^{-1}CPPU (applied at fruit set) and 0, 300, or 600 mg·L^{-1} ABA (applied at veraison) in 2005 and with 0, 5, 10, 15, or 20 g·ha^{-1} CPPU and 0, 200, 400, or 600 mg·L^{-1} ABA in 2006. Both plant growth regulators (PGRs) increased berry mass, but grapes treated with CPPU were as firm, or firmer than non-treated grapes, whereas those treated with ABA were of similar or lesser firmness. Treatment with CPPU generally reduced soluble solids and red berry color, whereas treatment with ABA reduced titratable acidity and increased red colour. The PGRs did not interact to affect any of the fruit quality variables measured, so beneficial effects of CPPU or ABA were apparent whether the grapes were treated with either or both PGRs. Thus, the combined use of CPPU and ABA may be a desirable cultural practice for 'Flame Seedless'.

Put, GA$_3$, SA, CPPU and CaCl$_2$ used at pre-harvest application also help in enhancing the shelf life as they increased berries firmness and decreased the percentage of unmarketable berries after keeping at ambient temperature for seven days after harvest.

Additional post harvest formulations also evaluated worldwide to improve the keeping quality in grapes. Aliabdollahi *et al.* (2012) reported use of postharvest spraying of essential oils from sweet basil (*Ocimum basilicum*), fennel (*Foeniculum vulgare*), summer savory (*Satureja hortensis*) and thyme (*Thymus vulgaris*) on fungal decay and quality parameters of the 'Thompson seedless' table grape stored at 0±1°C for 60 days were evaluated. Results showed that the essential oils, especially of thyme and fennel, have a good inhibitory effect on the development of fungal decay in Thompson table grapes. In addition, essential oils reduced weight loss, berry and rachis browning and had no considerable adverse effect on the flavour of the fruits. Comparable results were obtained with 'Fantac' formulation containing a liquid concentrate of amino acids, vitamins and proteins. It was used after April as well as October pruning in 0.5 –2.0 ml/l at different intervals. Significant differences were recorded due to variable application of Fantac in berry length, TSS and pulp content. The Berry length was recorded higher in recommended doses of plant growth regulators (PGRs) as per the standard schedule. The results on shelf life revealed that the significant differences were recorded on 7th day in shelf and the minimum loss of weight was recorded in recommended PGRs treatment as compared to control and on par with Fantac treatments. Hence, it can be concluded that Fantac has a potential over the untreated control in increasing the berry length, increasing the pulp content in the berries and increasing the shelf life.

Lurie *et al.* (2008) utilize modified ethanol atmosphere to control decay of table grapes. The Ethanol dip or impregnated papers were used for covering the grapes. Results showed that the taste of berries was not impaired by any of the ethanol applications but the taste of Thompson Seedless grapes stored for 8 weeks in modified atmosphere storage was affected by CO_2 concentrations above 7 per cent. Narayana Reddy *et al.* (2008) studied the effect of post harvest treatment with AOX inhibitors alone and in combination with polyamines on the shelf life when stored at 4°C. Result reported the PLW increased gradually during storage in grapes but the AOX-inhibitors and polyamines effectively reduced the PLW and berry shatter during storage (Spermine + n-PG) 1.0 mM, (Spermidine + n-PG) 1.0 mM recorded higher TSS. The shelf life was higher with SHAM 1.0 mM and 47 days with n-PG 1.0 mM and in the combination of spermidine + SHAM @ 1.0 mM gave the highest shelf life of 51 days followed by spermine 1.0 mM + SHAM 1.0 mM (45 days) and rest of the treatments were at par with each other. Narayana Reddy *et al.* (2008) explained the involvement of polyamines in the shelf life of grapes. Thompson Seedless and Sonaka recorded significantly lowest PLW, berry shatter and highest TSS and acidity. Putrescine was the major polyamine in grapes followed by spermidine and spermine. There was a general decline in the intensity of polyamine, and particularly spermine declined to undetectable levels during the storage in the varieties or treatments with shorter shelf life.

9. Conclusion and Recommendations

The present work sum up all the progress made for standardizing judicious use of bioregulators in different table grape cultivars to improve its productivity and quality with respect to Indian and abroad market. Considering the economic importance of Thompson Seedless grapes, the studies were presented with recommended practices followed in India. The potential of hydrogen cyanamide in bud break was extensively explained with its ability of uniform bud break induction in various table grape cultivars. The importance of nutrients and endogenous growth hormone levels at canopy development was studied with the use of growth retardants like CCC. The results obtained showed that by controlling excess vigour, the production of fruitful buds was possible with the application of growth retardants, among which CCC is extensively used. Maturity of cane indicating good food storage can support higher bunch load which adds in productivity. Further, the crucial stage of cluster development was influenced in each cultivar differently. The timing and microclimatic conditions at this stage varies with the vineyard and hence the concentration and stage of application of bioregulator changes. GA_3 is well known bioregulator used for improving berry characteristics and supported by other cytokinin like CPPU and homobrassinolides. CPPU enhances Pre- and Post-harvest quality of Seedless and

Seeded grapes. Different formulations of GA_3 in liquid, tabular and WG forms with the different adjuvant were tested to check its bioefficacy. Combinations of different bioregulators were standardized with phenological stage of development and mode of application. All these formulations of bioregulators were evaluated for quality parameters considering berry, bunch characteristics and yield/vine. Residues of these bioregulators also checked for consideration of its bioefficacy.

Along with optimised field practices for quality and productivity of grapes, storage and shelf life of grapes is also essential factor which influence market acceptability of fresh fruits. Various pre and post harvest spraying practices were carried out depending coloured and none colour grape verities. Use of calcium bioregulators for improving firmness of berries in bunch were explained for improving shelf life of grapes under cold storage. Finally, to improve the market acceptability of final product, pre and post-harvest application of bioregulators and bio stimulants are very much important which were briefly discussed. This systematic study explained the efforts made to improve the understanding of judicious use of bioregulators for the production of exportable quality table grapes.

10. Future Strategies

Now-a-day food safety is very much important and consumers are very much aware about not only for pesticide residues but also the residues of the plant hormones. Hence future line of work will be majorly focused on this angle and pre harvest interval of each plant hormone will be developed and based on this advice will be given to grape growers and application of these plant hormones/bioregulators/bio stimulants will be made accordingly.

References

Abbashassani, A., Irajbernousi, Y., Hadimeshkatalsadat, M., Seyed, R., (2012). Evaluation of Essential Oils for Maintaining Postharvest Quality of Thompson Seedless Table Grape. *Natural Product Research*, 26(1), 77-83.

Banerjee, K., Ramteke, S, Somkuwar, R., (2000). Terminal residues of hydrogen cyanamide in grape (*Vitis vinifera* L.) berries. *Indian Journal of Agricultural Sciences*. 70(7), 481.

BenArie, R., Sarig, P., CohenAhdut, Y., Zutkhi, Y., Sonego, L., Kapulonov, T., Lisker, N., (1998).Cppu and GA3 Effects on Pre- and Post-harvest Quality of Seedless and Seeded Grapes. *Acta Horticulturae*, No. 463.

Böttcher, C., Boss, P K., Davies, C. (2013).Increase in cytokinin levels during ripening in developing *Vitis vinifera* cv. Shiraz berries *Am. J. Enol. Vitic* 64(4), 527-531.

Carreno, J., Faraj, S., Martinez, A., (1999).The effects of hydrogen cyanamide on budburst and fruit maturity of 'Thompson Seedless' grapevine *Journal of Horticultural Science and Biotechnology*; 74 (4), 426-429.

Chadha, K., (2008). Indian viticulture scenario *Acta Horticulturae*, No. 785(3), 59-68.

Chougule, R., Tambe T., Kshirsagar D. (2008).Effect of canopy management on yield and quality attributes of Thompson seedless grapes. *ISHS Acta Horticulturae* 785(22), 183-190.

Chougule, R., Tambe, T., Kshirsagar, D. (2010). Biochemical changes associated with hydrogen cyanamide induced bud break in grapes. *J. Maharashtra agric. Univ.*, 35(3): 470-474.

Claudia, B., Rafael, R. (2010). Effects of foliar and soil calcium application on yield andquality of table grape cv. 'Thompson Seedless' *Journal of Plant Nutrition*, 33(3), 299- 314.

Cooper, T., Retamales, J., Bangerth, F. (1993) Berry drop occurrence as affected by GA3and promalin applications in Thompson seedless grapes. *Acta Horticulturae*, No. 329(24), 124-136.

ElDeen, S., Ashour, N.E. (2009). Decline N-fertilizer utilized for enhancing yield and fruit qualities of Thompson seedless grape using some physical mutants of *Saccharomyces cerevisiae. American-Eurasian Journal of Sustainable Agriculture*, 3(3), 321-327.

Fayed, T A. (2010) Effect of some antioxidants on growth, yield and bunch characteristics of Thompson Seedless grapevine. *American-Eurasian Journal of Agricultural and Environmental Sciences*, 8(3).

Fidelibus, M., Cathline, K., (2010). Dose and time dependent effects of methyl jasmonate on abscission of Grapes. *Acta Horticulturae*, No. 884.

Giacosa, S., Torchio, F., Giust, M., Tomasi, D., Gerbi, V., Rolle, L., (2014). Selection of a mechanical property for flesh firmness of table grapes in accordance with an OIV Ampelographic Descriptor *Am. J. Enol. Vitic*, 65(2), 206-214.

González Herranz, R., Cathline, K. A., Fidelibus, M W., Burns, J K., (2009). Potential of methyl jasmonate as a harvest aid for 'Thompson Seedless' grapes: concentration and time needed for consistent berry loosening. *Hort Science*, 44(5), 1330–1333.

Holzapfel, B., Smith, J. (2012). Developmental Stage and Climatic Factors Impact More on Carbohydrate Reserve Dynamics of Shiraz than Cultural Practice. *Am. J. Enol. Vitic.* 63: 333-342.

Kumar, H., Chohan, G. S., (1990). Effect of cycocel on floral drop, yield and fruit quality in Thompson seedless cv. of grapes. *Journal of Research-Punjab Agricultural University*, 27(2), 241-243.

Lombard, P.J., Cook, N.C., Bellstedt, D.U. (2006). Endogenous cytokinin levels of table grape vines during spring budburst as influenced by hydrogen cyanamide application and pruning. *Scientia Horticulturae*, 109(1), 92.

Lurie, S., Lichter, A., Zutahy, Y., Kaplonov, T. (2008) Modified ethanol atmosphere to control decay of table grapes. *Acta Horticulturae* No. 768.

Martínez, M. M., Ortega, R., Angulo, J., Janssens, M., (2014). Effect of different carbon rates on table grapes ('Thompson Seedless') in a controlled experiment. *Acta Horticulturae*, No. 1018.

Marzouk, H. A., Kassem, H. A., (2011). Improving yield, quality, and shelf life of Thompson Seedless grapevine by preharvest foliar applications. *Scientia Horticulturae*, 130(2)14: 425-430.

Narayana Reddy Y., Ramprasad V., Reddy M G D M. (2008) Studies on extension of shelf life of grape through alternative oxidase inhibitors alone and in combination with polyamines. *Acta Horticulturae*, No. 785.

Narayana Reddy, Y., Bhavani Shankar., V., Padmalatha, V. (2008) Studies on the possible involvement of polyamines in the shelf life of grapes (*Vitis vinifera* L.). *Acta Horticulturae*, No. 785.

Norrie, J., Branson, T., Keathley, P. E., (2003). Marine plant extracts impact on grape yield and quality. *Acta Horticulturae*, No. 594.

Norrie, J., Keathley, J. P., (2006). Benefits of *Ascophyllum Nodosum* marine-plant extract applications to 'thompson seedless' grape production. *Acta Horticulturae*, No. 727.

Oulkar, D., Banerjee, K., Ghaste, M. S., Ramteke, S. D., Naik, D. G., Patil, S. B., Jadhav, M. R., Adsule, P G (2011). Multiresidue analysis of multiclass plant growth regulators in grapes by liquid chromatography/tandem mass spectrometry. *J AOAC Int.*, 94 (3): 968-977.

Paranjpe, S. S., Ferruzzi, M., Morgan, M. T. (2012).Effect of a flash vacuum expansion process on grape juice yield and quality. *LWT Food Science and Technology*, 48(2), 147-155.

Peppi, M. C., Fidelibus, M. W. (2008). Effects of forchlor fenuron and abscisic acid on the quality of 'Flame Seedless' grapes. *Hort Science*, 43(1), 173-176.

Pérez, F., Vergara, R., Rubio, S. (2008). H2O2 is involved in the dormancy-breaking effect of hydrogen cyanamide in grapevine buds Plant Growth Regulation *An International Journal on Plant Growth and Development*, doi: 10.1007/s10725-008-9269-4.

Ramteke, S. D., Somkuwar, R. G. (2005). Effect of homobrassinolide on yield, quality and storage life in Thompson Seedless grape. *Indian Journal of Plant Physiology*, 10(2), 179-181.

Ramteke, S. D., Somkuwar, R. G. (2005). Effect of homobrassinolide on yield, quality and storage life in Thompson Seedless grape. *Indian J. Pl. Physiol.*, 10(2), 179-181.

Ramteke, S. D., Somkuwar, R. G. (2005). Effect of Quantum on increasing growth, yield and quality of grapes *Karnataka J. Agric. Sci.*, 18(1), 13-17.

Ramteke, S. D., Somkuwar, R. G., Adsule, P. G. (2008). Effect of CPPU on bunch and berry development in Thompson Seedless grapes. *Acta Hort.* 785, 213-216.

Ramteke, S. D., Somkuwar, R. G., Adsule, P. G. (2008). Use of bioregulators to improve the quality of Sharad Seedless grapes. *Acta Hort.* 785, 225-227.

Ramteke, S. D., Somkuwar, R. G. (2005). Effect of Cycocel sprays on increasing growth and yield of Tas-A-Ganesh grapes grafted on Dogridge rootstock. *Karnataka J. Agric. Sci.* 18(1), 18-20.

Ramteke, S. D., Somkuwar, R. G., (2006). Effect of stage and concentration of GA3 and carbaryl on cluster compactness in Tas-A-Ganesh on 1613C rootstock. *Journal of Maharashtra Agric. Univ.*, 31 (2), 364-365.

Ramteke, S. D., Somkuwar, R. G., (2010). Biochemical changes associated with hydrogen cyanamide induced bud break in grapes. *J. Mah. Agril. Uni.*, 35(3), 470-474.

Ramteke, S.D., Somkuwar, R.G., Shikhamany, S.D. and Banerjee, K. (2003).Cumulative effect of hydrogen cyanamide on growth, yield and quality of Tas-A-Ganesh grapes. *Ann. Plant. Physiol.* 17(1), 6-11.

Ramteke, S.D., Somkuwar, R.G., Shikhamany, S. D., Satisha, J. (2002). Growth regulators in increasing pedicel thickness and shelf life in Tas-A-Ganesh grapes (*Vitis vinifera*) grafted on 1613 C rootstock. *Indian J. Agric. Sci.* 72(1), 3-5.

Retamales, J., Rivas, A., Pinto, M., (1998) A novel mixture of gibberellins can replace both GA3 and CPUU on Thompson Seedless grapes. *Acta Horticulturae*, No. 463.

Retamales, J., Bangerth, F., Cooper, T., (1995) Effects of CPPU and GA3on fruit quality of Sultanina table grape. *Acta Horticulturae*, No. 394.

Rolle, L., Giacosa, S., Gerbi, V., Novello, V., (2011). Comparative study of texture properties, color characteristics, and chemical composition of ten white table-grape varieties *Am. J. Enol. Vitic*, 62(1), 49-56.

Salunkhe, N. M., Tambe, T. B., Kadam, J. H., (2008). Influence of pruning times and various chemicals on yield and qualityof Thompson Seedless grapes *Acta Horticulturae*, No. 785.

Shulman, Y., Nir, G., Fanberstein, L., Lavee, S., (1983).The effect of cyanamide on the release from dormancy of grapevine buds. *Scientia Horticulturae*, 19 (1–2), 97-104.

Singh, P. V., Sharma, P. K. (1996).Effect of ethrel on ripening and quality of Thompson Seedless grapes (*Vitis vinifera* L.). *Indian Journal of Horticulture*, 53(3), 202-205.

Singh, S., Bindra, S. S., Dhillon, W. S., Sandhu, S. S., (1992). Fruit quality improvement in 'Thompson Seedless' grapes. *Acta Horticulturae*, No. 321.

Somkuwar, R.G., Ramteke, S. D., Shikhamany, S D. (2002). Effect of ripeness on shelf life in Thompson Seedless grape.*Indian J. Hort.*, 59(3), 230-232.

Somkuwar, R.G., Ramteke, S.D., (2006). Effect of different rootstocks on fruitfulness in Thompson Seedless (Vitis vinifera L.) grapes. *Asian Journal of Plant Sciences.*, 5 (1), 150-152.

Somkuwar, R.G., Satisha, J., Ramteke, S. D., (2006). Effect of different rootstocks on fruitfulness in Thompson Seedless grapes. *Asian J. Plant Sciences*, 5(1), 150-152.

Somkuwar, R. G., Satisha, J., Sharma, J., Ramteke, S. D., (2008). Effect of cluster clipping and berry thinning on yield and quality of Thompson Seedless grapes. *Acta Hort.*, 785: 229-231.

Somkuwar, R. G., Satisha, J., Ramteke, S. D., Sharma, J. (2008). Effect of rootstocks and pre-harvest treatments on storage life of Thompson Seedless grapes. *Acta Horticulturae*, No. 785.

Somkuwar, R.G., Ramteke, S.D., (2007). Fruitfulness in relation to bud position in Tas-A-Ganesh grapes grafted on dogridge rootstock. *The Asian J. Hort.* 2(2), 87-88.

Tang, D., Wang, Y., Cai, J., Zhao, R., (2009). Effects of exogenous application of plant growth regulators on the development of ovule and subsequent embryo rescue of stenospermic grape (*Vitis vinifera* L.) *Scientia Horticulturae*, 120(1), 51-57.

Thorat, V.S., Singh, S., and Ghule, B. A. (2012). Dynamics of area, production and Export of grapes in India. Lambert Academic Publishing, 132.

Ugare, B, Banerjee, K., Ramteke, S., Pradhan, S., Oulkar, D., Utture, S., Adsule, P. (2013) Dissipation kinetics of forchlorfenuron, 6-benzyl aminopurine, gibberellic acid and ethephon residues in table grapes (*Vitis vinifera*). *Food Chemistry*, 141 (4), 4208-4214.

Vasconcelos, M., Greven, M., Winefield, C, Trought, M., Raw, V., (2009). The Flowering process of *Vitis vinifera*: A review *Am. J. Enol. Vitic.* 60(4), 411-434.

Violeta, D., Violeta, I., Fidanka, I., Elenica, S., (2011). Influence of bioregulator gibberellic acid on some technological characteristics of cluster and berry from some seedless grape varieties. *Journal of Agricultural Science and Technology*, 1: 1054-1058. ISSN 1939-1250.

Warusavitharana, A.J., Tambe, T.B., Kshirsagar, D.B. (2008). Effect of cytokinins and brassinosteroid with gibberellic acid on yield and quality of Thompson Seedless grapes *Acta Horticulturae*, No. 785.

Weaver, R. J., AbdelGawad, H. A., Martin, G. C. (1972). Translocation and persistence of 1, 2-14C-(2-Chloroethyl)-phosphonic Acid (Ethephon) in Thompson Seedless grapes. *PhysiologiaPlantarum*, 26(1), 13-16.

Williams, L. E., Ayars, J. E., (2005). Water use of Thompson Seedless grapevines as affected by the application of gibberellic acid (GA3) and trunk girdling – practices to increase berrysize. *Agricultural and Forest Meteorology*, 129(1-2), 85-94.

Zoffoli, J., Latorre, B., Naranjo, P., (2009). Preharvest applications of growth regulators and their effect on postharvest quality of table grapes during cold storage. *Postharvest Biology and Technology*, 51(2), 183-192.

Recent Advances in Crop Physiology Vol. 2 (2016)
Editor: Dr. Amrit Lal Singh
Published by: DAYA PUBLISHING HOUSE, NEW DELHI

Pages 203–224

Chapter 6

Physiological Basis of Iron Toxicity and its Management in Crops

K.K. Baruah * *and Ashmita Bharali*

Department of Environmental Science,
Tezpur (Central) University, Napaam, Tezpur, Assam

1. Introduction

Iron toxicity is one of the most widely distributed nutritional disorders in lowland rice soils derived from excessive amount of water soluble Fe^{2+} ions. Iron as an essential element for all plants, has many important biological roles in processes as diverse as photosynthesis, chloroplast development and chlorophyll biosynthesis (Sahrawat, 2003; Backer and Asch, 2005; Baruah *et al.*, 2007; Olaleye *et al.*, 2009). It is a major constituent of the cell redox systems like heme proteins including cytochromes, catalase, peroxidase, Fe-S proteins like ferredoxin, aconitase and superoxide dismutase (SOD) (Marschner, 1995). The Fe (II)/Fe (III) redox couple plays an important role in plant growth by enhancing the enzymatic redox reaction (Gill and Tuteja, 2010). Higher concentration of Fe^{2+} ion within plant cell may accelerate many redox reactions where Fe^{2+} acts as an electron donor (Baruah and Nath, 2001; Sahrawat, 2003). But lowland rice frequently suffers from iron toxicity and is reported to be an important yield limiting factor in terms of productivity. Since the first report of its occurrence (Ponnamperuma *et al.*, 1955), iron toxicity in rice has been reported in several countries in Asia, South America, West and Central Africa, and Uganda (Sahrawat 2004). Injuries due to iron toxicity are estimated to reduce overall rice grain yield by 30–60 per cent (Sahrawat 2000; Majerus *et al.*, 2007b). Toxicity at

* *Corresponding Author:* E-mail: kkbaruah@tezu.ernet.in

seedling and early vegetative stages can strongly affect plant growth, and may result in complete yield loss (Backer and Asch, 2005). Many cultivars are hypothesized to employ tolerance rather than avoidance or exclusion mechanisms (Becker and Asch 2005; Sahrawat, 2010). Storage in the apoplast and vacuole, and detoxification of Fe-induced reactive oxygen species (ROS) by antioxidant enzymes are the mostly reported tolerance mechanisms (Dufey *et al.*, 2009; Saikia and Baruah, 2012).

Various water, crop and nutrient management options have been reported to alleviate the negative effects of Fe toxicity on lowland rice performance. The most prominent and easily adoptable strategy to address Fe toxicity at field level is the development and use of tolerant rice germplasm. Breeders have proposed a wide array of genotypes with various degrees of adaptation. However, depending on the region, the soil type, the cropping season, and the severity and duration of Fe-toxicity occurrence, rice genotypes strongly differ in their response patterns to excess amount of Fe^{2+}. Hence, the toxicity impact can be reduced by using Fe tolerant genotypes, soil, water and nutrients management practices (Mandal *et al.*, 2004; Baruah *et al.*, 2007; Silveira *et al.*, 2007; Olaleye *et al.*, 2009; Saberi *et al.*, 2010). A range of agronomic management interventions have been advocated to reduce the Fe^{2+} concentration in the soil or to foster the rice plants' ability to cope with excess iron in either soil or the plant. In addition, the available rice germplasm contains numerous accessions and cultivars which are reportedly tolerant to excess Fe^{2+} (Qu *et al.*, 2005; Nozoe *et al.*, 2008; Sahrawat, 2010; Saberi *et al.*, 2010). However, none of those options is universally applicable or efficient under the diverse environmental conditions where Fe toxicity is expressed. Thus, cultivars that reportedly showed Fe^{2+} tolerance in soil-environmental conditions of one region frequently submit to iron toxicity in another region. Based on the available literature, this review focuses iron toxic environments, the steps involved in toxicity expression in rice, the current knowledge of crops' adaptation mechanisms and management practices and future research approaches.

2. Iron in Soil

Iron is the fourth most common element in the earth crust and exists either as Fe^{2+} (plant favourable form/at low soil pH) or Fe^{3+} (aerobic soil conditions). Apart from the seasonal changes due to the rainfall pattern or drainage, irrigation and changes in Eh, the Fe^{2+} concentrations are found on a small scale horizontally in the soil profile and vertically between the bulk soil and the rhizosphere. Three compartments have been differentiated and include a thin oxic surface layer, the reduced puddled bulk soil and the oxicrhizosphere soil and rhizoplane (Backer and Asch, 2005). The general characteristics of most Fe-toxic soils are high amounts of reducible Fe, low pH, and low CEC and exchangeable K content (Parent *et al.*, 2008) which is also associated with P and Zn deficiencies and H_2S toxicity (Kirk, 2004). Paddy rice soils are subjected to periodic changes between oxic and anoxic conditions. Since diffusion rate of oxygen in air is about 103–104 times faster than in water or in water-saturated soils, oxygen is depleted rapidly by the respiration of soil microorganisms and plant roots in waterlogged soils (Pradeet *et al.*, 1993). With the depletion in oxygen, NO_3^-, Mn^{4+}, Fe^{3+}, and SO_4^{2-} can act as electron acceptors for microbial respiration and sequentially become reduced in waterlogged rice field. The amount of extractable Fe^{2+} increases with the quantity of decomposable organic matter, the temperature, and the amount of available redox buffers and low initial soil pH (Backer and Asch, 2005; Sahrawat, 2010; Wang *et al.*, 2013).

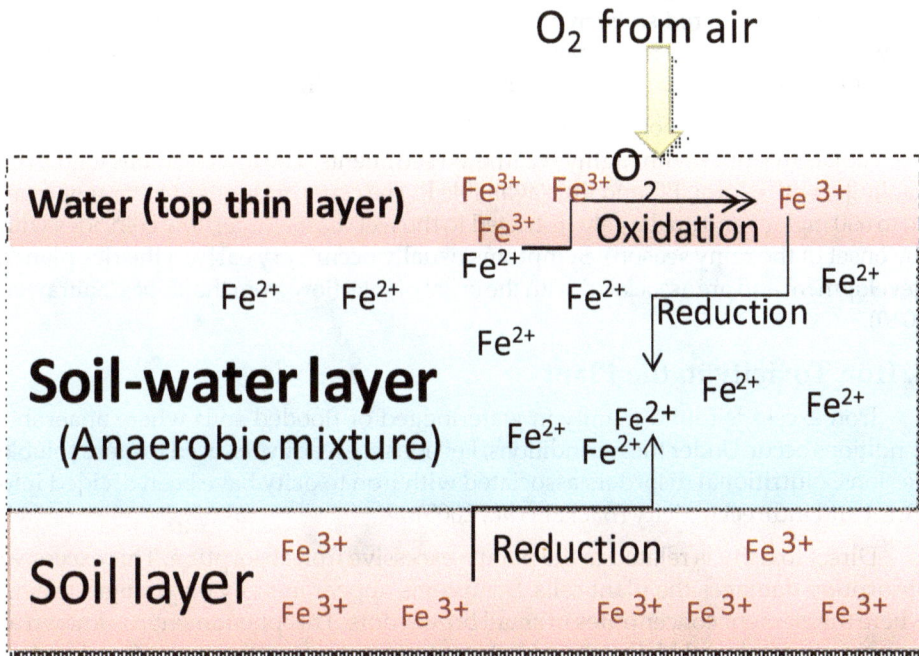

Figure 6.1: Iron Reduction Reactions In Waterlogged Soil
(Backer and Asch, 2005).

3. Soil Environment and Iron Toxicity

The severity of iron-toxicity expression in rice has been related to a number of soil factors. These involve most prominently (1) the content and type of the clay minerals (2) the amount of exchangeable soil Fe (3) the soil pH, and (4) the presence of "stress factors". The concentration of soil Fe^{2+} is reportedly less in clay than in sandy soils (Das *et al.*, 1997). Clay was found to control the content and distribution of iron in both Alfisols and Vertisols (Rajkumar *et al.*, 1997). Clay content affected the Fe dynamics primarily *via* Fe retention on clay-mineral surfaces due to cation-exchange capacity. There is a large inconsistency within the published literature on iron toxicity with regard to soil Fe content ranging from 20–5000 mgkg⁻¹and leaf Fe concentrations (300–2000 mg kg⁻¹), time of toxicity occurrence from 2 weeks after rice transplanting to the late reproductive phase, the distribution of toxicity symptoms in the field, and the observed yield losses (Backer and Asch, 2005). Conditions associated with iron toxicity can be categorised in three different phases.

Phase 1 refers to acid sulfate soils, in which extremely large concentrations of Fe^{2+} in the soil solution arise as a result of the soils' unusual mineralogy. The soil Fe^{2+} can range from 500 to over 5000 mgkg⁻¹, and tissue concentrations are generally between 500 and 2000 mg kg⁻¹(Backer and Asch, 2005). The symptoms on plants can occur throughout the rice-growing period, and yield losses reportedly range from 40 per cent –100 per cent.

Phase 2 is associated with more clayey acid, iron rich soils in sediments derived from highly weathered soils. Symptoms on rice appear during the late vegetative growth stage, and yield losses rarely exceed 30 per cent. However, yield losses of 35 per cent–50 per cent can occur when rice is grown during the dry season.

Phase 3 refers to poorly drained sandy soils in valleys receiving interflow water from adjacent slopes with highly weathered sediments. The lowland soils tend to be kaolinitic with a low CEC and little available P. The concentrations of Fe^{2+} range from 20 to 600 $mgkg^{-1}$ and are largely restricted to the wet season (interflow coincides with the onset of the rainy season). Symptoms usually occur very early in the rice plant's development and are associated with the onset of interflow from the slopes (Sahrawat, 2010).

4. Iron Toxicity in the Plant

Iron excess is found mainly in waterlogged or flooded soils where anaerobic conditions occur.Under these conditions, Fe^{3+} ions are readily reduced to more soluble Fe^{2+}ions. Nutritional disorders associated with iron toxicity have been divided into direct and indirect toxicity (Backer *et al.*, 2005).

Direct toxicity is related to the plant's excessive iron absorption. This excessive absorption damages the plant cells. Symptoms appear initially on younger leaves, where the element concentrates in small brown dots. This phenomenon is known as bronzing (Baruah and Nath, 1996). Under extreme situation, the leaves may become chlorotic and at advanced toxicity stages necrosis may occur, *i.e.* the leaves dry and eventually die. Although the degree of bronzing has been suggested as a good way to measure the degree of toxicity (Baruah and Nath, 1996), the mechanism underlying the bronzing is not well understood (Sahrawat, 2004; Briat *et al.*, 2007). Iron's low mobility is likely due to its precipitation in older leaves as insoluble oxides or phosphates, or to the formation of complexes with phytoferritin, an iron binding protein. Iron precipitation decreases the metal's subsequent mobilization inside the phloem.

Indirect toxicity results from the limited absorption of several nutrients such as calcium, magnesium, potassium, phosphorous and iron itself, due to iron precipitation on rice root epidermis. The formation of an oxide-hydroxide Fe^{3+} layer on the root blocks nutrient absorption, resulting in multiple nutritional deficiencies. Symptoms of this deficiency include plant atrophy, tillering reduction, orange leaves, and the covering of roots by red layers of iron oxides. Toxicity symptoms are usually correlated with iron deposition in the roots (Backer and Asch, 2005; Baruah *et al.*, 2007; Silveira *et al.*, 2007; Sahrawat, 2010). At the cellular level, it is not only insolubility, but iron's high reactivity that can cause severe damage. Tissue iron may catalyse the generation of active oxygen species via the iron catalysed Haber-Weiss reaction (Fenton reaction) (Figure 6.2). Without iron this reaction is reported to be slow in the medium (Halliwell, 2006). Excess Fe^{2+} irons accelerate the production of ROS comprising O_2^-, H_2O_2, $^1O_2$, HO_2, OH, OH^- and RO which are highly reactive and toxic, cause a multitude of damage to protein, lipid, carbohydrate and DNA leading to cell death (Bode *et al.*, 1995; Arora *et al.*, 2002).

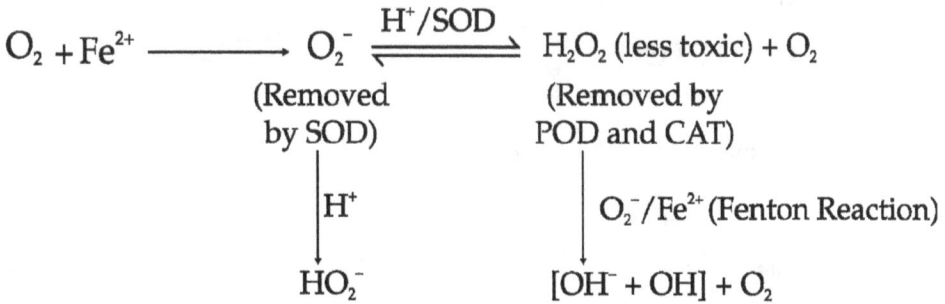

$$O_2 + Fe^{2+} \longrightarrow O_2^- \xrightleftharpoons{H^+/SOD} H_2O_2 \text{ (less toxic)} + O_2$$

$$\underset{\text{by SOD)}}{\text{(Removed}} \qquad\qquad \underset{\text{POD and CAT)}}{\text{(Removed by}}$$

$$\downarrow H^+ \qquad\qquad\qquad \downarrow O_2^-/Fe^{2+} \text{ (Fenton Reaction)}$$

$$HO_2^- \qquad\qquad\qquad [OH^- + OH] + O_2$$

Figure 6.2: Formation of ROS in Plant Cells due to Excessive Uptake of Fe²⁺ in Iron Toxic Soil (Arora *et al.*, 2002).

5. Iron Absorption and Transport Mechanism in the Plant

Rice plants have the tendency of taking up more iron than most other plants, and Fe^{2+} is the only iron species prevailing in paddy fields under flooded environments. After uptake into the root cortex, the reduced Fe^{2+} can enter the xylem after a symplastic passage through the casparian strip. However, a large allocates of Fe^{2+} may enter the xylem directly *via* an apoplastic bypass (Backer and Asch, 2005; Sperotto *et al.*, 2012). In the xylem, Fe^{2+} follows the transpiration stream in the acropetal long-distance transport. This is in contrast to the iron transport in upland crops or in plants that do not grow under iron-toxic conditions, where the long distance transport is dominated by Fe^{3+} complexes with citrate or peptide carbohydrate compounds (Schmidt, 1999). Reaching the leaf apoplastic space, Fe may re-enter the symplast. The exact mechanism by which Fe is taken up by leaf cells is not yet well understood. In rice, this uptake is likely to occur in the form of Fe^{2+}, being the physiologically active form of iron in the symplast (Sahrawat, 2004). Plants have developed sophisticated and tightly regulated mechanism for acquiring Fe from soil. The soluble Fe^{2+} ions are transported across the root plasma membrane through *IRT1* (Bashir *et al.*, 2010). The expression of genes like *OsNAS1-2*, *Os NAAT1* and *OsDMAS1* work together to produce DMA (Deoxymugineic Acid) for Fe acquisition and translocation (Bashir *et al.*, 2010). Among the Fe transporters, *OsYSL15* is predominantly expressed in phloem cells under Fe sufficiency (Inoue *et al.*, 2009). Besides secreting DMA, it also absorbs Fe^{2+}, which is more abundant than Fe^{3+} under the submerged conditions to which rice is well adapted (Ishimaru *et al.*, 2012). Although two homologs of ferric-chelate reductase are present in rice, the expression of these genes is not observed under Fe deficient or Fe-sufficient conditions, and the level of Fe^{3+} per cent chelate reductase activity is very low compared to that in other plants (Ishimaru *et al.*, 2006). As rice plants are adapted to grow under submerged conditions where Fe^{2+} is abundant, they may have lost the ability to reduce Fe^{3+} through the development of a functional Fe^{2+} regulated transporter, but not ferric-chelate reductase (Ishimaru *et al.*, 2006). *OsIRT1* and *OsIRT2* expression is observed only in Fe-deficient roots (Ishimaru *et al.*, 2006).

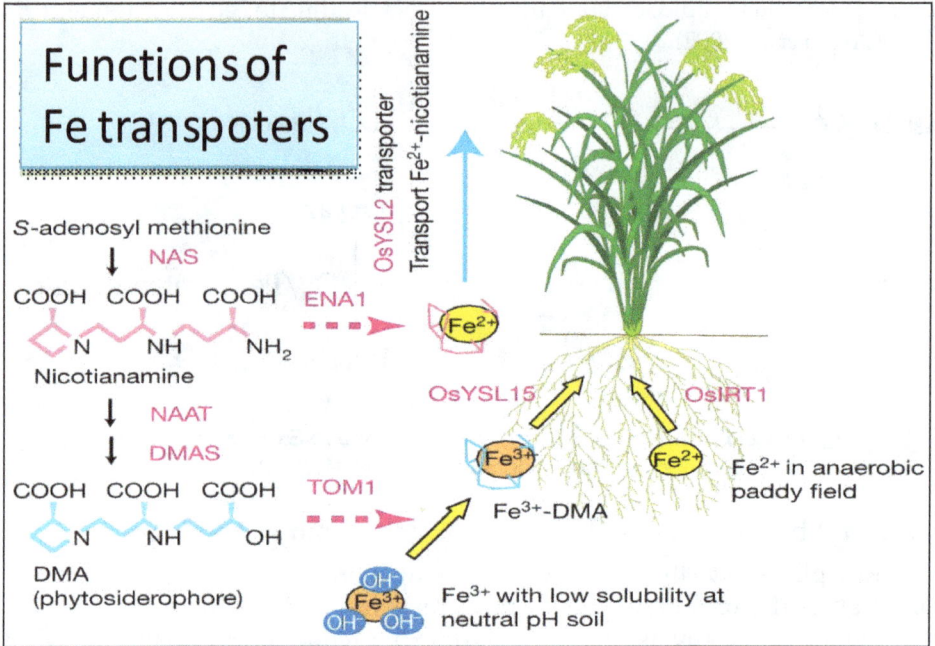

Figure 6.3: Fe Transport Phenomenon in Plant.
NAS: Nicotinanamine synthase; NAAT: Nicotinanamine aminotransferase;
DMAS: Deoxymugineic Acid synthase (Bashir *et al.*, 2010).

The existence of strategies I and II in rice allows it to utilize both ferrous and ferric iron, depending on their availability and environmental conditions. Fe transport from root to shoot and grain is essential for normal plant growth. In rice, Fe can be transported in various forms through the xylem and phloem, including Fe-citrate, DMA–Fe(III), and NA–Fe(II). Among the 18 putative *YSL* family genes in rice, *OsYSL5–7*, *-14*, and *-17* are expressed in the epidermis, cortex, and stele of Fe-sufficient and Fe-deficient roots. On the other hand, the expression of *OsYSL1–4, 9–11*, and *-18* was not observed in roots, irrespective of their Fe status (Bashir *et al.*, 2010). *OsYSL12* is expressed in the cortex and stele under Fe sufficient and Fe-deficient conditions, whereas *OsYSL16* is expressed in the epidermis/endodermis and cortex under Fe-sufficient condition and in the epi-/endodermis, cortex and stele under Fe-deficient condition (Inoue *et al.*, 2009). *OsYSL15* promoter-driven GUS expression was not only observed in leaf tissue but also at the flowering stage (Inoue *et al.*, 2009). These results indicate that *OsYSL15* is involved in Fe transport to rice grains. Recently *OZT1* was constitutively expressed in various rice tissues. The *OZT1* expression was significantly induced both in the seedlings of japonica rice Nipponbare and Indica rice, IR26 in response to Zn^{2+} and Cd^{2+} treatments. Besides, *OZT1* expression was also increased when exposed to toxic level of Fe^{2+} and Mg^{2+}.

6. Iron Toxicity Tolerance

6.1 Nutrients Uptake

Although lot of research have been conducted worldwide to identify adaptive responses of different rice genotypes still rate of nutrients absorption (ionic competition for absorption) and their availability (in favourable oxidation states) under higher iron concentrations is a matter of debate. Under waterlogged conditions O_2 release from rice roots, oxidise Fe^{2+} to polymeric oxy-hydroxide which coats on roots surface preventing the uptake of Fe^{2+}, Mn^{2+} and also acts as P reservoir. Many workers suggest that excess Fe^{2+} may result in lower uptake of other essential nutrients due to chemical interactions in soil. Resistance to Fe overload in rice plants can be a consequence of Fe avoidance and/or tolerance to high internal Fe concentrations (Sahrawat, 2004). Sahrawat (2010) has reported the possibility of "pseudo Fe toxicity" (Fe toxicity symptoms induced by nutrients deficiency) and "true Fe toxicity" (caused by excessive Fe^{2+} uptake) in rice grown at higher iron concentration. The Fe^{2+} formed by the reduction of Fe oxides shows antagonistic interactions with other cationic nutrients, mainly manganese, zinc, and potassium. Consequently, with increasing Fe concentrations, Mn, Zn and K deficiencies are likely to occur (Fageria, 2008). Thus, large amounts of reduced Fe adversely affected the uptake of Mn. Vice versa, Mn application can suppress iron desorption, delay the reduction of Fe oxides, and subsequently reduce Fe uptake by rice (Silveira *et al.*, 2007). Iron oxides are known to have a strong zinc-binding capacity. When soils are flooded, Zn becomes available in the process of iron oxide reduction (Mandal *et al.*, 2004). On the other hand, reduced Fe can also exert a direct antagonistic effect on Zn uptake (Baruah and Nath, 2001). Furthermore, the plaque formation resulting from Fe re-oxidation around the rice root can reduce the concentration of soluble Zn in the rhizosphere by forming sparingly soluble $ZnFe_2O_4$ (Backer and Asch, 2005; Silveira *et al.*, 2007; Sahrawat, 2010). When the iron plaque exceeds 25 g, the plaque becomes a physical barrier for the uptake of Zn (Zhang *et al.*, 2011)). Similar to Zn, the uptake of K is affected by excess Fe^{2+} in the soil solution (Baruah and Nath, 2001). The increase of Fe^{2+} concentrations in the growth medium has a synergistic impact on root and shoot nitrogen contents (Baruah *et al.*, 2007: Olaleye *et al.*, 2009; Sahrawat, 2010). Saikia and Baruah (2013) observed in an experiment on varieties grown at higher iron level in the soil that rice variety Ranjit and SiyalSali accumulated higher tissue iron in the shoot than relatively iron tolerant variety Mahsuri. Thus iron accumulation in the shoot may be related to exclusion/avoidance of excess Fe^{2+} from root, an adapting mechanism of iron toxicitytolerant rice variety (Backer and Asch, 2005; Silveira *et al.*, 2007). Ranjit and SiyalSali recorded a decreasing trend of Cu, Zn and Mn concentrations in root and shoot under higher Fe^{2+}while in Mahsuri micronutrients in roots and shoots were not affected by added Fe^{2+} concentrations.

6.2 Cell Biochemical Components

Iron toxicity is characterized by "bronzing" or "yellowing" of oldest rice leaves and formation of ROS in cells which affects the synthesis of chlorophyll, protein, leaf free amino acid nitrate reductase activity. Higher Fe uptake by plant is reported to reduce the protein synthesis in the leaf (Baruah *et al.*, 2007; Silveira *et al.*, 2007; Saikia and Baruah, 2012). Ferritin is considered crucial for iron homeostasis. It consists of multimeric spherical protein called phytoferritin, able to store upto 4500 Fe atoms inside its cavity in non toxic form (Briat *et al.*, 2006, 2007). It functions as a cellular Fe buffer. Plant ferritin is found mainly in plastids and also in mitochondria (Zancani *et al.*, 2007) is synthesized in responses to various environmental stresses including photo inhibition and iron overloading (Murgia *et al.*, 2001, 2002). Phytoferritin gene expression in plants is dually regulated by ABA and by antioxidants. Phytoferritins, therefore function as front-line defence mechanism against free iron-induced oxidative stress (Ravet *et al.*, 2009). The major function of plant ferritins is not to store and release iron, as previously reported, but to scavage free reactive iron and prevent oxidative damage (Ravet *et al.*, 2009). A resistant variety may accumulate more amount of phytoferritin which forms complex with Fe, reducing iron toxicity damage. Moreover, large amount of Fe^{2+} in plant can give rise to the formation of oxygen radicals, which are highly phytotoxic and responsible for protein degradation (Dobermann and Fairhurst, 2000). For sensitive variety, higher Fe^{2+}concentrations lead to rapid increase in lipid peroxidation accompanied with growth retardation (Fang *et al.*, 2001) and shift the balance of free radical metabolism towards production of active oxygen species that impair the cellular structure irreversibly and damage membrane proteins (Arora *et al.*, 2002; Dorlodot *et al.*, 2005).

Higher iron in the growth medium also affects the total leaf chlorophyll content in iron sensitive varieties (Baruah and Nath, 1996). It has been established that the appearance of iron toxicity in rice involves an excessive uptake by the roots and its acropetal translocation into the leaves where an elevated production of ROS can damage cell structural components and impair physiological processes (Backer and Asch, 2005). Moreover higher PPO activity which is involved in release of oxidized polyphenols under Fe toxic condition may accelerate browning reactions in leaves (Mayer, 2006). Such iron induced detrimental effects may be the reason of lower chlorophyll content in iron sensitive varieties. Dordas *et al.* (2003) reported the induction of non symbiotic haemoglobins (class-1) under hypoxic stress. A tolerant variety might have been able to maintain a higher level of haemenzyme activity in the leaf tissues even at higher level of iron in the medium and hence can maintain higher chlorophyll content through chloroplast development (Dordas *et al.*, 2003).

A positive correlation of higher soil Fe^{2+} concentration and nitrogen dynamic in resistant plants is reported. The leaf nitrate reductase (NR) activity and leaf free amino acid contents of the tolerant genotypes are not affected by Fe overloading (Baruah and Nath, 2001). Mandal *et al.* (2004) reported that the higher Fe uptake reduces the protein synthesis in leaf and subsequently inhibits or suppresses the normal NR activity. Iron tolerant rice cultivars Budumoni and CR 683 are reported to have higher chlorophyll content, proteins and NR activity at 7.12 mM Fe in the growth medium (Baruah *et al.*, 2007).

6.3 Genotypic Variations

Although range of agronomic management interventions have been advocated to reduce the Fe^{2+} concentration in soil, genetic diversity of rice plays an important role in sustainable development and food security, as it allows the cultivation of crops in the presence of various biotic and abiotic stresses. The major emphasis in breeding for iron toxicity tolerance in rice is to identify differences that are associated with resistance and connect them for genetic improvement. Onaga *et al.* (2013) had analyzed thirty accessions, including IRRI gene bank accessions for sensitivity to iron toxicity and genetic diversity using morphological and SSR markers. Two genotypes, IR61612-313-16-2-2-1 and Suakoko 8 showed significantly high resistance with an average score of d" 3.5 on 1 - 9 scale. Similarly Sahrawat *et al.* (1996) evaluated 20 lowland rice cultivars for tolerance of iron toxicity at iron toxic site on the basis of grain yield and toxicity scores. Mandal *et al.* (2004) evaluated Pokkali, IR50, IR2152-190-3 and IR2153-96-1 as Fe tolerant genotypes and reported two different genes to govern Fe tolerance in rice. Suakokos 8 harbours a dominant gene while Gossi 27 possesses a recessive gene for imparting tolerance to Fe toxicity (Mandal *et al.*, 2004). Nozoe *et al.* (2008) cultivated IR64 (check variety) and four lines of rice (*Oryza sativa* L.) developed at IRRI in a field with iron (Fe) toxicity at Iloilo City, Philippines, and also under normal soil conditions at IRRI farm.

6.4 Yield Components

Fe toxicity symptoms can occur at different growth stages and may affect rice at the seedling stage, during the vegetative growth, and at the early and late reproductive stages. In the case of toxicity occurring during seedling stage, the rice plants remain stunted with extremely limited tillering. Toxicity during the vegetative stage is associated with reduced plant height and dry-matter accumulation (Baruah *et al.*, 2007), with the shoot being more affected than the root biomass (Backer and Asch, 2005). Both the tiller formation and the share of productive tillers can be severely reduced. When iron toxicity occurs during the late vegetative or early reproductive growth phases, it is associated with fewer panicles per hill (Olaleye *et al.*, 2009), an increase in spikelet sterility and delayed flowering and maturity by up to 20–25 d. In highly susceptible cultivars, no flowering at all will occur (Backer and Asch, 2005).There is some correlation between the severity of iron-toxicity symptom expression and yield. This relationship can vary within the cropping season as well as between seasons and years. Average reported yield losses due to iron toxicity are in the range of 12 per cent –35 per cent (Sahrawat, 2010). Reduction in rice productivity has been reported to be directly proportional to concentration of Fe^{2+} in the solution and the tolerance of the cultivar type (Audebert *et al.*, 2006). Higher filled grain (per cent), 1000 grain weight and grain yield of a variety indicate its tolerance capacity to toxic iron concentration (Baruah *et al.*, 2007). Effect of higher Fe^{2+} iron in the growth medium is less prominent on yield and yield attributing parameters of a tolerant variety as reported by several research workers (Mandal and Roy 2005; Baruah *et al.*, 2007; Olaleye *et al.*, 2009; Saberi, 2010).

7. Mechanisms of Tolerance

Evidently, rice plants have developed morphological and physiological avoidance and/or tolerance mechanisms to cope with and survive adverse iron-toxic soil conditions and large amounts if iron in the plant. These mechanisms are important in the selection of tolerant or adapted rice genotypes. However, problems in selection of rice genotypes for tolerance to iron toxicity still relate to the inadequacy

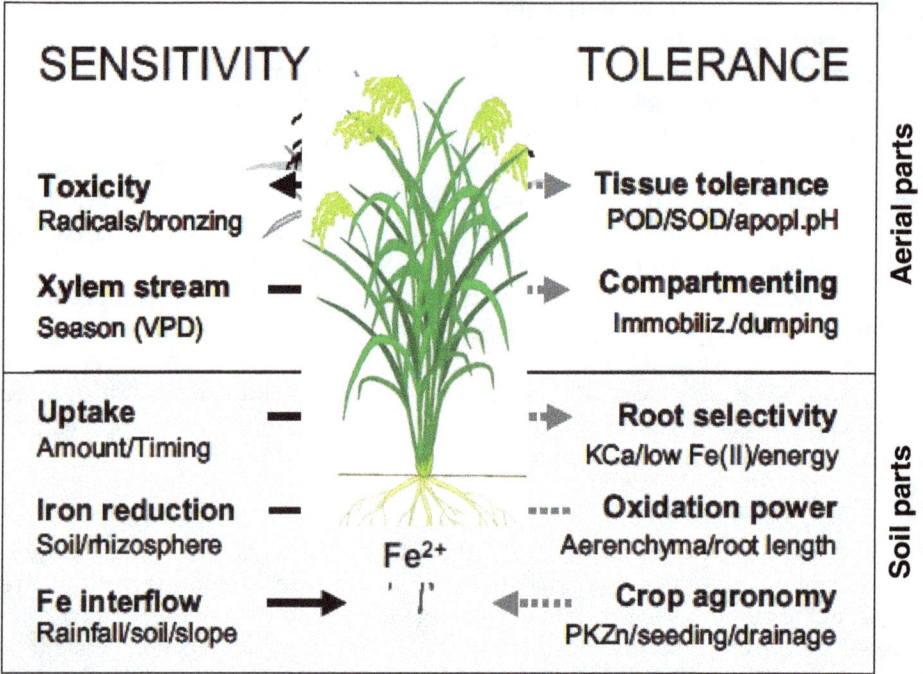

Figure 6.4: Schematic Representation of some Tolerance and Sensitivity Characters of Crops at Iron Toxic Environment (Backer and Asch, 2005).

of knowledge on physiological mechanisms of tolerance (Nishiuchi *et al.*, 2012). The relative importance of the avoidance mechanisms or tolerance mechanisms to iron (toxic) resistance depends on several factors. One important factor is the duration of iron toxic stress. For short-term stress situations tolerance mechanisms are probably adequate. For long-term stress conditions tolerance alone may not be sufficient and the plant may have to adopt avoidance mechanisms. Another distinction can be made between plants which exclude iron through oxidation of the rhizosphere or exclusion mechanisms and those which include high amounts of iron and subsequently resort to enzymatic adaptation or inactivation mechanisms of iron in the plant tissues (Marschner, 1995). The following mechanisms have been found to be relevant in rice plants in coping with excess iron concentrations. (1) Oxidation of iron at the root surface (Ando, 1983), (2) Exclusion of iron at the root surface (Tadano, 1976) and (3) Retention of iron in the root tissue (Tadano, 1976).

Three major types of adaptation strategies can be differentiated and comprise "include" and "excluder" strategies as well as "avoidance" and "tolerance" mechanisms (Becker and Asch, 2005). Plants employing strategy 1 (exclusion/ avoidance) exclude Fe^{2+} at the root level and hence avoid of Fe^{2+} damage to the shoot tissue (rhizospheric oxidation and root iron selectivity). With strategy ll (inclusion/ avoidance), Fe^{2+} is taken up into the rice root, but tissue damage may be avoided by either compartmentation (immobilization of active iron in dumping sites, *e.g.*, old leaves of photosynthetically less active leaf sheath tissue) or exclusion from the symplast (immobilization in the leaf apoplast). Strategy lll plants (inclusion/ tolerance) actually tolerate elevated levels of Fe^{2+} within leaf cells, probably via enzymatic "detoxification" in the symplast. Whereas Fe^{2+} exclusion by oxidation in the rhizosphere and the detoxification of leaf cells well established Fe-tolerance mechanisms of rice, the other mechanisms are not yet well understood and to date are not considered in rice breeding or screening for iron tolerance (Becker and Asch, 2005)

8. Strategies of Adaptation to Iron (Fe) Toxicity

8.1 Aerenchyma Formation

Formation of aerenchyma is essential to the survival and functioning of plants subjected to waterlogging. The aerenchyma contributes to O_2 supply from shoots to roots and to the ventilation of gases (*e.g.* CO_2 and methane) from roots to shoots (Colmer 2003a; Evans 2003). The ventilation of gases in aerenchyma is mainly caused by gas diffusion in rice, but in some wetland species with 'through-flow pathways' (*e.g.* along rhizomes), gas flows can also occur by humidity and Venturi-induced pressure flows (*e.g. Phragmites australis*) (Armstrong *et al.*, 1996). The aerenchyma may provide a photosynthetic benefit by concentrating CO_2 from root respiration and transporting it to the leaf intercellular spaces in some wetland plant species. In general, aerenchyma can be classified into two types: (i) schizogenous aerenchyma, which develops by cell separation and differential cell expansion that creates spaces between cells, in *e.g. Rumex palustris* and (ii) lysigenous aerenchyma, formed by the death and subsequent lysis of some cells, *e.g.*, in rice (Jackson and Armstrong, 1999), maize (Drew *et al.*, 2000), wheat and barley (Nishiuchi *et al.*, 2012). Fully developed aerenchyma, which is observed on the basal parts of roots, separates the inner stele from the outer cell layers (*i.e.*, sclerenchyma, hypodermis/exodermis, and epidermis) of the roots (Kozela and Regan, 2003; Ranathunge *et al.*, 2003). Strands of remaining cells and cell walls separate gas spaces in the cortex, forming radial bridges, which are important for the structural integrity of the root and for both apoplastic and symplastic transport of nutrients. During aerenchyma formation in rice root, cell death begins at the cells in the mid-cortex and then spreads out radially to the surrounding cortical cells. The epidermis, hypodermis/exodermis, endodermis and stele are unaffected, indicating that lysigenous aerenchyma formation occurs by closely controlled mechanisms (Yamauchi *et al.*, 2011). By contrast, in non-wetland plant species such as maize, wheat and barley root lysigenous aerenchyma does not form under well-drained soil conditions but it may be induced by poor aeration (McDonald *et al.*, 2001).

8.2 Formation of a Barrier to Radial Oxygen Lost (ROL)

Oxygen molecules diffusing longitudinally through aerenchyma towards the root tips may be either consumed by respiration or diffused radially to the rhizosphere (Colmer 2003a). ROL, the flux of O_2 from the aerenchyma to the soil, is determined by the concentration gradient, the physical resistance to O_2 diffusion in a radial direction, and consumption of O_2 by cells along this radial diffusion path (Colmer, 2003a). ROL aerates the rhizosphere and is therefore considered to be of adaptive significance in plants growing in waterlogged soil (Blossfeld *et al.*, 2011; Colmer, 2003a; Neubauer *et al.*, 2007). However, ROL reduces the supply of O_2 to the root apex and thereby causes a decrease in root length in anaerobic soil (Colmer 2003a; Armstrong *et al.*, 2000). The roots of many wetland species, including rice, have the ability to prevent ROL to the rhizosphere by forming a barrier in the root peripheral cell layers exterior to the aerenchyma (McDonald *et al.*, 2002; Visser *et al.*, 2000). This adaptive trait enhances longitudinal O_2 diffusion through the aerenchyma towards the root apex by diminishing losses of O_2 to the rhizosphere, thereby enabling the roots to elongate into anaerobic substrates. The roots of some wetland species have constitutively present barriers to ROL (Visser *et al.*, 2000), whereas in other species such as rice and *Hordeum murinum* the barrier to ROL is induced during growth under anaerobic conditions (Colmer, 2003b; Garthwaite *et al.*, 2003; Kotula *et al.*, 2009a). Analysis of the spatial patterns of ROL along rice roots has revealed that O_2 leakage from the basal regions of the long roots under stagnant conditions is quite low, but there are large amounts of O_2 flux from the root apexes and numerous short lateral roots that appear near the base of the main axes (Colmer, 2003b).

8.3 Root Morphological Responses to Fe^{2+} Toxicity

Toxicity of Fe^{2+} is one of the major constraints for lowland rice production in tropical and subtropical areas. However the mechanism of Fe-induced inhibition of root growth and the reasons for the spatial variations in Fe^{2+} sensitivity among the apical root zones are still poorly understood. The root tip is a primary site of Fe^{2+} toxicity in rice. The root border cells (BC), which originated from the root cap meristem by mitosis, can separately carry out metabolism and resist adverse stress through a series of distinct responses (Zhang *et al.*, 2009). Zhang *et al.* (2011) studied the effect of higher Fe^{2+} on root tip cells of two genotypes of rice (*Oryza sativa* L.), Azucena (iron tolerant) and IR64 (iron sensitive) to investigate the numbers and survival rates of root border cells (namely, in situ border cells) in plants that were exposed to excess iron (Fe^{2+}). Additionally, they examined the changes in the root tip cell morphology and activities of protective enzymes in response to Fe^{2+} toxicity. The seedlings of Azucena and IR64 known to vary in Fe^{2+} resistance at the whole-root level, were treated with Fe^{2+} -containing solutions at 0 (control), 50, 100, 200 or 400 µmol/litre for 24 h, and their responses to Fe^{2+} were studied. Fe^{2+} toxicity was found to inhibit the development of BC (root border cells). However, compared to IR64, Fe^{2+} at 100-200 µmol/litre was propitious to the development of BC in Azucena. With increase in Fe^{2+} concentration, the viability of the rice BC became lower, the cell wall of root tip outermost cells became thicker, and some characteristics of programmed cell death were observed in the cells (Fe-sensitive variety). The results of higher POD, SOD and CAT activities indicated that under Fe^{2+}toxicity, the root tip can resist Fe^{2+} toxicity by increasing BC and thickening the cell wall of resistance varieties (Zhang *et al.*, 2011).

8.4 Function of Anti-oxidative Enzyme Activities

Tissue iron in plants may catalyze the generation of active oxygen species via the iron catalyzed Haber-Weiss reaction (Fenton reaction) as shown in Figure 6.2. Without iron this reaction is reported to be slow in the medium (Halliwell, 2006).Under waterlogged situation, excess of water soluble iron in growth medium and its translocation into plant cells retards crop growth and causes oxidative damage within the cells (Stein *et al.*, 2009; Zhang *et al.*, 2011). Excess (Fe^{2+}) ion accelerates the production of ROS comprising O_2^-, H_2O_2, $^1O_2$, HO_2^-, OH, OH$^-$ and RO which are highly reactive and toxic, cause a multitude of damage to protein, lipid, carbohydrate and DNA leading to cell death (Arora *et al.*, 2002; Bode *et al.*, 1995). A great deal of research has established that the induction of the cellular antioxidant machinery is important for protection against Fe toxicity (Zhang *et al.*, 2011; Saikia and Baruah, 2012). The components of antioxidant defence system are enzymatic and non-enzymatic antioxidants. Enzymatic antioxidants include SOD, CAT, APX, MDHAR, DHAR and GR and non-enzymatic antioxidants are GSH, AA (both water soluble), carotenoids and tocopherols (lipid soluble) (Gill *et al.*, 2010). To protect themselves against the toxic oxygen intermediates, plant cells and its organelles like chloroplast, mitochondria and peroxisomes employ antioxidant defence systems (Agarwal *et al.*, 2006). Saikia and Baruah (2012), has reported that the variety Mahsuri (tolerant to higher Fe^{2+}) recorded efficient antioxidant activities in terms of better SOD, POD and CAT activities compared to Ranjit and Siyal Sali (sensitive to Fe toxicity).

9. Management Approaches

9.1. Genotypic Approaches for Reducing Iron Toxicity

Rice cultivars differ in their tolerance for iron toxicity and selection of rice cultivars with superior iron tolerance can be considered as an important biological mitigation strategy for reducing iron toxicity. Genetic differences in adaptation to and tolerance for iron toxic soil, conditions have indeed been exploited for developing rice cultivars with tolerance for iron toxicity (Sahrawat *et al.*, 2000; Sahrawat 2004; Baruah *et al.*, 2007; Nozoe *et al.*, 2008; Saikia and Baruah, 2012). Research on screening of rice cultivars to grow in Fe toxic soil has also been conducted under controlled conditions in nutrient solution experiments to identify iron tolerant cultivars as well as to establish the role of various factors involved in tolerance mechanism to Fe toxicity (Mandal *et al.*, 2004; Qu *et al.*, 2005; Saberi *et al.*, 2010; Zhang *et al.*, 2011). The selection of Fe tolerant rice cultivars is preferably better expressed in the field and is greatly influenced by the length of the growing season (Sahrawat 2004; Nozoe *et al.*, 2008). Breeding and screening efforts at the IRRI (Philipines) and WARDA (Africa) have identified a number of rice cultivars with improved tolerance to Fe toxicity for growing in iron prone soils. Nayak *et al.* (2008) evaluated 65 rice genotypes for growing in iron-toxic soil of Odisha, India.

9.2 Nutrient Management Practices

A high concentration of iron can cause nutrients imbalance through antagonistic relation, effects on the uptake of nutrients like P, K, Zn, Mn (Sahrawat, 2004; Baruah *et al.*, 2007; Fageria *et al.*, 2008). To study the involvement of other plant nutrients in the occurrence of and tolerance to Fe toxicity, Sahrawat (2004) determined the elemental compositions Fe tolerant (CK4) and susceptible (Bouake 189) lowland rice cultivars grown on an iron toxic Utisol under irrigated conditions in the field with and without the application of N, P, K and Zn. For both the rice varieties, there were no significant differences in elemental compositions. Silveira *et al.* (2007) found that except for Mn, the uptake of other nutrients was not impaired by iron toxicity. Several reports show that the applications of nutrients reduce iron toxicity maintaining improved growth rate and yield parameters. Especially application of K fertilizer is one of the best effective ways in reducing Fe toxicity in rice (Sahu *et al.*, 2001; Baruah *et al.*, 2007). K nutrition reduces Fe toxicity through enhanced root activity to exclude or retain iron in the root rhizosphere of rice plants (Sahrawat, 2004; Nayak *et al.*, 2008). Rice roots deficient in the ability to uptake K, P, Mn or Si are more susceptible to Fe toxicity than a healthy plant efficient with these nutrients (Backer and Asch, 2005; Sahrawat, 2010). Moreover, the combined applications of N, P, K and Zn significantly increase the rice grain and biomass yield in Fe toxic soil condition (Audebert *et al.*, 2006; Olaleye *et al.*, 2009).

9.3 Soil and Water Management

In situations where iron toxicity in the lowland is determined or aggravated by the influx of reduced iron from adjacent slopes, management options must consider factors that determine the extent of down slope water and solute movement. Such factors may include the amount and intensity of rainfall, the parent material and soil texture of the slopes, physical slope characteristics. In high-rainfall environments, water infiltration and percolation through the soil profile and up-welling of Fe-rich subsurface flow in the hydromorphic valley fringe is likely to be enhanced. Thus, the vast majority of Fe-toxic lowlands in West Africa are located in the high-rainfall humid-forest zone of Guinea, Liberia, Sierra Leone, and Cote d'Ivoire (WARDA, 2002). Also the severity of Fe toxicity in highland valleys in Madagascar appears to be more on the high-rainfall eastern than the moderate-rainfall western side of the island (Balasubramanian *et al.*, 1995).

Irrespective of the origin of the reduced iron, a broad range of soil-water and crop management interventions that either avoid the build-up of Fe^{2+} or reduce its negative effects on crop growth may be considered at plot level. These include measures that avoid a rapid drop in Eh in the bulk soil and the rhizosphere, the removal or re-oxidation of Fe in the bulk soil and the rhizosphere, the avoidance of an uncontrolled Fe influx into the root, and the application of mineral nutrients that strengthen the rice plants' defence mechanisms or act as "competing ions". Flush irrigation reduced the uptake of Fe compared to flood irrigation. Delayed flooding and shortened submergence periods have also been shown to reduce iron toxicity. Mid-season drainage of the field can remove some of the reduced iron together with the flood water and result in the temporary reoxidation of the topsoil thus substantially reducing the level of toxic Fe^{2+} and increasing crop yields (Backer and Asch, 2005; Sahrawat, 2010). When such drainage coincides with critical growth stages of rice (*e.g.*, seedling establishment, panicle initiation and flowering), the negative effects of excess Fe on crop yield can be significantly reduced (Olaleye *et al.*, 2009). The timeliness of such drainage and irrigation measures requires a near-perfect control of the irrigation water, a situation usually limited to fully irrigated environments and to the dry-season crop. A dense transplanting of rice with a high root oxidation power was hypothesized to avoid the accumulation of excess iron by increasing the share of the oxic rhizosphere compared to anoxic bulk soil (Backer and Asch, 2005).

9.4 Seed Cultivation Practices

There are indications that direct-seeded rice suffers less from Fe toxicity than transplanted rice. This may partially be related to the fact that the production of directly seeded rice is usually linked to systems with good irrigation-water control and thus the possibility to manage Fe by drainage. On the other hand, it is conceivable that root injury that particularly older seedlings endure in the process of pulling before transplanting presents an opening for uncontrolled Fe influx. This root damage is considerably more severe in seedlings grown in aerobic compared to flooded or water-saturated seedbeds.

The uncontrolled Fe^{2+} influx may be particularly severe when the injured root section is planted into the soil layer with highest Fe^{2+} concentrations (2–15 cm) (Backer and Asch, 2005). Transplanting depth could thus be used to avoid high soil- borne iron concentrations during the sensitive seedling stages of rice.

9.5 Fertilizer Management

Iron toxicity is associated with situations where the redox potential declines rapidly upon soil flooding. High levels of reducible iron combined with large amounts of organic matter hasten the occurrence and increase the severity of Fe-toxicity symptoms. Thus, during the initial stages of inundation, up to 60 per cent –80 per cent of the easily decomposable organic matter of rice soils is mineralized by micro organisms using Fe^{3+} as electron acceptor (Baruah *et al.*, 2007). In addition, organic acids can form complexes with Fe^{2+} and > 40 per cent of the total Fe^{2+} in the soil can be present in chelated form (Bashir *et al.*, 2010). Addition of large quantities of easily decomposable organic matter (*e.g.*, green manure) should be avoided, particularly in alluvial clays and acid sulfate soils (Clusters 1 and 2). The metabolic activity of the rhizoflora is regulated by the root membrane, which determines the efflux of organic exudates. Increased root permeability as a result of K deficiency can lead to an enhanced exudation of low-molecular weight metabolites, which stimulate the activity of Fe-reducing bacteria in the rhizosphere. Application of K fertilizer was shown to reduce the uncontrolled exudation of low-molecular-weight substances and hence to limit microbial iron reduction in the rhizosphere (Ramirez *et al.*, 2002; Sahrawat, 2004; Baruah *et al.*, 2007).

10. Research Needs

The extent of Fe-toxicity problems in tropical rice-based systems and the lack of widely adapted and adoptable solutions highlight the need to match crop adaptation mechanisms with site-specific factors associated with Fe toxicity and, in situations where adaptation is insufficient, the specific targeting of site-specifically appropriate agronomic management practices. In addition, the functioning and the genetic variability of adaptation mechanisms need to be understood for the successful targeting of cultivars to clearly defined environments (severity, time of occurrence, and duration of Fe-toxic stress). The processes and mechanisms will first be examined thereafter, following the pathway of Fe^{2+} from the rhizosphere to the leaf cells.

Although significant progress has been made in different aspects of Fe toxicity in a large number of plant species, most of the studies were conducted under laboratory conditions. Therefore, to get more realistic information we need to move from the laboratory to actual field conditions by using various approaches like QTL mapping inhibition, identification of functional genes, transgenic approaches and molecular approaches for iron toxicity tolerance (Shimizu, 2009).The targeting of the technical options needs to be strengthened with progress in rice breeding. Efforts to identify QTLs have focused on easily measurable traits such as leaf bronzing index, shoot and root dry weight, tiller number, plant height, root length, and Fe accumulation in the shoot (Dufey *et al.*, 2009). Simple sequence repeats (SSR) are the most widely-used DNA marker type to characterize germplasm collections of crops (Van Inghelandt *et al.*, 2010).

An improved understanding of crop adaptation for a range of iron-stress situations needs to be translated into repeatable and strong tools for the screening of improved rice cultivars (Onaga *et al.*, 2013). Recently, a few genes responsible for acquired tolerance to metal stress have been identified. Ectopic expression of these genes and its metabolizing enzymes in different species plant showed promising results (Hossain *et al.*, 2012). However, very few transgenic plants showed resistant reactions over a range of Fe toxicity Therefore, concerted efforts by various research domains will further increase our understanding of the fundamental mechanisms involved in hyperaccumulation processes that naturally occur in metal hyper accumulating plants.

11. Summary and Conclusion

Fe toxicity is a serious problem for sustainable crop productivity in low land rice soils and in acid soils of India and other parts of the world. Higher concentrations of Fe were found to be toxic to crop plants, particularly rice and upland crops. The last decade has seen rapid progress in our understanding of Fe toxicity mechanisms and plants' tolerance to soil acidity. Development of tolerant varieties of crops for Fe toxicity for sustainable agriculture in toxic soils can be the best solution to these physiological problems.Understanding both morphological and genetic diversity of a germplasm collection provides a basis for improvement of crops and development of resistant varieties. Finally, these cultivars must be targeted to the Fe toxic environment in which their adaptation mechanism will actually be translated into higher or more robust yields and may need to be supplemented with the appropriate agronomic management interventions. Only such an integrated approach seems to effectively address the problem of Fe toxicity in lowland rice. Research on genetic engineering, transgenic and implementation of molecular markers to facilitate breeding for Fe^{2+}tolerance/resistance may bring out complete solution to the problem of yield loss and sustainability of lowland and acid soil agriculture. Physiological response of crop varieties especially at field conditions in iron toxic soils will probably shed light on other associated plant nutrition related problems and can give new directions for future research.

Fe toxicity is a wide spread problem of rice cultivation in India and South-East Asia. Despite of current knowledge of the processes and mechanisms involved, the problem of Fe toxicity remains as an important constraint to rice production in wetland soils primarily because of the complex nature of the chemistry of iron toxicity. Although some attempts have been made to ameliorate such diverse conditions, no sustainable and economically feasible technology has been available for increasing productivity in the affected areas. Identification of rice genotypes with physiological tolerance to iron toxicity and further improvement by breeding may be an effective means for ameliorating iron toxicity problem. The breeding effort could be facilitated by a better understanding of the genetic background and physiological mechanisms related to tolerance of Fe toxicity of a genotype. Fe toxicity in lowland rice has been more complex due to Fe toxicity induced multiple nutritional disorders caused by deficiency of other nutrients like K, P, Ca and Mg. Several other factors such as hydrogen sulphide and organic acids in the soil are associated with Fe toxicity. Soil pH, soil organic matter, soil redox potential and cation exchange capacity also affect the intensity of toxicity. A range of agronomic management interventions have been advocated to reduce the Fe^{2+} concentration in the soil which contribute to the mechanism of ameliorating the toxicity problem even if sensitive varieties are planted in the field with high Fe concentration. However, more knowledge on different management practices and their relation with soil-water chemistry, seed cultivation practices and fertilizer application are required so that toxicity impact of higher Fe can be minimised in farmers' field.

References

Agarwal, S., Sairam, R. K., Meena, R. C., Tyagi, A and Srivastava G. C. (2006). Effect of excess and deficient levels of iron and copper on oxidative stress and antioxidant enzymes activity in wheat. *Journal of Plant Science*.1: 86-97.

Ando, T. (1983). Nature of oxidizing power of rice roots. *Plant Soil*.72: 57–71.

Armstrong, J., Armstrong, W., Beckett, P.M., Halder, J.E., Lythe, S., Holt, R and Sinclair, A. (1996). Pathways of aeration and the mechanisms and beneficial effects of humidity-and Venturi-induced convections in *Phragmites australis* (Cav.) Trinex Steud. *Aquatic Botany*. 54: 177–197.

Armstrong, W., Cousins, D., Armstrong, J., Turner, D.W and Beckett, P.M. (2000). Oxygen distribution in wetland plant roots and permeability barriers to gas-exchange with the rhizosphere: a microelectrode and modelling study with *Phragmites australis*. *Annals of Botany*. 86: 687–703.

Arora, A., Sairam, R. K. and Srivastava, G. C. (2002). Oxidative stress and antioxidative system in plants. *Current Science*. 82: 1227-1338.

Audebert, A. (2006). Iron partitioning as a mechanism for iron toxicity tolerance in lowland rice. In: Audebert A, Narteh LT, Kiepe P, Millar D, Beks B. (eds) Iron Toxicity in Rice-based Systems in West Africa, Africa Rice Center (WARDA), Print Right, Ghana. pp. 34-46.

Backer, M and Asch, F. (2005). Iron toxicity in rice-conditions and management concepts. *Journal of Plant Nutrition and Soil Science*. 168: 558-573.

Balasubramanian, V., Rabeson, R., Razafinjara, L and Ratsimandresy, J. (1995). Rice soil constraints and fertility management in the highlands of Madagascar, in IRRI: Fragile Lives in Fragile Ecosystems. The International Rice Research Institute, Manila, Philippines, p. 313–324.

Baruah, K.K., Nath, B.C. (1996). Changes in growth, ion uptake and metabolism of rice (*Oryza sativa* L) seedlings at excess iron in growth medium. *Indian Journal of Plant Physiology*. 2: 122-125.

Baruah, K.K., Nath, B.C. (1997). Ion uptake, metabolism and yield of rice (*Oryza sativa* L.) at excess iron in the growth medium. Ando, T *et al.*, eds, *Plant Nutrition for Sustainable Food Production and Environment*, In Kluwer Academic Publishers, Tokyo pp. 403-404.

Baruah, K.K. and Nath, B.C. (2001). Physiological and biochemical traits of rice associated with tolerance to iron toxicity. In: *Stress and Environmental Plant Physiology*. Bora, K.K. *et al.*, eds, In Pointer Publisher, Jaipur. pp. 296-306.

Baruah, K. K., Nath, B. C and Gogoi, N. (2001). Physiological and biochemical trait of rice (*Oryza sativa* L) genotypes associated with tolerance to iron toxicity. In W. J. Horst *et al.*, eds. Plant *Nutrition-Food Security and Sustainability of Agro-ecosystem*. Kluwer Academic Publishers, Dordrecht, pp. 476-477.

Baruah, K. K., Das, S and Das, K. (2007). Physiological disorder of rice associated with high levels of iron in growth medium. *Journal of Plant Nutrition*. 30: 1871-1883.

Bashir, K., Ishimaru, Y and Nishizawa, N. K. (2010). Iron uptake and loading into rice grains. *Rice*. 3: 122-130.

Blossfeld, S., Gansert, D., Thiele, B., Kuhn, A. J and Losch, R. (2011). The dynamics of oxygen concentration, pH value, and organic acids in the rhizosphere of *Juncus* spp. *Soil Biology and Biochemistry*. 43: 1186–1197.

Briat, J. F., Celllier, F and Gaymard, F.(2006), Ferritins and iron accumulation in plants tissues, In. Barton, LL; Abadia, J. (Edt), *Iron Nutrition in Plants and Rhizospheric Microorganisms*. New York: Springer, Chapter 17: 341-357.

Briat, J. F., Curie, C and Gaymard, F. (2007). Iron utilization and metabolism in plants. *Current Opinion in Plant Biology*. 10: 276-282.

Colmer, T. D (2003a). Long-distance transport of gases in plants: a perspective on internal aeration and radial oxygen loss from roots. *Plant, Cell and Environment*. 26: 17–36.

Das, K. N., Bordoloi, P. K and Bora, N. (1997). Tolerance level of iron in irrigation water for rice crop. *International Journal of Tropical Agriculture*.15: 59–166.

Dobermann, A and Fairhurst, T.H. (2000). Rice: Nutrient disorder and nutrient management. Oxford Graphic Printers Pte Ltd, IRRI, Manila, Philippines.

Dordas, C., Rivoal.J and Hill, R. D. (2003). Plant haemoglobins, nitricoxide and hypoxic stress. *Annals of Botany*. 91: 172–178.

Dorlodot, S., Lutts, S and Bertin, P. (2005). Effect of ferrous iron toxicity on the growth and mineral composition of interspecific rice. *Journal of Plant Nutrition*. 28: 1-20.

Drew, M. C., He, C. J and Morgan, P. W. (2000). Programmed cell death and aerenchyma formation in roots. *Trends in Plant Science*. 5: 123–127.

Dufey, I., Hakizimana, P., Draye, X., Lutts, S and Bertin, P. (2009). QTL mapping for biomass and physiological parameters linked to resistance mechanisms to ferrous iron toxicity in rice. *Euphytica.* 167: 143-160.

Evans, D. E. (2003). Aerenchyma formation. *New Phytologist.* 161: 35–49.

Fageria, N. K., Santos, A. B., Filho, M.P. and Guimaries, C. M. (2008). Iron toxicity in lowland rice. *Journal of Plant Nutrition.* 31: 1676-1697.

Garthwaite, A.J., Von Bothmer, R., Colmer, T.D. (2003). Diversity in root aeration traits associated with waterlogging tolerance in genus Hordeum. *Functional Plant Biology.* 30: 875–889.

Halliwell, B. (2006). Reactive species and antioxidants. Redox biology is a fundamental theme of aerobic life. *Plant Physiology.* 141: 312-322.

Hossain, M. A., Piyatida, P., Jaime, A., da Silva, T and Fujita, M. (2012). Molecular mechanism of heavy-metal toxicity and tolerance in plants: Central role of glutathione in detoxification of reactive oxygen species and methylglyoxal and in heavy-metal chelation. *Journal of Botany* (Article ID 872875).

Inoue, H., Kobayashi, T., Nozoye, T., Takahashi, M., Kakei, Y., Suzuki, K., Nakazono, M., Nakanishi, H., Mori, S and Nishizawa, N. K. (2009). Rice OsYSL15 is an iron-regulatediron(III)-deoxymugineic acid transporter expressed in the roots and is essential for iron uptake in early growth of the seedlings. *Journal of Biological Chemistry.* 284: 3470-3479.

Ishimaru, Y., Suzuki, M., Tsukamoto, T., Suzuki, K., Nakazono, M., Kobayashi, T., Wada, Y., Watanabe, S., Matsuhashi, S., Takahashi, M., Nakanishi, H., Mori, S and Nishizawa, N.K. (2006). Rice plants take up iron as an Fe^{3+}-phytosiderophore and as Fe^{2+}. *Plant Journal.* 45: 335–346.

Ishimaru, Y., Takahashi, R., Bashir, K., Shimo, H., Senoura, T., Sugimoto, K., Ono, K., Yano, M., Ishikawa, S., Arao, T., Nakanishi, H and Nishizawa, N. (2012). Characterizing the role of rice NRAMP5 in manganese, iron and cadmium transport. *Scientific Reports.* 2: 286.

Jackson, M. B and Armstrong, W. (1999). Formation of aerenchyma and the processes of plant ventilation in relation to soil flooding and submergence. *Plant Biology.* 1: 274–287.

Kozela, C and Regan, S. (2003). How plants make tubes. *Trends in Plant Science.* 8: 159–164.

Majerus, V., Bertin, P., Swenden, V., Fortemps, A., Lobréaux, S and Lutts, S. (2007b). Organ-dependent responses of the African rice to short-term iron toxicity: ferritin regulation and antioxidative responses. *Biol. Plant.* 51: 303-312.

Mandal, A. B and Roy, B. (2005). Increased Fe- toxicity tolerance in rice calli and modulation in isonzyme profiles. *Indian Journal of Biotechnology.* 4: 65-71.

Mandal, A. B., Basu, A. K., Roy, B., Sheeja, T. E and Roy, T. (2004). Genetic management for increased tolerance to aluminium and iron toxicities in rice-A review. *Indian Journal of Biotechnology.* 3: 359-368.

Marschner, H. (1995). Mineral nutrition of higher plants. Academic Press, London, 313-323.

Mayer, A. M. (2006). Polyphenol oxidase in plants and fungi: Going place? A review. *Phytochem*. 67: 2318-2331.

McDonald, M. P., Galwey, N.W and Colmer, T.D. (2001). Waterlogging tolerance in the tribe triticeae: the adventitious roots of *Critesion marinum* have a relatively high porosity and a barrier to radial oxygen loss. *Plant, Cell and Environment*.24: 585–596.

McDonald, M.P., Galwey, N.W and Colmer, T.D. (2002). Similarity and diversity in adventitious root anatomy as related to root aeration among a range of wetland and dryland grass species. *Plant, Cell and Environment*. 25: 441–451.

Murgia, I., Briat, J. F., Tarantino, D and Soave, C. (2001). Plant ferritin accumulates in response to photoinhibition but its ectopic overexpression does not protect against photoinhibition. *Plant Physiology and Biochemistry*. 39: 797–805.

Murgia, I., Delledonne, M and Soave, C. (2002). Nitric oxide mediates iron-induced ferritin accumulation in Arabidopsis. *Plant Journal*. 30: 521–528.

Neubauer, S.C., Toledo-Duran, G.E., Emerson, D and Megonigal, J.P. (2007). Returning to their roots: iron-oxidizing bacteria enhance short-term plaque formation in the wetland-plant rhizosphere. *Geomicrobiology Journal*. 24: 65–73.

Nishiuchi, S., Yamauchi, T., Takahashi, H., Kotula, L and Nakazono, M. (2012). Mechanisms for coping with submergence and water logging in rice, *Rice* 5: 2 (http://www.thericejournal.com/content/5/1/2).

Nozoe, T., Agbisit, R., Fukuta, Y., Rodriguez, R and Yanagihara, S. (2008). Characteristics of iron tolerant rice lines developed at IRRI under field conditions. *JARQ*. 42: 187—192.

Olaleye, A. O., Ogunkunle, A. O., Singh, B. N., Akinbola, G. E., Tabi, F. O., Fayinminu, O. M and Iji, E. (2009). Ratios of nutrients in lowland rice grown on two iron toxic soils in Nigeria. *Journal of Plant Nutrition*.32: 1-17.

Onaga, G., Egdane, J., Edema, R and Abdelbagi, I. (2013). Morphological and genetic diversity analysis of rice accessions (*Oryza sativa* L.) differing in iron toxicity tolerance. *Journal of Crop Science and Biotechnology*. 16 (1): 53—62.

Ponnamperuma, F.N., Bradfield, R and Reech, M. (1955). Physiological disease of rice attributable to iron toxicity. *Nature*. 175: 265.

Parent, C., Capelli, N., Berger, A., Crevecoeur, M and Dat, J. F. (2008). An overview of plant responses to soil waterlogging. *Plant Stress*. 2: 20-27.

Pradeet, K., Ottow, J. C. G and Jacq, V. A. (1993). Excessive iron uptake (iron toxicity) by wetland rice (*Oryza sativa* L.) on an acid sulphate soil in the Casamance/ Senegal. International Institute for Land Reclamation and Improvement, Wageningen. The Netherlands. *ILRI Public*. 44: 150–162.

Qu, I. Q., Yoshihara, T., Ooyama, A., Goto, F and Takaiwa, F. (2005). Iron accumulation does not parallel the high expression level of ferritin in transgenic rice seeds. *Planta*. 222: 225-233.

Ramirez, L.M., Claassen, N., Ubiera, A.A., Werner, H and Moawad, A.M. (2002). Effect of phosphorous, potassium and zinc fertilizers on iron toxicity in wetland rice (*Oryza sativa* L). *Plant and Soil* 239: 179—206.

Ravet, K., Touraine, B., Boucherez, J., Briat, J. F., Gaymard, F and Cellier, F. (2009). Ferritins control interaction between iron homeostasis and oxidative stress in Arabidopsis. *Plant Journal* 57: 400–412.

Saberi, H. K., Rengel, Z., Wilson, R and Setter, T.L. (2010). Variation for tolerance to high concentration of ferrous iron (Fe^{2+}) in Australian hexaploid wheat. *Euphytica*. 172: 275-283.

Sahrawat, K. L. (2003). Iron toxicity in wetland rice: Occurrence and Management through integration of genetic tolerance with plant nutrition. *Journal of the Indian Society of Soil Science*. 51: 409-417.

Sahrawat, K. L. (2004). Iron toxicity in wetland rice and role of other nutrients. *Journal of Plant Nutrition*. 27: 1471-1504.

Sahrawat, K. L. (2010). Reducing iron toxicity in lowland rice with tolerant genotypes and plant nutrition. *Plant Stress*. Global Science Books.

Sahrawat, K.L., Diatta, S and Singh, B. N. (2000). Reducing iron toxicity in lowland rice through an integrated use of tolerant genotypes and plant nutrient management. *Oryza* 37: 44-47.

Sahrawat, K. L.(2000). Elemental composition of the rice plant as affected by iron toxicity under field conditions. *Comm. Soil Sci. Plant Anal.*, 31: 2819-2827.

Sahu, S.K., Sandha, B and Dev, G. (2001). Relationship between applied potassium and iron toxicity in rice. *International Rice Research Notes*, 26: 52-53.

Saikia, T and Baruah, K.K.(2012). Iron toxicity tolerance in rice (*Oryza sativa*) and its association with anti-oxidative enzyme activities. *Journal of Crop Science*. 3: 90-94.

Schmidt, W.(1999). Mechanisms and regulation of reduction-based iron uptake in plants. *New Phytol*. 141: 1–26.

Shimizu, A. (2009). QTL analysis of genetic tolerance to iron toxicity in rice (*Oryza sativa* L.) by quantification of bronzing score. *J. New Seeds*. 10: 171-179.

Silveira, V. C., Oliveira, A. P., Sperotto, R. A., Espindola, L. S., Amaral, L., Dias. J. F., Cunha, J. B and Fett, J. P. (2007). Influence of iron on mineral status of two rice (*Oryza sativa* L.) cultivars. *Brazilian Journal Plant Physiology*. 19: 127-139.

Sperotto, R. A., Vasconcelos, M. W., Grusak, M.A and Fett, J. P. (2012). Effect of different Fe supplies on mineral partitioning and remobilization during the reproductive development of rice (*Oryza sativa* L.). *Rice*. 5: 27.

Stein, R.J., Duarte, G.L., Spohr, M.G., Lopes, S.I.G and Fett, J.P. (2009). Distinct physiological responses of two rice cultivars subjected to iron toxicity under field conditions. *Annals of Applied Biology*.154: 269-277.

Tadano, T. (1976). Studies on the methods to prevent iron toxicity in the lowland rice. Mementos of the Faculty of Agriculture HokkaidoUniversity 10: 22-68.

Van Inghelandt, D., Melchinger, A.E., Lebreton, C and Stich, B. (2010). Population structure and genetic diversity in a commercial maize breeding program assessed with SSR and SNP markers. *Theor. Appl. Genet.* 120: 1289-1299.

Visser, E.J.W., Colmer, T.D., Blom, C.W.P.M., Voesenek, L.A.C.J. (2000). Changes in growth, porosity, and radial oxygen loss from adventitious roots of selected mono and dicotyledonous wetland species with contrasting types of aerenchyma. *Plant, Cell and Environment*, 23: 1237-1245.

Wang, Y. P., Yu-Huan, W. U., Peng, L. I. U., Zheng, G. H., Zhang, J. p. and Xu, G. D. (2013). Effectsof potassium on organic acid metabolism of Fe-sensitive and Fe-resistant rices (*Oryza sativa* L.), *AJCS*. 7: 843-848.

Yamauchi, T., Rajhi, I and Nakazono, M. (2011). Lysigenous aerenchyma formation in maize root is confined to cortical cells by regulation of genes related to generation and scavenging of reactive oxygen species. *Plant Signaling and Behavior* 6: 759-761.

Zancani, M., Peresson, C., Tubaro, F., Vianello, A., Macri, F. (2007). Mitrochondrial ferritin distribution among plant organs and its involvement in ascorbate-mediated iron uptake and release. *Plant Science*. 173: 182-189.

Zhang, Y., Zhang, G.H., Liu, P., Song, J,M., Xu, G.D., Cai, M.Z. (2011). Morphological and Physiological responses of root tip cells to Fe2+toxicity in rice. *Acta Physiologia plantarum*. 33: 683-689.

Recent Advances in Crop Physiology Vol. 2 (2016)   *Pages* **225–236**
Editor: **Dr. Amrit Lal Singh**
Published by: **DAYA PUBLISHING HOUSE, NEW DELHI**

Chapter 7

Can Water Deficit be Useful in Potato?–Some Issues

Devendra Kumar[1]* and J.S. Minhas[2]

[1]*ICAR-Central Potato Research Institute Campus,*
Modipuram, Meerut – 250 110, Uttar Pradesh
[2]*ICAR-Central Potato Research Station,*
PO Model Town, PB No. 1, Jalandhar – 144 003, Punjab

1. Introduction

Potato (*Solanum tuberosum* L.) is regarded as sensitive to water-stress (Salter and Goode, 1967) with yields frequently constrained by drought in most of the environments. The effects of water stress on a plant depend on the timing of the stress, its duration and severity. Transient water-stress, induced by high evaporative demand, may have acute, but only temporary effects on plant growth. Such stress commonly occurs even in well watered crops, particularly in hot, sunny conditions when evaporative demand exceeds the capacity of roots to supply water to shoots. In contrast, prolonged drought with gradual depletion of soil moisture affords greater opportunity for adaptation by the plant and may also induce chronic effects on plant growth which lead to premature senescence. Adaptations to deficit water stress involve variations in several morphological, biochemical and physiological traits and most of them are negative for the plant productivity. Nevertheless there are instances, across the crops, where water stress as an independent event has altered some traits/ constituents which attract attention and hopefully can be subjected to exploitation/ utilization. In the present article, attempt has been made to cite some of the responsive

* *Corresponding Author:* E-mail: devkmalik@gmail.com

traits which were adaptive for plant and appeared useful, depending upon the targeted needs.

In rose bush, plant branching, which results from axillary bud burst, governs plant architectural development, shape and visual quality. Bud burst is known to be sensitive to environmental conditions, including long-term changes in water status or light intensity. Sabine *et al.* (2013) showed that temporary water or light restrictions promote the branching and development of rose bush axes. They suggest that sugar metabolism and hormonal regulation may be involved in stimulating branching after the release of these two types of constraint. In sweet potato, Ekanayake and Collins (2004) demonstrated that the use of drought resistant genotypes and better water management practices can enhance its' root quality and yield. Sweet potato is sensitive to water deficit stress. The response of sweet potato genotypes were evaluated to 4 seasonal irrigation regimes, during 3 summer crop seasons in Lima, Peru. The significant genotypic (G) differences, irrigation (I) effects, and G x I interaction effects were observed for leaf water potential (y) indicating different internal tissue stress levels in various irrigation treatments. Drought stress significantly reduced nitrogenous compounds and root yield. The lower content of nitrogenous compounds and higher dry matter are preferred traits from angle of product consumption. In contrast, root dry matter increased as water stress increased and was the most sensitive root quality trait. It was advocated that higher root dry matter is the best indicator for drought resistance and therefore breeding efforts to select newly improved sweet potato cultivars with high dry matter content is warranted in drought-prone tropics. FarshidVazin (2013) investigated the effects of different levels of water deficit (I_{100}, I_{50} and I_0) on yield, fatty acids and essential oil yield and composition of Cumin and reported that moderate water deficit increased the essential oil yield. The main essential oil constituents were p-Menta-1,3-diene-7-al, cuminaldehyde, γ-Terpinene, p-Cymene and β-Pinene which showed a decreasing of their contents under very high stress. These results demonstrated that the essential oil monoterpene hydrocarbons compounds increased with moderate water deficit. It was highlighted that water deficit decreased primary metabolites but it increased secondary metabolites.

2. Candidate Event in Potato: Tuberization

Higher proportion of seed potatoes is indispensable for profitable production system(field crop, net house, aeroponics). The process of tuber initiation and its development is a very complex event in potato. Tuberization is immensely influenced by genetic back ground, biotic (viruses etc.) and abiotic factors, environmental (photoperiod, temp, light) and edaphic (mineral nutrients, water availability). These environmental factors act by the synthesis, destruction, transport and activation of growth substances. Water availability also alter growth–substance balance in plants hasten tuberization in *Solanum* potatoes (Ito and Kato,1951). Cessation of shoot and leaf growth under water stress leads to initiation of tubers much earlier but only grow slowly because of smaller productive area.Cavagnaro*et al.* (1971) observed that although more tubers initiated when plants were exposed to water shortage during tuberization, fewer were harvested because of an increase in the number of tubers or incipient tubers that were reabsorbed.

However, Minhas and Bansal (1991) did not found varietal differences in tuber number when water stress was imposed at stolon initiation, tuber initiation, early tuber bulking and late bulking growth phases in three potato cultivars *i.e.*, Kufri Lauvkar, Kufri Chandramukhi and Kufri Jyoti.

Studies on water stress at Modipuram during 2004-07 showed that number of tubers was influenced differently in various potato varieties and by saving one irrigation, the number of small tubers increased in 2, 16 and 20 genotypes in 2004-05, 2005-06 and 2006-07, respectively (Kumar and Minhas, 2013). The number of total tubers increased in 7, 5 and 4 genotypes in 2004-05, 2005-06 and 2006-07, respectively (Table 7.1). Variation in number of tubers under stress also varied in different years which signifies and underlines the role of other climatic factors mainly temperature regime and light.

Table 7.1: Influence of Water Deficit on Number of Small Size (<25g) Tubers in various Genotypes of 90 Days Duration (Average of three years) at Meerut (⁺more tubers)

S.N.	Varieties/Genotypes	2004-05		2005-06		2006-07		Potential/ Frequency
		Small	Total	Small	Total	Small	Total	
1.	K. Alankar							
2.	K. Anand*			+	+	+		3/6
3.	K. Ashoka					+		
4.	K. Badshah*			+	+	+		3/6
5.	K. Bahar							
6.	K. Chamatkar							
7.	K.Chandramukhi*		+	+		+		3/6
8.	K. Chipsona-1					+		
9.	K. Chipsona-2*	+	+			+	+	4/6
10.	K. Deva							
11.	K. Giriraj			+		+		
12.	K. Jawahar					+	+	
13.	K. Jyoti			+				
14.	K. Kanchan							
15.	K. Khasigiri			+				
16.	K. Kuber*			+	+		+	3/6
17.	K. Kumar*			+	+	+		3/6
18.	K. Kundan			+				
19.	K. Lalima		+			+		
20.	K. Lauvkar							
21.	K. Megha					+		
22.	K. Muthu					+		

Contd...

Table 7.1–*Contd...*

S.N.	Varieties/Genotypes	2004-05		2005-06		2006-07		Potential/ Frequency
		Small	Total	Small	Total	Small	Total	
23.	K. Naveen			+		+		
24.	K. Neela					+	+	
25.	K. Pukhraj*		+	+	+	+		4/6
26.	K. Red							
27.	K. Safed			+				
28.	K. Sheetman		+	+				
29.	K. Sherpa		+					
30.	K. Sinduri							
31.	K. Sutlej							
32.	K. Swarana			+		+		
33.	K. Surya							
34.	HT/93-707		+			+		
35.	HT/94-203			+		+		
36.	K. Arun					+		
37.	Kathadin			+				
38.	LT-2					+		
39.	LT-7							
40.	Desiree	+						
	Total	**2**	**7**	**16**	**5**	**20**	**4**	**7**

These results indicate the influence of water stress on actual numbers of tuber *per se* and also on the shifting of tubers among different size categories. Data on minimum temperature particularly in the year 2005-06 and 2006-07 showed that least temperature dip in the month of December and January (Table 7.2) has added impact on numbers of small tuber. Considering the frequency of increase in tuber numbers under water stress, a total of 7 genotypes/varieties indicated some potential to exploit this response to customer advantage. Noticeable among them are K. Anand, K. Badshah, K. Chipsona-2 and K. Pukhraj with highest frequency (Table 7.2). These studies suggest that water stress event can be attempted/experimented as an external stimulus for triggering the tuberization in selected variety to enhance the proportion of seed size tubers. Successful exploitation in either of the production system *viz.* field, glass house, controlled environment etc can be replicated for seed production technologies through net house, aeroponics and so on.

Four potato cultivars differing in their precocity and contrasted for their drought tolerance were investigated by Ouiam and Jean (2005) in the field and two of them in the greenhouse. They were subjected to two water treatments, well irrigated and droughted. The evolution of stolon number, stolon length, mass of roots and the adventitious roots which are formed on stolons were monitored to check the

Table 7.2: Monthly Meteorological Parameters during Crop Seasons at Meerut

Months	2004-05			Monthly BSS* (hr)	2005-06			Monthly BSS* (hr)	2006-07			Monthly BSS* (hr)
	Monthly Mean Air Temperature (°C)				Monthly Mean Air Temperature (°C)				Monthly Mean Air Temperature (°C)			
	Max	*Min*	*Mean*		*Max*	*Min*	*Mean*		*Max*	*Min*	*Mean*	
Oct	34.0	13.5	23.8	6.05	33.5	13.0	23.3	7.7	36.0	10.0	23.0	7.6
Nov	28.6	9.5	19.1	6.06	31.0	6.2	18.6	7.9	31.0	6.5	18.8	6.9
Dec	28.0	3.0	15.5	4.26	27.0	2.0	14.5	6.2	27.0	3.0	15.0	5.8
Jan	24.0	3.2	13.6	4.49	26.0	1.0	13.5	6.3	27.0	1.5	14.3	5.6

* Bright sunshine hours.

response of these characters to water shortage throughout the cycle, and examine if they could have, especially the adventitious roots on stolons, contributed to drought tolerance. On expected line, they noted that all physiological and agronomical studied parameters were sensitive to drought. Maximum root dry mass was reduced by drought. Stolon number was enhanced by drought but the total length of stolons, was reduced. The number of the adventitious roots on stolons was reduced by drought and negatively related to root dry mass of the plant. Fresh tuber yield was significantly correlated to root dry mass in the field and drought tolerance index was significantly associated to root depth in the field, but in stressed conditions only.

3. Production of 'Baby' Potatoes

Another area for exploitation of water stress as an external stimulus for tuberization could be the production of baby potatoes. Nutritionally, baby potatoes (< 40mm) are low in starch and high in fiber (with skin on) and therefore considered healthier than the large size potatoes (Jha, 2005).

Moreover, these 'little luxury' are being preferred for high-end retail and food service markets (Figure 7.1). Recently a promising hybrid has been identified for this segment under the sub project "Value chain on potato and potato products" funded by NAIP. Evaluation of field trial indicated that number of baby size and total tubers (812 and 1247 '000' ha^{-1}, respectively) were significantly higher in potato hybrid HT/03-704 than control variety Kufri Himsona (593 and 850 '000' ha^{-1}) in 60 days crop. Likewise baby grade tuber yield (11.5 t ha^{-1}) was better by 10.6 per cent against control (10.4 t). These yields (10-12 ha^{-1}) in baby size potatoes are very low and need more attention. Short crop duration (60 days) has been used to have more number of small tubers. The crop duration can be extended further provided more tubers are initiated through water stress event. Some genotypes are known to

Figure 7.1: Baby Potatoes in a NC Mall.

induce more tubers even when they have a history of drought stress exposure (Pankaj *et al.*, 2013). Under such situation productivity of this segment can be improved and instead of two crops of 60 days in one season, farmers may opt for this new segment.

In a green house study (17 per cent RH) to examine the resistance mechanism in potato genotypes to stress acclimation, resistance and recovery, Pankaj *et al.,* 2013 observed that among the three contrasting genotypes (Russet Burbank, U 1002 and Fu 12) genotype U1002 having treatments with a pervious drought stress history induced more tubers in 5-50 g category in the subsequent generation compared to treatments with no drought stress history. Russet Burbank, Fu 12 and U 1002 showed high resistance, moderate resistance and sensitivity, respectively.

4. Acclimatization of TPS Seedlings

Prior exposure to mild stress in seedlings enhance establishment and yield of true potato seeds (TPS). Seed priming is the induction of a particular physiological state in plants by the treatment of natural and synthetic compounds to the seeds before germination. In plant defence, priming is defined as a physiological process by which a plant prepares to respond to imminent abiotic stress more quickly or aggressively. In recent years, seed priming has been developed as an indispensable method to produce tolerant plants against various stresses such as water deficit, high salinity, extreme temperature, submergence, etc. Jisha *et al.* (2013) has recently reviewed various studies in which seed priming is practiced and various seed-priming methods and their effects were discussed. Plants raised from primed seeds showed sturdy and quick cellular defence response against abiotic stresses. Priming for enhanced resistance to abiotic stress obviously is operating via various pathways involved in different metabolic processes.

The seedlings emerging from primed seeds showed early and uniform germination. Moreover, the overall growth of plants is enhanced due to the seed-priming treatments. In potato, botanical seed true potato seeds (TPS) can be used for commercial cultivation mainly by two methods: (i) raising the seedlings in nursery beds and transplanting them directly in field and (ii) *In situ* tuber production in nursery beds *i.e.* seedling tubers for raising crop in the next season. Apart from genotypic variability in TPS lines, the time lag between seedling uprooting and their transplanting in field also influence seedling survival, vigour and tuber yield. Kumar and Pandey (2006) conducted experiments during 2005-06 with three TPS lines *viz.* HPS I/13, C-3 and 92-PT-27 to find out: (a) the physiological changes in the seedlings during different time periods (0, 12 and 24 hr) between uprooting and transplanting and (b) the influence of resultant physiological changes on seedling vigour/hardiness and their productivity.

Evaluation of seedlings under *in vitro* water stress (PEG 40 per cent) showed that injury to cell membrane in root tissues increased progressively with the increase in time periods (0, 12 and 24 hr) of up-rooting. Proline contents declined in root tissues in 12 and 24 hr treatments (140 and 120 μg/g FWas against 0 hr, (180 μg/g FW) which probably caused higher injury to root tissues in the former treatments. Seedlings, however, did not show any adverse effect of being susceptible to water stress in root tissues because, as a practice, proper watering was ensured during their establishment. Consequently, the seedling survival was >90 per cent in all treatments. Leaf proline was increased in 12 and 24 hr treatments (140-170 μg/g FW) as compared to freshly uprooted seedlings (120 μg/g FW). It appears that dehydration of a few roots in 12 and 24 hr treatments may generate a chemical response that moves to the shoot, even if the flux of water has not yet been reduced as the foliage of uprooted seedlings were wrapped with a piece of wet hessian cloth. Enhanced proline content in leaf tissues in 12 and 24 hr treatments appears to impart tolerance to mid-day water stress subsequently during growth and development in the transplanted crop as against control. Mid-day water stress is commonly observed even in well-watered normal crops raised from seed-tubers in sandy soils. Plants were more vigorous and tuber fresh wt. was significantly higher (34.0 g/3plants) even as early as 45 day stage in 12 and 24 hr treatments than the 0 hr treatment (19.4 g/3plants). Leaf area was also higher in 12 and 24 hr treatments (1000-1100 cm^2/3plants) as compared to control (851 cm^2/3plants). Besides, stomatal resistance (RS) was less in 12 and 24 hr (0.48 - 0.55 s cm^{-1}) treatment than in 0 hr (0.68 s cm^{-1}). The higher leaf area and low RS in plants developed from the seedlings uprooted 12 or 24 hr before transplanting suggests the possibility of higher photosynthates. The morpho-physiological superiority in crop grown from seedlings transplanted after 12 and 24 hr of up-rooting was sustained up to 90 days. Marketable (>25g tuber) as well as total tuber yield at 90 days were significantly higher in 12 and 24 hr treatments (121-129 q/ha) than 0 hr. (108 q/ha). Higher tuber yield in crop raised from seedlings uprooted 12 or 24 hr before transplanting could be attributed to higher leaf proline, less RS and development of more vigorous canopy. These responses where prior exposure to stress event infuse hardiness in the plant system holds merit for further exploitation.

5. Adaptation and Resilient Productivity to Partial Root Drying/ Progressive Soil Drying

Progressive soil drying during mild stress has been shown to offer plant opportunity for chemical regulations. Lieu *et al.* (2005) showed that ABA regulates stomatal opening and photosynthetic water use efficiency in potato during progressive soil drying. Glasshouse experiments were conducted to investigate plant water relation and leaf gas exchange characteristics during progressive soil drying at two developmental stages (tuber initiation and tuber bulking) of potatoes. Leaf relative water content (RWC), leaf water potential (Ψ_l), root water potential (Ψ_r), stomatal conductance (g_s), photosynthesis (A), and xylem sap abscisic acid (ABA) concentration ($[ABA]_{xylem}$) were determined in well-watered (WW) and drought-stressed (DS) plants. At both stages, RWC and Ψ_l were hardly affected, significant decreases of the two parameters occurred only at severe soil water deficits; however, g_s decreased much early at 2 and 1 days after imposition of stress (DAIS) at tuber initiation and tuber bulking, respectively, and coincided with decrease of Ψ_r and increase of $[ABA]_{xylem}$; while A decreased 2 days later than g_s at each stage. Analyses of the pooled data of the two stages showed that g_s was linearly correlated with $[ABA]_{xylem}$ at mild soil water deficits (*i.e.* Ψ_r > -0.3 MPa); photosynthetic water use efficiency, *viz.* A/g_s, increased linearly with decreasing g_s until the latter reached 0.2 mol m^{-2} s^{-1}, below this point, it decreased sharply. The results suggest that at mild soil water deficits, g_s of potato is seemingly controlled by xylem-borne ABA. As a consequence of A being less sensitive than g_s to soil drying, photosynthetic water use efficiency, *i.e.* A/g_s, is increased at mild soil water deficits. Therefore, irrigation management practices involving mild water stress event can be employed and exploited to attain resilient yields.

6. Tuber Quality: Enhancement of Dry Matter and Storability

Tuber dry matter concentration is an important component of tuber yield which dependent on both the dry matter accumulated into tubers and tuber water content.

Tuber dry matter tends to show a progressive increase from the time of tuber initiation until maturity except when stress induces a remobilization of starch (Jefferies, 1995).Tuber water potential is related to soil water potential (Epstein and Grant 1973) and exhibit diurnal fluctuations. Under water stress conditions, tuber dry matter concentration increases because tuber water content is affected to a greater extent than dry matter accumulation. As long as tubers are attached to an active root system the re-watering of water stressed plants results in a decrease in dry matter concentration as the uptake of water is increased (MacKerron and Jefferies, 1985).

Cultivars showing greater change in tuber dry matter are drought sensitive in contrast to drought tolerant cultivars that show the least change in tuber dry matter concentration during drought. Selection of genotypes for stability in dry matter concentration over a range of soil moisture conditions may contribute towards increased drought tolerance. Drought increased tuber dry matter concentration in all the genotypes examined, the effect differed between genotypes with increasing tuber dry matter concentration by 12.9 per cent in clone 8906 AC-11 but by 8.3 per cent in cultivar up-to-date. Tourneux *et al.* (2003) also found that tuber dry matter content of cultivar Lucky and Clone CIP 382171.10 increase to drought by + 15 per cent and + 20 per cent in R_1 (Progressive drought after tuberisation, with a recovery period) and R_2 (Progressive drought after tuberisation without recovery) treatments respectively. Lahlou et al. (2003) studied four potato cultivars in the field and in green house with well watered and stressed treatment. They found 10 per cent increase in tuber dry matter in field and 8 per cent in the green house for Remarka under water stress while Desiree was not affected in field but its dry matter content increased by 11 per cent under drought in the green house and Nicola and Monalisa had on average increase of 25 per cent. Drought induced elevation in dry matter content can be aptly harnessed in processing sector such as production of starch or other dry matter based finished products. Considering such aspects, potato cultivar Atlantic was adjudged best for producing starch from the dry-land potatoes grown in western Nebraska in US (Alexander, 2013).

Apart from advantage of higher tuber dry matter *per se*, such tubers are expected to have good storability (due to lesser water content), compact and shining skin and possibly concentrated typical potato flavor.

7. Production of Desired Chemicals *viz.* Antioxidants for Use in other Fields (Pharmaceuticals, Medicines)

A better understanding of tuber secondary metabolites in potatoes could facilitate production of a more nutritious crop. Tubers having inherently higher antioxidants, are gaining ground in the market (Figure 7.2). Attempt can be made to further enrich such variety through mild water stress events. Phenylpropanoids have important role in potato physiology and influence the nutritional value. Phenylpropanoids metabolism is strongly affected by development (Roy *et al.*, 2013). Immature tubers were found to have more active phenylpropanoid metabolism, including variation in genotypes (up to three- fold) and responsive to environmental variations. Further, Roy *et al.* (2013) reported that a human feeding study with immature potatoes has clearly indicated a reduction in blood pressure.

8. Future Thrust

The candidate events such as tuberization, accumulation of some secondary metabolites, acclimatization in seedlings, elevated levels of antioxidants and involved enzyme systems, need carefully drawn roadmap for their scientific exploitation. Further, the judicious application of water (creating stress event through reduced use) is likely to offer indirect benefit to soil properties (aeration, texture, nutrient retention etc). In the process, a larger database will be developed for understanding the cross tolerance among other abiotic stresses such as heat, frost, high light intensity etc.

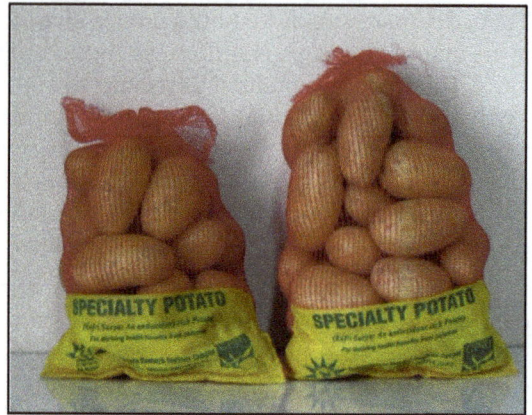

Figure 7.2: Antioxidant-Rich Potato Variety Kufri Surya.

9. Conclusion and Recommendations

Potato is regarded as sensitive to water-stress, but the effects depend on the timing of the stress, its duration and severity as the process of tuber initiation and its development is a very complex event and influenced by genetic back ground, biotic (viruses etc.) and abiotic factors, environmental (photoperiod, temp, light) and edaphic (mineral nutrients, water availability). Higher proportion of seed potatoes is indispensable for profitable production system. Water availability alter growth–substance balance in plants hasten tuberization in *Solanum* potatoes. Cessation of shoot and leaf growth under water stress leads to initiation of tubers much earlier but grow slowly because of smaller productive area and by saving one irrigation,the number of small tubers increased. Nutritionally, baby potatoes (< 40mm) are low in starch and high in fiber (with skin on) and considered healthier than the large size potatoes. Thus water stress can be used as an external stimulus for tuberization and production of baby potatoes.

Recently a promising hybrid (HT/03-704) has been identified with more number of baby size and total tubers with baby grade tuber yield (11.5 t ha^{-1}). However, these 10-12 t ha^{-1} yields in baby size potatoes are very low and need more attention using short duration (60 days) varieties which have more number of small tubers

Botanical seed true potato seeds (TPS) can be used for commercial cultivation mainly by two methods: (i) raising the seedlings in nursery beds and transplanting them directly in field and (ii) *In situ* tuber production in nursery beds *i.e.* seedling tubers for raising crop in the next season. The time lag between seedling uprooting and their transplanting in field influence seedling survival, vigour and tuber yield, but exposure to mild stress in seedlings enhance establishment and yield of true potato seeds (TPS).

Irrigation management practices involving mild water stress event can be employed and exploited to attain resilient yields. Tuber yield tends to show a progressive increase from the time of tuber initiation until maturity except when stress induces a remobilization of starch. Drought induced elevation in dry matter content can be aptly harnessed in processing sector such as production of starch or other dry matter based finished products. Considering such aspects, potato cultivar Atlantic was adjudged best for producing starch from the dry-land potatoes. Cultivars showing least change in tuber dry matter during drought are drought tolerant and the one showing greater change in tuber dry matter are sensitive. Such tubers are expected to have good storability (due to lesser water content), compact and shining skin and possibly concentrated typical potato flavor. A human feeding study with immature potatoes has clearly indicated a reduction in blood pressure.

References

Alexander, D. Pavlista (2013) Organic, Dryland potatoes for starch in western Nebraska. *Am. J.Potato Res.* 90: 144 pp.

Cavagnaro, J.B., de-Lis, B.R. and Tizio, R.M. (1971). Drought hardening of the potato plant as an after-effect of soil drought conditions at planting. *Potato Res.,* 14: 181-192.

Ekanayake, IJ and Collins,W (2004). Effect of irrigation on sweet potato root carbohydrates and nitrogenous compounds. *Journal of Food Agriculture and Environment* 2(1): 243-248.

Epstein, E. and Grant, W.J. (1973). Water stress relations of the potato plant under field conditions. *Agron. J.,* **65**: 400-404.

FarshidVazin (2013) Water stress effects on Cumin (*Cuminumcyminum* L.) yield and oil essential components. *Scientia Horticulturae.*151: 135-141.

Ito, H and Kato T (1951).the physiological foundation of the tuber formation of potato. Tohoku J. Agric. Res., 2: 1-14.

Jefferies, R.A. (1995) Physiology of crop response to drought. Pp. 61-74 In: A.J. Haverkort and D.K.L. MacKerron (ed.). *Potato Ecology and Modelling of Crops under Conditions Limiting Growth,* Kluwer Academic Publishers, Printed in Netherlands.

Jha DK (2005). Himalya bags mega order for baby potatoes from US. *Business Standard,* Mumbai, November 30, 2005.

Jisha, KC, K Vijayakumari and JT Puthur (2013). Seed priming for abiotic stress tolerance: an overview. *Acta Physiologiae Plantarum.* 35 (5): 1381-1396.

Kumar Devendra, J S Minhas. 2013. Evaluation of indigenous potato varieties, advanced clones and exotic genotypes against water deficit stress under sub-tropical environment. *Indian J. Plant Physiology* 18(3): 240-249.

Kumar Devendra and PCPande (2006). Studies on physiological attributes imparting hardiness in true potato seed (TPS) seedlings. (Abstr.) *In* National Seminar on *"Physiological and Molecular Approaches for the Improvement of Agricultural,*

Horticultural and Forestry Crops" organized by The Indian Society for Plant Physiology, IARI, New Delhi during November 28-30th 2006 at Kerala Agricultural University, Thrissur, Kerala. pp. 143.

Lahlou, Q, Ouattar S and Jean-François Ledent (2003).The effect of drought and cultivar on growth parameters, yield and yield components of potato. Agronomie **23**, 3 (2003) 257-268.

MacKerron, D.K.L. and Jefferies, R.A. (1985). Observations on the effects of relief of late water stress in potato, *Potato Res.,* 28: 349-359.

Minhas JS and KC Bansal (1991). Tuber yield in relation to water stress at different stages of growh in potato (*Solanumtuberosum L*). *J. Indian Potato Assoc.,* 18: 1-8.

OuiamLahlou; Jean-François Ledent, 2005. Root Mass and Depth, Stolons and Roots Formed on Stolons in Four Cultivars of Potato under Water Stress. *European Journal of Agronomy.* 22 (2)159-173.

Pankaj, Banik, HalenTaiand Karen Tanini. (2013). Resistance of potato genotypes to drought stress. *Am. J. Potato Res.* 90: 124 pp.

Roy, Navarre, RajaPayyavula, Shyam Kumar, Joe Vinson Rick Knowles, Joe, Kuhl and Alberto Pantoja. (2013) Secondary Metabolism and nutritionally important compounds in potatoes. *Am. J. Potato Res.* 90: 141 pp.

Sabine Demotes-Mainard, Lydie Huché-Thélier, Philippe Morel, Rachid Boumaza, Vincent Guérin and Soulaiman Sakr (2013). Temporary water restriction or light intensity limitation promotes branching in rose bush. *Scientia Horticulturae.* 150: 432-440.

Salter, P.J. and Goode, J.E. (1967).Crop responses to water at different stages of growth. *Research Reviews of the Commonwealth Bureau of Horticulture, East Malling* 2, pp. 93-97.

Tourneux, C., Devaux, A., Camacho, M.R., Mamani, P. and Ledent, J.F. (2003). Effect of water shortage on six potato genotypes: morphological parameters, growth and yield in the highlands of Bolivia (I). *Agronomie,* 23: 169-179.

Recent Advances in Crop Physiology Vol. 2 (2016)
Editor: Dr. Amrit Lal Singh
Published by: DAYA PUBLISHING HOUSE, NEW DELHI

Pages 237–259

Chapter 8

Bioregulators Ameliorate Water Deficit Stress in Wheat

*Sushmita and Pravin Prakash**

Department of Plant Physiology,
Institute of Agricultural Sciences, Banaras Hindu University,
Varanasi – 221 005, Uttar Pradesh

Introduction

Wheat (*Triticum aestivum* L.), a member of the family Graminae, is an important staple food crop of the entire world. It is the second most important cereal crop after rice in India. The cultivation of wheat dates back to more than 5000 years during the era of Indus valley civilization where the original species was *Triticum sphaerococcum* popularly known as Indian wheat, which has now been replaced by present day species- *Triticum aestivum* or the common bread wheat, *Triticum durum* or the macaroni wheat and the *Triticum dicoccum* or the emmer wheat. It occupies a unique position in human life as it is the main source of food and energy with a large number of end use products like chapati, bread, biscuits, pasta etc.

In the year 2012-13, wheat production was 93.5 million tonnes out of total food grain production of 257million tonnes. This record has been estimated to be broken by the wheat production in the year 2013-14 which is around 96 million tonnes and India is likely to produce 264 million tonnes of food grain. This improvement in wheat productivity (along with rice production) has made India a self sufficient nation in terms of foodgrain production. However, a major part of wheat growing belt falls under rainfed condition. Life cycle of wheat (*T. aestivum* and *T. durum*) in

* *Corresponding Author:* E-mail: prakashpbhu@gmail.com

India covers the period from Oct-Nov to Mar-Apr, during which rainfall is negligible and inconsistent. This makes water availability a limiting factor at critical stages of wheat production and results in reduced grain yield and size than expected. Thus, to meet the requirement of the growing population, water availability at crucial stages of wheat growth needs to be ensured along with the other management practices. Although, India is not a water poor country, due to growing human population, severe neglect and over-exploitation of this resource, water is becoming a scarce commodity. India is more vulnerable because of the growing population and in-disciplined lifestyle. This calls for immediate attention by the stakeholders to make sustainable use of the available water resources.

Stress is an altered physiological condition caused by factors that tend to disrupt the equilibrium (Gaspar *et al.*, 2002). Plants, while growing naturally, are frequently exposed to many stresses such as drought, salinity, high and low temperature and oxidative stress etc. Water deficit refers to temporary unavailability of water to the plants for their normal growth and development. Water deficit occurs when the available water in the soil is reduced and atmospheric conditions cause continuous loss of water by transpiration or evaporation. The stress is manifested in the plant as loss of turgidity, stomatal closure, decreased cell expansion and restricted growth. Severe water stress may result in the arrest of photosynthesis, disturbance of metabolism and finally the death of plant (Jaleel *et al.*, 2008). Removal of water affects membrane fluidity that results in enhanced porosity of the bilayer. This results in accumulation of cellular electrolytes that disrupt the ionic balance and hence the cellular metabolism.

A common effect of drought stress is the disturbance between the generations and quenching of reactive oxygen species (ROS) which are highly reactive and in the absence of effective protective mechanism can seriously damage plants by lipid peroxidation, protein degradation, breakage of DNA and cell death (Beligni and Lamattina, 1999). Plant cells can tolerate ROS by endogenous protective mechanisms involving non-enzymatic as well as enzymatic systems (Asada, 1994).Plants have an array of enzymes that scavenge ROS (like SOD, CAT, Peroxidases, Glutathione reductases), products that restrict lipid peroxidation and a network of low molecular mass antioxidants (like Ascorbate, Glutathione, Phenolic compounds and Tocopherols). These defensive strategies are endogenous to the plants which are generated as a result of oxidative stress. Oxidative stress is a common outcome of all the stresses.

Most plants undergo osmotic adjustment as an acclimation strategy when exposed to water deficit. Osmotic adjustment is an active process in which plants adjust osmotically to a decline in soil water potential, by accumulating a high concentration of low molecular weight, neutrally charged organic solutes which are highly soluble and do not interfere with the enzymes function and hence cellular metabolism. Such solutes are termed as compatible osmolytes. The examples of compatible osmolytes include proline, glycine betaine, polyhydric alcohols, sugars and polyamines. These osmolytes maintain a low water potential in the compartments where these are accumulated so that water from the extracellular space enter across an osmotic gradient and maintain the turgidity of the cell and prevent it from being dessicated during water deficit.

In addition to above defense strategies, an extensive cross talk takes place between different hormones *viz.*, Auxin, Cytokinin, Gibberellins, Ethylene, Strigolactones and ABA, during water deficit (or any other abiotic stress). Out of all the phytohormones, ABA is said to be the "Stress hormone" and plays crucial roles in stomatal movements and thereby regulates water balance of plants. ABA has also been found to be the primary signal for the gene expression of LEA proteins, which are low molecular weight proteins, involved in protecting plant cells from dessication under water deficit conditions.

Apart from the major classes of phytohormones, some bioregulators which regulate plant growth and impart defensive properties against abiotic stresses are also observed to be crucial for abiotic stress signaling in plants. These include salicylic acid, hydrogen peroxide, polyamines, brassinosteroids and sodium nitroprusside (NO donor). These bioregulators act as signaling molecules which mediate an array of reactions that ultimately elicit an altered response to protect plants during stress situations. A detailed review on amelioration of water deficit stress in wheat, which occurs under the influence of bioregulators is thus presented in this book chapter.

2. Water Deficit and Defense Mechanism in Plants

Drought is the meteorological term characterized by a period of insufficient precipitation resulting in water deficit in plants. Soil water content and the relative humidity of the atmosphere are the primary determinants of water status of the plant. At field capacity, soil water potential becomes close to zero and the plant may have access to optimum water availability depending on the relative humidity of the atmosphere. In case the relative humidity is very low, a significant amount of water is lost from the plants even if sufficient water is available in the soil and the plant is subjected to water deficit. If the relative humidity is optimum, even then the plant may suffer from water deficit if the soil water potential approaches permanent wilting point (*i.e.* -15 MPa). At this point the plant is unable to recover from water deficit even at night, when the transpirational losses are minimized and the plant cells get damaged due to loss of turgidity.

The effect of water deficit in plants may be classified as primary and secondary. The primary effects include: reduced cell water potential, loss of turgidity, decreased cell expansion that ultimately leads to reduced growth. All of these parameters are interrelated and these can be expressed by the equation:

$G = m (\Psi p - Y)$

where, G is growth rate, m is wall extensibility, Ψp is turgor pressure and Y is yield threshold.

The secondary effects that result from primary effects, include on toxicity due to greater evapotranspirational losses, stomatal closure, deeper root penetration, reduced leaf area, reduced photosynthetic rate, trichomes.

Plants resist water deficit by different strategies categorized as drought escape, avoidance and tolerance. Escape is characterized by completion of life cycle before the onset of drought stress. Such plants have greater vigour and complete their life cycle when favourable conditions prevail and become dormant during stress. Dehydration avoidance refers to ability to maintain tissue hydration even during water deficit, and thus a high water potential either by minimizing the water loss or maximizing the water uptake. Water loss is minimized through various strategies like stomatal closure, reduced leaf area, wax deposition, trichomes etc. and uptake is maximized through deeper root penetration to reach the receding water table. Drought tolerance can be explained as ability of a plant to function even at a low water potential. This may be accomplished through osmotic adjustment, increased permeability of aquaporins for the entry of water.

3. Impact of Water Deficit Stress on Physiology of Wheat

The growth stages of wheat are generally categorized into growth stage 1 (GS1) from emergence to double ridges; growth stage 2 (GS2) from double ridges to anthesis and growth stage 3 (GS3) from anthesis to maturity. If water deficit occurs at growth sensitive stages, it may result in a complete crop failure. Among various stages of wheat growth, critical stages for water deficit include mainly, crown root initiation (CRI), spikelet differentiation, anthesis and grain filling period (*i.e.* milk stage, soft dough and hard dough stage).A brief explanation of different phenophases of wheat and the impact of water deficit on these stages are presented as follows:

3.1 Emergence to Double Ridges (GS1)

The minimum water content required in the grain for wheat germination is 35 to 45 per cent by weight (Evans *et al.*, 1975). Seminal roots are the first to grow during germination which is followed by coleoptile growth. Seminal roots support the growing seedlings till they become independent. Before emergence, seedling establishment is dependent on stored food reserves, with larger seeds having greater proportion and thus, faster ability of seedling establishment. As the first leaf gets exposed to green light, the plant becomes photosynthetically independent and autotrophic. Crown root initiation occurs at 21 DAS (days after sowing) from the leaf node and supports the growth of tillers. Tillering starts on the main stem after the 3 leaves have fully expanded and the 4th leaf is emerging. Tillering is generally found to be optimum at a temperature of 25°C. Water deficit at CRI stage adversely affects the number of productive tillers which in turn affects spike number. Tillering is almost halved if conditions are dry enough (Peterson *et al.*, 1984; Rickman *et al.*, 1983). The most sensitive stage to water stress is leaf expansion (Acevedo *et al.*, 1971). As the water potential becomes more negative, leaf growth gets further restricted and that adversely affects leaf area index the most. Water deficit prior to anthesis causes reduction in number of spikelets per spike (Oosterhius and Cartwright, 1983).

3.2 Double Ridges to Anthesis (GS2)

Double ridge stage occurs approximately 28 days after sowing (DAS) which is marked by the transition of growing apex from vegetative to the reproductive stage. At this stage, the plant bears 4-8 leaves in the main shoot. The stage is said to be double ridge as there are two parts, the lower ridge is the leaf primordium whose development is suppressed and the upper ridge eventually differentiates into spikelet. The spikelet further differentiates into structures like: glumes, lemma and palea, lodicules, stamens and carpels. The formation of terminal spikelet determines the final number of spikelets which is usually 20 (Allison and Daynard, 1976; Kirby and Appleyard, 1984). At this point, the last initiated primordial develops into glume and floret primordia and this marks the formation of a complete spike or ear. The central part of each spikelet consists of 8-12 floret primordia and the basal and distal spikelets have 6 to 8 floret primordia. Anthesis is observed to be completed in less than half of the florets while the rest abort before getting fertilized (Kirby, 1988; Kirby and Appleyard, 1987; Hay and Kirby, 1991). Florets become completely sterile at a temperature above 30°C (Owen, 1971; Saini and Aspinall, 1982).Number of spikelets are stimulated, from double ridges to terminal spikelet, under short day conditions (Rawson, 1971; Rahman and Wilson, 1978).

GS2, being a period of very active plant growth, even mild water deficits during this period will decrease cell growth and leaf area with consequent decrease of photosynthesis per unit area. Under more intense situation of water deficit, partial stomatal closure causes greater reduction in net photosynthesis (Acevedo *et al.*, 1991). Stomatal closure has been observed to be started at leaf water potential of -1.5 MPa (Kobata *et al.*, 1992; Palta *et al.*, 1994).

Water deficit prior to anthesis leads to reduced number of spikelets; while if the water deficit occurs after anthesis, then the grain number gets affected. Water stress developing from 10 days before spike emergence stage also decreases spikelets per spike of fertile tillers (Hochman, 1982; Moustafa *et al.*, 1996) and causes death of the distal and basal florets of the spikes (Oosterhuis and Cartwright, 1983). Water stress also reduces carbon and nitrogen availability which are required for spike growth.

3.3 Anthesis to Physiological Maturity (GS3)

Each spikelet has between three and six potentially fertile florets (Kirby and Appleyard, 1984) which get self pollinated as wheat is a self pollinated crop. Anthesis begins in the central part of the spike and continues towards the basal and apical parts during a 3 to 5 day period (Peterson, 1965). About 20 to 30 per cent of the grain filling period is involved in formation of endosperm cells and amyloplasts, which is the result of a rapid cell division following floret fertilization. The embryo is formed during endosperm growth (Jones *et al.*, 1985). After this lag phase of grain growth, linear growth phase (50 to 70 per cent) in grain filling occurs that corresponds to starch deposition in the endosperm.

Water deficit close to anthesis has been observed to decrease the linear phase of grain growth (Nicholas and Turner, 1993) as the remobilization of assimilates from different parts to the grains get reduced. Remobilization of total non-structural carbohydrates from wheat leaves and stems that contribute significantly to grain growth is also reduced (Bidinger *et al.*, 1977; Richards and Townley-Smith, 1987; Kiniry, 1993; Palta *et al.*, 1994). After anthesis, water deficit causes reduction in total number of spikelets that bear florets. In case there is water deficit during grain filling stage then the test weight of grain gets reduced. This ultimately leads to the reduction in the overall economic yield in terms of reduced grain size as well as weight. Water stress during grain filling does not affect the number of fertile tillers; grain weight is, however, reduced (Hochman, 1982; Kobata *et al.*, 1992). Reduced grain filling duration is observed because senescence of different plant parts begins under stress (mostly under the influence of ABA), due to which lesser parts contribute to assimilate partitioning towards grain and shortened life cycle for plant's survival.

4. Phytohormones and their Cross Talk

The hormones act as signals that elicit a cascade of events that ultimately generate an altered physiological response of a plant to stress. The degree of drought tolerance varies with developmental stages in most plant species (El-Far and Allan 1995; Reddy *et al.*, 2004; Rassaa *et al.*, 2008). Plants have a variety of mechanisms through which they perceive stress (abiotic or biotic) and respond to such stress by generating a series of reactions. These may involve complex interactions between different hormones (ABA, JA, SA and Ethylene) via synergistic and antagonistic actions and regulating the defensive responses, which are referred to as signaling crosstalk (Fujita *et al.*, 2006).

ABA is the primary hormone that regulates the plant responses to abiotic stresses including drought, salinity, low temperature. In response to biotic stress *i.e.* pathogen infection, key role is played by SA, JA and Ethylene to impart defensive response in plants. Researches have shown that response generated in plants to abiotic stress confers some effects that impart altered response to pathogens also. In many cases, ABA acts as a negative regulator of disease resistance (Narusaka *et al.*, 2004). The abiotic stress especially water deficit is the major concern of this chapter, hence the hormonal crosstalks involved in this stress is explained in detail.

Stomatal closure is a rapid response to drought stress and is regulated by a complex network of signaling pathways. During drought stress, abscisic acid (ABA) is considered to be the primary phytohormone that triggers short-term responses such as stomatal closure. ABA controls longer-term growth responses through the regulation of gene expression that favors maintenance of root growth, which optimizes water uptake (Zhang *et al.*, 2006). ABA synthesis is one of the fastest responses of plants to abiotic stress, triggering ABA-inducible gene expression (Yamaguchi and Shinozaki, 2006) and causing stomatal closure, thereby reducing water loss via transpiration (Wilkinson and Davies, 2010) and eventually restricting cellular growth.

As part of the regulation of drought stress responses, ABA may interact with jasmonic acid (JA) and nitric oxide (NO) to stimulate stomatal closure, while its regulation of gene expression includes the induction of genes associated with response to ethylene, cytokinin or auxin. ABA regulates stomatal opening during stress, however, recent studies suggest that other hormones such as CK, ethylene, BR, JA, SA, andNOalso affect stomatal function (Acharya and Assmann, 2009). While ABA,BR, SA, JA, and NO induce stomatal closure, CK and IAA promote stomatal opening. NO operates as a key intermediate in the ABA-mediated signaling network that regulates stomatal closure (Ribeiro *et al.*, 2009). ABA is also a regulator of strigolactones biosynthesis, as shown using tomato ABA-deficient mutants of different steps in the ABA biosynthetic pathway and specific inhibitors for different carotenoid cleaving enzymes (Lo'pez-Ra'ez *et al.*, 2010).

4.1 Association of H_2O_2 and NO with ABA in the Regulationof Stomata

An increase in oxidative stress is a common result of most abiotic stress conditions, including drought (Jaspers and Kangasjarvi, 2010), and is often associated with an increase in NO production (Neill *et al.*, 2008). While there is considerable evidence for the roles of H_2O_2 and NO in ABA-mediated stimulation of stomatal closure, their roles in signaling associated with drought stress remain unclear (Neill *et al.*, 2008; Sirichandra *et al.*, 2009). Studies of ABA action demonstrate that ABA mediated regulation of stomatal closure requires NO and H_2O_2 (Bright *et al.*, 2006). In these studies, both H_2O_2 and ABA treatments stimulated NO synthesis within 25 min after drought induction and resulted in stomatal closure in leaf tissue within 2.5 h (Bright *et al.*, 2006). ABA-stimulated stomatal closure was reduced when NO was removed, and in mutants with impaired NO biosynthesis. Likewise, mutants that lack a key enzyme for H_2O_2 production demonstrated reduced NO production and reduced stomatal closure. In addition, an ABA-induced SnRK2 activates a guard cell NADPH oxidase that releases H_2O_2 (Sirichandra *et al.*, 2009; Hubbard *et al.*, 2010). These results demonstrate both the complex interactions between these molecules and that ABA regulation requires the production of H_2O_2 to stimulate NO production.

4.2 Maintenance of Primary Root Growth

Under drought conditions, the region near the root tip continues to grow, while regions away from the tip are inhibited or cease to grow (Sharp, 2002). The tip region of the primary root is also the area where ABA and ROS accumulate and where a decrease in ethylene production is noted 3.5 h after drought stress (Yamaguchi and Sharp, 2010). Therefore, during drought stress, through cross-talk between phytohormone signaling pathways an increase in ABA concentration inhibits ethylene production and down regulates genes that respond to the presence of ethylene, thus maintaining root growth (Sharp, 2002).

Cytokinin is considered to be a negative regulator of root growth and branching,and root-specific degradation of cytokinin may also contribute to the primary root growth and branching induced by drought stress (Werner *et al.*, 2010). Werner *et al.* (2010) demonstrated that an increase in cytokinin degradation in the roots results in an increase in both primary root length and lateral root formation during drought conditions. Thus, increase in the root-to-shoot is associated with drought conditions and with increased drought tolerance.

4.3 Regulation of Lateral and Adventitious Root Development

While auxin is considered to be the primary hormone involved in the initiation and growth of lateral and adventitious roots, ABA may play a role in regulating lateral root growth under stress conditions (De Smet *et al.*, 2006). During drought conditions, *Arabidopsis* produces specialized, short lateral roots that remain in adormant or nongrowing condition while the plant is under stress (Wasilewska *et al.*, 2008). These roots replace dehydrated lateral roots once drought conditions are relieved. Therefore, for *Arabidopsis*, the additional ABA produced in response to drought stress inhibits the outgrowth of lateral roots from existing meristems (De Smet *et al.*, 2006).

Several mRNAs that are associated with auxin signaling are upregulated by drought or osmotic stress and may play important roles in the regulation of root architecture. A microarray analysis supports the overall down regulation of genes involved in auxin-regulated responses as a consequence of root dehydration (Huang*et al.*, 2008).

Recently, the role of auxins in drought tolerance was postulated; *TLD1/OsGH3.13*, encoding indole-3-acetic acid (IAA)-amido synthetase, was shown to enhance the expression of LEA (late embryogenesis abundant) genes, which correlated with the increased drought toleranceof rice seedlings (Zhang *et al.*, 2009). The expression of many other genes associated with auxin synthesis, perception, and action has been shown to be regulated by ethylene (Stepanova and Alonso, 2009). Among them, are the auxin-responsive factors *ARF2* and *ARF19* (Li*et al.*, 2004;Li and Zhao, 2006), the auxin transporters *PIN1*, *PIN2*, *PIN4*, *AUX1* (Ruzicka *et al.*, 2007), and genes encoding auxin biosynthetic enzymes (*ASA1/WEI2/TIR7*, *ASB1/WEI7,TAA1/SAV3/WEI8*) (Stepanova *et al.*, 2005; Stepanova *et al.*, 2008). Conversely, auxin was found to affect ethylene biosynthesis. Several members of the 1-aminocyclopropane-1-carboxylate synthase (ACS) gene family, encoding rate-limiting enzymes in ethylene biosynthesis, were shown to be regulated by auxin treatment (Tsuchisaka and Theologis, 2004). Recently, CK was also shown to be a positive regulator of auxin biosynthesis, and it was postulated that a homeostatic feedback regulatory loop involving both CK and IAA signaling acts to maintain appropriate CK and IAA concentrations in developing root and shoot tissues.

5. Bioregulators as Vital Signaling Molecules during Water Deficit Stress

5.1 Salicylic Acid (SA)

Salicylic acid (SA), a phenolic compound is known as an endogenous signal molecule and plays a crucial role in inducing defense mechanisms against abiotic and biotic stresses in plants (Yalpani *et al.*, 1994 and Szalai *et al.*, 2000). In higher plants SA derives from the shikimate-phenylpropanoid pathway (Sticher *et al.*, 1997). The precursor for SA biosynthesis has been recognized as phenylalanine, where phenylalanine after getting converted to Cinnamic acid undergoes: (1) hydroxylation to form o-coumaric acid followed by oxidation of the side chain or (2) the side chain of CA is initially oxidized to give benzoic acid which is then hydroxylated in the ortho position. SA has also been found to be biosynthesized via chorismic acid and isochorismic acid. This pathway was originally described in bacteria but has recently been shown to take place in the chloroplast in plants (Wildermuth *et al.*, 2001; Wildermuth, 2006).

The SA plays key roles in a number of processes. The most widely studied role comprises of enhancing cyanide-resistant respiration and also increasing the expression of alternative oxidase, where, the electron transport instead of moving to the terminal electron acceptor, *i.e.* oxygen, bypasses the pathway and releases heat instead of ATP during respiration (Raskin *et al.*, 1987). An increased expression of alternative oxidase has been found under the influence of SA (Rhoads and McIntosh, 1992).SA affects the generation of reactive oxygen species (ROS) in photosynthetic tissues of *Arabidopsis thaliana* during salt and osmotic stresses, thus participating in the development of stress symptoms (Borsani *et al.*, 2001).

An experiment was conducted to ascertain the effect of salicylic acid on the growth and metabolic profile of wheat seedlings under water stress. Irrespective of the SA concentration (1–3 mM) and water stress, SA treated plants showed in general, a higher moisture content, dry mass, carboxylase activity of Rubisco, superoxide dismutase (SOD) activity and total chlorophyll compared to those of untreated seedlings. SA treatment, under water stress, protected nitrate reductase (NR) activity and maintained especially at 3 mM SA concentration, the protein and nitrogen content of leaves compared to water sufficient seedlings. Results signify the role of SA in regulating the drought response of plants and suggest that SA could be used as a potential growth regulator for improving plant growth under water stress (Singh and Usha,2003).

Spraying wheat leaves with 1 mM SA increased antioxidant enzyme activities, chlorophyll and relative water content, and the membrane stability index and decreased H_2O_2 and lipid peroxide levels under moderate water stress (Agarwal *et al.*, 2005).

Thus, SA is a key signaling molecule which upon exogenous application leads to increase in endogenous ROS, especially H_2O_2. This ROS generation elicits the activation of the entire antioxidant system within the treated plant that encourages the defense system to alleviate biotic as well as abiotic stress.

5.2 Sodium Nitroprusside (SNP)

Nitric oxide (NO) is a relatively stable free radical gas that acts as a signaling molecule in mediating a number of processes like seed germination, stomatal closure, hypersensitive responses and root development (Neill *et al.*, 2003; Duan *et al.*, 2007; Delledonne *et al.*, 2001). Mata and Lamattina (2001) showed that exogenous NO (applied as SNP) reduced transpiration and induced stomatal closure in several species such as *Vicia faba*, *Salpichroa* and *Tradescantia* spp. and NO was indicated to be a component of ABA signaling pathways in ABA-induced stomatal closure. Zhao *et al.* (2001) found that NO scavenger PTIO or nitric oxide synthase inhibitor, L-NNA, strongly blocked the stress induced accumulation of ABA in root tips of wheat seedlings, and exogenously NO supplied also induced ABA accumulation. NO is expected to interact with various signals affecting guard cells and thereby regulates stomatal movement and metabolism. NO has also been found to regulate ROS metabolism and regulate the responses to a variety of abiotic and biotic stresses (Neill *et al.*, 2003; Del Rio *et al.*, 2004).

Effects of SNP (NO donor) on germination and metabolism of ROS, were being surveyed in wheat (*Triticum aestivum* L.) seeds. Germination of wheat seeds and even the elongation of radical and plumule were dramatically promoted by SNP treatment during germination under osmotic stress. Activity of amylase and endosperm was enhanced leading to degradation of storage reserves in seeds. In addition, activities of CAT and SOD and proline content were observed to be increased simultaneously, but activities of LOX were inhibited, both of which were beneficial for improving the antioxidant capacity during germination of wheat seeds under osmotic stress (Zhang *et al.*, 2003).

NO is also being observed to be involved in hypersensitive response, in which cells immediately surrounding the infection site die rapidly, depriving the pathogen of nutrients and preventing its spread. An increase in both NO and ROS is required for the activation of the hypersensitive response. NO is synthesized by NO synthase which is activated by a rapid spike in cytosolic calcium concentration during hypersensitive response.

Studies of ABA action demonstrate that ABA mediated regulation of stomatal closure requires NO and H_2O_2 (Bright *et al.*, 2006). In these studies, both H_2O_2 and ABA treatments stimulated NO synthesis within 25 min after drought induction and resulted in stomatal closure in leaf tissue within 2.5 h (Bright *et al.*, 2006). ABA-stimulated stomatal closure was reduced when NO was removed, and in mutants with impaired NO biosynthesis. Likewise, mutants that lack a key enzyme for H_2O_2 production demonstrated reduced NO production and reduced stomatal closure. In addition, an ABA-induced SnRK2 activates a guard cell NADPH oxidase that releases H_2O_2 (Sirichandra *et al.*, 2009; Hubbard *et al.*, 2010). These results demonstrate both the complex interactions between these molecules and that ABA regulation requires the production of H_2O_2 to stimulate NO production.

5.3 Hydrogen Peroxide

Hydrogen peroxide (H_2O_2) is the most stable species of ROS that plays a signaling role in plant response to stresses, such as mediating abscisic acid (ABA) regulated stomatal closure (Pei *et al.*, 2000; Zhang *et al.*, 2001). H_2O_2 is thought to affect target protein activities through modification of thiol groups of Cys residues (Hancock *et al.*, 2005).

H_2O_2 was originally viewed mainly as a toxic by-product of aerobic cellular metabolism. However, recent evidence has suggested that H_2O_2 is an important signal molecule that regulates various developmental processes such as PCD, stomatal movement, senescence, and cellular differentiation and plant morphogenesis (Neill *et al.*, 2002). A recent work has reported that exogenously applied H_2O_2 and endogenous H_2O_2 accumulation alleviated the symptoms of drought stress in plants (Ishibashi *et al.*, 2011).

Chloroplasts, mitochondria and peroxisomes are often included among the major sites of intracellular H_2O_2 production. However, under stress conditions much attention is focused on plasmalemma and apoplast compartments, in which increases in H_2O_2 could be associated with the activities of NADPH oxidases and peroxidases (Frahry and Schopfer, 1998; Bolwell *et al.*, 2002; Bindschedler *et al.*, 2006). Apart from its role in plant responses to abiotic and biotic stresses (Neill *et al.*, 2002), this ROS is emerging as a key regulator of plant development as it participates in cell growth and controls plant form and hormone action (Gapper and Dolan, 2006; Kwak *et al.*, 2006).

There are two important aspects to the function of H_2O_2 production and its diffusion to its sites of action. The production of H_2O_2 by photosynthetic electron transport and its ability to diffuse through the chloroplast envelope membranes investigated using spin trapping electron paramagnetic resonance spectroscopy and H_2O_2-sensitive fluorescence dyes indicated that even at low light intensity, a portion of H_2O_2 produced inside the chloroplasts can leave the chloroplasts thus escaping the effective antioxidant systems located inside the chloroplast. The production of H_2O_2 by chloroplasts and the appearance of H_2O_2 outside chloroplasts increased with increasing light intensity and time of illumination. The fact that H_2O_2 produced by chloroplasts can be detected outside these organelles is an important finding in terms of understanding how chloroplastic H_2O_2 can serve as a signal molecule (Maria *et al.*, 2010).

The effects of nitric oxide (NO) and hydrogen peroxide (H_2O_2) on adventitious rooting in marigold (*Tagetes erecta* L.) under drought stress showed that the promoting effect of NO or H_2O_2 on rooting under drought stress was dose-dependent, with a maximal biological response at 10 mM NO donor sodium nitroprusside (SNP) or 600 mM H_2O_2 indicating that endogenous NO and H_2O_2 may play crucial roles in rooting and H_2O_2 may be involved in rooting promoted by NO under drought stress. NO or H_2O_2 treatment attenuated the destruction of mesophyll cells ultra structure by drought stress. Similarly, NO or H_2O_2 increased leaf chlorophyll content, chlorophyll fluorescence parameters (Fv/Fm, FPS II and qP), and hypocotyls soluble carbohydrate and protein content, while decreasing starch content. Results suggest that the protection of mesophyll cells ultra structure by NO or H_2O_2 under drought conditions improves the photosynthetic performance of leaves and alleviates the negative effects of drought on carbohydrate and nitrogen accumulation in explants, thereby adventitious rooting being promoted (Liao *et al.*, 2012).

5.4 Polyamines

Polyamines (PAs) are low molecular weight aliphatic nitrogen compound positively charged at physiological pH (Groppa and Benavides, 2008) that are found in a wide range of organisms from bacteria to plants and animals (Alcázar *et al.*, 2006). They bind strongly *in vitro* to negatively charged nucleic acids (Feurstein and Marton,1989), acidic phospholipids (Tadolini *et al.*, 1985) and many types of proteins, including numerous enzymes whose activities are directly modulated by polyamine binding (Carley *et al.*, 1983). They are essential cell components and growth regulators in plants as well as other living organisms, play a critical role in different biological processes, including cell division, growth, somatic embryogenesis, floral initiation, development of flowers and fruits (Davies, 1995). Also, they are effective retardant of senescence (Couee *et al.*, 2004; Tang *et al.*, 2004).

The diamine putrescine (Put), the triamine spermidine (Spd) and the tetramine spermine (Spm) are the main PAs found in all living cells. In plants, Put is synthesized by the decarboxylation of either arginine catalysed by arginine decarboxylase or ornithine catalyzed by ornithine decarboxylase, respectively. This reaction is followed by formation of Spermidine and Spermine under the catalytic effect of spermidine synthase and spermine synthase respectively. Spermine, spermidine and putrescine all reduce level of superoxide radicals generated by senescing plant cells (Drolet *et al.,* 1986). The catabolism of polyamines produces hydrogen peroxide that acts as a signal to activate antioxidant defense response through signal transduction.El-Bassiouny (2004) and El-Bassiouny and Bekheta (2005) reportedthat, exogenous application of Putrescine on *Pisum sativum* and wheat, respectively, increased the amount of endogenous promoters (indole acetic acid, gibberellins and cytokinins) and decreased abscisic acid. Polyamines play an important role in the photosynthetic system. Their potential role was illustrated in the studies of HuiGuio *et al.* (2006) where, polyamines prevent chlorophyll degradation in leaf discs of several higher plants. Other researchers added that, polyamines could prevent chlorophyll loss by stabilizing thylakoid membrane structure of oat leaf (Besford *et al.,* 1993), or by redistributing pigment protein complexes (Iordanov and Goltsev, 1990).

In a study to investigate the effect of exogenous PAs on the membrane status and proline level in roots of water stressed cucumber (*Cucumis sativus* cv. Dar) seedlings, it was found that water shortage resulted in an increase of membrane injury, lipoxygenase (LOX) activity, lipid peroxidation and proline concentration in cucumber roots during progressive dehydration and PA pretreatment resulted in a distinct reduction of the injury index, and this effect was reflected by a lower stress-evoked LOX activity increase and lipid peroxide levels at the end of the stress period. In contrast, PA-supplied stressed roots displayed a higher proline accumulation. The presented results suggest that exogenous PAs are able to alleviate water deficit-induced membrane permeability and diminish LOX activity. Observed changes were accompanied by an accumulation of proline, suggesting that the accumulation of this osmolyte might be another possible mode of action for PAs to attain higher membrane stability, and in this way mitigate water deficit effects in roots of cucumber seedlings (JanKubis *et al.,* 2014).

5.5 Brassinosteroids (BRs)

Brassinosteroids biosynthesis takes place via two alternative pathways *i.e.* Early C-6 oxidation and Late C-6 Oxidation pathway. The precursor is Campesterol and the end product is Brassinolide and shows the highest biological activity among all other BRs. In general, BRs with a 6,7-lactone functionality, as exhibited by brassinolide, possess higher biological activity than 6-keto steroids, and BRs lacking a B-ring oxygen function.

BRs application also resulted in enhancement of seedling growth which was evident in terms of seedling length, seedling fresh and dry weights of all the three varieties of sorghum (*Sorghum vulgare*) *viz.* CSH-14 and ICSV-745 (susceptible to water stress) andM-35-1 (resistant towater stress) under osmotic stress (Vardhini and Rao, 2003). Furthermore, osmotic stress resulted in considerable reduction in the protein contents in all the three varieties of sorghum. However, BRs not only restored but also stimulated the level of protein and free proline (Vardhini and Rao, 2003).

Homobrassinolide also had a stimulatory effect on the growth of drought tolerant (C306) and drought-susceptible (HD2329) wheat (*Triticum aestivum*) varieties under stress conditions. Application of homobrassinolide resulted in increase relative water content, nitrate reductase activity, chlorophyll content and photosynthesis under both conditions. It also improved membrane stability (lower injury).These beneficial effects resulted in higher leaf area, biomass production, grain yield and yield related parameters in the treated plants. However, homobrassinolide showed a higher activity in drought tolerant wheat variety under water stress conditions. Increased water uptake and membrane stability and higher carbon dioxide and nitrogen assimilation rates in BR-treated plants under stress were correlated with BR-induced drought tolerance (Sairam, 1994).

The effect of 28-homobrassinolide and 24-epibrassinolide on the activities of four oxidizing enzymes (superoxide dismutase, glutathione reductase, IAA oxidase and polyphenol oxidase) and two hydrolyzing enzymes (protease and ribonuclease) in the seedlings of four varieties of sorghum *viz.*, CSH-15R, CSH-14 and ICSV-745 (susceptible to water stress) and M-35-1 (resistant to water stress) under PEG – imposed water stress was studied. Supplementation of both the brassinosteroids resulted in enhanced superoxide dismutase and glutathione reductase but lowered IAA oxidase, polyphenol oxidase, protease and ribonuclease activities indicating the alleviating ability of brassinosteroids on water stress in the drought sensitive as well as tolerant varieties of sorghum seedlings (Vardhini *et al.*, 2011).Li *et al.* (2012) reported that EBL induces changes in antioxidative enzyme activities and content of antioxidants and so improve plant growth under drought stress. Leaf wilting, reduction in growth, and complete drying of some seedlings are frequently observed in untreated, but are considerably reduced in EBL-treated seedlings.

6. Conclusion and Future Research Strategies

The deficit in plants have a primary and secondary effects. The primary effects include: reduced cell water potential, loss of turgidity, decreased cell expansion that ultimately leads to reduced growth. All of these parameters are interrelated and these can be expressed by the equation: $G = m (\Psi p - Y)$; where G is growth rate, m is wall extensibility, Ψp turgor pressure and Y is yield threshold. The secondary effects result from primary effects, include ion toxicity due to greater evapotranspirational losses, stomatal closure, deeper root penetration, reduced leaf area, reduced photosynthetic rate, trichomes. Plants resist water deficit by various strategies known as drought escape, avoidance and tolerance. Escape is characterized by completion of life cycle before the onset of drought. Dehydration avoidance refers to ability to maintain tissue hydration even during water deficit, and thus a high water potential either by minimizing the water loss or maximizing the water uptake. Drought tolerance can be explained as ability of a plant to function even at a low water potential.

Among various stages of wheat growth, critical stages for water deficit include mainly crown root initiation (CRI), spikelet differentiation, anthesis and grain filling period. Water deficit at CRI stage, results in reduction in number of tillers per plant; water deficit prior to anthesis causes reduction in number of spikelets per spike. The grains per spikelet is reduced when the water deficit occurs after anthesis, however, water deficit at grain filling stage reduces the size of the grain test weight. The crops phenophases are regulated through an extensive cross talk of different phytohormones. During stresses, these hormones generate an array of signaling cascade that alters physiological response which in turn helps the plants to withstand the stress. The

ABA is primary hormone regulating the plant responses to abiotic stresses including drought, salinity and low temperature. Stomatal closure is a rapid response to drought stress mediated by ABA. Long term growth responses are controlled by ABA through the regulation of gene expression that favors maintenance of root growth to optimize water uptake. The ABA also interacts with jasmonic acid and nitric oxide to stimulate stomatal closure, while its regulation of gene expression includes the induction of genes associated with response to ethylene, cytokinin or auxin.

There are a number of bioregulators (like salicylic acid, sodium nitroprusside, hydrogen peroxide, polyamines and Brassinosteroids) that have been observed to influence growth under stress. These act as signaling molecules that mediate defense mechanisms so that the plant could withstand the stress situation. SA is a key signaling molecule of the exogenous application leads to increase in endogenous ROS, especially H_2O_2 which elicits the activation of the entire antioxidant system that encourages the defense system to alleviate biotic as well as abiotic stress.

Sodium nitroprusside (SNP) acts as the NO donor, nitric oxide (NO) is a relatively stable free radical gas that acts as a signaling molecule in mediating a number of processes like seed germination, stomatal closure, hypersensitive responses and root development. An increase in NO and ROS is required for the activation of the hypersensitive response. NO is synthesized by NO synthase activated by a rapid spike in cytosolic calcium concentration during hypersensitive response. The ABA mediated regulation of stomatal closure requires NO and H_2O_2. The H_2O_2 is an important signal molecule that regulates PCD, stomatal movement, senescence and cellular differentiation and plant morphogenesis and affects target protein activities through modification of thiol groups of Cys residues. Polyamines are low molecular weight aliphatic nitrogen compound positively charged at physiological pH bind strongly *in vitro* to negatively charged nucleic acids, acidic phospholipids and many types of proteins and modulate the activities. They play a critical role in cell division, growth, somatic embryogenesis, floral initiation, development of flowers and fruits and also are effective retardant of senescence. Brassinosteroid biosynthesis takes place via two alternative pathways. Early C-6 oxidation and Late C-6 oxidation pathway. The EBL(24- epibrassinolide) induces changes in antioxidative enzyme activities and content of antioxidant, and improves plant growth under drought stress by reducing leaf wilting, growth and drying of seedlings.

Plants have a number of phytohormones being endogenous to them which mediate a variety of responses to stress situations, especially water deficit which is the most predominant. However, these responses are comparatively slow and not much significant in their expression. The application of bioregulators, being exogenous (either through seed treatment or through foliar spray) generates a rapid response by mediating a number of signaling pathways that ultimately exaggerate the effects of phytohormones (either acting synergistically or antagonistically), and elicit different defense mechanisms in response to stress. Thus application of bioregulators exogenously may help in strengthening the plant defense strategies in response to various stress situations.

There exists a number of bioregulators affecting the plants and their interactions with the environmental stresses, which may act synergistically or antagonistically with the endogenous phytohormones to impart defense response to crop plants under stress situations. An extensive cross talk occurs at various levels, activating a series of reactions that ultimately elicits an altered response so that the plant survives the stress situations and minimized yield reduction. However, these bioregulators are beneficial within a narrow range of concentrations beyond which they may affect adversely. As the drought stress is of major concern and the crops are mostly grown under rainfed situation, these bioregulators should be applied in a form or combination that alleviate the damage taking place during the critical growth stages. In case of wheat, water deficit at different growth stages has different outcomes. The occurrence of drought during critical stages of crop growth affects the most to both, yield components and the yield. Thus, researches on concentration, timing of application, their coordination with other phytohormones need to be much better analyzed before being advocating their exogenous application on crops under stress. The bioregulators are required to act in a proper coordination with the endogenous phytohormones so that the adverse reactions do not occur which may otherwise lead to even more intense damage to the crop and the crop fails to survive.

Abbreviations

ABA: Absissic acid

ACS: Aminocyclopropane -1- carboxylate synthase

BL: Brassinolide

BR: Brassinosteroid

CA: Cinnamic acid

CAT: Catalase

Ci: Internal carbon dioxide concentration

CK: Cytokinin

EBR: 24-Epibrassinolide

g_s: Stomatal conductance

H_2O_2: Hydrogen peroxide

IAA: Indole-3- acetic acid

JA: Jasmonic acid

LEA: Late embryogenesis abundant protein

LOX: Lipoxygenase

NO: Nitric oxide

PCD: Programmed cell death

PEG: Polyethylene glycol

ROS: Reactive oxygen species

RWC: Relative water content

SA: Salicylic acid

SNP: Sodium nitroprusside

SnRK2: Sucrose non-fermenting related protein kinase 2

SOD: Superoxide dismutase

References

Acevedo, E., Harris, H. and Cooper, P.J.M. (1991). Crop architecture and water use efficiency in Mediterranean environments. *In* H. Harris, P.J.M. Cooper and M. Pala eds., Soil and Crop Management for Improved Water Use Efficiency in Rainfed Areas, pp.106-118.

Acevedo, E., Hsiao, T.C. and Henderson, D.W. (1971). Immediate and subsequent growth responses of maize leaves to changes in water status. *Plant Physiol.*, 48: 631-636.

Acharya, B. and Assmann, S. (2009). Hormone interactions in stomatal function. *Plant Mol. Biol.*, 69: 451-462.

Agarwal, S., Sairam, R.K., Srivastava, G.C. and Meena, R.C. (2005).Changes in antioxidant enzymes activity and oxidative stress by abscisic acid and salicylic acid in wheat genotypes. *Biol Plant*, 49: 541– 550.

Alcázar, R., Marco, F., Cuevas, J.C., Patron, M., Ferrando, A., Carrasco, P., Tiburcio, A.F. and Altabella, T. (2006).Involvement of polyamines in plant response to abiotic stress. *Biotechnol Lett*, 28: 1867–1876.

Allison, J.C.S. and Daynard, T.B. (1976).Effect of photoperiod on development and number of spikelets of a temperate and some low-latitude wheats.*Ann. Appl.Biol.*, 83: 93-102.

Asada, K. (1994). Production and action of active oxygen species in photosynthetic tissues.*In*: Foyer, C.H., Mullineaux, P.M. (eds.), Causes of Photooxidative Stress and Amelioration of Defense System in Plants. pp. 77-103, CRC Press, Boca Raton.

Beligni, M.V. and Lamattina, L. (1999). Nitric oxide counteracts cytotoxic processes mediated by reactive oxygen species in plant tissues. *Planta*,208: 337-344.

Besford, R.T., Richardson, C.M., Campos, J.L. and Tiburcio, A.F. (1993).Effect of polyamineson stabilization of molecular complexes in thylakoid membranes of osmotically stressed oat leaves. *Planta*, 189: 201–206.

Bidinger, F.R., Musgrave, R.B. and Fischer, R.A. (1977). Contribution of stored preanthesis assimilates to grain yield in wheat and barley. *Nature*, 270: 431-433.

Bindschedler, L.V., Dewney, J. and Blee, K.A. (2006). Peroxidase dependent apoplastic oxidative burst in Arabidopsis required for pathogen resistance. *The Plant Journal*, 47: 851–863.

Bolwell, K.A., Bindschedler, L.V., Blee, K.A., Butt, V.S., Davies, D.R., Gardner, S.L., Gerrish, C. and Minibayeva, F. (2002). The apoplastic oxidative burst in response to biotic stress in plants: a three component system. *Journal of Experimental Botany*, 53: 1367–1376.

Borsani, O., Valpuesta, V. and Botella, M.A. (2001). Evidence for a role of salicylic acid in the oxidative damage generated by NaCl and osmotic stress in Arabidopsis seedlings. *Plant Physiol.*, 126: 1024–1030.

Bright, J., Desikan, R., Hancock, J.T., Weir, I.S. and Neill, S.J. (2006). ABA induced NO generation and stomatal closure in Arabidopsis are dependent on H_2O_2 synthesis. *Plant J.*, 45: 113–122.

Carley, E., Wolosiuk, R.A. and Hertig, C.M. (1983). Regulation of the activation of chloroplast fructose-1,6-bis phosphatase (E.C.3.1.3.11). Inhibition by spermidine and spermine, *Biochem.Biophys. Res. Commun.*, 115: 707–710.

Couee, I., Hummel, I., Sulmon, C., Gouesbet, G. and El-Amrani, A. (2004). Involvement of polyamines in root development.*Netherlands*, 76(1): 1-10.

Davies, P.J., 1995. Plant Hormones: *Physiology and Biochemistry and Biology*, pp. 159.

De Smet, I., Zhang, H.M., Inze´, D. and Beeckman, T. (2006). A novel role for abscisic acid emerges from underground. *Trends Plant Sci*, 11: 434–439.

Del Rio, L.A., Corpas, F.J. and Barroso, J.B. (2004). Nitric oxide and nitric oxide synthase activity in plants. *Phytochemistry*, 65: 783–792.

Delledonne, M., Zeier, J., Marocco, A. and Lamb, C. (2001). Signal interaction between nitric oxide and reactive oxygen intermediates in the plant hypersensitive disease resistance response. *Plant Biol*, 98: 13454–13459.

Drolet, G., Dumbroff, E.B., Legge, R.L. and Thompson, J.E. (1986). Radical scavenging properties of polyamines, *Phytochemistry* 25: 367–371.

Duan, X., Su, X., You, Y., Qu, H., Li, Y. and Jiang, Y. (2007). Effects of nitric oxide on pericarp browning of harvested longan fruit in relation to phenolic metabolism. *Food Chem.*, 104: 571–576.

El-Bassiouny, H.M.S. (2004).Increasing thermotolerance of *Pisum sativum* L. plants through application of putrescine and stigmasterol. *Egypt. J. Biotech.*, 18: 93-118.

El-Bassiouny, H.M.S. and Bekheta, M.A. (2005). Effect of salt stress on relative water content, lipid peroxidation, polyamines, amino acids and ethylene of two wheat cultivars. *Int. J. Agric. Biol.*, 7(3): 363-368.

El-Far, I.A. and Allan, A.Y. (1995).Responses of some wheat cultivars to sowing methods and drought at different stages of growth.*Assuit J. Agric. Sci.*,26: 267–277.

Evans, L.T., Wardlaw, I.F. and Fischer, R.A. (1975).Wheat. *In* L.T. Evans, eds. *Crop Physiology*, pp. 101-149.Cambridge University Press. USA.

Feurstein, B.G. and Marton, L.G. (1989). Specificity and binding in polyamine: nucleic acid interactions, in: U. Bachrach, Y.M. Heimer (Eds.). *The Physiology of Polyamines*, CRC Press, Boca Raton, FL, pp. 109–120.

Frahry, G. and Schopfer, P. (1998).Hydrogen peroxide production by roots and its stimulation by exogenous NADH.*Physiologia Plantarum*, 103: 395–404.

Fujita, M., Fujita, Y., Noutoshi, Y., Takahashi, F., Narusaka, Y., Yamaguchi-Shinozaki, K. and Shinozaki, K. (2006). Crosstalk between abiotic and biotic stress responses: a current view from the points of convergence in the stress signaling networks. *Current Opinion inPlant Biology*, 9: 436–442.

Gapper, C. and Dolan, L. (2006).Control of plant development by reactive oxygen species.*Plant Physiology*, 141: 341–345.

Gaspar, T., Franck, T., Bisbis, B., Kevers, C., Jouve, L., Hausman, J.F. and Dommes, J. (2002). Concepts in plant stress physiology. Application to plant tissue cultures.*Plant Growth Regul.*, 37: 263–285.

Groppa, M.D. and Benavides, M.P. (2008). Polyamines and abiotic stress: recent advances. *AminoAcids*,34: 35–45.

Hancock, J.T., Henson, D., Nyirenda, M., Desikan, R., Harrison, J., Lewis, M., Hughes, J., and Neill, S.J. (2005).Proteomic identification of glyceraldehyde 3-phosphate dehydrogenase as an inhibitory target of hydrogen peroxide in Arabidopsis.*Plant Physiol. Biochem.*, 43: 828–835.

Hay, R.K.M. and Kirby, E.J.M. (1991).Convergence and synchrony-A review of the coordination of development in wheat.*Aust. J. Agric. Res.*, 42: 661-700.

Hochman, Z.V.I. (1982). Effect of water stress with phasic development on yield of wheat grown in a semi-arid environment.*Field Crop Res.*, 5: 55-67.

Huang, D., Wu, W., Abrams, S.R., Adrian, J. and Cutler, A.J. (2008). The relationship of drought-related gene expression in Arabidopsis thaliana to hormonal and environmental factors. *J Exp Bot* 59: 2991–3007.

Hubbard, K.E., Nishimura, N, Hitomi, K., Getzoff, E.D. and Schroeder, J.I. (2010). Early abscisic acid signal transduction mechanisms: newly discovered components and newly emerging questions. *Genes Dev.*, 24: 1695–1708.

HuiGuo, D., Y. Shu, L. WenJuan, X. DeHui, Q. DongHong, l. HouGuo and HongHui, (2006). Effects of exogenous spermidine on photosystem II of wheat seedlings under water stress. *J. Integrative Plant Biol.*, 45 (8): 920 – 927.

Iordanov, I. and Goltsev, V. (1990).The protective effect of some polyamines on functioning of thylakiod membranes. *Fiziologiya na Rasteniyata.*, 16(4): 42 – 51.

Ishibashi, Y., Yamaguchi, H., Yuasa, T., Iwaya-Inoue, M., Arima, S. and Zheng, S.H. (2011). Hydrogen peroxide spraying alleviates drought stress in soybean plants. *J. Plant Physiol.* 168: 1562-1567.

Jaleel, C.A., Manivannan, P., Lakshmanan, G.M.A., Gomathinayagam, M. and Panneerselvam, R. (2008). Alterations in morphological parameters and photosynthetic pigment responses of *Catharanthus roseus* under soil water deficits. *Colloids Surf. B: Biointerfaces*, 61: 298–303.

Jan Kubi[, Jolanta Floryszak-Wieczorek, and Magdalena Arasimowicz-Jelonek (2014). Polyamines induce adaptive responses in water deficit stressed cucumber roots. *J.Plant Res.*, 127(1): 151–158.

Jaspers, P. and Kangasjarvi, J. (2010).200 Reactive oxygen species in abiotic stress signaling.*Physiol. Plant.*, 138: 405–413.

Jones, R.J., Roessler, J. and Quattar, S. (1985). Thermal environment during endosperm cell division in maize : Effects on number of endosperm cells and starch granules. *Crop Sci.*, 25: 830-834.

Kiniry, J.R. (1993). Nonstructural carbohydrate utilisation by wheat shaded during grain growth. *Agron. J.*, 85: 844-849.

Kirby, E.J.M. (1988). Analysis of leaf stem and ear growth in wheat from terminal spikelet stage to anthesis. *Field Crop Res.*, 18: 127-140.

Kirby, E.J.M. and Appleyard, M. (1984).Cereal development guide. Arable Unit, National Agriculture Centre, Stoneleigh, Kenilworth. England. 95p.

Kirby, E.J.M. and Appleyard, M. (1987).Development and structure of the wheat plant. *In* F.G.H. Lupton, ed. *Wheat Breeding*, p. 287-311. Chapman and Hall.London.

Kobata, T., Palta, J.A. and Turner, N.C. (1992). Rate of development of post anthesis water deficits and grain filling of spring wheat. *Crop Sci.*, 32: 1238-1242.

Kwak, J.M., Nguyen,V. and Schroeder, J.I. (2006). The role of reactive oxygen species in hormonal responses.*Plant Physiology*, 141: 323–329.

Li, H., Johnson, P., Stepanova, A., Alonso, J.M. and Ecker, J.R. (2004).Convergence of signaling pathways in the control of differential cell growth in Arabidopsis.*Developmental Cell*, 7: 193-204.

Li, J., Dai, X. and Zhao, Y. (2006). A role for auxin response factor 19 in auxin and ethylene signaling in Arabidopsis. *Plant Physiology*, 140: 899-908.

Li, Y.H., Liu, Y.J., Xu, X.L., Jin, M., An, L.Z. and Zhang, H. (2012). Effect of 24-epibrassinolide on drought stress-induced changes in *Chorispora bungeana.Biol.Plant.*,56: 192-196.

Lo´pez-Ra´ez, J.A., Kohlen, W., Charnikhova, T., Mulder, P., Undas, A.K., Sergeant, M.J., Verstappen, F., Bugg, T.D.H., Thompson, A.J. and Ruyter- Spira, C. (2010). Does abscisic acid affect strigolactone biosynthesis? *New Phytologist*, 187: 343-354.

Mubarakshina, M.M., Ivanov, B.N., Naydov, I.A., Hillier, W., Murray R. Badger, M.R. and Krieger-Liszkay, A. (2010). Production and diffusion of chloroplastic H_2O_2 and its implication to signaling.*Journal of Experimental Botany*, 61 **(13)**: 3577–3587.

Mata, C.G. and Lamattina, L. (2001). Nitric oxide induces stomatal closure and enhances the adaptive plant responses against drought stress. *Plant Physiol* 126: 1196–1204.

Moustafa, M.A., Boersma, L. and Kronstad, W.E. (1996). Response of four spring wheat cultivars to drought stress.*Crop Sci.*, 36: 982-986.

Narusaka, Y., Narusaka, M., Seki, M., Umezawa, T., Ishida, J., Nakajima, M., Enju, A. and Shinozaki, K. (2004). Crosstalk in the responses to abiotic and biotic stresses in Arabidopsis: analysis of gene expression in cytochrome P450 gene superfamily by cDNA microarray. *Plant Mol Biol*, 55: 327-342.

Neill, S., Barros, R., Bright, J., Desikan, R., Hancock, J., Harrison, J., Morris, P., Ribeiro, D. and Wilson, I. (2008). Nitric oxide, stomatal closure, and abiotic stress. *J. Exp Bot.* 59: 165-176.

Neill, S., Desikan, R. and Hancock, J. (2003).Nitric oxide signaling in plant. *New Phytol* 159: 11–35.

Neill, S.J., Desikan, R., Clarke, A., Hurst, R.D. and Hancock, J.T. (2002).Hydrogen peroxide and nitric oxide as signalling molecules in plants.*J. Exp. Bot.* 53: 1237-1242.

Nicholas, M.E. and Turner, N.C. (1993). Use of chemical desiccants and senescing agents to select wheat lines maintaining stable grain size during postanthesis drought. *Field Crop Res.*, 31: 155-171.

Oosterhuis, D.M. and Cartwright, P.M. (1983). Spike Differentiation and floret survival in semidwarf spring wheat as affected by water stress and photoperiod. *CropSci.*, 23: 711-716.

Owen, P.C. (1971). Responses of a semi-dwarf wheat to temperatures representing a tropical dry season. II. Extreme temperatures. *Exp. Agric.*, 7: 43-47.

Palta, J.A., Kobata, T., Turner, N.C. and Fillery, I.R. (1994).Remobilization of carbon and nitrogen in wheat as influenced by postanthesis water deficits. *CropSci.*, 34: 118-124.

Pei, Z.M., Murata, Y., Benning, G., Thomine, S., Klüsener, B., Allen, G.J., Grill, E. and Schroeder, J.I. (2000). Calcium channels activated by hydrogen peroxide mediate abscisic acid signalling in guard cells. *Nature* 406: 731–734.

Peterson, C.M., Klepper, B., Pumphrey, F.B. and Rickman, R.W. (1984).Restricted rooting decreases tillering and growth of winter wheat. *Agron. J.*, 76: 861-863.

Peterson, R.F. (1965). Wheat: Botany, cultivation and utilisation. Leonard Hill, London, pp. 448.

Rahman, H.M. and Wilson, J.H. (1978).Determination of spikelet number in wheat. III Effect of varying temperature on ear development. *Aust. J. Agric. Res.*, 28: 575-581.

Raskin, I., Ehmann, A., Melander, W.R. and Meeuse, B.J.D. (1987). Salicylic acid: a natural inducer of heat production in *Arum lilies*. *Science*, 237: 1601-1602.

Rassaa, N., Ben Haj Salah, H. and Latiri, K. (2008). Thermal responses of Durum wheat Triticum durum to early water stress. Consequence on leaf and flower development.*C. R.Biologies*, 331: 363–371.

Rawson, H.M. (1971). An upper limit for spikelet number per ear in wheat as controlled by photoperiod. *Aust. J. Agric. Res.* 22: 537-546.

Reddy, A.R., Chaitanya, K.V. and Vivekanandan, M. (2004).Drought-induced responses of photosynthesis and antioxidant metabolism in higher plants.*J. Plant Physiol.* 161: 1189–1202.

Rhoads, D.M. and McIntosh, L. (1992).Cytochrome and alternative pathway respiration in tobacco.Effects of salicylic acid.*Plant Physiol,* 103: 877–883.

Ribeiro, D.M., Desikan, R., Bright, J.O., Confraria, A.N.A., Harrison, J., Hancock, J.T., Barros, R.S., Neill, S.J. and Wilson, I.D. (2009).Differential requirement for NO during ABA-induced stomatal closure in turgid and wilted leaves.*Plant, Cell and Environment,* 32: 46-57.

Richards, R.A. and Townley-Smith, T.F. (1987). Variation in leaf area development and its effects and water use, yield and harvest index of droughted wheat. *Aust.J. Agric. Res.,* 38: 983-992.

Rickman, R.W., Klepper, B.L. and Peterson, C.M. (1983). Time distribution for describing appearance of specific culms of winter wheat. *Agron. J.,* 75: 551-556.

Ruzicka, K., Ljung, K., Vanneste, S., Podhorska, R., Beeckman, T., Friml, J. and Benkova, E. (2007). Ethylene regulates root growth through effects on auxin biosynthesis and transport-dependent auxin distribution. *Plant Cell,* 19: 2197-2212.

Saini, H.S. and Aspinall, D. (1982). Abnormal sporogenesis in wheat (*Triticum aestivum* L.) induced by short periods of high temperature. *Ann. Bot.,* 49: 835-846.

Sairam, R.K. (1994). Effects of homobrassinolide application on plant metabolism and grain yield under irrigated and moisturestress conditions of two wheat varieties. *Plant Growth Regul.*14 : 173–181.

Sharp, R.E. (2002). Interaction with ethylene: changing views on the role of abscisic acid in root and shoot growth responses to water stress. *Plant Cell Environ,* 25: 211–222.

Singh B. and Usha, K. (2003).Salicylic acid induced physiological and biochemical changes in wheat seedlings under water stress.*Plant Growth Regul.,*39: 137–141.

Sirichandra, C., Wasilewska, A., Vlad, F., Valon, C. and Leung, J. (2009). The guard cell as a single-cell model towards understanding drought tolerance and abscisic acid action. *J Exp Bot,* 60: 1439–1463.

Smirnoff, N. (1998). Plant resistance to environmental stress.*Curr.Opin.Biotechnol.*9: 214-219.

Stepanova, A.N. and Alonso, J.M. (2009). Ethylene signaling and response: where different regulatory modules meet. *Current Opinion in Plant Biology,* 12: 548-555.

Stepanova, A.N., Hoyt, J.M., Hamilton, A.A. and Alonso, J.M. (2005).A Link between ethylene and auxin uncovered by the characterization of two root-specific ethylene-insensitive mutants in Arabidopsis.*Plant Cell,* 17: 2230-2242.

Stepanova, A.N., Robertson-Hoyt, J., Yun, J., Benavente, L.M., Xie, D.Y., Dolezal, K., Schlereth, A., Ju¨ rgens, G. and Alonso, J.M. (2008).TAA1-mediated auxin biosynthesis is essential for hormone crosstalk and plant development. *Cell*, 133: 177-191.

Sticher, L., MauchMani, B. and Metraux, J.P. (1997). Systemic acquired resistance. *Annu Rev Plant Pathol*, 35: 235–270.

Szalai, G., Tari, I., Janda, T., Pestenácz, A. and Páldi,E. (2000). Effects of cold acclimation and salicylic acid on changes in ACC and MACC contents in maize during chilling.*Biol. Plant.*, 43: 637-640.

Tadolini,B., Cabrini,L., Varani, E. and A.M. Sechi, A.M. (1985). Spermine binding and aggregation of vesicles of different lipid composition, *Biol. Amines*, 3: 87–92.

Tang, W., R.J. Newton, R.J. and V. Outhavong, V. (2004). Exogenously added polyamines recover browning tissues into normal callus cultivars and improve plant regeneration in pine. *Physiol. Plant*, 122(3): 386-395.

Tsuchisaka, A and Theologis, A (2004). Unique and overlapping expression patterns among the Arabidopsis 1-amino-cyclopropane-1- carboxylate synthase gene family members. *Plant Physiology*, 136: 2982-3000.

Vardhini, B.V. and. Rao, S.S.R (2003).Amelioration of osmotic stress by brassinosteroids on seed germination and seedling growth of three varieties of sorghum.*Plant Growth Regul.*, 41: 25–31.

Vardhini, B.V., Sujatha, E. and Rao, S.S.R. (2011). Brassinosteroids: Alleviation of Water Stress in Certain Enzymes of Sorghum Seedlings. *Journal of Phytology*, 3 (10): 38-43.

Wasilewska, A., Vlad, F., Sirichandra, C., Redko, Y., Jammes, F., Valon, C., Frey, N.F. and Leung, J. (2008).An update on abscisic acid signaling in plants and more.*Mol. Plant*, 1: 198–217.

Liao W.B., Gao-Bao Huang, Ji-Hua Yu and Mei-Ling Zhang (2012). Nitric oxide and hydrogen peroxide alleviate drought stress in marigold explants and promote its adventitious root development. *Plant Physiology and Biochemistry*, 58: 6-15.

Werner, T., Nehnevajovaa, E., Kollmera, I., Nova´kb, O., Strnadb, M., Kramerc, U. and Schmullinga, T. (2010). Root-specific reduction of cytokinin causes enhanced root growth, drought tolerance, and leaf mineral enrichment in Arabidopsis and Tobacco. *Plant Cell*, 22: 3905–3920.

Wildermuth, M.C. (2006). Variations on a theme: synthesis and modification of plant benzoic acids. *Curr Opin Plant Biol*, 9: 288–296.

Wildermuth, M.C., Dewdney, J., Wu, G. and Ausubel, F.M. (2001). Isochorismate synthase is required to synthesize salicylic acid for plant defence. *Nature,*414: 562–565.

Wilkinson, S. and Davies, W.J. (2010). Drought, ozone, ABA and ethylene: new insights from cell to plant to community. *Plant, Cell and Environment*, 33: 510-525.

Yalpani, N., Enyedi, A.J., León, J. and Raskin, I. (1994). Ultraviolet light and ozone stimulate accumulation of salicylic acid and pathogenesis related proteins and virus resistance in tobacco. *Planta*, 193: 373-376.

Yamaguchi, M. and Sharp, R.E. (2010). Complexity and coordination of root growth at low water potentials: recent advances from transcriptomic and proteomic analyses. *Plant CellEnviron*, 33: 590–603.

Yamaguchi-Shinozaki K and Shinozaki K (2006).Transcriptional regulatory networks in cellular responses and toletance to dehydration and cold stresses.*Annual Review of Plant Biology*, 57: 781-803.

Yuan G.F., Cheng-Guo Jia, Zhen Li, Bo Suna, Li-Ping Zhanga, Na Liua and Qiao-Mei Wang (2010). Effect of brassinosteroids on drought resistance and abscisic acid concentration in tomato under water stress.*Scientia Horticulturae*, 126: 103–108.

Zhang Hua, Shen Wen-Biao and Xu Lang- Lai (2003).Effect of Nitric Oxide on Germination of Wheat Seeds and Its Reactive Oxygen Species Metabolisms under Osmotic Stress.*Acta Botanica Sinica*, 45 (8): 901-905.

Zhang, J., Jia, W., Yang, J. and Ismail, A.M. (2006). Role of ABA in integrating plant responses to drought and salt stresses.*Field Crops Res*, 97: 111–119.

Zhang, M., Zhai, Z., Tian, X., Duan, L. and Li, Z. 2008. Brassinolide alleviated the adverse effect of water deficits on photosynthesis and the antioxidant of soybean (*Glycine max* L.). *Plant Growth Regul.*,56: 257-264.

Zhang, S.W., Li, C.H., Cao, J., Zhang, Y.C., Zhang, S.Q., Xia, Y.F., Sun, D.Y. and Sun,Y. (2009). Altered architecture and enhanced drought tolerance in rice via the down-regulation of indole-3-Acetic Acid by TLD1/OsGH3.13 activation. *Plant Physiology*, 151: 1889-1901.

Zhang, X., Zhang, L., Dong, F., Gao, J., Galbraith, D.W., and Song, C.P. (2001). Hydrogen peroxide is involved in abscisic acid-induced stomatal closure in Vicia faba. *Plant Physiol.*, 126: 1438–1448.

Zhao, Z., Cheng and Zhang, C. (2001). Interaction between reactive oxygen species and nitric oxide in drought-induced abscisic acid synthesis in root tips of wheat seedlings. *AustJ Plant Physiol*, 28: 1055–1061.

Recent Advances in Crop Physiology Vol. 2 (2016)   *Pages* 261–293
Editor: **Dr. Amrit Lal Singh**
Published by: **DAYA PUBLISHING HOUSE, NEW DELHI**

Chapter 9

Phenology and Productivity of Forest Flora of Gujarat

R.N. Nakar[1] *, B.A. Jadeja[2] and A.L. Singh[3]

[1]*Assistant Professor,*
Seth P.T. Science College, Godhra – 389 001, Gujarat
[2]*Associate Professor,*
M.D. Science College, Porbandar – 360 575, Gujarat
[3]*Principal Scientist,*
ICAR-Directorate of Groundnut Research, Junagadh – 362 001, Gujarat

1. Introduction

Forest is the main source of economy for any country or state as it deals directly with plant wealth, food, fodder, livestock and are involved with human lives one or other way. It provides pure oxygen and give many ethnobotanically useful plants being used by humans since ancient times. Gujarat has many important forests such as Girnar and Barda forest in Saurashtra region, Dang and Aravalli forest in south and north Gujarat. In addition, Kachchh has important diversity of xerophytic community. There are 2,198 species of higher plants belonging to 902 genera and 155 families in Gujarat which represent 12.91 per cent flora of the country (GEC, 1996). All these wealths provide great opportunity to study with various plants in forest areas. The floristic, vegetational and diversity studies have received much attention in the later part of the nineteenth century in Gujarat, when it was bifurcated from Bombay state, and information was gathered on plant wealth of the state. Earlier most of work done was on species diversity however, very few studies were on

* *Corresponding Author:* E-mail: rupeshnakar@gmail.com

phenological and physiological aspects. Seed, the main source of future forest, is not studied in depth with about germination, viability, morphology, ecology, and embryology. There is further need to study the phenology, physiology and productivity of these plants particularly on, emergence and fall of leaf, flowering and fruiting, and seed biology. Role of ecology is also very vital in understanding forest plants, density, frequency and abundance of species and their correlation with other traits. Finally, demarcate forest areas based on their species distribution, habitats and utility of plants and other flora with much detail.

In current chapter, an effort was made to highlight the current scenario of forest in Gujarat and their phenological and physiological studies.

2. Forests of Gujarat

In Gujarat, there are three distinct geomorphologic divisions, *viz.* (1) Gujarat main land (2) Saurashtra peninsula and (3) Kutch peninsula.

Gujarat main land rises from estuarine tracts between Narmada and Tapi rivers, and goes North side, about 400 km and merges into desert of Rajasthan and Runn of Kachchh. In Eastern side it is surrounded by Aravalli, Vindhya, Satpura and Sahyadri.

Saurashtra peninsula slopes towards all direction forming an elevated table land.

Kutch peninsula is greatly isolated by great Rann on the north and east, and little Rann, on the south east side. During November it is barren tract of dry bed of salt encrusted mud, presenting aspects of inconceivable desolation. While during other half, it is flooded with water of rivers that are held back owing to rise of sea by south west monsoon.

Geographically forests of Gujarat can be divided into:

1. The tropical dry moist deciduous forest- Hilly regions in south parts in Bulsar, Dangs and Surat districts.
2. Dry deciduous forest- subdivided into (a) Dry teak forest (Rajpipla) and (b) Non dry teak forest (Chota Udepur, Panchmahal, Sabarkantha in North Gujarat).

While in Gujarat, scrub forests occur only in Kachchh where mangrove forest is mostly found along sea. There are different important forest areas in Gujarat which need to be surveyed at regular intervals. Photographs of some forest areas have been shown in Figure 9.1a-b.

Rodgers and Panwar, (1988) divided these forests into four major bio-geographic regions (1) Biogeographic zones (2) The biotic province (3) Region and (4) The biome. India has total 10 bio-geographic regions, (1) Trans-Himalayan with one province (2) The Himalayan with one province (3) The Indian Desert with one province (4) The Semi-arid zone with two provinces (5) The Western Ghats with two provinces (6) The Deccan peninsula with two provinces (7) The Gangetic plain with two provinces (8) North-east India with two provinces (9) The Islands with three provinces (10) The Coasts with two provinces. Thus, Gujarat has total four bio-geographic zones: Indian

A View of Girnar Reserve Forest, Junagadh (Nakar, 2013)

Kutchchh Region Forest, Near Greater Rann (Joshi, 2013)

Figure 9.1a: Different Forest Sites from Gujarat.

Desert-Province 3A-Kutch, The semi arid zone- Province 4B-Gujarat- Rajwada, The Western Ghat 5A- Malbar coast, Province 5B-Western Ghat Mountain.

Floristic Diversity of Forest in Dang (Tadvi, 2014)

A View of Tapkeswari Hills (Joshi, 2012)

Figure 9.1b: Different Forest Sites from Gujarat.

There are eight agro-climatic zones in Gujarat: (1) Southern Hills (2) Southern Gujarat (3) Middle Gujarat (4) North Gujarat (5) North west (6) North Saurashtra (7) South Saurashtra (8) Bhal and coastal.

3. Floristic Diversities

The angiosperm flora of Gujarat mostly varied in extent and composition, representing 12.91 per cent of the flora of the country. Floristic studies have long roots in Gujarat. Flora of the Presidency of Bombay (1901-1908) was studied for Gujarat flora along with Sind flora, and 450 plants species were identified for Gujarat and 176 species for Sind. However, flora of Kuchchh and Saurashtra was studied later (Blatter, 1908-09; Thakar, 1910, 1926). South Gujarat region has been intensively studied and by and large, the central part, part of North Gujarat, West Saurashtra and Eastern Kachchh remain virtually unexplored. Work of Santapau (1958) is well known among all researchers who had published survey based book, flora of Saurashtra (Part-1), which had covered Rannunculaceae to Rubiaceae family. Although work was incomplete, later Santapau and Janardan (1966) gave checklist of plants occurring in Saurashtra with 1,136 species under 591 genera of 126 families.

During 1926, Santapau came with 'Kutch ane Saurashtra ni vanaspatio ane teni Upyogita' (Plants of Kutch and Saurashtra along with their utility) which had 10 chapters along with picture outlines. During second phase of sanctuary, Botanical survey of Nawanagar (now Jamnagar) was carried out. Further research was done by adding 71 new species, published as "Vanaspati Shastra, Barda dungar ni jadibuti teni parikhsa ane upyog" by Thaker.

Shah (1978) published Flora of Gujarat (Two volumes), listing 684 species. Since last decades floristic studies increased showing increasing interest among workers. Sixty-two species of plants have been given different conservation status, majority of them are distributed in the semi arid regions of Kachchh. In addition, reforestation, afforestation, deforestation depend directly on physiology of any forest hence traits like photosynthesis, transpiration, water use efficiency along with morphological traits are very useful to study.

The recent studies of Jadeja (1999) for Barda Hills, Punjani (2002) for Aravalli Hills, Nagar (2005) for Barda Hills and surroundings, Vediya and Kharadi (2011) for Sabarkantha region, Nair (2011) for Dadra and Nagar Haveli, Modi and Dudani (2013) for North Gujarat, Kumar *et al.* (2011) for North Gujarat, Bhatt (2013) for Jamnagar area, Sharma and Sikarwar (2013) for Diu region and Mankad (2007) for Girnar forest enrich further our knowledge. Owing to importance of forest and its vast scope more people should work on forest diversity, conservation of endangered plants, phenology of new foliage, leaf fall, flowering, pollination and fruiting, germination and viability of seed, density and ecological aspects, (Figure 9.2). List of PhD thesis and other publications on forest surveys is given in Table 9.1.

Due to increased human population, for the last decade demand of plant species is increased and many species from the forest are wiped out. Data shows that about 53 species have become rare and are listed for immediate conservation (mostly from Kuchchh region). Different authors have suggested different categories for these plants (Table 9.2).

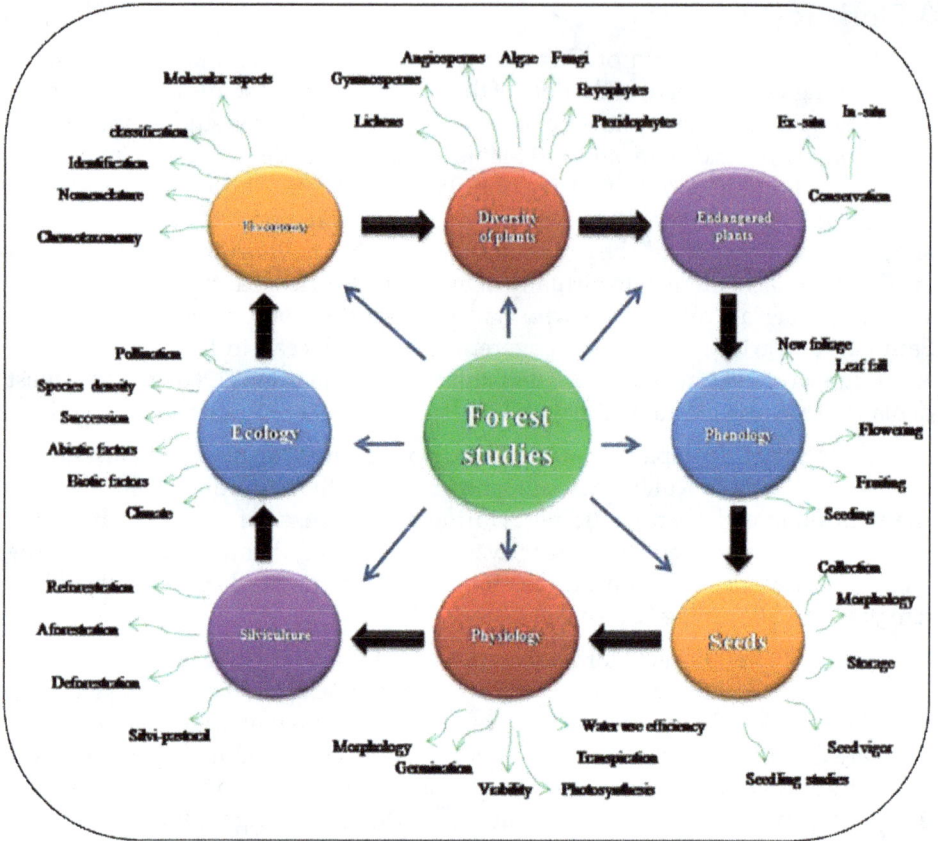

**Figure 9.2: Scopes of Inter Disciplinary Studies on Forest
Directly Helping in Economy of any Country.**

Table 9.1: Major Floristic Studies of Gujarat Region

Sl.No.	Contents	References
1.	Flora of Saurashtra	Santapau, 1952
2.	Contribution to the Flora of the Gir forest in Saurashtra	Santapau and Raizada, 1954
3.	Iter Kathiwarense: Being notes on a botanical tour in Nawanagar State	Santapau, 1945
4.	A contribution to the flora of North Gujarat	Yogi, 1970
5.	Common Trees.-Book	Santapau, 1971
6.	Working plant for Junagadh forest division.	Sinha *et al.,* 1972
7.	Studies on the botany of Jamnagar District.	Malahotra and Wadhawa, 1973
8.	A contribution to the flora of Bansada Forest	Desai, 1976
9.	Flora and vegetation of Saurashtra	Pandya, 1976
10.	Flora of Gujarat State, Vol I and II	Shah, 1978

Contd...

Table 9.1–*Contd...*

Sl.No.	Contents	References
11.	Floristics and phytosociological studies on some parts of Saurashtra	Menon, 1979
12.	Grasses of Western India	Toby and Patrick, 1982
13.	Observation on some rare or endangered endemics of Southern Kutch: Assessment of threatened plants of India	Sabnis and Rao,1983
14.	Rare species with restricted distribution in South Gujarat	Shah, 1983
15.	Floristic phytosociology and Ethno botanical study of Vapi and Umargaon area in South Gujarat.	Contractor, 1986
16.	Contribution to the flora of Surat district, Surat	Mac, 1986
17.	Flora of Dharampur Forest	Reddy, 1987
18.	Flora of Saurashtra. Part II and III	Bole and Pathak, 1988
19.	Forest wealth of Kutch, its Utility and Potencials	Pandya, 1989
20.	Red data book of Indian plants, Vol-I to III	Nayar and Shastry, 1990
21.	Taxonomical and ecological studies of the flora of Bhavangar	Oza, 1991
22.	Vegetational and wildlife studies in Gir	Chavan, 1993
23.	Studies on endangered and endemic desert taxa	Raole, 1993
24.	Flora of Gir-Report submitted to Department of forest, Gujarat	Ansari, 1994
25.	Flora of Girnar-Report submitted to Department of Forest, Gujarat	Ansari, 1994
26.	Biological diversiy of Gujarat-Current knowledge.	Pilo and Pathak, 1996
27.	Plants used by the tribe Rabari in Barda Hills of Gujarat.	Jadeja, 1999
28.	Biodiversity of Barda Hills	Nagar, 2000
29.	*Stylosanthes hamatus* (Linn.) Taub.(Papilionaceae), A New record to the flora of Gujarat.	Nagar and Pandya, 2002
30.	Addition to the flora of Saurashtra, Gujarat, India.	Nagar and Pandya, 2002
31.	*Cleome scaposa* D.C. (Capparidaceae). A rare species of Saurashtra	Nagar and Pandya, 2002
32.	Rediscovery of *Tephrosia jamnagarensis* (Fabaceae), an endangered and narrow endemic plant species of Saurashtra.	Nagar *et al.,* 2003
33.	Floristic diversity of Barda hills and their surroundings, Saurashtra, Gujarat.	Nagar and Pandya, 2003
34.	Status of some regionally threatened species of Saurashtra, Gujarat, India.	Nagar, 2004
35.	Studies on flora along the river bank of the Saraswati River from Mukteshwar to Patan District with ethnobotanical aspect	Patel, 2004
36.	Floristic diversity of Barda Hills and its surroundings	Nagar, 2005

Contd...

Table 9.1–*Contd...*

Sl.No.	Contents	References
37.	Floristic and ethnobotanical studies of Ambaji and Gorakhnath mountains of Girnar	Makad, R.S., 2007
38.	Phenological observations on some dry deciduous Forest trees at Barda hills, Gujarat	Jadeja *et al.*, 2008
39.	Study on medicinal plant diversity in Arboretum	Tadvi, D., 2009
40.	Plant Biodiversity: Collection, characterization and conservation of Motibaug Botanical Garden, Junagadh)	Dhaduk *et al.*, 2010
41.	Floristic diversity of Gujarat	Nagar and Daniel, 2010
42.	Floristic status and its conservation in the forests of North Gujarat Region, Gujarat	Rajendra Kumar, 2010
43.	Pictorial floristic diversity of grasses and associated vegetation from three grasslands of Randhikpur forest range, Dahod, Gujarat	Tyagi *et al.*, 2010
44.	A soujoun to the herbal treasures of MSU., M.S. University	Daniel and Nagar, 2010
45.	Phenological studies of some tree species from Girnar Reserve forest, Gujarat.	Jadeja and Nakar, 2010
46.	Floristic diversity of the stream and riparian zone of the Vishwamitri river Baroda.	Patale and Parikh, 2010
47.	Vegetational diversity in the land of Bhey (Garbada).	Gandhi *et al.*, 2010
48.	Taxoethnobotanical significance of some plants from Ambaji Forest (North Gujarat).	Dave *et al.*, 2010
49.	Observation on tree species of Danta Range forest of North Gujarat.	Patel *et al.*, 2010
50.	A review on some medicinal plants of Gujarat College campus, Ahmedabad, Gujarat.	Gondalia *et al.*, 2010
51.	Plant diversity and its life forms of Visnagar Taluka, North Gujarat.	Solanki *et al.*, 2010
52.	Traditional aboriginal knowledge of the flora in the Girnar holy hills areas, Gujarat.	Solanki *et al.*, 2010
53.	Floristic diversity of Isari zone, Megharj range forest District Sabarkantha, Gujarat, India	Vediya and Kharadi, 2011
54.	Floristic analysis of flora of Bhanvad, Jamjodhpur and LalipurTalukas of Jamnagar district, Gujarat	Jadeja *et al.*, 2011
55.	Floristic studies of Dadra and Nagar Haveli.	Nair, 2011
56.	In depth studies on some of the floristic components of Baroda and Panchmahal Districts with reference to Bioprospecting and speciation.	Gohil, 2012
57.	Floristic diversity of the Ahmadabad city	Jadeja and Patel, 2012
58.	Floral and avifaunal diversity of the Thol lake, wild life sanctuary of Gujarat State, India	Kariya, 2012
59.	Biological spectrum of Taluka Modasa, District Sabarkantha (Gujarat), India	Jangid, 2012

Contd...

Table 9.1–*Contd...*

Sl.No.	Contents	References
60.	Study of plant diversity in Vadali range forest district Sabarkantha, North Gujarat, India	Desai and Ant, 2012
61.	Floristical and ecological assessment of Joint forest Management Plantations and Natural Forests	Bariya, and Sandhyakiran, 2012
62.	Taxonomic status and utilization of halophytes in coastal Kachchh district, Gujarat.	Shah and Thivakaran, 2012
63.	Floristic studies on scared grooves situated near Utkantheswar, Kheda, Gujarat	Patel and Patel, 2012
64.	Floristic and ethnobotanical study of Palanpur and Dantiwada, Gujarat.	Patel and Patel, 2012
65.	Plant diversity assessment at land scape level in Jamnagar district, Gujarat using satellite remote sensing and geographic information system	Bhatt, 2013
66.	Plant richness modeling in South Gujarat using remote sensing and geographic information system	Bhatt *et al.*, 2013
67.	Floristic diversity and ecological studies in Forest areas of Idar and Vadali Talukas, district Sabarkantha	Desai, 2013
68.	Floristic study of Kaprada's hilly Forest in South Gujarat	Rao *et al.*, 2013
69.	Study on the floristic diversity of two newly recorded sacred groves from Kachchh district of Gujarat, India	Patel *et al.*, 2014
	Floristic diversity of Dangs forest, Gujarat	Tadvi, 2014
70.	Floristic diversity of Santrampur forest Range of Panchmahals, Gujarat	Chaniyara, 2014

4. Phenological Attributes of different Forest Areas

Phenology, the time of recurring phenomena is expressed in the form of flowering, fruiting, new foliage, leaf drop and germination of seedling like events in response to calendar and climate by different plants. There are only a few studies on phenological characteristics to see life cycle of plant forms in Gujarat. In earlier phenological studies on a limited flora of Gujarat gave information of only some phenological stages.

Study on phenology of Reserve forest, Bhavnagar carried out by Gohil (2005), reported that amenity plantation was needed to decrease the increased concern about environment conservation. She reported that flowering in tree species was during February-March followed by fruiting however, new foliage was characterized by June-July but leaf fall was set during August to December. On the basis of study in Victoria reserve forest, Bhavanagar and in GIDC polluted area, Dr. Gohil (2005) suggested plantation of plant species like *Cassia siamea* Lam, *Ceiba pentandra* L., *Delonix regia* (Boj.) Raf. *Gliricidia sepium* (Jacq.) Walp., *Millingtonia hortensis* L.f., *Peltophorum pterocarpum* (DC) Backer ex Heyne for amenity plantation in Urban area. The recent physiological studies are compiled in Table 9.3.

Phenological study on three species of Prosopis during 2008-09, showed that new foliage time for these species was from June to November while leaf fall was between

Table 9.2: Endangered Plants Species Listed by various Workers

Species	Family	Status	Distribution	Author
Ammannia desertorum Blatt. and Hallb.	Lythraceae	Rare	Jamnagar	Nayar and Sastri, 1988
Anogeissus sericea var. *nummuaria* King ex Duthie	Combretaceae	Rare	Ahemdabad	
Campylanthus ramosissimus Wt.	Scrophulariaceae	Rare	Kuchch, Bhuj	
Ceropegia odorata Nimmo ex Hook.f.	Asclepiadaceae	Rare	Pavagadh	
Chlorophytum borivilianum Sant et Fernand	Liliaceae	Rare	Dang,Ahwa	
Cyperus dwarakensis Sahni et Naithani	Cyperaceae	Rare	Dwarka, Jamnagar coast	
Helichrysum cutchicum (C.B. Cl.) Rolla Rao et Deshpande	Asteraceae	Rare	Kuchchh and North Saurashtra	
Indigofera coerulea Roxb. var. *monosperma* (Sant.) Sant.	Fabaceae	Rare	Kuchch, Saurashtra Bhadreswar	
Tephrosia jamnagarensis Sant.	Fabaceae	Rare	Jamnagar, Bhat bunt, Udhna	
Anticharis senegalensis (Walp.) Bhandari	Scrophulariaceae	Rare	Anjar, Kuchchh	Rao, K.S.S., 1981
Astragalus prolixus Sieb.	Fabaceae	Rare	Mundra, Bhadreswar	
Bouchea marrubifolia Schauer	Verbinaceae	Rare	Anjar	
Cassia holosericea Fresen	Caesalpiniaceae	Rare	Anjar, North Gujarat	
Cassia senna Linn.	Caesalpiniaceae	Rare	Anjar, Bhachau, Gandhidham	
Commiphora wightii (Arnott) Bhandari	Bursaraceae	Rare	Kuchchh, Bhuj	
Convolvulus stocksii Boiss.	Convolulaceae	Rare and threatened	Bhuj	
Corbichonia decumbens (Forsk.) Exell	Chenopodiaceae	Rare	Anjar	
Cyperus conglomeratus Rottb.	Cyperaceae	Rare	Near to Mandavi	
Dignathia hirtella Stapf	Cyperaceae	Rare	Kandla	

Contd...

Table 9.2–*Contd*...

Species	Family	Status	Distribution	Author
Ephedra foliata Boiss. and Kotschy ex Boiss.	Gnetaceae	Rare	Bharatpur	
Halopyrum mucronatum (Linn.) Stapf	Chenopodiaceae	Rare	Mandvi	
Haloxylon recurvum (Moq.) Bunge ex Boiss	Chenopodiaceae	Rare	Bhuj	
Heliotropium bacciferum Forsk. var. *suberosum* (Clarke) Bhandari	Boraginaceae	Rare	Mandavi	
Hibiscus obtusifolius Garcke	Malvaceae	Rare	Bhuj	
Hyphaene indica Beccari	Fabaceae	Rare	Mundra and Mandavi	
Indigofera argentea Burm. f.	Fabaceae	Rare	Mandavi North west Kuchchh	
Indigofera coerulea Roxb. var. *monosperma* Sant.	Fabaceae	Rare	Bhadreswar, Saurashtra and Kuchchh	
Ipomoea kotschyana Hochst. ex Choisey	Convolulaceae	Rare	Eastern Kuchchh	
Launaea resedifolia (Linn.) Druce	Asteraceae	Rare	Mandavi	
Limonium stocksii (Boiss.) O. Ktze.	Rutaceae	Rare	Bhadreswar	
Micrococca mercurialis Linn.	Euphorbiaceae	Rare	Anjar	
Monsonia senegalensis Guill. Perr.	Geraniaceae	Rare	Bhuj	
Pavonia arabica Steud.	Malvaceae	Rare	Bhuj	
Pavonia ceratocarpa Mast.	Malvaceae	Rare	Bhuj	
Polycarpaea spicata Wt. and Arn.	Caryophyllaceae	Rare	Bhuj	
Premna resinosa Schau	Verbinaceae	Rare	Bhuj	
Psoralea plicata Del.	Acanthaceae	Rare	Bhadreswar Bhuj	
Sarcostemma acidum (Roxb.) Voigt	Acanthaceae	Endangered	Bhuj	
Schweinfurthia papilionacea (Linn.) Merrill	Scrophulariaceae	Rare	Anjar, Kandla and Gandhidham	

Contd...

Table 9.2–*Contd...*

Species	Family	Status	Distribution	Author
Schweinfurthia pterosperma A. Braun	Scrophulariaceae	Rare	Bhuj, Gandhidham	
Senra incana Cav.	Malvaceae	Rare	Bhuj	
Solanum incanum Linn.	Solanaceae	Rare	Bhuj	
Sporobolus virginicus (Linn.) Kunth	Gramineae	Rare	Mandavi	
Sterculia urena Roxb.	Sterculiaceae	Rare and threatened	Chauduva and Kirgaria Hillocks	
Tecomella undulata (Sw.) Seem.	Bignoniaceae	Rare and threatened	Anjar, Chauduva Hills	
Capparis cartilaginea	Capparaceae	Rare	Naliya and Narayan Sarovar	Bhatt, J. B.,1993
Fagonia indica Burm.f.	Zygophyllaceae	Rare	Westrn Kuchchh	
Sedera latifolia Hocht. Ex Steud	Convolulaceae	Rare	Western Kuchchh	
Solanum albicaule Kotschy ex Dunal	Solanaceae	Rare	Nakhtrana	
Chascanum marrubifolium Fenzel ex Walp	Verbinaceae	Rare	Nakhtrana, Kuchchh	
Aeluropus lagopiodes (Linn.) Trin.	Poaceae	Rare	Kuchchh, coastal Saurashtra	
Urochondra setulosa (Trin.) Hubb.	Poaceae	Rare	Kuchchh near India Bridge	

Table 9.3: Phenological and Physiological Studies of Forest Trees in Gujarat

Contents	Study Areas	References
The improvement of germination of some Forest species by acid scarification	Bhavanagar forest range	Zapade, 1991
Influence of seed traits on germination of *Prosopis cineraria* L.	Rajasthan, Haryana and Gujarat	Manga and Sen, 1995
Study on soil properties and their influence on vegetation in western region of Gujarat, India.	Different ecosystems of Gujarat	Panchal and Pandey, 2002
Phenological studies of Reserve Forest (Victoria Park) near Bhavanagar and GIDC area	Bhavanagar and its GIDC area	Gohil, 2005
Effect of storage on seed viability and germination of certain Leguminous trees	Gujarat Forests	Arya and Arya, 2006
Phenological observations on some dry deciduous Forest trees at Barda hills, Gujarat.	Barda Hills, Porbandar	Jadeja *et al.*, 2008
Salinity tolerance of *Aegicersa corniculatum* (L) Blaco. From Gujrat coast of India	Saurashtra University, Rajkot	Patel and Pandey,2009
Study on morphology, ethnobotany and phenology of *Prosopis* in Girnar Forest Junagadh, Gujarat.	Girnar Reserve forest, Junagadh	Nakar and Jadeja, 2009
Germination and growth of *Commiphora whitii* and *Acacia senegal* under different salinity from arid coast of Kachchh	Kuchchh region	Shah and Thivakaran, 2010
Methods to break seed dormancy in *Andrographis paniculata* (Burm f)-a medicinal herbs of tropical Asia	Vidhyanagar, Anand	Kumar *et al.*, 2010
Seed germination of selected taxa from Kuchchh desert	Kuchchh region	Raole *et al.*, 2010
Phenological studies of some tree species from Girnar Reserve Forest, Gujarat, India	Girnar Reserve forest, Junagadh	Jadeja and Nakar, 2010
Phenological studies of tree species of Satlasana range Forest (North Gujarat)	Satlasana Range forest, Sabarkantha	Desai and Patel, 2010
Comparision of phenology of *Cassia siamea* Lam. of Girnar Reserve Forest and GIDC area Junagadh	Girnar Reserve forest and GIDC, Junagadh	Nakar and Jadeja, 2013a
Studies on phenology of some shrubs from Girnar Reserve Forest, Gujarat.	Girnar Reserve Forest, Junagadh	Nakar and Jadeja, 2013b
Studies on Phenological patterns of Girnar Reserve Forest, near Junagadh, Gujarat	Girnar Reserve forest, Junagadh	Nakar, 2013
Floristic diversity and ecological studies in forest areas of Idar and Vadali Talukas, Sabarkantha	Vadali area, Sabarkantha	Desai, 2013

Contd...

Table 9.3–*Contd...*

Contents	Study Areas	References
Phenological studies of two bombacacean members from Girnar Reserve Forest, Junagadh, India	Girnar Reserve forest, Junagadh	Nakar and Jadeja, 2014a
Seed pattern, germination and viability studies on some forest tree species seeds from Girnar reserve Forest	Girnar Reserve forest, Junagadh	Nakar and Jadeja, 2014b
Standardization of wild Forest seeds for morphology and viability studies	Girnar Reserve forest, Junagadh	Nakar and Jadeja, 2014c
Phenological attributes of seven Caesalpiniaceae members in association with climate	Girnar Reserve forest, Junagadh	Nakar and Jadeja, 2014d
Flowering and fruiting phenology of some herbs, shrubs and undershrbs from Girnar Reserve Forest	Girnar Reserve forest, Junagadh	Nakar and Jadeja, 2015

September to May. Flowering and fruiting occurred during February to May in Girnar Reserve forest, (Nakar and Jadeja, 2009). Study at Satlasana range forest, north Gujarat on 19 tree species with 18 genera and 13 families for their month wise phenological behavior, exhibited that maximum leaf fall in month of January and minimum leaf fall in June to August, highest flowering was observed in March while lowest in November to December, maximum new flowering found from April to May (Desai and Patel, 2010).

The phenological behavior was studied on 10 woody tree species in 2008-09, in Girnar reserve forest of Saurashtra region, concluded that leaf fall in most of species occurred during January, new leaves started before monsoon in February to March. Fruiting activity was observed maximum in December. Duration of maturation of leaves was shortest but that of fruit ripening was longest in most of the species (Jadeja and Nakar, 2010). In other studies on *Cassia siamea* Lam. at two different locations in Girnar Forest and GIDC polluted area; the tree proved to be better option against pollution and recommended for polluted areas (Nakar and Jadeja, 2013a).

In other two years study of two tree species, of Bombaceae family of Girnar reserve forest, both species *viz. Adansonia digitata* L. and *Bombax ceiba* L showed new foliage in June-July, within 50 days, and leaf drop for 132 and 139 days during August to December for *Adansonia digitata* L., and August to January for *Bombax ceiba* L. However, flowering period was found between May to June with 505 and 581 days for both species followed by fruiting during June- July for 33 and 58.5 days, respectively during 2008 to 2010. Interestingly, duration for interphenophase phenology was 195 days for new foliage to leaf drop however; duration for flowering to fruiting was 66.5 for *Adansonia digitata* L. and, in *Bombax ceiba* L., new foliage to leaf fall behavior was observed for 335 days but flowering to fruiting was recorded for 115 days respectively. Moreover, positive significant correlation was observed between branches per tree and leaves per branch, as well as branches per tree and flowers per branch for both species (Nakar and Jadeja, 2014a).

Recent studies for three years from 2008 to 2011 on phenological pattern of 26 plant species including herbs, shrubs and undershrubs of Girnar reserve forest expressed that, for non woody species, dry season is best time for flowering and fruiting explained by the fact that seeds have to be remained in the soil to germinate in next rainy seasons. In this study, it was also concluded that perennials took highest flowers and fruiting days compared to annuals, and among comparison of herbs, shrubs and undershrubs, highest flowering duration was noted for shrubs followed by undershrubs and herbs. Though direct relation was not found between number of days for phenological events and climate but affected indirectly. In addition, species classification was proposed on basis of highest, medium and lowest flowering or fruiting duration containing species (Nakar and Jadeja, 2015).

In a study on 7 tree species including 6 Caesalpiniaceae trees and a member of Fabaceaea family Nakar and Jadeja, (2014d) found positive and negative correlation with number of phenological days as well as morpho-phenological traits in relation to climatic factors. In *Bauhinia purpurea* L., negative correlation was recorded between inflorescence per branch and maximum temperature (-0.67**) and also between inflorescence per branch and minimum temperature (-0.61**) however, *Caesalpinia crista* L. pertained positive correlation with wind speed (0.75**). *Cassia fistula* L. showed positive association between inflorescence per branch and wind speed (0.59**), flowers per branch and wind speed (0.56 **) but *Cassia siamea* Lam. exhibited positive correlation between fruiting per branch and wind speed (0.49*), and rain (0.74**). Interestingly, in Fabaceae member *Derris indica* (Lam.) Bennet, there was positive relation between inflorescence per branch and minimum temperature (0.64**). These results indicated that different species behaved differently against temperature and *Bauhinia purpurea* L. showed negative whereas *Derris indica* (Lam.) Bennet pertained positive correlations with temperature (Nakar and Jadeja, 2014d).

Most of the studies from Gujarat on phenology of forest trees indicated similar season for all phenological events such as new foliage, leaf fall, flowering and fruiting as there is very small difference between different places of semi arid region. Climate has its impact on phenological sequence, furthermore occurs increase or decrease in inflorescence or flowering if temperature increases or decreases. Climatic factors such as rain, wind speed and temperature had direct effect on phenological characteristics of Girnar Reserve forest area in seven tree species. Although comprehensive studies are not available, these studies are in right direction for further research. Phenological aspects can be used in determining proper collection stage of plant material. Different stages of phenology indicate different response against climate at that particular time (Figure 9.3). In a one year study on life cycle of 10 major tree species, it was observed that peak time for new foliage in plants is April-May and for leaf fall is January-February. However, maximum flowering was observed during January-February followed by peak season for fruiting during March-April.

5. Physiological Attributes

The physiological attributes and traits such as germination, viability and regeneration have direct effect on phenology, productivity and floristic composition of any forest. Soil moisture study indicates that previous season is important in triggering flowering of any species. Differences in soil moisture level can produce ecotypes for short period. Germination is awakening of dormant embryo, or coming out of radical after imbibitions and absorption process. It is vital for any species; hence efforts have been made to improve germination percentage for species.

5.1. Germination

Nakar and Jadeja (2014b), in a seed germination study of nine tree species of six families of Girnar reserve forest, found average germination, 27.2, 55.6 and 69.4 per cent after 5, 10 and 20 days respectively for all species at room temperature. Highest germination was in two Bombacaceae species, *viz. Bombax ceiba* with 95 per cent germination, followed by *Ceiba pentandra* species with 30 per cent germination after 5 days of incubation. Average seed vigor index increased from 59, 579 and 865 after 5, 10 and 20 days. Average viability was 91 per cent with highest value of 94 per cent for *Albizia lebbeck*. In comparison of germination and viability, viability was high as compared to germination in most of the species studied.

Figure 9.3: Phenological Events in 10 Tree Species of Girnar Reserve Forest (Jadeja and Nakar, 2010). The A, B, C and D are new foliage, leaf fall, flowering and fruiting phenophases, respectively.

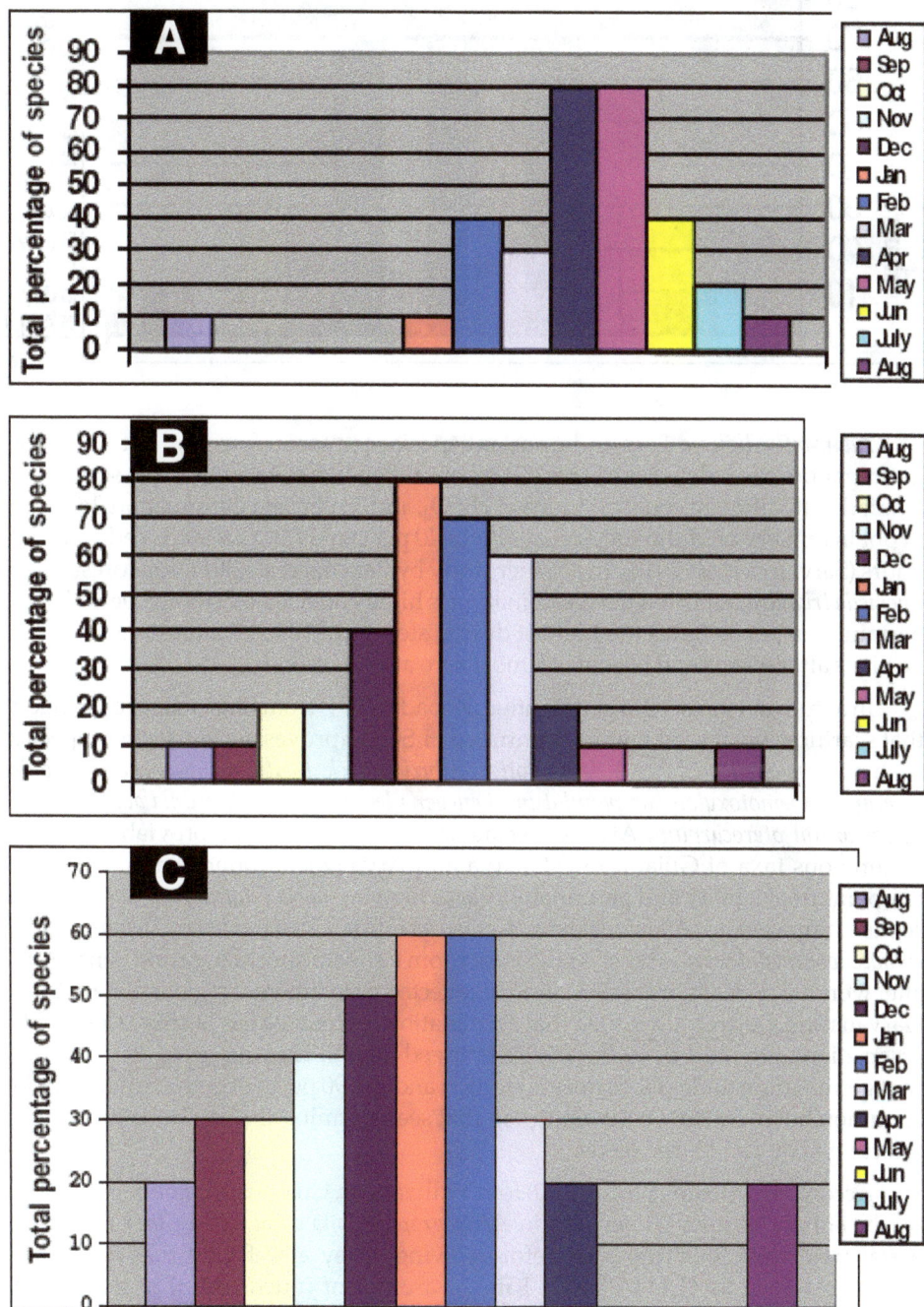

Contd...

Figure 9.3—*Contd...*

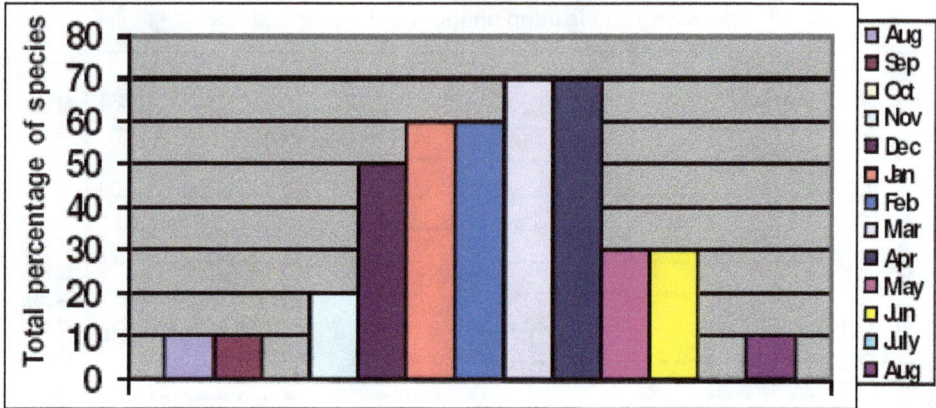

Adansonia digitata L. is an important species for social forestry, but had poor germination due to hard seed coat. So, germination behavior of this tree seeds was studied with different concentrations of H_2SO_4, and 50 per cent germination (highest) in pot in nursery at Motibaug, Junagadh for 40 per cent H_2SO_4 treated seeds after 48 hours (Sarvaliya *et al.*, 2010). In another study by Panchal *et al.* (2011), on forest grass *Themeda triandra*, reported that germination is highly affected by storage period, that germination per cent reaches highest during storage of 10 to 15 months however, it declines after words and becomes almost zero after 22 months.

In a study on forest seed germination, Zopade (1991) from Bhavanagar concluded that scarification of seed with concentrated H_2SO_4 improves germination in species such as *Acacia leucophloea, Acacia nilotica, Albizia lebeck, Beutia monosperma, Prosopis cineraria, Haemotoxylon companulatum, Leucaena lecocphala, Parkinsonia aculeate* and *Peltophorum pterocarpum*. After six years study from 1997-2002 on viability of 23 leguminous taxa of Gujarat forests Arya and Arya (2006) found that at different temperature, viability and germinability was high in *Acacia nilotica* and *Cassia fistula* seeds compared to *Albizia lebbeck, Cassia javanica, Cassia siamea, Delonix regia, Pithocellobium dulce* and *Prosopis juliflora* at room temperature. Manga and Sen (1995), after working on seeds of *Prosopis cineraria* collected from different regions of Rajasthan, Haryana and Gujarat, advocated that germination percentage has positive significant association with most of seed traits except seed length. They reported 30 to 100 per cent germination under laboratory condition and 7 to 90 per cent germination in field condition however they further argued that seed germination can be improved by selecting heavy and large seeds.

According to Kumar *et al.* (2010) fom Vidhyangar, Gujarat germination capacity can be enhanced upto 93 per cent in *Andrographis paniculata*, wild plant using hot water treatment for 5 minutes before sowing. They also found that hormonal treatments such as GA, IAA and Kinetin are not of much useful in increasing germination for this species.

Raole *et al.* (2010) worked on germination of 11 wild rare species of Kutch region committed that all the wild seeds are dormant due to thick seed coat hence most of the seeds require different storage period to get mature. Among all species, only *Capparis cartilaginea* germinated without prior treatment, however growth regulators had increased germination in few taxa singly and in some cases with combinations. In another study from Kuchchh region on two endangered plant species *Commiphora whitii,* and *Acacia nilotica* for adaptation and germination against salinity, Shah and Thivakaran (2005) found that *Commiphora whitii* can grow upto 3 to 5 dSm^{-2} however, after that growth of the plant is stunted and plant starts to shrink but in other species *Acacia nilotica,* survival limit for plant is 6.5 dSm^{-2}. Thus, *Acacia nilotica* is more salinity tolerant plant species in Kuchchh region.

With weather of Saurashtra, a green house study on germination of *Aegicersa cornicalatum* (L) Patel and Pandey (2009) found optimum growth of seedling with salinity of 24 dSm^{-2} where proline and Na concentration increased, but K, Ca, N and P decreased due to excess salinity. In another study on four different ecosystems namely, forest, grassland, degraded land and desert in Gujarat, Panchal and Pandey (2002) observed that, there is positive correlation between decreased clay content, water holding capacity, field capacity, organic carbon, phosphorus and number of plant species or species density. In addition, they also opined that, calcium, magnesium and sodium have negative correlation between herbaceous biomass and deteriorated ecosystem.

The germination and viabilities of 9 forest species (1) *Aegle marmelos* (L.) Corr., (2) *Albizia lebbeck* (L.) Bth., (3) *Bombax ceiba* L., (4) *Cassia siamea* Lam., (5) *Ceiba pentandra* L., (6) *Delonix regia* (Boj.) Raf., (7) *Leucaena leucocphala* (Lam) de Wit, (8) *Peltophorum pterocarpum* (DC) Backer ex Heyne, (9) *Sterculia urens* Roxb.) after 20 days of collection are given in Figures 9.4 and 9.5 (Nakar and Jadeja, 2014b). Overall viability was high in all of the species, but only two Bombacacea members *viz. Bombax ceiba* L. and *Ceiba pentandra* exhibited more germination percentage than that of viability.

5.2. Studies on Seed

Seed, which is a matured or ripened ovary formed after fertilization process, in forest is essential to understand as it deals with its ecosystem functioning. Seed collection is major aspect for any nursery plantation. Furthermore, local forest seed can adapt easily in local climate hence process such as collection, identification and classification of seeds for any local forest area is necessary for getting planting material as nursery stock and as well as to maintain species diversity of any forest.

Seed morphological qualitative and quantitative traits, studied for 97 species including trees, herbs, shrubs and climbers in Girnar Reserve forest showed variation in length, width, thickness, weight and volume. Highest seed volume was recorded for trees followed by climbers, shrubs and herbs. Overall, *Terminalia arjuna* showed biggest seed volume with value of 21.4 cm^3. In qualitative traits such as shape, colour, surface and hilum shape, different shapes such as, obovate, sub obovate, ovate, trigonal pyramidal, wedge shaped, kidney shaped, reniform, subreniform, tubular, elliptical, narrow elliptical etc. for different species whearas, different colors such as brown, brownish black, brownish yellow, yellowish white, yellowish brown were found.

Figure 9.4A-F: Germination, Viability and Seedling Vigour Index among 9 Forest Species (1 *Aegle marmelos* (L.) Corr., 2 *Albizia lebbeck* (L.) Bth., 3 *Bombax ceiba* L., 4 *Cassia siamea* Lam., 5 *Ceiba pentandra* L., 6 *Delonix regia* (Boj.) Raf., 7 *Leucaena leucocphala* (Lam) de Wit, 8 *Peltophorum pterocarpum* (DC) Backer ex Heyn, 9 *Sterculia urens* Roxb.). (Nakar and Jadeja, 2014b).

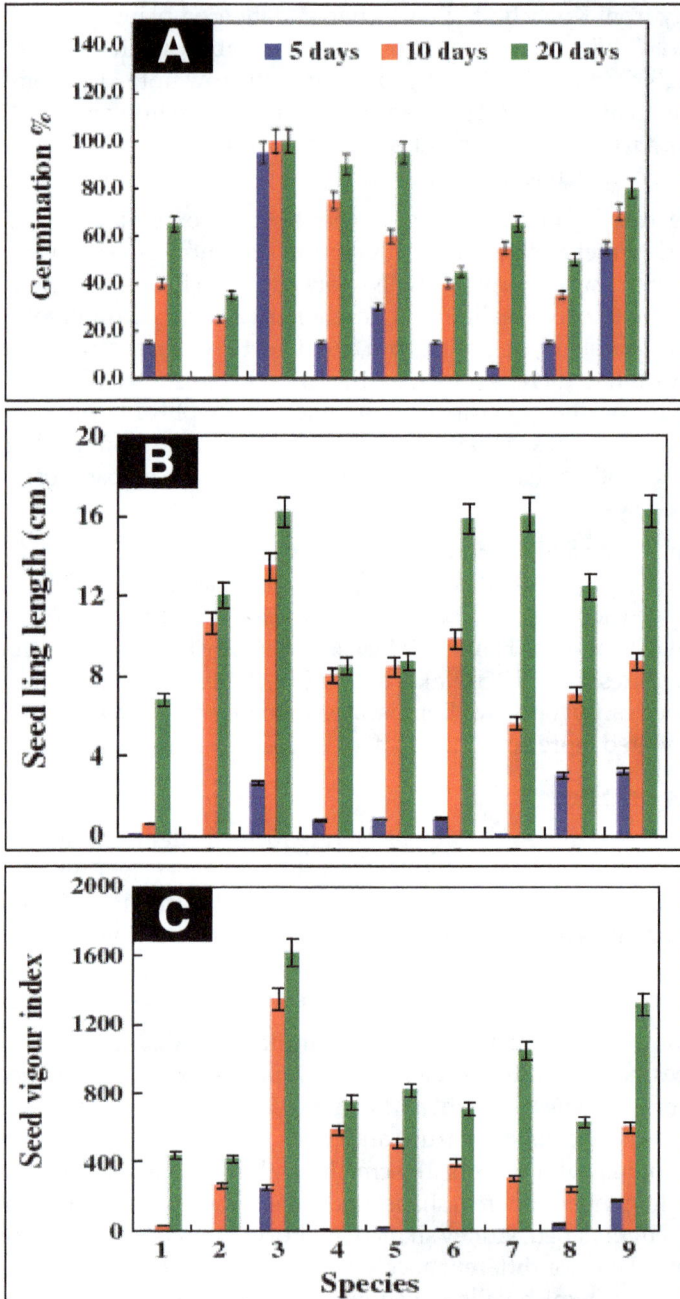

Contd...

Figure 9.4–*Contd...*

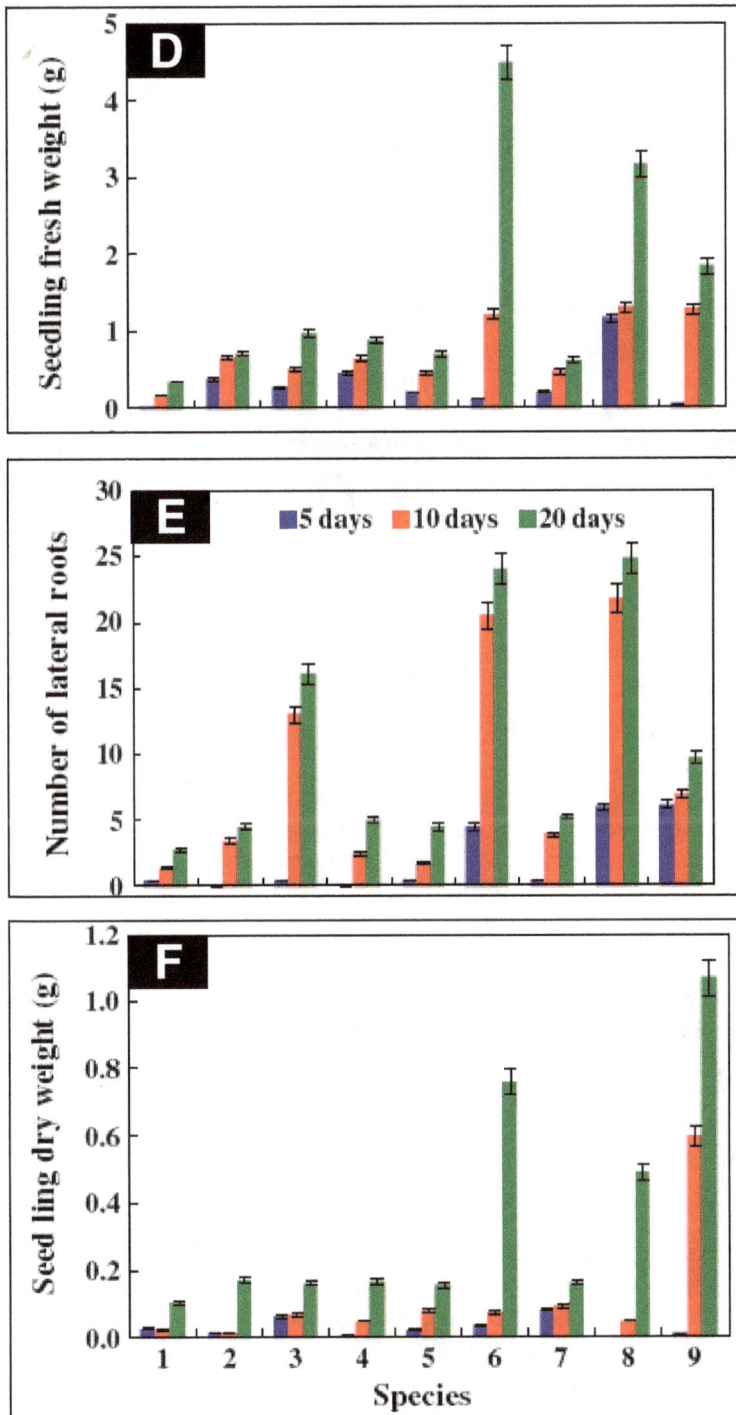

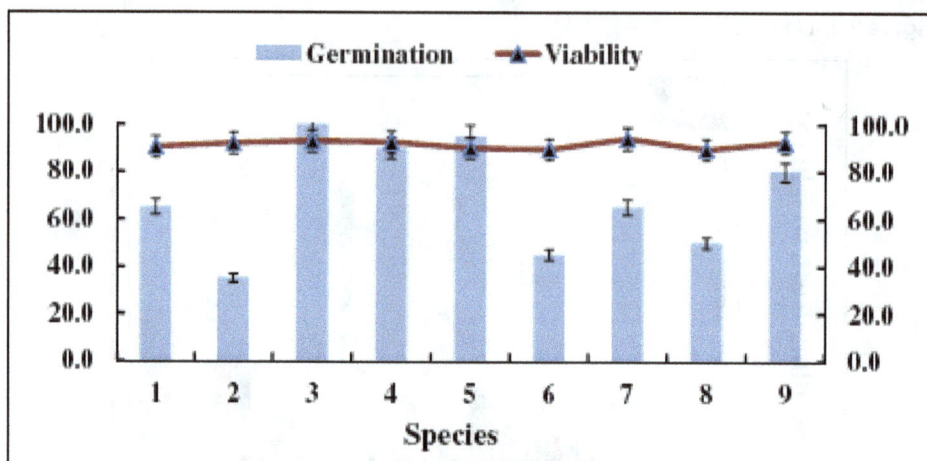

Figure 9.5: Germination and Viabilities of 9 Forest Species. 1: *Aegle marmelos* (L.) Corr., 2: *Albizia lebbeck* (L.) Bth., 3: *Bombax ceiba* L., 4: *Cassia siamea* Lam., 5: *Ceiba pentandra* L., 6: *Delonix regia* (Boj.) Raf., 7: *Leucaena leucocphala* (Lam) de Wit, 8: *Peltophorum pterocarpum* (DC) Backer ex Heyne, 9: *Sterculia urens* Roxb.) after 20 Days of Collection. (Nakar and Jadeja, 2014b).

There were different surfaces of seeds like rugose, glabrous, granular, however, Hilum is a part of attachment of seed with fruit, varied in different shapes such as circular, elliptic, broad elliptic etc (Nakar, 2013).

The seed atlas of Gujarat by Gavit and Parabaia (1989), reports that morphological features of seed are of immense use in forest studies. Gavit (1990) studied morphological features of seeds of plant in South Gujarat area and committed that morphological features are very useful. Recently Gohil (2005) studied morphological attributes of seeds for Reserve forest of Bhavanagar and reported variation in qualitative and quantitative traits for species of trees, herbs, shrubs and climbers.

Recently Nakar and Jadeja, (2014c) worked on seed morphology and viability of 97 species of Girnar reserve forest using X-ray visualization technique and standardized different doses related with four different groups of plants including trees, herbs, shrubs and climbers for proper visualization.

5.3. Morphological Traits

The morphophenological traits studied by Nakar and Jadeja, (2014a) in two Bombacean plants of Girnar reserve forest, Junagadh are given in Figure 9.6.

5.3.1 Productivity and Carbon Sequestration

Pandit *et al.* (1993) in a study reported that leaf litter production of dry deciduous forest ecosystem of Gir alone contributed major portion to the total leaf production of 3.68 t ha^{-1} and species like *Acacia catechu, Acacia latifolia, Butea monosperma* and *Diospyros melanoxylon* provided 10 to 17 tonne ha^{-1} annual contribution of biomass. Among most species tree density affected litter production but in *Acacia latifolia* inspite of high density leaf litter production was less compared to other tree species.

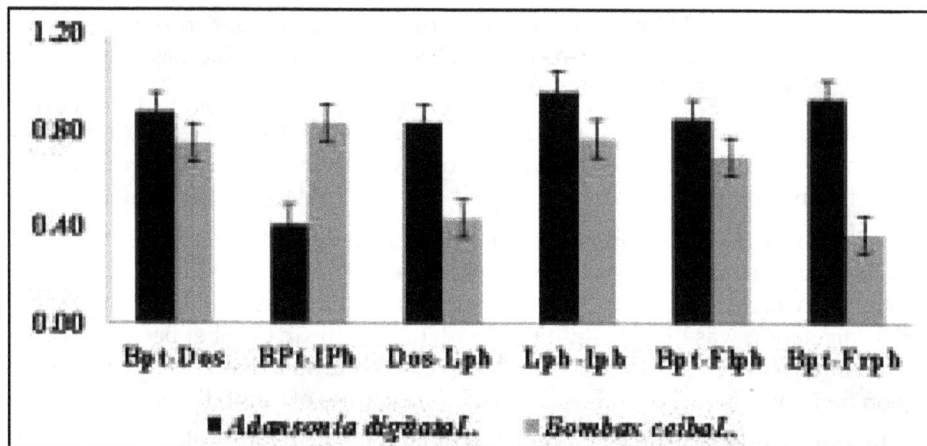

**Figure 9.6: Morphophenological Traits in Two Bombacean Plants of
Girnar Reserve Forest (Nakar and Jadeja, 2014a). The Bpt is
branches per tree, Ipb is Inflorescence per branch, Lpb is leaves per branch,
Flpb and Frpb are flowers per branch and fruits per branch respectively.**

In Dangs forest net production of five tree species was measured by Pandeya *et al.* (1967) where *Ougeinia oojenensis* had higher production followed by *Dalbergia latifolia, Garuga pinnata, Terminalia cranulata* and *Tectona grandis*. The net primary productivity was proportional with annual growth rings with increasing trend of concentration of growth in wood tissue along with increasing size of plant.

Parwani and Jasrai (2013) found seasonal variation in *Cenchrus cilliaris,* a common grass in Gujarat and reported that the above ground and below ground biomasses were, 155 and 62 kg ha^{-1}, respectively during dry season (February to May), and 256 and 76 kg ha^{-1}during monsoon (June to September). But during winter season (October to January), these were 115 and 54 kg ha^{-1}, respectively. This grass exhibits maximum biomass during monsoon season and could be grown in mixture with *Dichanthium annulatum,* another common grass.

Bhatt *et al.* (2013) used Resource sat-1 Satellite data to interpret forest cover and land map of South Gujarat, and reported that forest areas contribute 17.4 per cent to the total area and highest biomass was of the teak (15 t ha^{-1}) followed by other forest of mangroves, mangroves scrub, riverine forest, reverine thorn forest, degraded forest and grassland. Using satellite data, Baria (2003) in a study established that due to increase in green cover of Rajpipla net productivity has increased and recommended leaving certain percentage of litter and foliage on the ground to enhance physical and chemical properties of soil due to high nutrient contents of foliage.

In a study, the above ground biomass was estimated using micro-wave remote sensing technique with DLR-ESAR air borne data at Shool Paneshwar Wild life Sanctuary, Narmada district, Gujarat where, L bands (Longer wave lenth bands) have higher ability to penetrate into forest canopy and can estimate biomass for low frequency for 160 mg ha^{-1} to 200 mg ha^{-1} whereas P-band is capable of estimating from 100 mg ha^{-1} to 200 mg ha^{-1}, while in deep, with increase in biomass, back scattering co-efficient increases, and C-band ESAR can estimate biomass upto 70 mg ha^{-1} (Nizalpur *et al.*, 2010)

In another study, Pandey and Pandey (2013) estimated that total of 8.116 m tonne carbon has been sequestered by mangroves of Gujarat at 30 cm depth as a result of mangrove soil contribute more carbon compared to mangrove plants.

In Surat district, a study carried out at 42 different sites using LLSS-III and MODIS data to measure biomass of the area gave range of biomass from 6-389 t ha^{-1} (where area weight formula of 250 x 250 m, measured biomass at 5.5 t ha^{-1} to 134 t ha^{-1}) and correlation was highly significant with high value in October between above ground biomass (AGB) and band (R) of MODIS satellite data (Patil *et al.*, 2010). Mean carbon density was 19.4 t carbon ha^{-1} in the forest areas of Surat district.

In a study ISRO scientists argued that biomass densities in Gujarat were 142.8, 138 and 94 t ha^{-1} for the respective crown densities, 70 per cent and above, 40 to 70 per cent and 10 to 40 per cent. Accordingly, average growing stock density was 74.4 m^3 ha^{-1} whereas standing biomass (Above and below ground biomass) was estimated with value of 8683.7 million tone. The mean biomass density of total Indian forest was 135.6 t ha^{-1} however it varied with 27.4 t ha^{-1} in Punjab and 252 t ha^{-1} in Jammu-Kashmir. They also argued that these kinds of studies could be used in deforestation studies as well as in carbon pool studies.

In a recent study of Dediyapada taluka, using remote sensing technique for forest phenology and biomass estimation, resulted with useful information and remote sensing method, could be useful to aid foresters in order to prepare recovery programme for forest. Details on tree damage gave idea regarding dominant tree species such as *Tectona grandis, Butea monosperma, Dalbergia sisoo, Terminalia cranulata* and *Madhuca indica*. The phonological changes in these trees were in relation to abiotic factors such as rain, wind speed and temperature using Remote sensing data. They concluded that, technique with use of polarimetric Radarsat-2 and ENVISAT ASAR data are more practical for forest monitoring due to their better mapping acquisition capability.

Using data by non destructive or algometric method, carbon storage of 25 tree species with similar diameter girth, was evaluated by Pandya *et al.* (2013) found maximum biomass in *Tamarindus indica* with value of 56 t ha^{-1} followed by *Terminalia arjuna* 45 t ha^{-1}. Lowest carbon in the study, was rendered by *Emblica officinalis* with value of 1.77 t ha^{-1}.

Forest biomass is most cruisal step in quantification of carbon.), the total carbon stock in India's forest biomass was 3325 to 3161 million tone (Seikh *et al.*, 2011). In West Coast of India, the total biomass was 300 million tone in 2003. In study on carbon sequestered by 10 major trees from Gandhinagar forest area (Attarsumba range), it was recorded that total carbon stock of that forest was 46 t and of this total carbon stock, maximum carbon 55 per cent was contributed by *Acacia tortilis* however *Cordia perrottetii* and *Pheonix sylvestris* had storage potential of 25 and 56 per cent, respectively. Other species such as *Azadirechta indica, Acacia nilotica, Acacia senegal* and *Prosopis cineraria* contributed 3981, 3421, 4457 and 1329 t carbon. According to Rathore and Jasrai (2013), trees with high woody density should be selected because they can grow at high level of CO_2 in the atmosphere, with maximum biomass with huge canopy.

Social, cultural, organizational and market factors in addition with to physical factors are important in the production criteria of forest management programme. There is 1 per cent increased dependence of user group on forest which supply produce upto 2.92 per cent with forest canopy cover by 0.67 per cent (Mishra and Shashikant, 2003). The joint forest management (JFM) can be used effectively in conservation of any forest biodiversity. In Rajpipla forest the dominant families were Mimosaceae followed by Anacardiaceae, Combretaceae, Bursaraceae, Verbinaceae, Rubiaceae, Meliaceae Euphorbiaceae and Sapotaceae (Baria, 2003). Compared with natural forest, the area adopted by joint forest management showed more biomass, tree density (440 to 1990 ha^{-1} in JFM plot and 330 to 930 ha^{-1} in natural forest) and better regeneration potential of plant species, *Tectona grandis, Butea monosperma and Diospiros melanoxylon* (Baria, 2003).

7. Summary, Conclusion and Future Aspects

The Girnar, Dang, Jambughoda, Aravali are the main forests wealth of Gujarat and xerophytic forest in Kachchh housing 2,198 species of higher plants of 902 genera and 155 families representing 12.91 per cent flora of India. These forest wealths provide great opportunity to study, but the plant species are not studied in detail. The floristic, vegetational and diversity studies of these forest of Gujarat have received attention only in the later part of the nineteenth century and information on plant wealth of the state was gathered. Most of work done is on species diversity and only a few studies are on phenological and physiological aspects.

Much time has been passed, since last flora was published and a lot of progress has been made. There is an urgent need to revise the same. For that, various research institutes, universities, forest departments should come forward with initiative and surveys with floristic diversity for different districts and taluka level.

Seed, with its germination, viability, morphology, ecology, is the main source of future forest. Physiological parameters like photosynthesis, transpiration, vapor pressure deficit and radiation use efficiency are important traits to be studied for the plant species of Gujarat forests. Molecular aspects are another area of research. Based on information gathered from various sources from different regions of Gujarat, there is need to identify missing links.

Plants are disappearing rapidly and there is direct need of conserving them by *In- situ* and *Ex- situ* techniques, failing which, we may lose some of very important plant species from the forest in future. Unfortunately, due to lesser employment opportunities, researchers on taxonomy and classical botany are decreasing. However, with such a huge wealth of plants, there is bright future ahead with lots of scopes in the changing climatic scenario for which the government departments should identify institution with the postings to work out the productivity of forest flora of Gujarat.

References

Arya, C. and Arya, A. (2006). Effect of storage on seed viability and germination of certain leguminous trees. *Indian Forester* **132** (5): 125-131.

Asari, R.A. (1994). Gir Forest-report submitted to Gujart Government. Govt. of India.

Asari, R.A. (1994). Girnar Forest-report submitted to Gujart Government. Govt. of India.

Baria, P. (2003). Ecological assessment of vegetation in JFMP adopted and natural forest villages. M.Sc. Thesis, M.S. University, Vadodara.

Bariya, P. and Sandhyakiran, G.(2012). Floristical and Ecological assessment of Joint Forest Management Plantations and Natural Forests. *Bulletin of Environmental and Scientific Research*, **1**: 11-17 [ISSN: 2278-52105].

Bhatt G.D. (2013). Plant diversity assessment at land scape level in Jamnagar District, Gujarat using satellite remote sensing and geographic information system. *International Journal of advancement in earth and environmental sciences*, **1**(1): 23-25.

Bhatt, G.D., Khushwaha, S.P.S., Nandy, S., Bargali, K., Tadvi, D., Nagar, P.S. and Daniel, M. (2013). Plant richness modeling in South Gujarat using remote sensing and geographic information system. *Indian Forester*, **139**(9): 757-768.

Bhatt, G.D., Kushwaha, S.P.S., Nandy, S., Bargali, K. (2013). Vegetation types and land uses maping in South Gujarat using remote sensing and geographical information system. *International Journal of Advancement in Remote sensing, GIS and Geography*, **1**(1): 20-31 [ISSN: 2321-8355].

Bhatt, J.B. (1993). Studies on the flora of Western Kachchh, PhD Thesis, Maharaja Sayaji Rao University, Baroda.

Blatter, E. (1908). Flora of Bombay presidency; Statistical and biological notes. *Journal of Bombay Natural Histroy Socety*. **18**, 562-571.

Bole and Pathak (1988). *Flora of Saurashtra*. Part II and III. Botanical survey of India, Culcutta, pp. 545.

Chabra, A., Palria, S., Dadhwal, V.K. (2002). Growing stock based forest bio mass estimate for India. *Biomass and Bioenergy*, 22: 187-194.

Chaniyara, H.B. (2014). Floristic diversity of Santrampur Forest Range, of Panchmahals District, Gujarat, PhD Thesis, HNGU North Gujarat University, Patan.

Chavan, S.A. (1993). Vegetation and wildlife studies in Gir Forest, PhD Thesis, Maharaja Sayaji Rao University, Baroda.

Contractor, B.J. (1986). Floristic phytosociology and Ethno botanical study of Vapi and Umargaon area in South Gujarat.

Daniel, M. and Nagar, P.S. (2010). A soujoun o he herbal treasures of MSU., Maharaja Sayaji Rao University, Arihant offset Printers, Karelibaugh, Baroda.

Dave, M., Patel, R.S. and Patel, K.C. (2010). Taxoethnobotanical significance of some plants from Ambaji Forest (North Gujarat). *International Journal of Bioscience Reporter*, **8**(1): 67-70.

Desai, P.B. and Patel, N.K. (2010). Phenology of tree species of Satlasana Range Forest (North Gujarat). *Life Science Leaflets*, **3**: 41-61 [ISSN: 0976-1098].

Desai, R. (2013). Floristic diversity and ecological studies in Forest areas of Idar and Vadali Talukas, District Sabarkantha. PhD Thesis. HNGU North Gujarat University, Patan.

Desai, R.K. and Ant, H.M. (2012). Study of plant diversity in Vadali range Forest District Sabarkantha, North Gujarat, India. *Life science leaflets*, **1**: 32-43 [ISSN: 2277-4297].

Dhaduk, H.L., Chovatia, V.P. and Mandavia, C.K. (2011). *Aushdhiya vanaspatini Kheti* (Gujarati). [Farming of medicinal plants]. Metro Offset Press, Junagadh.

Dhaduk, H.L., Dhruj, I.U., Mandavia, C.K. and Chovatia, V.P.(2010). Plant Biodiversity: Collection, characterization and conservation. In souvenir of National Conference on Biodiversity Conservation, M.S.University, Baroda, pp.110.

Gandhi, D., Susy, A, Pandya, N. and Panchal, K.(2010). Vegetational diversity in the land of Bhey (Garbada). *International Journal of Bioscience Reporter*. **8**(1): 45-50.

Gavit, N.C. (1990). A contribution to the study of Systematic Seed Morphology of South Gujarat Plants.Ph. D. Thesis, South Gujarat University, Surat.

Gavit, N.C. and Parabia, M.H. (1989). Seed Atlas of Gujarat-1: Papilionaceae.In proceeding of all India Symposium on the Biologyll and utility of wide plants.South Gujarat University, Surat.

Gohil Deepa (2012). In depth studies on some of the floristic components of Baroda and Panchmahal districts with reference to Bioprospecting and speciation. PhD Thesis. Maharaja Sayaji Rao University, Baroda.

Gohil., K. (2005). Phenological studies of Reserve Forest (Victoria park). near Bhavnagar. Ph.D.Thesis. Bhavnagar University, Bhavnagar.

Gondaliya, H., Patel, R.S. and Patel, K.C. (2010). A review on some medicinal plants of Gujarat College campus, Ahmedabad, Gujarat. *International Journal of Bioscience Reporter*, **8**(1): 71-73.

Jadeja, B.A. (1999). Plants used by tribe rabari in Barda Hills of Gujarat, Ph.D. Thesis, Bhavanagar University, Bhavanagar.

Jadeja, B.A. and Nakar, R.N. (2010). Phenological studies of some tree species from Girnar Reserve Forest, Gujarat, India. *Plant Archieves*, **10**: 825-828.

Jadeja, B.A. and Patel, N.A.(2012). *Floristic diversity of the Ahmadabad city*. Lap Lambert Academic Publishing GmbH and Co, Germny [ISBN: 978-3-8484-2097-1].

Jadeja, B.A., Kanjaria, K.V. and Odedara, N.K.(2011). Floristic analysis of flora of Bhanvad, Jamjodhpur and Lalipur Talukas of Jamnagar district, Gujarat. *Bio Nano frontiers*. **4**: 294-296.

Jadeja, B.A., Odedara, N.K. and Singh,R.S.(2008). Phenological observations on some dry deciduous Forest trees at Barda hills, Gujarat. *Journal of economic and taxonomic botany*, **32**: 51-56.

Jaiswal, D.J., Maheta, V.R., Patel, Y.B., Pandya, H.A. (2014). Carbon stock estimation of major tree species in Attarsumba range, Gandhinagar forest division, India. *Annals of Biological Research.* 5(9): 46-49.

Jangid, M.S. (2012). Biological spectrum of Taluka Modasa, District Sabarkantha (Gujarat), India. *Life science leaflets,* **12**: 77-98 [ISSN: 2277-4297].

Joshi, P.N, Joshi, E., Jain, B.K. (2010). Ecology and conservation of threatened plants in Tapkeswari Hills, Kachchh island, Gujarat, India. *Journal of threatened taxa,* 4(2): 2390- 2397 [ISSN: 0974-7907].

Kariya, J.P. (2012). Floral and avifaunal diversity of the Thol lake, wild life sanctuary of Gujarat State, India. Biodiversity enrichment in the diverse world. Intech Publications, pp.1-34.

Kumar R. S., Joshi, P.N., Joshua, J., Sunderraj, S.F.W., Kalawathy, S., Raghunathan, P. (2011). Importance and conservation values of disturbed lands of North Gujarat region (NGR), Gujarat, India. *Plant Sciences Feed* 1(8): 121-141 [ISSN: 2231-1971].

Kumar, R.N., Chakraborty, S., Nirmal Kumar, J.I. (2010). Methods to break seed dormancy of Andrographis paniculata (Burm.f. Nees). – An important medicinal herb of tropical Asia. *Advances in bio Research,* 1(2): 35-39 [ISSN: 0976-4586].

Mac R.N.(1986). A contribution to the flora of Surat district, Surat.

Makad, R.S. (2007). Floristic and ethnobotanical aspects of Ambaji and Gorakhnath Hills, PhD Thesis, Bhavangar University, Bhavanagar.

Malhotra, S.K. and Wadhava, B.M. (1978). Studies on Botany of Jamnagar District. M.V.M. Patrika 1978, **8**: 3-23.

Manga, V.K. and Sen, D.N. (1995). Influence of seed traits on germination in *Prosopis cineraria* (L). *Journal of Arid Environments,* 31 (5): 371-375.

Menon, A.R.R. (1979). Floristics and phytosociological studies of some parts of Saurashtra. PhD Thesis, S.P. University, Vallabh Vidhyanagar.

Mishra, D., Shashikant (2003). Production functions for multiple outputs of joint forest management. XII th World Forestry Congress, Quebec City, FAO Publication.

Modi, N.R. and Dudani, S.N. (2013). Biodiversity conservation through urban green spaces: a case study of Gujarat university campus in Ahmedabad. *International Journal of Conservation Science,* 4: 189-196.

Murlidhar, A.N. (2013). Forest studies using optical and microwave remote sensing data. PhD Thesis. M.S. University, Vadodara.

Nagar, P. S. and Pandya, S. M. (2002). *Stylosanthes hamatus* (Linn.). Taub. (Papilionaceae), A New record to the flora of Gujarat. (With one text figure). *Journal of Bombay Natural History Society,* **99**: 363-364.

Nagar, P. S. and Pandya. S. M. (2002). Addition to the flora of Saurashtra, Gujarat, India. *Journal of Economic and Taxonomic Botany.* **26**: 75-77.

Nagar, P. S. and Pandya. S. M. (2002). *Cleome scaposa* D.C. (Capparidaceae). A rare species of Saurashtra, Gujarat. *Journal of Bombay Natural History Society,* **99:** 543-544.

Nagar, P. S. and Pandya. S. M. (2003). Floristic diversity of Barda Hills and their surroundings, Saurashtra, Gujarat. *Journal of Economic and Taxonomic Botany,* **27:** 1166-1180.

Nagar, P. S., Sata, S. J. and Pathak, S. J. (2003). Rediscovery of *Tephrosia jamnagarensis* (Fabaceae), an endangered and narrow endemic plant species of Saurashtra Side. *Journal of Economic and Taxonomic Botany,* **20:** 1701-1705.

Nagar, P.S. (2000). Floristic Diversity of Barda Hills and its surroundings. PhD Thesis, Saurashtra University, Rajkot.

Nagar, P.S. (2005). Floristic Diversity of Barda Hills and its surroundings. Scientific Publishers, Jodhpur, pp. 325.

Nagar, P.S. and Daniel, M. (2010). Floristic diversity of Gujarat. In National conference on Biodiversity conservation. Maharaja Sayaji Rao University, Baroda, pp. 63-77.

Nagar, P.S.(2004). Status of some regionally threatened species of Saurashtra, Gujarat, India. *Journal of Current Biosciences.*1: 238-247.

Nair, R. (2011). Floristic studies of Dadra and Nagar Haveli. *Life science leaflets,* **20:** 898-903 [ISSN: 0976-1098].

Nakar, R.N. (2013). Studies on phonological patterns of Girnar Reserve Forest, near Junagadh, Gujarat, PhD Thesis, Saurashtra University, Rajkot.

Nakar, R.N. and Jadeja, B.A. (2009). Study on morphology, ethnobotany and phenology of *Prosopis* in Girnar Forest Junagadh, Gujarat. In Proceedings of the National Symposium on Prosopis: ecological, economic significance and management challenges. Gujarat Institute of Desert Ecology, Bhuj, pp. 51-53.

Nakar, R.N. and Jadeja, B.A. (2013a). Comparision of phenology of *Cassia siamea* Lam. from Girnar Reserve Forest and GIDC polluted area from Junagadh, Gujarat. In UGC sponsored State level seminar on conservation of biodiversity and environment-A challenge, KKSJ Maninagar Sci.College, Ahmadabad. pp: 41.

Nakar, R.N. and Jadeja, B.A. (2013b). Studies on phenology of some shrubs from Girnar Reserve Forest, Gujarat. In National conference on 'Medicinal and aromatic plants for rural development and prosperity', Ananad agricultural University, Ananad, pp.18.

Nakar, R.N., Jadeja, B.A. and Dhaduk, H.L. (2014a). Phenological studies of two bombacacean members from Girnar Reserve Forest, Junagadh, Gujarat, India. *Indian Forester,* **140** (1): 59-64.

Nakar, R.N. and Jadeja, B.A. (2014b). Seed pattern, germination and viability of some tree species seeds from Girnar Reserve Forest of Gujarat. *Indian Journal of Plant Physiology,* **19**(1): 57-64.

Nakar, R.N. and Jadeja, B.A. (2014c). Standardization of wild Forest seeds for morphology and viability studies. *Plant archives*, **14**(1): 111-114.

Nakar, R.N. and Jadeja, B.A. (2014d). Phenological attributes of seven Caesalpiniaceae members in association with climate from Girnar Reserve Forest. In proceedings of National Symposium on 'Strategies to understand sustainable utilization of plant wealth (SUSUP-2014)', Department of Botany, Gujarat University, Ahmedabad, pp.41.

Nakar, R.N. and Jadeja, B.A. (2015). Flowering and fruiting phenology of some herbs, shrubs and undershrubs from Girnar Reserve Forest, Gujarat, *Current Science* 108: 111-118.

Nayar, M.P. and Sastry, A.R.K. (1987-1990). Red Data Book of Indian Plants (3 vols). Botanical Survey India, Kolkata.

Nayar, M.P. and Shastry, A.R.K. (1987-1988). Red Data Book of Indian Plants. Botanical Survey of India, Calcutta.

Nizalpur, V., Jha, C.S., Madugundu, R. (2010). Estimation of above ground biomass in Indian tropical forested areas using multi frequency DLR-ESAR data. *International Journal of Geomatrics and Geosciences*, 1(2): 167-178 [ISSN: 0976-4380].

Oza, R.A. (1991). Taxonomical and ecological studies of the flora around Bhavnagar, PhD Thesis, Bhavnagar University, Bhavangar.

Panchal K., Pandya N., Albert S. and Gandhi D. (2011). Effect of storage period on the germinability of Kangaroo grass (*Themeda triandra*). *Bioscience guardian*, 1 (2): 291-295.

Panchal, N.S. and Pandey, A.N. (2002). Study on soil properties and their influence on vegetation in western region of Gujarat, India. In Proceeding of 12ᵗʰ ISCO Conference, Beijing, pp: 610-615.

Pandey, C.N., Pandey, R. (2013). Carbon sequestration by mangroves of Gujarat, India. *International Journal of Botany and Research*. 3(2): 57-70 [ISSN: 2277-4815].

Pandeya, S.C., Kuruvilla, K., Gosai, G.N., Pandit, B.R., Trivedi, K.A. (1967).Net production relations of five important tree species at Waghai range, Dang forest, Gujrat. *Proceedings of National Academy of Science, Section-B*, 66(1): 25-36.

Pandit, B.R., Subramanyam, S.V.S., Rao R.V.K. (1993). Leaf litter production of the dry deciduous forest ecosystem of Eastern Gir (Gujarat). *Indian Forester*, 119 (5): 375-380.

Pandya, I., Salvi, H., Chahar, O., Vaghela, N. (2013). Quantitative analysis on carbon storage of 25 valuable tree species of Gujarat, Incradable India. *Indian Journal of Scientific Reserch*. 4(1): 137-141.

Pandya, S.M. (1989). Forest wealth of Kutch, its utility and potentials. In All India symposium. Biology and Utility of Wild Plants, Surat, pp. 175-189.

Parwani, R., Mankad, A. (2012). Seasonal impact on biomass production of *Cenchrus cilliaris* grass. *Life Science Leaf Lets*, 1: 348-354.

Patale, V. and Parikh, P. (2010). Floristic diversity of the stream and riparian zone of the Vishwamitri river Baroda. *Bioscience guardian*, 1: 201-207.

Patel, A.M. and Patel, K.C.(2012). Floristic studies on scared groves situated near Utkantheswar in Kheda District, Gujarat State. In National Conference on Plant Science: Changing pathways, Changing lives, Talod, pp. 41.

Patel, H.K., Patel, K.C., Patel, R.S. (2010). Plants used in preparation of musical instruments and agricultural impliments of Ambaji Forest in Banashkantha District (North Gujarat). *International Journal of Bioscience Reporter*, 8(1): 81-63.

Patel, M.K., Patel, P.K. (2012). Floristic and ethnobotanical study of Palanpur and Dantiwada, Gujarat. Lap-Lambert Academic Publishing, Germany, pp: 1-292 [ISBN: 978-3-659-12553-9].

Patel, N.T. and Pandya, A.N. (2009). Salinity tolerance of *Aegiceras corniculatum* (L.). Blanco. from Gujarat coast of India. *Analesde Biologia*, 31: 93- 104.

Patel, P.K. (2004). Studies on the flora along the river bank of Saraswati River from Mukteshwar to Patan district with ethnobotanical aspects, PhD Thesis, HNGU North Gujarat University, Patan.

Patel, R., Roy, A.K. and Patel, Y.S. (2014). Study on the floristic diversity of two newly recorded sacred groves from Kachchh District of Gujarat, India. *Indian Journal of Plant Science*, 1(1): 34-36.

Patil, P., Singh, S., Dadhwal, V.K. (2012). Above ground forest phytomass assessment in Southern Gujarat. *Journal of Indian Society of Remote Sensing*. 40(1): 37-46.

Pilo, B. and Pathak, J. (1996). Biological diversity of Gujarat- Current knowledge. Gujarat Ecology Commission, Baroda, pp.329.

Punjani, B.L. (2002). Ethnobotanical aspects of some plants of Arravalli Hills in North Gujarat. *Ancient Life Sciencce*, 21: 268-280.

Rajendra Kumar (2010). Floristic status and its conservation in the Forests of North Gujarat Region, Gujarat India. Ph.D. Thesis. Bharathidasan University, Tiruchirappalli.

Rao, K.S.S. (1981). Flora of South eastern Kachchh, PhD Thesis, Maharaja Sayaji Rao University, Baroda.

Rao, V.H., Gohil, T.G. and Thakor, A.B. (2013). Floristic study of Kaprada's hilly Forest in South Gujarat *International Journal of Plant Science*. 8(1): 100-102.

Raole, V.M. (1993). Studies on endangered and endemic desert taxa, PhD Thesis, Maharaja Sayaji Rao University, Baroda.

Raole, V.M., Joshi, A.G., Garge, S.K., Desai, R.J. (2010). Seed germination of selected taxa from Kachchh desert, Gujarat, India. Notulae Scientia Biologieae, 2(2): 41-45.

Rathore, A., Jasrai, Y.T. (2013). Urban green patches as carbon sink: Gujarat University Campus, Ahmedabad. *Indian Journal of Fundamental and Applied Life Science*, 3(1): 208-213.

Reddy, A. S. (1987). Flora of Dharampur Forest, Ph. D. Thesis, S. P. University, Vallabh Vidhyanagar, Anand.

Rodgers, W.A. and Panwar, H.S. (1988). Planning a wildlife protected area network in India. Wildlife Institute of India, Dehra Dun.

Sabnis, S.D. and Rao, K.S.S.(1983). Observation on some rare or endangered endemics of Southern Kutch. An assessment of threatened plants of India. Botanical Survey of India, Culcutta. pp.71-77.

Santapau, H. (1952b). Report of the work done under the auspices of the Saurashtra Research Society, for the botanical exploration of Saurashtra. Annual Report of Saurashtra Research Society pp. 9-20.

Santapau, H. (1956). La esploracien botanica de Saurashtra, India. *Anal. Inst. Bot. Cavanilles, Madrid*, **13:** 23-454.

Santapau, H. (1958). The floristic study in India. *Mem. Indian Bot. Soc.* 1: 117- 121.

Santapau, H. (1971). *Common trees.* National book trust of India, New Delhi.

Santapau, H. and Janardan (1966). The flora of Saurashtra. Bulletin of Botanical Survey of India, 8: 1-58.

Santapau, H. and Raizada, M.B. (1954). Contribution to the flora of Gir Forest in Saurashtra. *Indian Forester*, 80: 379-389.

Santapau, H.(1950). Inter Kathiwrense: Being notes on a botanical tour in Nawagar State, Oct Nov 1945. *Gujarat Research Society.* **11-12**, 226-237.

Santapau, H.(1952a). A note on the Jay Krishna Indraji Thaker collections of plants preserved in the Herbarium, Agricultural College, Poona. Agriculture College Magazine, Poona, **42:** 210-233.

Santapau, H.(1952c). Report of the second field season in the botanical exploration of Saurashtra. Rajkot, pp. 16.

Santapau, H.(1952d). Report of the third field season in the botanical exploration of Saurashtra. Rajkot, pp. 16.

Santapau, H.(1952e). Plants of Saurashtra. A preliminary list. Rajkot, pp. 16.

Sarvalia, V.M., Chovatia, V.P., Nakar, R.N. and Meta, H.R. (2010). Seed germination and seedling establishment studies in *Adansonia digitata* Linn. *Bioscience guardian*.1 (1): 77-79.

Seikh, M.A., Kumar, M., Bussman, R.W., Todaria, N.P. (2011). Forest carbon stock and fluxes in physiographic Zones of India. *Carbon balance and Management*, 6: 1-15.

Shah, G.L. (1978). Flora Gujarat State, Vol 1 and 2, Saradar Patel Univeristy, Vallabh Vidhyanagar. pp.1067.

Shah, G.L. (1983). Rare species with restricted distribution in South Gujarat. In: S.K. Jain and R.R. Rao (eds.). An assessment of threatened plants of India. Botanical survey of India. Cucultta. pp. 50-54.

Shah, J.P and Thivakar, G.A. (2010). Germination and growth performance of *Commiphora whitii* and *Acacia Senegal* under different salinity from arid coast of Kachchh, Gujarat. Abhinav, 2(7): 51-61 [ISSN: 2277-1174].

Shah, J.P. and Thivakaran, G.A. (2012). Taxonomic status and utilization of Halophytes in coastal Katch District, Gujarat. In National Conference on Plant Science: Changing pathways, Changing lives, Talod, pp. 20.

Sharma, P.P. and Sikarwar, R.L.S. (2014). Floristic diversity of Diu Island, Gujarat. Uttar Pradesh Biodiversity Board.

Sinha, S.K., Pinto, A.G. and Patel, R.I.(1972). Working plant for Junagadh Forest division. Printed at the Government Press, Baroda, pp. 167.

Solanki, H.A., Bhatt, K.J., Naghera, B.D., Goswami, N.D. and Sirdhvad, K.N. (2010). Traditional aboriginal knowledge of the flora in the Girnar Holy hills areas, Gujarat. *International Journal of Bioscience Reporter*, 8(1): 91-96.

Solanki, H.A., Dabgar, Y.B., Mali, M.S. and Khokhariya, B.P. (2010). Plant diversity and its life forms of Visnagar Taluka, North Gujarat. *International Journal of Bioscience Reporter*. 8(1): 43-48.

Tadvi, D. (2009). Study on medicinal Plant diversity in Arboretum. M.Sc. Thesis. Maharaja Sayaji Rao University, Baroda.

Tadvi, D. (2014). Floristic diversity of Dangs, Gujarat, PhD Theis, Maharaja Sayaji Rao University, Baroda.

Thakar, J.I. (1910). Vanaspati sastraane Barda dungarni jadibuti teni parikhsa ane upyog. (A compete and comprehensive account of the flora of Barda Mountain (Kathiyavad). Gujrati printing Press). Bombay, pp. 717.

Thakar, J.I. (1926). Katchni sawathani Vanaspatiyo ane tani upyogita. (Plants of Kutch and their utility- An elaborate treatise containing ten chapters and litho figure). Gujarati printing press and Nirnaya- Sagar Press. Bombay, pp.200.

Thakar, J.I. (1966). Flora of Baroda, Sastu Sahitya Press, Ahmedabad.

Toby, H. and Patricia, H. (1982). Grasses of Western India. Bombay Natural History Society, Cornell University.

Tyagi,S. N., Padhiar, A, Albert, S., Pandya, N.,Gandhi, D.and Panchal, K. (2010). Pictorial Floristic diversity of grasses and associated vegetation from three grasslands of Randhikpur Forest Range, Dahod, Gujarat. *Indian Forester*, 136: 1581-1592.

Vediya, S.D. and Kharadi, H.S. (2011). Floristic diversity of Isari zone, Megharj range Forest District Sabarkantha, Gujarat, India. *International Journal of Pharmacy and Life Science*, 2: 1033-1034 [ISSN: 0976-7126].

Yogi, D. V. (1970). A contribution to the flora of North Gujarat. Ph.D. Thesis, S.P.University, Vallabh Vidyanagar.

Zopade, S.T. (1991). The improvement of germination of some Forest species by acid scarification. *Indian Forester*, 117(1): 61-66.

Recent Advances in Crop Physiology Vol. 2 (2016)
Editor: **Dr. Amrit Lal Singh**
Published by: **DAYA PUBLISHING HOUSE, NEW DELHI**

Pages **295–314**

Chapter 10

Radiotracer Use in Understanding Mineral Nutrition of Crop Plants

Bhupinder Singh[1]*, Prashant Kumar Hanjagi[2],
Manoj Shrivastava[1], Achchelal Yadav[1],
Sumedha Ahuja[1] and Rinki[2]

[1]*Nuclear Research Laboratory, CESCRA*
[2]*Division of Plant Physiology,*
ICAR-Indian Agricultural Research Institute, New Delhi – 110 012

1. Introduction

The capability of living cells to take up substances from the environment and use them for synthesis of their own cellular components or as energy source is an outstanding feature of life. Plant mineral nutrition involves studies on nutrient uptake, translocation, cellular partitioning and utilization. The other important factors controlling plant mineral nutrition are (re)-translocation of the nutrients once they are taken by the plant, an important adaptive mechanism under condition of nutrient deficiency. Use of nuclear and allied techniques in agricultural research and development provides excellent opportunity. Isotopes have been exceptionally useful in the effort to understand various aspects of ion uptake, translocation and utilisation or assimilation. Their use has helped in characterization of ion influx into apoplasm, understanding the phenomenon of active uptake through insight into the interactions of ions, efflux process and foliar uptake and understanding long distance transport processes including radial transport, upward, downward and lateral transport,

* *Corresponding Author:* E-mail: bsingh@iari.res.in, bhupindersinghiari@yahoo.com

exchange adsorption in xylem, resorption from xylem (apoplast) into living cells and long distance transport and understanding the process of nutrient utilisation/ assimilation. Isotopes yield an analytical advantage that they can be easily traced and measured in minute quantities and analysis remains uncomplicated by the presence of large cellular pools of the ions in question. Infact radioisotopes can be used effectively in developing modern agro-techniques for better understanding of the mineral fertilizers, water-use and biological nutrient use efficiency. This chapter collates and critically analyzes the research on the role of radioisotopes and stable isotopes in boosting our understanding on nutrient dynamics in soil-plant-environment continuum.

Understanding the science of mineral nutrition is immensely important for crop plants but and also for humans and animal health. Complexities in the field of plant mineral nutrition arise mainly due to variability in the type of soil, crops and cropping systems. Add to these, the abiotic stress factors that influence plant nutrition and growth either directly or indirectly. Further, we have serious issues of nutrient deficiencies, toxicities and imbalance, all of which inhibit grain/crop yields and threaten human nutrition and health.

Green Revolution in India gave over emphasis on NPK. However, intensive agricultural practices have pressured other resources leading to the depletion of soil quality mainly sulfur and micronutrient levels of our soil. The practice in the past has led to widespread nutritional imbalance in the soil and has resulted in malnutrition. Zn, Fe and B deficiencies are presently predominant in our soils and require a more focused and targeted research as scientists are working on the idea of producing nutrient rich crops *i.e.*, biofortification approach. The Plant Physiologists need to work for enabling a higher accumulation of nutrients in seed/grain as not only the capacity for uptake of nutrients but also efficiency for accumulation of nutrients in the seed is crucial for biofortification. It is important to determine the translocation efficiency of micronutrients. For example, Fe content of the Fe inefficient plants is low despite the roots accumulating 8-9 times more Fe than the shoot. This clearly indicates a limitation at the level of root to shoot Fe translocation.

Also application and strategies need to be planned in line with the stage specific nutritional requirement of sorghum, maize, rice, wheat crops which differ greatly for the critical limits of deficiency and sufficiency at different growth stages. All these parameters put together make the task of the plant scientists, in general, and plant nutritionists in particular very- very difficult. Key question arises *i.e.*, how much fertilizer one should apply for a crop on a soil to achieve maximum yield and nutritional quality together high resource use efficiency. The above challenges can be partially addressed by exploiting the genetic variability in resource use in different species. A very high degree of variability for Zn deficiency tolerance among monocots such as durum and bread wheat, triticale and rye and Fe and Zn deficiencies tolerance in peanut has been reported. Low fertilizer nutrient use efficiency (NUE) is puzzling say for N, we are working at a very low nutrient use efficiency (30-33 per cent). Use of neem coated urea was shown to improve the N efficiency by slow release of the N. The plant traits with considerable effect on nutrient use efficiency are rooting attributes, root uptake, root–shoot nutrient translocation and also nutrient retranslocation.

Root response to the variable supply of the nutrients is another gray area where the role of hormones in relation to root development needs to be deciphered (GuoHua *et al.*, 2010). There is a lot more to roots than the mass. There is a need to understand the change in root architecture under individual or combined nutrient sufficiency/ deficiency condition. Further a change in root pattern may influence root exudation in crops and consequently the nutrient availability, uptake and plant growth. A lot needs to be done in the area of nutrient sensing and phytohormone signaling. Later may also be playing a significant role in nutrient homeostasis or nutrient balance under condition of low nutrient availability. Nutrient balance is important at the rhizosphere level, and at the tissue, cell and intracellular level to optimally regulate the rate of metabolic and catabolic reactions in the cytosol and its embedded organelles like the mitochondria and the chloroplast. Competition between the mineral nutrients may also influence nutrient balance.

2. Radioisotopes, Stable Isotopes and their Applications

Natural abundance of isotopes of many elements having the same atomic number but different atomic mass, is known. These isotopes may be either stable or radioactive. Stable isotopes, are defined as those that are energetically stable and do not decay; thus, they are not radioactive. An isotope tends to be stable when the number of neutrons and the number of protons are quite similar. A radioactive tracer is a chemical compound in which one or more atoms have been replaced by a radioisotope so by virtue of its radioactive decay and emits energy in the form of alpha or beta particles or electromagnetic radiations such as gamma rays. Radiations emitted by the radioisotopes are discrete, measurable and quantifiable. They dissipate the associated energy while passing through the matter and ionize it to alter the properties of the matter. Exploiting these characteristics, researchers have explored the mechanism of chemical reactions by tracing the path that a radioisotope follows from the substrate to the end product. There are roughly 300 stable isotopes, over 1200 radioactive isotopes, and only 21 elements that are known to have only one isotope (Hoefs, 1997).

Just as early man harnessed fire to improve his life, society in the last century was able to harness radiation. The development of nuclear technology is one of the most significant achievements of the 20th century. From the time of the first documented practical application of a radioisotope (George de Hevesy, 1911), the radionuclides have been exploited to understand the physical, chemical and biological processes. Probably the application with the biggest impact has been in the medical field where radionuclides have been incorporated into biologically active molecules and used to diagnose a wide variety of diseases and to treat many disorders. Stable isotopes can be measured on an isotope ratio mass spectrometer (IRMS) which can be configured for analysis of either the lighter elements (*e.g.*, H, C, O, N, and S) or heavier elements such as Fe. Among stable isotopes the most useful as biological tracers are the heavy isotopes of carbon and nitrogen. These two elements are found in the earth, the atmosphere, and all living things.

Table 10.1: Terrestrial Abundances of Stable Isotopes of major Elements

Elements	Isotopes	Abundance (per cent)
Hydrogen	1H	99.985
	2H (D)	0.015
Carbon	12C	98.89
	13C	1.11
Nitrogen	14N	99.63
	15N	0.37
Oxygen	16O	99.75
	17O	90.037
	18O	0.204
Sulfur	32S	95.00
	33S	0.76
	34S	4.22
	36S	0.014

Stable isotopes are commonly used to identify source of pollutants to a stream, infer processes such as heterotrophic nitrification, estimate rates of soil C turnover and other processes, determine proportional inputs for example percent contribution of a particular prey item to a predator's diet and confirm, reject, or constrain models derived from the use of other techniques. On the other hand, the variety of uses for radiotracers in science are almost boundless depends only upon ones imagination. Today the largest number of applications of radiotracers is in biology and medicine. Radionuclides have been largely used in three major areas in life science research as tracers of metabolic processes, as analytical reagents and as diagnostic and therapeutic agents in nuclear medicine. Immense applications of gamma radiation in agriculture have been found for reducing post-harvest losses through suppressing sprouting and contamination, eradication or control of insect pests, reduction of food-borne diseases and in extension of shelf life and breeding of high-performance well adapted and disease resistant agricultural crop varieties (Andress *et al.*, 1994, Emovon, 1996).

3. Radiotracers in Plant Mineral Nutrition Research

Agriculture is dependent on soil for a variety of reasons, one of the most important being the fact that soil contains the elements which are essential to plant growth and can store nutrient elements (fertilizers) added from outside. Study of soil characteristics and processes by which plants take up nutrients from the soil is therefore extremely valuable in devising effective methods of farming. Radioisotopes have greatly facilitated such investigations and are now being widely used in soil-plant nutrition research. Several studies on use of radioisotopes in soil-plant nutrition studies have showed the varied ways in which isotopes can contribute to agricultural production by helping to investigate soil characteristics and soil-plant relationships. Processes involved in the uptake and translocation of nutrient elements by plants are being investigated with radioisotopes. Radiotracers offer a unique advantage in terms of high sensitivity, their simplicity and small expense (compared to competing technologies such as mass spectrometry). In a well-designed experiment, the presence of radiotracers does not affect the system under study and any analysis is non-destructive. Radiotracers can be used to study the dynamic biological processes like the ion transport across cell membranes, turnover, intermediary metabolism or translocation in plants that were earlier being studied indirectly with a greater degree of analytical error. Radiotracer technology also aids in identification of the rate determining step in any biological process. Details of some of the applications of radiotracers in the area of plant mineral nutrition are highlighted in the following sections.

3.1. Ion Influx, Water Free Space and Donnan Free Space

The term 'free space' refers to the fraction of the volume of plant tissue readily accessible to diffusion of an externally applied solute dissolved in water. Free space in primary root consists of all the regions occupied by primary cell walls and intercellular spaces in cortical tissues and is bounded by endodermis and plasma membrane. Movement of low molecular weight solutes (ion, organic acids, amino acids, sugars etc.) from the external solution into the free space of roots is a non-metabolic passive process driven by diffusion or mass flow. However, the cell walls can interact with solutes and thus, may facilitate or restrict further movement to the uptake sites of the plasma membrane of individual cells of roots. Free space is a functional concept. Its dimensions can be measured by physiological experiments described below. Because of the negative charges in the cell walls of the apoplasm, Hope and Stevens (1952) introduced the term apparent free space (AFS). This comprises the water free space (WFS) which is freely accessible to ions, and the Donnan Free Space (DFS), where cation exchange and anion repulsion take place.

The magnitude of free space in tissue can never be measured directly. It can, however, be measured indirectly, more conveniently if the ion used is radioactive. The entry of solute into free space can be visualized by immersing a plant tissue into a solution and following the time course of uptake. The time course of uptake of a cation (radioactive strontium) by excised root tissue shows a very rapid uptake of solute immediately on immersion of the tissue in the solutions. After about 30 min the period of rapid absorption ceases and is replaced by a period of slower absorption. To examine the desorption process the plant tissue is removed from the radioactive strontium solution and placed in water some of the radioactive strontium diffuses out of the tissue. The amount of radioactive strontium coming out can be easily quantified, and measures the water free space. If the tissue is then transferred to solution containing non-radioactive strontium, more radioactive strontium comes out. These radioactive strontium ions were adsorbed to cation adsorption sites in cell walls and require exchange with other cations for non-radioactive strontium. Estimation of this fraction gives a measurement of Donnan Free Space. Even after this, not all radioactive strontium is removed. About 10-15 per cent is retained by the roots. This indicates that about 10-15 per cent is not free to diffuse out of the tissue. This fraction of the strontium ion can be presumed to have crossed the plasma membrane of the root cells and to have entered the 'inner space'.

3.2. Nutrient Uptake Characteristics

Radioisotopes are ideally suited as tools for the investigations of nutrients. Important plant nutrients, such as calcium, phosphorus, iron, potassium, copper, sodium, sulfur, and zinc have radioisotopes with appropriate half lives and decay characteristics to be used as tracers. These elements can be incorporated in fertilizers and applied to the soil to determine the effect on plant utilization of fertilizer composition or the method of application. Plant uptake of the activated fertilizer can be readily measured and can be distinguished from the uptake of the same compound already present in the soil. Movement of water and ions via the apoplastic pathway can occur through the walls of the cortex until restricted by the impermeable Casparian strips on the endodermal cells. At the Casparian strip further progress of solutes toward the xylem is controlled by the plasma membranes of the endodermal cells or nearby cortical cells. These membranes help control the rates of ion absorption and the kind of solutes absorbed. Furthermore, as dissolved ions move along the walls of the epidermal and cortical cells, a certain percentage of each of these move into the cytosol and thus enters the symplastic pathway. Transport across the membranes requires energy and is called active transport.

Radiotracers are widely used for the kinetics of nutrient uptake. The most widely used tracer for studies in tracer kinetics in plants is N-13. In spite of its relatively short half-life (<10 min) a wide variety of studies have been undertaken to understand the incorporation of nitrogen into plant systems. These studies have had a wide impact on understanding the adaptive abilities of plant systems associated with changing environmental conditions to monitor the nitrogen content of genetically modified rice in attempts to increase the protein content of rice species as the primary protein food around the world (Britto, 2004). The process is characterized by selectivity, kinetic constants like K_m, V_{max}, and also influenced by temperature, O_2, carbohydrate content of the tissue etc. Such characterization of the uptake process has been possible, more easily, with the use of isotopes. Differences in nutrient uptake can be studied by nutrient depletion method in a short term experiment using labelled source. Pandey and Singh (2003) compared uptake rates of two maize genotypes using the nutrient depletion technique. In this experiment, plants were incubated in a known concentration of a nutrient cold (KH_2PO_4) and hot (here ^{32}P- phosphoric acid @ 2000 Bq/ml) and a known volume is collected every 20 minutes over 2 hours and the depletion in label in the solution over time is recorded. The data can be used to work out the kinetics of P uptake system in terms of K_m, V_{max}, and C_{min}. In another study Hart *et al.* (1998) studied the concentration dependent uptake of $^{65}Zn^{2+}$ in intact bread and durum wheat roots and the kinetics of zinc uptake in two wheat types was worked out. Durum wheat had a high K_m as well high V_{max} than bread wheat genotype which shows that the difference in zinc efficiency of bread (efficient) and durum (inefficient) wheat is not due to difference Zn^{2+} uptake capacity and probably some other factors may be responsible.

3.3. Nutrient Interactions

Rubidium, is an isotope not discriminated at the plant level can be used to characterize potassium. Consequently ^{86}Rb has often been used as a radiotracer for investigation of K^+ absorption by excised tissues of plants and is more convenient to use than ^{42}K because the half life of Rb is 18.6 days, compared with 12.47 h for ^{42}K. The K and Rb are mutually competitive in the presence of Ca, *i.e.*, the binding sites (carriers) at the plasma membrane of root cells do not seem to distinguish between these two cations. Uptake characteristics of K as affected by Ca and Na have been investigated using Rb. The absorption of Rb by excised barley roots from solution containing calcium is a strictly linear function of time upto 1 hour, and is temperature sensitive throughout; there is no evidence for an initial non-metabolic exchange phase of uptake. The Rb absorbed reached concentrations many times the external concentrations without any slackening of the rate of absorption-evidence for a high degree of irreversibility of the overall absorption process. In the absence of Ca, Rb absorption from a 1 mM solution of RbCl is progressively diminished by increasing concentration of NaCl. In the presence of Ca, the effect of Na is quite different. There is a slight inhibition of absorption at 1 to 2 mM Na. Between 9 -10 mM Na, absorption of Rb or K rises, and at still higher Na concentration declines slightly. This marked indifference of K and Rb absorption to even high concentration Na, in the presence of Ca is in marked contrast to the pronounced inhibition of K absorption by Rb and Rb absorption by K. Similarly labelled Rb was used to assess the effect of varying the Ca concentration on K. Rb absorption from 1 mM solution is reduced to 2.1 µmole g^{-1} fr. wt. h^{-1} in the absence of Ca and there is no effect from Ca at 10^{-6} M. At higher Ca concentration, Ca becomes increasingly effective in reversing the inhibition due to sodium, Rb absorption rising most steeply at about 1×10^{-4} MCa and levelling off at 3×10^{-4} to 1×10^{-3} MCa. In the absence of Ca, Na interferes with K^+ or Rb absorption. In the presence of Ca, there is little or no further inhibition of K absorption.

3.4. Cation Competition

Another example of competition between cations is that between Mg^{2+}, Mn^{2+}, and Ca^{2+}, the affinity of the hydrated Mg^{2+} for binding sites at root plasma membrane seems to be particularly low. Other cations, Mn^{2+} and Ca^{2+} in particular, therefore, compete quite effectively with Mg^{2+} and the uptake rate of $^{28}Mg^{2+}$ is thus strongly depressed. This strong competition is in agreement with the observation of Mg deficiency induced in crop plants by extensive application of K^+ and Ca^{2+} fertilizers.

3.5. Nutrient Efflux

Another area where isotope has been used is to study the process of ion efflux from tissues. An example is that of nitrate efflux studied with the help of ^{15}N. When seedlings of maize and millet were initially fed with $^{15}NO_3^-$ and transferred to a solution containing $^{14}NO_3$, it could be seen that nitrate efflux out and the rate of nitrate efflux was more in the dark than in the light in millet while the rate remained constant both in light and dark in the case of maize. Because of the small quantity of NO_3^- effluxing out, its detection was possible only with the help of the isotope. Also in natural systems, such efflux process can not be detected without labelling, as the surrounding soil solution will also have nitrate.

3.6. Rhizospheric Changes

Interactions between plants and microorganisms in the rhizosphere are complex and varied. They include the general transfer of nutrients and specific interactions mediated by the release of signalling molecules from plant roots. Until recently, understanding the nature of these interactions was limited by a reliance on traditional, cultivation-based techniques. Recently developed $^{13}CO_2$ pulse labelling and stable isotope probing (SIP) methods offer the potential to track ^{13}C-labelled plant photosynthate into phylogenetic groups of microbial taxa in the rhizosphere, permitting an examination of the link between soil microbial diversity and carbon flow in situ (Griffiths *et al.*, 2004 and Prosser *et al.*, 2006).

Changes in the rhizosphere influence nutrient availability and uptake. These modifications are effected or induced not only by the attributes constituting the rhizosphere and accompanied pH changes but also by the plant roots themselves. Roots of almost all plant species release certain high (mucilage) or low molecular weight solutes (organic acids, sugars, phenolics and amino acids). The production of root exudates is higher under stress. Pandey and Singh (2003) showed a variation in the release of root exudates based on ^{14}C label in the washings of the rhizosphere of the two genotypes and their ^{32}P mobilization capacity. This reliable radiotracer technique was documented recently (Pandey *et al.*, 2013).

The availability of nutrients especially that of micronutrients is affected by the release of certain non proteineous amino acids, the phytosiderophores that are released mainly under Fe stress but recent reports now suggest that they act as general metallochelatins and can mobilise not only Fe but also Zn, Cu, Mn etc (Figure 10.1). Studies to ascertain the rates of ^{59}Fe phytosiderophore uptake in barley as affected by deficiency of various micronutrients (Fe, Zn, Mn) supply reveal a higher ^{59}Fe uptake rates in Fe deficient plants suggesting that Fe-PS uptake system is active mainly under Fe deficiency. Von wiren *et al.* (1998) studied the kinetics of Fe association from ^{55}Fe-phytosiderophore (DMA) to root plasma membrane vesicles of wild type Fe efficient maize and a mutant deficient in Fe-phytosiderophore uptake, in the presence of various amount of free phytosiderophore (DMA). Both genotypes showed a similar kinetics of ^{55}Fe incorporation in the isolated vesicles.

In another study von Wiren *et al.* (1995) ascertained the concentration of ^{59}Fe and ^{14}C labeled DMA in roots and shoots of the Fe efficient maize cultivar WF9 and Fe inefficient maize YS1 supplied with double labeled DMA for 30 min. They first incubated the plant to $^{14}CO_2$ and collected the ^{14}C–DMA released by the roots in hydroponics and later labeled the ^{14}C-DMA with ^{59}Fe. A similar concentration of ^{59}Fe to ^{14}C in roots of both genotypes suggests that Fe_DMA is taken up by the roots with any reduction. Radiotracers were effectively used to prove that graminaceous roots take up not only Fe-phytosiderophore but also Zn-phytosiderophore (DMA) and that the bread (Zn efficient) and durum (Zn inefficient) genotypes do not differ in the rates of uptake of ^{65}Zn-DMA when it is available to them in similar concentration in the rhizosphere (Figure 10.2).

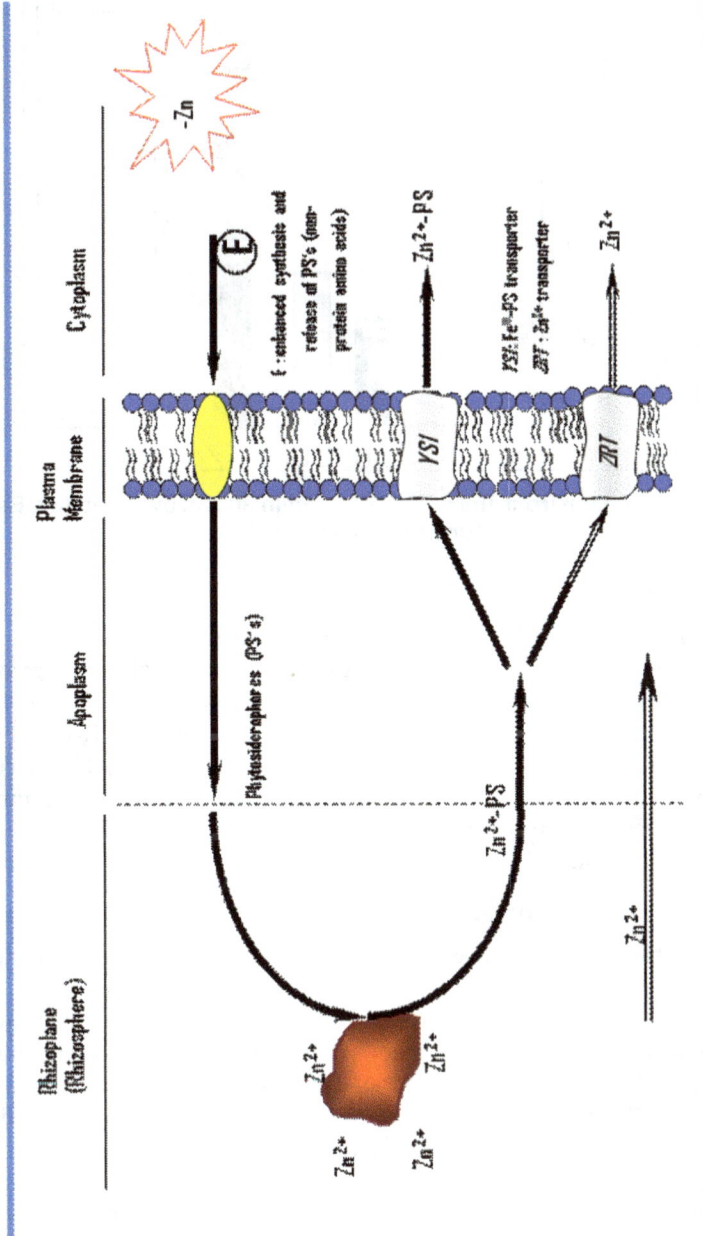

Figure 10.1: Role of Phytosiderophore in Micronutrient Mobilization in Crops.

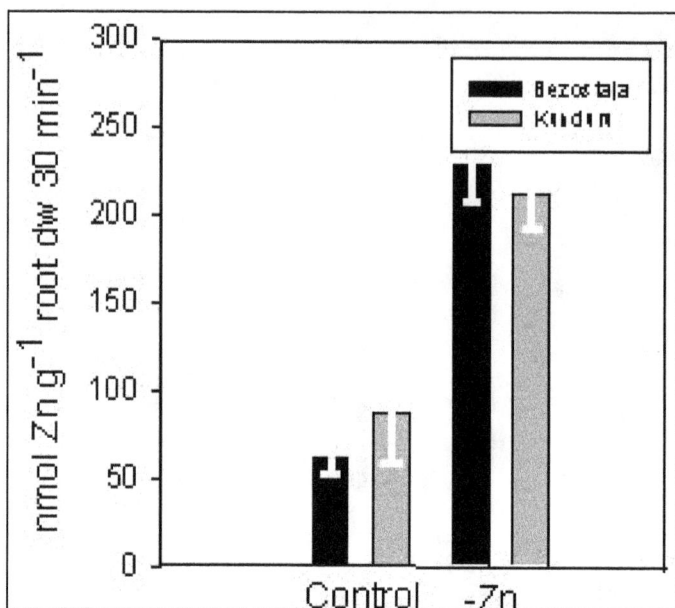

Figure 10.2: Zn Uptake from 1mM ^{65}Zn-DMA within 30 Min by Bread (cv. Bezostaza) and Durum (cv. Kunduru) Wheat.

3.7. Foliar Uptake

Another area where isotope has been useful to detect and quantify foliar absorption of nutrients, either from the atmosphere or from solutions sprayed on the foliage (*e.g.* foliar uptake of ammonia from atmosphere). When maize plants were exposed to air containing ammonia labelled with nitrogen-15 (1, 10 and 20 ppm) for 24 hrs, the 15N content of the plant showed that at one ppm, 43 per cent of the ammonia was absorbed whereas at 10 and 20 ppm, 30 per cent of the ammonia was absorbed. The results demonstrated that growing plants may be a natural sink for 'atmospheric ammonia.

3.8. Radial Transport of Nutrients

There are two parallel pathways of movement of ions (solutes) and water across the cortex towards the stele : one passing through the apoplasm (cell walls and intercellular spaces) and another passing from cell to cell in the symplasm through the plasmodesmata. For mineral nutrients the symplastic pathway plays a key role, either beginning at the rhizodermis and root hair, at the exodermis or at the endodermis. The mechanism of simplastic transport of solutes seems to be chiefly diffusional, facilitated by radial water flux. During the radial transport of ions competition occurs between accumulation in the vacuoles and transport in the symplasm. This competition depends on the mineral nutrient and its concentration in the vacuoles of the root cells along the pathway. This is elegantly shown by an experiment with ^{42}K. At low internal concentrations (low salt root) accumulation in the vacuoles competes very effectively with the simplastic transport. In short term studies this competition is reflected in a typical delay in the translocation of ions from the roots to the shoots of plants which originally had low internal concentration of the ion being investigated. As a result of this competition, when the supply of nutrient is suboptimal, the roots usually have higher concentration of the particular nutrient than the shoot (restricted translocation). In the long term studies, this regulatory mechanism is in part responsible for a shift in the relative growth rates of roots and shoots in favour of the roots.

3.9. Long Distance Transport in the Xylem and Phloem

The long distance transport of water and solutes takes place in the vascular system of xylem and phloem. Long distance transport from the roots to the shoots occurs predominantly in the non-living xylem vessels. In contrast to the xylem, long distance transport in the phloem takes place in the living sieve tube cells and is bi-directional. Mineral elements can also enter into the phloem in the roots and thus be translocated bi-directionally.

Radioactive tracers have provided an extremely powerful tool for measuring velocities, rates and paths of translocation. The most informative type of experiment has been that which separated the phloem from the xylem in the stem (the tissue breaks easily, and the bark separates along the natural shear line of the cambium). This can be done by making an incision in the bark, then inserting a sheet of wax-paper or some material so as to form a cylindrical barrier around the stem between the xylem and phloem to prevent any lateral translocation between the two tissues. Radioactive tracers are then applied either above or below the operated area, and either xylem or phloem may be interrupted independently if necessary. After a suitable period of time the plant is cut into small segments and the distribution of labelled nutrient is determined in plant tissues above or below the operated zone. A comparison of the result for intact plants (control), with phloem cut, and plants with xylem cut will provide information on the pathway or tissue of upward and downward movement of various substances. In addition internal transport from one tissue to another can be studied. The translocation of phosphate was studied using radioactive phosphorus (^{32}P), supplied to a leaf where results indicate that the phosphate moves downward in the phloem, but not in the xylem and is also able to move laterally from the phloem to xylem. Similarly radiotracers can also be used to study the pattern of nutrient absorption along the root length.

3.10. Exchange, Adsorption and Resorption of Nutrients in Xylem

Along the pathway of solute transport in the non-living xylem vessels (*i.e.* in the apoplasm), from the roots to the leaves important interactions take place between solutes and both the cell walls of the vessels and the surrounding xylem parenchyma cells. The major interactions are exchange adsorption of polyvalent cations in the cell walls and resorption (uptake) and release of mineral elements and organic solutes by surrounding living cells (xylem parenchyma and phloem). The interactions between cations and the negatively charged groups in the cell walls of the xylem vessels are similar to those in the apparent free space of the root cortex. This cation exchange adsorption is not restricted to xylem vessels; the cell walls of the surrounding tissue also take part in these exchange reactions. The degree of retardation of cation translocation depends on the valency of cations and complexing agents, the charge density of the negative groups (dicots > monocots), the diameter of xylem vessels and the pH of the xylem sap. Cation exchange in long distance transport through the stem can be demonstrated through the following example. When only $^{45}CaCl_2$ is supplied to the cut stem of de-rooted bean plants, the acropetal transport of ^{45}Ca is severely retarded. The addition of other cations strongly facilitates acropetal ^{45}Ca transport; the effect being similar to that seen with the exudate (xylem sap) of decapitated bean plants. Exchange adsorption reduces the speed of the long distance transport of polyvalent cations such as calcium.

Solutes are resorbed from the xylem (apoplast) into the living cells (cytoplasm and vacuole) along the pathway of xylem sap from the roots to the leaves. The concentration and composition of the xylem sap change, along the pathway. In some plant species, the resorption of certain mineral elements from the xylem sap is very pronounced and can have important consequences for the mineral nutrition of these plants. This is most evident in the so called natrophilic plant species. In natriophobic plant species (*e.g.* bean), Na^+ is retained mainly in the roots and lower stem, whereas in natrophilic species (*e.g.* Sugar beet) translocation into the leaves readily occur as shown in an experiment with $^{22}NaCl$. Resorption of Na^+ from the xylem sap is therefore an effective mechanism of restricting translocation to the leaf blades.

3.11. Retranslocation of Nutrients

Erenoglu *et al.* (2002) showed that the relative distribution of ^{65}Zn in the treated part of older leaf, remainder of shoot and roots 4 and 8 days after application of label to first leaf of two bread and two durum wheats. They mooted that durum wheat translocates more Zn to the root while bread wheat preferentially accumulates more Zn in the shoots especially by 8 days of growth. In another experiment Hart *et al.* (1998) used the ratio of ^{65}Zn translocation in shoot to root as an indicator of Zn efficiency differences in durum and bread wheat. They also observed that bread wheat translocates more Zn from root to shoot than durum wheat.

3.12. Nutrient Utilization

Different nutrient elements once taken up by roots and transported to the demanding sinks are utilised or assimilated into various compounds or incorporated into enzymes either as a cofactor or a prosthetic group. Requirement of Mg in chlorophyll synthesis is well known. Isotopes have been used successfully to relate the differences in nutrient efficiency among crop species to its utilisation and also as a tracer to determine the activity of carbon fixing enzymes like Rubisco (in C3 plants) and PEP carboxylase (in C_4 plants and in legume root nodules) using $^{14}CO_2$ (as $Na_2H^{14}CO_3$) and label measured using Liquid scintillation counter (Singh and Singh, 2000).

Similarly dark carbon fixation in legume root nodules, a beneficial process as upto 25 per cent of the nodules requirement of carbon for N_2 fixation can be met by refixation of the bacteroid respired CO_2 can be measured using radiolabel. *In vivo* fixation of $^{14}CO_2$ by nodules of seven legume species was measured by exposing the nodules to labelled carbon in a plexiglass chamber and the incorporated label was determined, where (Figure 10.3) clusterbean was most efficient carbon fixer compared to other legumes and a significant correlation between nodule nitrogen and carbon fixation was obtained for all the investigated legume species.

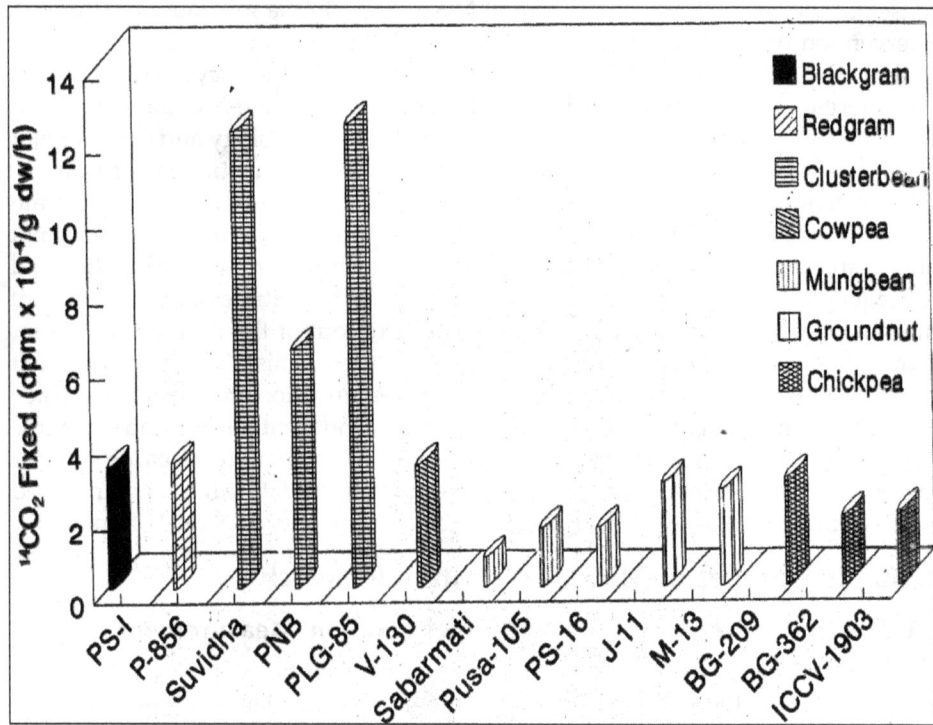

Figure 10.3: Inter Species Variation in Dark Carbon Dioxide Fixation in Legumes.

In another study, ^{14}C glucose was fed to the cut roots either with unlabelled methionine or histidine to ascertain whether histidine or methionine acts as a precursor for phytosiderophore (MA) biosynthesis. The results reveal that in case of a the label from glucose is readily incorporated in methionine and MAs but in case of b the label is not incorporated in histidine indicated that synthesis of phytosiderophores (MA) occurs from glucose via methionine and not via histidine. Fughio *et al.* (2001) conducted a real time study with ^{11}C methionine to prove conclusively that methionine is a precursor for the synthesis of phytosiderophores. They showed this by estimating the translocation of ^{11}C, fed to roots, from root to shoot. A and C depict the normal iron deficient and sufficient plants, respectively while B and D denote the autoradiograph of the Fe deficient and sufficient plants. Most of the labelled methionine is retained in the roots of Fe deficient plants which are also the organ for the biosynthesis of phytosiderophore while Fe sufficient plants which do not release any phytosiderophore show translocation of label to the shoot.

3.13. Nutrient Sensing and Signalling

Stable and radioisotopes are the most dependable tools to validate the data on nutrient sensing and signaling through accurate uptake (label) and molecular tools where labelled nucleotides (^{32}P) are used. GuoHua *et al.* (2010) recently showed the localized production of roots in response to nitrate N availability in soil pockets. Plants are very intelligent as they have the mechanisms by which they can detect/ sense pockets where a higher amount of N is present in the soil and then, through intervention of growth hormone, produce roots in the direction/location of the available N. Another important area of research on N-use efficiency is to understand the differential crop response to different forms of N *i.e.*, nitrate-N or ammonia-N. It has been shown that different N forms can alter leaf morphology and overall plant growth. Krouk *et al.* (2010) provided evidence for dual functioning of nutrient transporters. NRT-1.1 was shown to act as a nitrate transporter as well as a nitrate sensor. Imagine a transporter localized on the root performing a dual function of not only transporting N (in general, nutrients) but also sensing N availability and subsequently altering, favorably, its root behavior. When nitrate is in the vicinity of the root, it senses and stops the transport of auxin out of the root and promotes growth of lateral root in the direction of sensed nutrient. This is one very important piece of evidence in favor of coordination between nutrition and hormones which will generate new interest and will boost research on nutrient sensing and signaling. Besides the above, radiotracers are also used in molecular research aid for identification of nutrient uptake transporters and for elucidating the genes involved in different metabolic processes that are affected by the nutrient status of a plant.

4. Stable Isotopes in Mineral Nutrition Research

4.1. ^{15}N Method of Biological Nitrogen Fixation Measurement

The ^{15}N isotopic method is used to study the contribution of biological nitrogen fixation (BNF) to the nitrogen economy of both soils and plants in forage and tree legumes, and Azolla, and also helped in identification of superior genotypes of grain legumes for nitrogen fixation (Hardarson, 1994). The N-15 isotope dilution technique showed that up to 80 per cent of N accumulated by Azolla may be derived from N_2 fixation, depending on the Azolla species and environmental conditions. Multi locations and years studies showed that Azolla is an effective source of N for rice (Kumarasinghe and Eskew, 1993; Singh and Singh 1989).

4.2. Stable Isotopes in Soil Organic Matter Studies

The organic matter contained in the earth's soil is a large reservoir of carbon (C) that can act as a sink or source of atmospheric CO_2 (Raich and Potter, 1995). The upper meter of the world's mineral soils contain 1500 Pg C (1 Pg=1015 g) (Batjes, 1996). Because human modifications of land use, such as the replacement of natural ecosystems with agroecosystems, have the potential to dramatically alter soil organic matter (SOM) dynamics in these soils (Schlesinger, 1984), estimating SOM pools and turnover rates in natural and human-influenced systems are fundamental to our ability to estimate fluxes of C between the soils and the atmosphere. Carbon isotopic techniques using stable (^{13}C) tracers are well suited to the study of soil C dynamics over the time scales ranging from a few years to several hundreds of years are relevant to understanding the consequences of human-induced land use change. 13C represents approximately 1.11 per cent atom of the earth's carbon (Craig, 1957), but biological material varies around this average value as a result of isotopic discrimination during biological, physical and chemical processes (Galimov, 1985).

4.3. ^{15}N in Soil and Fertilizer Nitrogen Dynamics

Isotopes are a very powerful tool for measuring nutrient uptake from various fertilizers and soil, for studying the processes that influence the efficiency of fertilizers and for assessing the fate of the non-efficient fraction, to minimize losses of nutrients and water from the agro ecosystem (Zapata and Hera, 1995). The use of N-15 labelled fertilizers enabled the study of fertilizer use efficiency and the carry-over or residual effect of applied fertilizer N to a subsequent crop in a cropping sequence or rotation. A great deal of information on the fate of fertilizer N in different agro-ecosystems was accumulated. The total recovery values in plant and soil varied according to the growing season, crop and specific soil/water and fertilizer management conditions. On the average, total fertilizer N recovery values were of the order of 70-90 per cent but also low values of 25-50 per cent were found in non-efficient systems leading not only to low yields but also environmental pollution (IAEA, 1980, 1984).

4.4. ^{13}C/^{12}C and D/H Discrimination as a Measure of Water and Nutrient Use Efficiency

Water use efficiency is the ratio of productivity to water loss by a plant. It is defined on an instantaneous basis as the molar ratio of photosynthesis to transpiration or over the long term as the ratio of biomass produced to water consumed (Farquhar *et al.*, 1989). A major limitation to progress in our understanding of water use efficiency had been adequate method for assessing this parameter on large numbers of plants. A major breakthrough occurred when Farquhar *et al.* (1982) showed that, on a theoretical basis, carbon isotopic composition (^{13}C/^{12}C) should be correlated with both short term and long term integrated estimates of water use efficiency. Leaf carbon isotope discrimination values are used to estimate long term water use efficiency (molar ratio of photosynthesis to transpiration). Soil moisture sources are estimated from D/H ratios of stem waters. Carbon isotope discrimination (^{13}C/^{12}C) varied significantly among species within the community with the highest discrimination values occurring in short lived perennials. Long lived perennials have the lowest carbon isotope discrimination values. Substantial variation also occurred in the xylem D/H ratios, indicates that groups of species are utilized soil water from different depths within the soil profile. Midday leaf water potential values and D/H ratios in xylem waters are not correlated across species, indicate that leaf water potential alone is not a reliable indicator of the depth in the soil from which a species is extracting water. The variation in carbon isotope discrimination, and thus presumably also in water use efficiency, is not related to the water source used by a species. Leaf type or leaf duration appeared also not to be correlated with either water source or carbon isotope discrimination. Rather it appeared that tree species and shrub species (irrespective of longevity) were utilizing different soil moisture regimes. The life expectancy (longevity) of a species was negatively correlated with its carbon isotope discrimination values, suggesting that longer lived species are more conservative in their use of water. In addition, the likelihood of a species utilizing summer precipitation was also inversely proportional to longevity, indicating the more opportunistic nature of shorter lived perennials (Ehleringer and Cook, 1991). There is no fractionation of isotopes during water uptake by roots and therefore the hydrogen isotope ratio (D/H) of stem waters is a reliable indicator of the water source(s) being utilized by a plant (White *et al.*, 1985).

Stable isotopic composition of water ($\delta\ ^2H$ and $\delta\ ^{18}O$) is modified by processes like evaporation and condensation, and hence the recharge water in a particular environment will have a characteristic isotope signature. The signature serves as a natural tracer for water movement (Craig, 1961). Fresh water, essential for the survival of the vegetation, is usually present near the soil surface and in the groundwater. Stable isotopes of water (^2H and ^{18}O) can be used to find which reservoir of water the vegetation uses preferentially according to the environmental conditions of climate and soil. The assumption underlying the stable isotope method is that no fractionation of the isotopes occurs during the transfer of water from the soil to the plant (Förstel and Hützen, 1982). Since water and nutrient uptake follows the same pattern physiologically. Carbon discrimination can be taken as a indirect measure or an indicator of nutrient uptake capacity of the plant.

4.5. Sulphur Movement and Transformations in Soil Plant System

Sulphur is present in the soil in two basic forms; inorganic and organic S. In the inorganic form (sulphate S), S is available for plant uptake, but organic S accounts for 95 per cent of the total sulphur in most soils. This is due to the close relation of organic S with organic C and total N. Organic sulfur has two forms: organic S that is not bonded directly to C, consisting largely of S in the form of ester sulfates (organic sulfates containing C-O-S linkages) and organic S that is directly bonded to C (C-S), consisting largely of S in the form of S-containing amino acids, such as methionine and cysteine (Havlin *et al.*, 2005). Of the known isotopes of sulfur, four are stable (Table 10.1). The first studies using the 34S isotopic tracer were developed by Hamilton *et al.* (1991) and Awonaike *et al.* (1993). Trivelin *et al.* (2002), used sulphate-34S in studies on sulphur in soil under rice and sunn hemp (*Crotalaria juncea*) and Zhao *et al.* (2001) used to study sulphur uptake and distribution in wheat plants.

5. Innovative Radiological Tools in Plant Nutrition Research

5.1. Positron Emission Tomography

Short-lived positron-emitting radiation (PER) provides time-dependent data that is critical for developing models of metabolite transport and resource distribution in plants and their microenvironments. Until recently these techniques were applied to measure radiotracer accumulation in coarse regions along transport pathways. The recent application of PET technique to plant research allows detailed quantification of real-time metabolite dynamics on previously unexplored spatial scales. PET provides dynamic information with millimeter-scale resolution on labeled C, N and water transport over a small plant-size field of view, because details at the millimeter scale may not be required for all regions of interest. Hybrid detection systems that combine high-resolution imaging with other radiotracer counting technologies offer the versatility needed to pursue wide-ranging plant physiological and ecological research. Recently Duke University developed a hybrid detection system that provides researchers flexibility to carry out measurements of the dynamic responses of whole plants to environmental change using short-lived radiotracers (Kiser *et al.*, 2008).

5.2. Biofortication

Biofortified food offers a potentially powerful tool against malnutrition, that targets the most vulnerable population *i.e.*, resource-poor women, infants and children. Using Fe-52 it has been shown that under Fe deficiency, Nicotianamine synthase is not only involved in the production of metal-chelating phytosiderophore but that these phytosiderophore-matal complexes are absorbed by roots and transported as a moiety upto the shoot and to the grain (Zhao and McGrath, 2009). Also important fact is that the total content of a given nutrient in a food might not be a useful indicator of its nutritional quality, because not all of the nutrients in foods are absorbed. Thus, it is important to assess the bioavailability of the biofortified foods (Hirschi, 2008). The absorption of elements in animals or humans can be studied using radiotracers or enriched stable isotopes (Patterson and Veillon, 2001). Radiotracers have the advantage of easy detection and minimal preparation before analysis. However, because of radioactive decay and associated biohazards, this approach is used only in animal feeding regimes. Meanwhile, stable isotopes, although often expensive, occur naturally and do not represent any risk when labeling foods for human consumption. Isotopes can be used to label foods intrinsically or extrinsically (Abrams, 2003).

5.3. Real Time Radioisotope Imaging and Nutrient Uptake

Imaging is now an indispensable analysis tool in understanding the biological activity of plants. Among the imaging methods, significant developments have occurred in the fluorescent imaging technique, which requires specific conditions (darkness) for signal acquisition and cannot be easily applied to real-time imaging of nutrient uptake since plants require light for growth and development. Nuclear magnetic resonance (NMR) is another candidate for real-time imaging of nutrient uptake in plants (Ishida *et al.*, 2000), however, control of environmental conditions, such as humidity, temperature and light is extremely difficult to achieve since the sample is prepared in a small test tube in a magnetic field. The use of radioisotopes offers a promising method and allows imaging under light and enables quantitative analysis of the image, as the imaging is based on radioactivity.

Radioisotope imaging techniques that overcome such restrictions and allow for real-time imaging of ionic movement have been developed (Kanno *et al.*, 2012). Systems like macro imaging was developed to visualize and measure ion uptake and translocation between organs at a whole-plant scale. Even the fluorescent microscopes have been modified to measure ion uptake at the microscopic level. Both the above systems allow numerical analysis of images and possess a wide dynamic range of detection because they are based on radioactivity.

6. Conclusions

It is amply clear that most of the developments in the field of plant nutrition have been achieved through the use of radiotracers and radioisotopes have been effectively used to improve our understanding of the processes in nutrient uptake, interaction, nutrient sensing, signaling, phyto-hormone regulation, nutrient use efficiency. State of the art radiation analytical tools have made the real time analysis of nutrients at the whole plant, the tissue and the cell level. There is no doubt that radiotracers have a great future and will continue to better our understanding of nutrient dynamics under challenging condition of multiple nutrient deficiencies and for addressing other important challenges in the area of plant mineral nutrition.

References

Abrams, S.A. 2003. Using stable isotopes to assess the bioavailability of minerals in food-fortification programs. *Forum Nutr.* **56**, 312–313.

Awonaike KO, Danso SKA, Zapata F 1993. The use of a double isotope (N-15 and S-34) labeling technique to assess the suitability of various reference crops for estimating nitrogen-fixation in *Gliricidia sepium* and *Leucaena leucocephala*. *Plant Soil*, 155/156: 325-328.

Batjes NH. 1996. Total carbon and nitrogen in the soils of the world. *Eur. J. Soil Sci.*, 47: 151–163.

Britto, D, T, Ruth, T. J., Lapi, S. and Kronzucker H, J. 2004. Cellular and whole-plant chloride dynamics in barley: insights into chloride-nitrogen interactions and salinity responses *Planta*, **218**, 615–22.

Craig H 1957. Isotopic standards for carbon and oxygen and correction factors for mass spectrometric analysis of carbon dioxide *Geochim. Cosmochim. Acta*, 12: 133–149.

Craig, H. 1961.Isotopic variations in meteoric water. *Science*, 133, 1702-1703.

Cregg. B. and Zhang, J. 2000. Carbon Isotope Discrimination as a Tool to Screen for Improved Drought Tolerance. *Metria* 11.

Ehleringer JR and Cook CS 1991 Carbon isotope discrimination and xylem hydrogen isotope ratios in desert plants. *IAEA Techdoc* no. IAEA-SM-313-71.

Epstein, E 1972. *Mineral Nutrition of Plants : Principles and Perspective*, Wiley, New York.

Farquhar GD, Oleary MH, Berry JA 1982. On the relationship between carbon isotope discrimination and the intercellular carbon dioxide concentration in leaves. *Aust. J. Plant Physiol.* 9: 121-137.

Farquhar GD, Ehleringer JR Hubick KT 1989. Carbon isotope discrimination and photosynthesis. *Annu. Rev. Plant Physiol. Plant Mol. Biol.* 40: 503-537.

Forstel H., Hutzen H 1982, Use of water with different 18O-content to study transport processes in plants. *Stable Isot.* 511-516.

Galimov, E.M., 1985. *The Biological Fractionation of Isotopes*. Academic Press, New York.

Glass, A.D.M. 1989. *Plant Nutrition: An Introduction to Current Concepts*. Jones and Bartlett, Boston.

Griffiths, R. I., Manefield M., Ostleb, N., McNamarab, N., O'Donnellc, A. G., Bailey, M. J. and Whiteley, A. S. 2004. $^{13}CO_2$ pulse labelling of plants in tandem with stable isotope probing: methodological considerations for examining microbial function in the rhizosphere. *Journal of Microbiological Methods* **58,** 119–129.

GuoHua Mi, FanJun Chen, QiuPing Wu, NingWei Lai, LiXing Yuan and FuSuo Zhang (2010) Ideotype root architecture for efficient nitrogen acquisition by maize in intensive cropping systems science. *China Life Sciences* 53 (12): 1369-1373.

Haminton S.D.; Chalk, P.M.; Undovich, M.J. and Smith J 1991 The measurement of fertilizer-S uptake by plants using radioactive and stable isotopes. *Appl. Radiat. Isot.*, 42: 1099-1101.

Havlin, JL, Beaton JD, Tisdale SL, Nelson WL 2005. *Soil Fertility and Fertilizers*. 7th ed. Upper Saddle River: Pearson Education, 515p.

Hardarson, G., 1994. International FAO/IAEA programmes on biological nitrogen fixation. In: P.H.Graham, M.J.Sadowsky and C.P. Vance (eds). Symbiotic Nitrogen Fixation. Kluwer Acad.Publ., Dordrecht, The Netherlands: 189-202.

Hirschi. K., 2008. Nutritional improvements in plants: time to bite on biofortified foods. *Trends in Plant Science*, **13,** 459-463.

Hoefs, J. 1997. *Stable Isotope Geochemistry*. Springer-Verlag, New York.

International Atomic Energy Agency, 1980. Soil nitrogen as fertilizer or pollutant. Panel Proccedings Series, IAEA, Vienna, Austria.

International Atomic Energy Agency, 1984. *Soil and Fertilizer Nitrogen*. Technical Report Series No.244, IAEA, Vienna, Austria.

Ishida, N., Koizumi, M. and Kano, H. 2000 The NMR microscope: a unique and promising tool for plant science. *Ann. Bot.* **86,** 259–278.

Kanno, S.,Yamawaki, M., Ishibashi, H., Kobayashi, N, I., Hirose, A., Tanoi, K., Nussaume, L. and Nakanishi, T, M. 2012. Development of real-time radioisotope imaging systems for plant nutrient uptake studies. *Phil. Trans. R. Soc. B, 367,* 1501-1508.

Kiser, M.R., Reid, C.D., Crowell, A.S., Phillips, R.P. and Howell, C.R. 2008. Exploring the transport of plant metabolites using positron emitting radiotracers. *HFSP Journal, 2,* 189-204.

Krouk G, Lacombe B, Bielach A, Perrine-Walker F, Malinska K, Mounier E, Hoyerova K, Tillard P, Leon S, Ljung K, Zazimalova E, Benkova E, Nacry P, Gojon A. (2010) Nitrate-regulated auxin transport by NRT1.1 defines a mechanism for nutrient sensing in plants. *Dev Cell*. 18 (6): 927-37.

Kumarasinghe, K.S. and Eskew, D.L., 1993. Isotopic studies of Azolla and nitrogen fertilization of rice. *Developments in Plant and Soil Sci.*Vol.51, Kluwer Acad.Publ., Dordrecht. The Netherlands.

Marschner, H (1995). *Mineral Nutrition of Higher Plant*. Academic Press London.

Patterson, K.Y. and Veillon, C. 2001. Stable isotopes of minerals as metabolic tracers in human nutrition research. *Exp. Biol. Med.* **226,** 271–282.

Pettigrew, W.T. 2008. Potassium influences on yield and quality production for maize, wheat, soybean and cotton. *Physiologia Plantarum,* **133**: 670-681.

Prosser, J.I., Rangel-Castro, J.I., Killham, K. 2006. Studying plant-microbe interactions using stable isotope technologies. *Curr Opin Biotechnol* **17**: 98–102.

Raich, JW, Potter CS 1995. Global patterns of carbon dioxide emissions from soils *Global Biogeochem. Cycles*, 9 : 23–26.

Schlesinger WH 1984. Soil organic matter: A source of atmospheric CO_2 SCOPE, 23 : 117–127.

White. J.W.C., Cook E.R. Lawrence. J.R. Broecker WS 1985. The D/H ratios of sap in trees: Implications for water sources and tree ring DIH ratios, *Geochim. Cosmachim. Acta* 49: 237-246.

Zapata, F. and Hera, C. 1995. Enhancing nutrient management through use of isotope techniques. In: IAEA Proc. *Int. Symp. Nuclear Techniques in Soil-Plant Studies for Sustainable Agriculture and Environmental Preservation.* Vienna, Austria: 83-105.

Zhao F.J.; Verkampen K.C.J.; Birdsey M, Blakekalff MMA, McGrath, SP 2001 Use of the enriched stable isotope 34S to study sulfur uptake and distribution in wheat. *J. Plant Nutr.*, 24: 1551-1560.

Zhao FJ, McGrath SP. (2009). Biofortification and phytoremediation. *Curr Opin Plant Biol.* 12(3): 373-80.

Recent Advances in Crop Physiology Vol. 2 (2016)   *Pages* 315–335
Editor: **Dr. Amrit Lal Singh**
Published by: **DAYA PUBLISHING HOUSE, NEW DELHI**

Chapter 11
Nutritional Quality of Wheat

Sewa Ram

ICAR-Directorate of Wheat Research, Karnal – 132 001, Harnaya

Introduction

Wheat is an important cereal crop, the main source of energy and nutrition in human diet. Nutritional value of food stuff is defined by its ability to provide nutrients required by the body. Proteins, carbohydrates, fats, minerals and vitamins are main ingredients in food stuffs. Wheat is an important source of these nutrients (Pomeranz, 1998; Ram and Mishra, 2010). The distribution of nutrients within wheat kernel is typical of many cereals. These components are concentrated in different parts of the grain. The germ is rich in vitamins B and E, high quality protein, unsaturated fats, minerals and carbohydrates. The bran consists mostly of the insoluble carbohydrates such as cellulose, protein, traces of B vitamins and minerals and anti-nutritional factors such as phytic acid. The endosperm is the largest part of the grain and consists mainly of the starch and protein, and trace amounts of vitamins and minerals. Despite all these nutritional qualities, wheat lacks adequate amounts of certain essential nutrients, vitamins (A, B_{12} and C), fats, micronutrients and the essential amino acid lysine. Therefore, increasing micronutrient density and lysine content and decreasing anti-nutritional factors such as phytic acid are important in wheat along with improving its processing quality. In addition, certain populations suffer from gluten enteropathy called coeliac disease and there is a need to understand clearly the gluten fragments responsible for this disease. This can be accomplished by better understanding of genetics of these traits and their molecular basis. Recently good amount of information has been generated on various genes related storage proteins, starch quality, mineral nutrition, lysine content and anti-nutritional factors.

2. Protein Quality Assessment

Protein quality relates to the efficiency with which various food proteins are used for synthesis and maintenance of tissue proteins. The composition of various proteins may be so unique that their influence on physiological function in the human body could be quite different. The quality of a protein is vital when considering the nutritional benefits that it can provide. The quality of a protein is determined by assessing its essential amino acid composition, digestibility and bioavailability of amino acids. The capacity of a protein to meet the amino acid requirements of an organism depends on several factors such as amino acid composition, adequacy of a diet as a whole, physiological, nutritional and health status of an animal or human consuming it. There are several measurement scales and techniques based on chemical and biological evaluations.

2.1. Rating Scales

Numerous methods exist to determine protein quality. These methods have been identified as protein efficiency ratio (PER), biological value (BV), net protein utilization (NPU), and protein digestibility corrected amino acid score (PDCAAS). Usually weanling rats, suckling pigs, mouse, chickens and adult and adolescent humans are used as experimental subjects.

2.2. Efficiency Ratio

The protein efficiency ratio determines the effectiveness of a protein through the measurement of animal growth. This technique requires feeding rats a test protein and then measuring the weight gain in grams per gram of protein consumed. The computed value is then compared to a standard value of 2.7, which is the standard value of casein protein. Any value that exceeds 2.7 is considered to be an excellent protein source. PER value of wheat has been reported around 50 per cent of casein control. However, this calculation provides a measure of growth in rats and does not provide a strong correlation to the growth needs of humans. The assumptions that weight increase is an index of protein synthesis and all proteins consumed resulted in tissue growth are not valid.

2.3. Biological Value

Biological value measures protein quality by calculating the nitrogen used for tissue formation divided by the nitrogen absorbed from food. This product is multiplied by 100 and expressed as a percentage of nitrogen utilized. The biological value provides a measurement of how efficient the body utilizes protein consumed in the diet. A food with a high value correlates to a high supply of the essential amino acids. Animal sources typically possess a higher biological value than vegetable sources due to the lack of one or more essential amino acids of vegetable sources. There are, however, some inherent problems with this rating system. The biological value does not take into consideration several key factors that influence the digestion of protein and interaction with other foods before absorption. The biological value also measures a maximal potential quality of protein and not its estimate at requirement levels.

2.4 Net Protein Utilization

Net protein utilization is similar to the biological value except that it involves a direct measure of retention of absorbed nitrogen. Net protein utilization and biological value both measure the same parameter of nitrogen retention; however, the difference lies in that the biological value is calculated from nitrogen absorbed whereas net protein utilization is calculated from nitrogen ingested. NPU= (N intake-fecal N-Urine N)*100/N intake + (metabolic N + endogenous N)*100/N intake. NPU of wheat has been reported to be around 60 per cent.

2.5 Protein Digestibility Corrected Amino Acid Score

Protein digestibility is calculated by the formula: True protein digestibility = PI – (FP – MFP)*100/PI; where PI = protein intake, FP = fecal protein and MFP = metabolic fecal protein. The amount of protein in the feces of ratsfed the protein-free diet is used as the estimate for MFP. However, in 1989, the Food and Agriculture Organization and World Health Organization (FAO/WHO) in a joint position stand stated that protein quality could be determined by expressing the content of the first limiting essential amino acid of the test protein as a percentage of the content of the same amino acid content in a reference pattern of essential amino acids (FAO/WHO, 1991). The reference values used were based upon the essential amino acids requirements of preschool-age children. The recommendation of the joint FAO/WHO statement was to take this reference value and correct it for true fecal digestibility of the test protein. The value obtained was referred as protein digestibility corrected amino acid score (PDCAAS). PDCAAS is calculated by the following formula: PDCAAS (per cent) = True protein digestibility X amino acid score or the lowest amino acid ratio. This method has been adopted as the preferred method for measurement of the protein value in human nutrition. Amino acid ratios [(mg of an essential amino acid in 1.0 g of test protein/mg of the same amino acid in 1.0 g of reference protein) × 100] for essential amino acids are calculated using a rat growth pattern of amino acid requirements (Sarwar *et al.*, 1985) and a human pattern of amino acid requirements FAO/WHO/UNU, (1985) suggested pattern of amino acid requirements for preschool children, 2-5 yrs.) as the reference proteins.

The PDCAAS method has been considered to be a simple and scientifically sound approach for routine assessment of dietary protein quality for humans. The PDCAAS is now a federally approved alternative method to the protein efficiency ratio (PER)rat bioassay procedure. Although the PDCAAS is currently the most accepted and widely used method, limitations still exist relating to over estimation in the elderly (likely related to references values based on young individuals), influence of ideal digestibility and antinutritional factors (Sarwar, 1997). Antinutritional factors may occur naturally or may be formed during heat processing. Some examples of naturally occurring antinutritional factors include glucosinolates in mustard andrapeseed protein products, trypsin inhibitors and hemagglutinins in legumes, phytates in cereals and oilseeds and gossypol in cotton seed protein preparations, which could adversely affect nutrient utilization and may contribute to growth depression in animals. In addition to that, food processing conditions may influence protein digestibility.

2.6 Amino Acids Composition and Protein Quality of Wheat

Wheat is an important in human diet containing proteins from 6.0 per cent to 21.0 per cent, though germ contains higher proportion of proteins in wheat, 72 per cent of the total amount of proteins is present in endosperm. The quality and quantity of proteins in relation to processing quality are described in other chapter in the book. Here an account of nutritional quality of wheat proteins will be provided. Based on solubility criteria, wheat proteins are divided into albumins (water soluble), globulins (salt soluble), gliadins (ethanol soluble) and glutenins (acid soluble). Glutenins and gliadins are main storage proteins constituting around 80 per cent of total proteins. Albumins and globulins are mainly confined in embryo and aleurone layer while gliadins and glutenins in endosperm. Comparatively glutenins and gliadins are rich in glutamine and proline and poor in lysine while globulins have more lysine and lower content of glutamine and proline. Because of higher content of glutenins and gliadins, wheat proteins are deficient in lysine which is one of the indispensable amino acids. Other indispensable amino acids are leucine, isoleucine, valine, threonine, methionine, phenylalanine and tryptophan (Table 11.1). Recently transgenic approach has shown promise for the improvement of lysine content.

Table 11.1: Composition of Amino Acids in Wheat, Flour and Protein Fractions (g/16g N) (Salunkhe *et al.*, 1986)

Amino Acids	Wheat	Flour	Albumin	Globulin	Gliadin	Glutenin
Tryptophan	1.5	1.5	1.1	1.1	0.7	2.2
Lysine	2.3	1.9	3.2	5.9	0.5	1.5
Histidine	2.0	1.9	2.0	2.6	1.6	1.7
Arginine	4.0	3.1	5.8	7.0	1.9	2.7
Aspartic acid	4.7	3.7	5.8	7.0	1.9	2.7
Threonine	2.4	2.4	3.1	3.3	1.5	2.4
Serine	4.2	4.4	4.5	4.8	3.8	4.7
Glutamic acid	30.3	34.7	22.6	15.5	41.1	34.2
Proline	10.1	11.8	8.9	5.0	14.3	10.7
Glycine	3.8	3.4	3.6	4.9	1.5	4.2
Alanine	3.1	2.6	4.3	4.9	1.5	2.3
Cystine	2.8	2.8	6.2	5.4	2.7	2.2
Valine	3.6	3.4	4.7	4.6	2.7	3.2
Methionine	1.2	1.3	1.8	1.7	1.0	1.3
Isoleucine	3.0	3.1	3.0	3.2	3.2	2.7
Leucine	6.3	6.6	6.8	6.8	6.1	6.2
Tyrosine	2.7	2.8	3.4	2.9	2.2	3.4
Phenylalanine	4.6	4.8	4.0	3.5	6.0	4.1

2.7 Lysine Deficiency

Protein malnutrition is a major problem in areas where cereal based diet is predominant. Wheat is deficient in essential amino acid lysine and needs improvement. In recent years, major progress has been made in understanding the metabolic pathways of essential amino acids, as well in the identification of regulatory and limiting steps in these metabolic pathways that could be overcome by metabolic engineering. As limited variability exists in wheat for lysine content, transgenic approach has a potential for enhancing lysine content. Two-pronged strategies can be adopted in improving lysine content. One is to manipulate biosynthetic and degrading pathway and second is to enhance content of seed proteins with higher lysine level for which lysine biosynthetic machinery is required to be enhanced and/ or degradative pathway is to be suppressed. During the last two decades the biochemical pathways of lysine biosynthesis and degradation have been studied extensively (Figure 11.1).

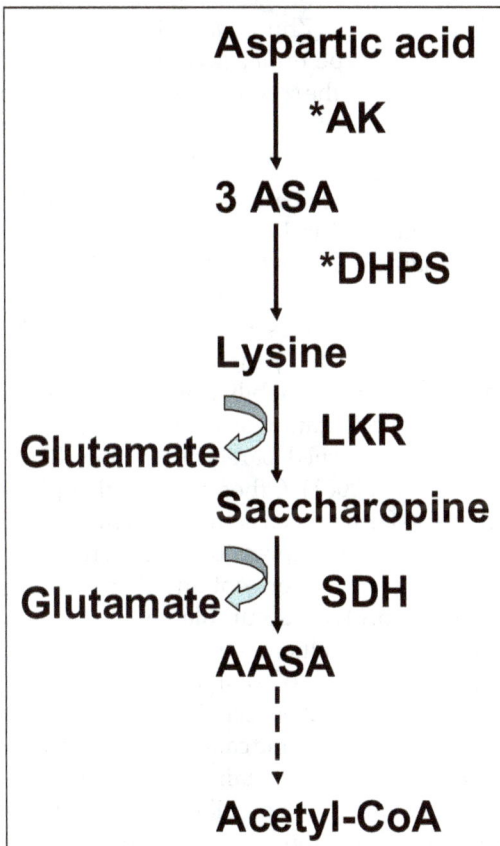

Aspartic acid

| *AK

3 ASA

| *DHPS

Lysine

Glutamate | LKR

Saccharopine

Glutamate | SDH

AASA

Acetyl-CoA

Figure 11.1: Main Steps Involved in the Biosynthesis and Catabolism of Lysine.

ASA: Aspartic semialdehyde; AASA: Amino adipicsemialdehyde; AK: Aspartate kinase; DHPS: Dihydrodipicolinate synthase; LKR: Lysine ketoglutaratereductase; and SDH: Saccharopine dehydrogenase. *: Feedback inhibited by lysine. Dashed arrow in the last step indicates many enzymes (Ram and Mishra, 2010).

Enzymes of lysine biosynthetic pathway and corresponding genes have been characterized. Two main enzymes of lysine biosynthesis such as aspartate kinase (AK) and dihydrodipicolinate synthase (DHPS) are feedback inhibited by lysine. Genetic mutations in plant DHPS genes to render them insensitive to lysine can cause the overproduction of lysine in all plant organs. Although an increase in protein lysine in seeds is beneficial, an increase in vegetative tissues is undesirable, because high concentrations of free lysine cause abnormal vegetative growth and flower development which, in turn, reduces seed yield. Targeted expression of transgenic DHPS in seeds of several crop plants by using seed-specific promoters eliminates its undesirable effects in vegetative tissues, resulting in plants with good growth characteristics that accumulate high concentrations of lysine in their seed proteins. Lysine ketoglutarate reductase (LKR) and saccharopine dehydrogenase (SDS) are main enzymes of lysine catabolism in plants and control lysine content, particularly in seeds. The significance of lysine catabolism has been demonstrated by knockout mutants of *Arabidopsis* with the *LKR/SDH* gene, encoding the first two linked enzymes of the lysine catabolism pathway (Zhu *et al.*, 2001). This knockout exhibited a morphologically normal phenotype under favorable growth conditions and also possesses enhanced lysine levels in the seeds. The regulatory role of lysine catabolism has also been confirmed in balancing lysine concentration in natural high lysine maize mutants (Azevedo *et al.*, 2003) and in *Arabidopsis* (Zhu and Gallili, 2004).

Another approach is to enhance native plant proteins rich in essential amino acids and are naturally synthesized and accumulated in the tissue of interest. Singh *et al.* (1993) isolated and characterized wheat triticin cDNA revealing a lysine-rich repetitive domain and subsequently used in transformation experiments for improving lysine content in wheat. The content of essential amino acids in these proteins may be further improved by inserting them into desirable places that will interfere minimally with the structural conformation and stability of the proteins. An example of such successful engineering is the modification of a native barley gene encoding a lysine-rich seed chymotrypsin inhibitor, termed 'barely high lysine' (BHL) protein, to increase its lysine codons (Forsyth *et al.*, 2005). Other successful application of engineered native lysine-rich proteins so far has been achieved in maize, where enhanced seed lysine levels were obtained by the expression of genetically engineered BHL or hordothionine (HT12) containing 24 per cent and 28 per cent lysine, respectively (Jung and Falco, 2000). These proteins accumulated in transgenic maize to 3–6 per cent of total grain proteins when introduced together with a bacterial lysine-insensitive DHPS, resulted in very high elevation of total lysine to over 0.7 per cent of seed dry weight compared to approximately 0.2 per cent in wild-type maize. Combining these traits with seed-specific reduction of lysine catabolism offers an optimistic future for commercial application of high-lysine maize. However, improvement of these metabolic pathways by overcoming such regulatory and limiting steps was generally associated with impairment of plant growth and productivity. This suggests that further comprehensive understanding of regulatory and metabolic networks as well as their interactions with plant growth and response to the environment is needed to optimally improve plant nutritional quality. This can be achieved by genomics approaches, such as gene expression profiling in micro arrays, proteomics, metabolic profiling and flux measurements.

3. Starch Quality (Resistant starch)

Carbohydrates constitute nearly 80 per cent of the total dry matter of wheat kernel. Starch is present in discrete granules within the cells of the endosperm. Starch content is available in the range of 65 per cent to 75 per cent and generally related to the protein content. Starch structure is a crucial determinant of quality where amylose/amylopectin ratio is a useful descriptor of starch structure. The rate of starch digestion varies from source to source. The starch in potatoes, cereals and baked goods is digested very rapidly. Other starchy foods such as beans, barley or long grained brown rice are digested more slowly and cause a much slower and lower blood sugar rise. Resistant starch goes all the way through the small intestine without being digested. In this way, it is more like fibre, and in some cases is classified and labelled as fibre.

Although starch can be digested(theoretically to completion) in the human small intestine, amylolysis varies by rate and extent for many foods. Reduction of the rate of starch digestion in the small intestine has the potential to lower the rate of entry of glucose into the circulation thus, reduces demand for insulin. This is measured as the glycemic index (GI), and lowering the GI is emerging as an important mechanism for managing the incidence and severity of type II diabetes. The fraction of the ingested starch escapes into the human large bowel is known as RS. Short-chain fatty acids (SCFA) are major end products of the fermentation of NSP and RS by the microflora, and they promote important aspects of large bowel function. SCFAs are also produced by soluble fibre and oligosaccharides - this is the reason why on some food labels, some fibre is shown as having calories associated with it, but these calories do not raise blood glucose. One of the principal SCFA, butyrate, may also play a role in promoting a normal phenotype in colonocytes and lowering the risk of colorectal cancer. It is emerging that RS contributes substantially to some of the benefits that have been ascribed solely to dietary fibre. Therefore, modification of the starch composition of wheat by raising its RS content presents an opportunity for a potentially large-scale improvement in public health. There is a case for increasing RS consumption as an effective means of improving nutrition for public health at the population level.The strategy of combining wheat bran (WB) with RS was effective at producing beneficial changes in fecal bulking, transit time and fermentation-dependent indexes that persisted to the distal regions of the human colon.Moreover, combining WB with RS may attenuate any potentially negative effects of RS. This work suggests that to maximize the health benefits of dietary fibre in the gastrointestinal tract, a combination of different fibre types may be required.

3.1. Amylose Content and Resistant Starch

Starches consist of glucose monomers polymerized through α-1,4and α-1,6 linkages into two classes of polymers, amylose and amylopectin. Amylopectin is a large highly branched polysaccharide [degree of polymerization (DP) >5,000], whereas amylose has infrequent α-1,6 linkages and a lower DP (<2,000). In cooked foods, amylase molecules re-associate rapidly on cooling, forming complexes that resist digestion, whereas amylopectin molecules re-associateslowly and are more readily digested. This difference explains the higher RS content of high-amylose products.

Therefore, altering the ratio of amylose to amylopectin in starch can led to higher resistant content having potential health benefits. This can be accomplished by altering the levels of different enzymes involved in starch biosynthesis. The synthesis of amylose and amylopectin occurs via two pathways (Figure 11.2). Amylosesynthesis requires an active granule-bound starch synthase (GBSS). Amylopectin is synthesized by a complex pathway involving, amongothers, several isoforms of starch synthase (SSI, SSII, SSIII), starch-branching enzymes (SBEI, SBEIIa, SBEIIb) and starch-debranching enzymes (DBE).

3.2 Development of Wheat with Enhanced Amylose Content (RS)

Theoretically, the suppression of the activities of starch synthase and starch branching enzymes can lead to enhance amylose content. Recently RNAi technology has been developed to silence expression of genes in plants. This has the benefits of silencing genes present in all the homeologous chromosomes at a time with obvious advantage in wheat being hexaploid. Silencing homeologous genes in wheat by other methods as mutation is difficult. Regina *et al.* (2006) used RNA interference technique to down-regulate the two different isoforms of starch-branching enzyme (SBE) II (SBEIIa and SBEIIb) in wheat endosperm to raise its amylose content. Suppression of SBEIIb expression alone had no effect on amylose content; however, suppression of both SBEIIa and SBEIIb expressions resulted in starch containing >70 per cent amylose. Rat nutritional studies were conducted using high amylose wheat generated and found positive effects on indices of gastrointestinal health. Indices of large-bowel fermentationwere all significantly higher in rats fed the high-amylose wheat, consistent with more RS, compared with those fed the standard wheat. Thus large-bowel digest a wet weight and SCFA pools and fecal SCFA excretion were all

100 per cent higher in rats fed with the novel wheat compared with those fed the control diet. pH values were also significantly lower, again consistent with greater fermentation. These data in rats fed the high-amylose wheat were generally similar to those noted previously in rats fed *Himalaya 292*, a barley cultivar with high RS arising through altered starch synthesis (Bird *et al.*, 2004). Interestingly, the molar ratios of the major SCFA in rats fed the novel wheat were rather different from *Himalaya 292* with considerably more butyrate in the investigation. This is of some interest in view of the apparent importance of butyrate in promoting large-bowel function. Collectively, the data supports the potential of the new high-amylose wheat to produce foods high in RS having a low glycemic index. The data justify further investigation of the health potential of high-amylose wheats, especially in processed foods, as an important additional mechanism to deliver significant health benefits to large numbers of consumers through their diet.

4. Wheat Gluten Ellergy, Intolerance and Enteropathy

Gluten is the viscoelastic complex mainly formed of glutenins and gliadins. Adverse reactions to wheat, as to any food, can be allergic (in this case, wheat allergy), intolerance (in this case, wheat intolerance, gluten intolerance and coeliac disease) or due to other naturally occurring constituents. Different mechanisms cause different adverse reactions. The resulting symptoms may be quite different from or confusingly similar to each other. The human body is able to mount a variety of defense mechanisms against proteins it regards as foreign or harmful. It is not clearly known why food proteins are regarded as harmful by the body. For example, wheat-sensitive allergic individuals typically produce IgE antibodies to the soluble grain proteins, but some develop gluten-specific IgE antibodies. Patients with coeliac disease develop gliadin-specific IgA and IgG antibodies. Wheat allergy refers specifically to adverse reactions involving immunoglobulin E (IgE) antibodies to one or more protein fractions of wheat, including albumin, globulin, gliadin and glutenin (gluten). The majority of IgE-mediated reactions to wheat involve the albumin and globulin fractions. Gliadin and gluten may also, rarely, induce IgE-mediated reactions. Allergy to wheat may occur in any individual, unlike Coeliac Disease, which is hereditary.

Allergic reactions to wheat may be caused by ingestion of wheat-containing foods or by inhalation of flour containing wheat (Baker's asthma). Clinical experience suggests that wheat allergy is relatively uncommon, but there are no accurate figures for prevalence. The allergy is more prevalent in certain groups: *e.g.*, wheat allergy is responsible for occupational asthma in up to 30 per cent of individuals in the baking industry. Allergic reactions to wheat (IgE-antibody mediated) usually begins within minutes or a few hours after eating or inhaling wheat. The more common symptoms involve the skin (urticaria [hives], eczema, angioedema [swelling due to allergy]), the gastrointestinal tract (abdominal cramps, nausea and vomiting, oral allergy syndrome) and the respiratory tract (asthma or allergic rhinitis). In association with exercise, reactions to gliadin or gluten can cause urticaria, angioedema or life-threatening anaphylaxis. As these proteins are present in other cereals, these symptoms may also occur in wheat-allergic individuals due to cross-reactivity.

Coeliac Disease (CD), also called Gluten Enteropathy has until recently been known as Gluten Intolerance is a hereditary disorder of the immune system in which eating gluten leads to damage of the mucosa (lining) of the small intestine (small gut). This results in malabsorption of nutrients and vitamins. CD is the result of IgA and IgG antibodies responses to gluten. It is important to differentiate between CD, mediated by IgA and IgG antibodies and wheat allergy, which is mediated by IgE antibodies. Coeliac Disease is one of the commonest life-long disorders in Western countries. CD is frequently under-diagnosed, particularly in adults, who may present with subtle symptoms. In some countries the incidence is as high as 1 in 200 (Sweden) or 1 in 10,000 (Denmark). The incidence in South Africa has not been ascertained, but is thought to be low, although the disease is most probably under-diagnosed. Recently cases of coeliac disease have been reported in India.

Typically CD presents at the age of 6-24 months with symptoms of intestinal malabsorption, impaired growth, abnormal stools, abdominal distension, muscle wasting, poor muscle tone (hypotonia), poor appetite or irritability, following the introduction of cereals into the diet. In adults, the symptoms of CD may be quite varied, from severe weight loss and diarrhoea and bulky, offensive stools to subtle complaints of cramps, abdominal bloating, flatulence and even constipation. These individuals are often mistakenly diagnosed as having Irritable Bowel Syndrome. Recent studies show that some individuals with CD present with no symptom but a form of ataxia. Recurrent oral aphthous ulcers are common and should arouse suspicion of the condition. Other symptoms may include persistent iron-deficiency anemia, folate deficiency anemia or a calcium metabolism disturbance. Dermatitis herpetiformis is a variant of Coeliac Disease in which clusters of itchy blisters occur, usually over the buttocks, knees and elbows. Doctors must have a low threshold of suspicion when seeing patients with symptoms such as those described above.

5. Wheat Improvement for Coeliac Patient

Medication is ineffective in treating this condition, hence the only treatment available is the complete removal of gluten from the diet. This usually entails life-long avoidance of all cereals containing gluten, (wheat, oats, rye and barley). Individuals on any avoidance diet are at risk of developing deficiencies of micro-nutrients (*e.g.*, thiamine, riboflavin, niacin, iron, selenium, chromium, magnesium, folacin, phosphorus and molybdenum). It is therefore essential that patients be managed in collaboration with a dietician. Information on gluten-free diets is becoming increasingly available worldwide. Other option would to develop varieties having lower immunogenicity. Gluten proteins from wheat can induce coeliac disease (CD) in genetically susceptible individuals. Specific gluten peptides can be presented by antigen presenting cells to gluten-sensitive T-cell lymphocytes leading to CD. During the last decades, a significant increase has been observed in the prevalence of CD. This may partly be attributed to an increase in awareness and to improved diagnostic techniques, but increased wheat and gluten consumption is also considered a major cause. Recent studies demonstrate that there is increase in the presence of the Gliα-9 epitope in the modern varieties, whereas the presence of the Gliα-20 epitope was lower, as compared to the landraces (Van der Broeck *et al.*, 2010). This suggests that modern wheat breeding practices may have led to an increased exposure to CD epitopes. On the other hand, some modern varieties and landraces have been identified that have relatively low contents of both epitopes. Such selected lines may serve as a start to breed wheat for the introduction of 'low CD toxic' as a new breeding trait. Large-scale culture and consumption of such varieties would considerably aid in decreasing the prevalence of CD.

6. Micronutrient Density and Phytic Acid

Wheat is the main source of starch, fibre and also contain significant amount of proteins, minerals and vitamins. Minerals and trace elements are essential nutrients for good health. For example Ca is required for formation and maintenance of skeleton, normal contraction of muscle, heart, nervous activity and blood clotting. Magnesium is implicated in the regulation of nervous system and prevents spasmophilia. Iron is incorporated in the haemoglobin molecule and plays a role in transport of oxygen. Zinc and copper act as constituents of a number of enzymes. The large part of the population in the world faces problem of micronutrient deficiency (especially Fe and Zn). This problem is more predominant in infants and pregnant women who require comparatively larger quantity of micronutrients. Though whole wheat is an important source of these minerals particularly Mg, Zn and Fe and contributes significantly to mineral consumptions; the daily requirements are not met by wheat based diets as shown in the Table 11.2.

Table 11.2: Mineral Content (mg/kg) of Wheat and Wheat Fractions and RDA Requirements

Wheat Sample	Fe	Zn	Cu	Mn
Wheat	18-31	21-63	1.8-6.2	24-37
Bran	74-103	56-141	8.4-16.2	72-144
Germ	41-58	<100-144	7.2-11.8	101-129
Flour	3.5-9.1	3.4-10.5	0.62-0.63	2.1-3.5
Maximum adult RDA	15 mg	15 mg	1.5-3.0 mg	2-5 mg

RDA: Recommended dietary allowances are the daily levels of intake of essential nutrients judged to be adequate to meet the known nutrients needs of practically all healthy persons (Grusak and Penna, 1999; Betschart, 1998).

The recommended daily requirements of these micronutrients are not met by consuming wheat because of lower bioavailability of micronutrients in wheat based diet. This is mainly caused by antinutritional factors such as phytic acid present in the grain. Phytic acid inhibits the release of Fe and Zn in the intestine and thus their absorption is reduced. Phytate in wheat grains is found predominantly in aleurone which remains attached to the pericarp during milling and therefore is concentrated in the bran fraction. Therefore, Fe and Zn deficiencies are considered major micronutrient deficiencies in people where cereals are consumed as major foods. The enhancement of their density and also their bioavailability by reduced antinutritional factors can lead to improve nutritional status of vast sections of the human beings throughout the world. Recent reports indicate that sufficient variability for the trait of Fe and Zn does not exist in cultivated hexaploids but wild species of wheat have wider variation of micronutrient content (upto 110 mg/kg) (Chhuneja *et al.*, 2006). Therefore wild progenitors of wheat can be used in biofortification for enhanced micronutrient density. In addition molecular tools are available to enhance both density and bioavailability of Fe and Zn in wheat, which are discussed in greater detail in other chapter of the book.

The bioavailability of these nutrients can be improved by reducing phytic acid content as well as increasing phytase activity in the grain. Therefore, identifying genotypes having low phytic acid content and high phytase activity along with higher Fe and Zn content is necessary before we proceed to improve nutritional quality of wheat. Phytic acid is the primary storage form of phosphorus (P) in seeds accounting for upto 85 per cent of the total seed phosphorus. The negatively charged phosphates in phytic acid bind strongly to metallic cations as K, Mg, Mn, Fe, Ca and Zn to form mixed salt called phytin which reduces bioavailability to humans. Phytin accumulates in seed protein bodies either dispersed or dense inclusions called globoids. The stable complexes of these minerals with phytic acid can lead to micronutrient deficiencies (*e.g.*, Fe and Zn) particularly in third world countries where diets are primarily seed based. Moreover, monogastric animals such as pig, poultry and fish cannot digest phytic acid. As a result animal feeds are supplemented with P, and a large fraction of phytic acid is excreted leading to accumulation in soil and water causing pollution. Development of low phytate crops may improve phosphorus availability in animal feeds and reduce phytic acid excretion, thus lessening the negative effect of animal waste on the environment. Another potential advantage of such crops is in an increase in the availability of Fe and Zn, which would significantly improve human nutrition.

In addition, high phytase levels in wheat can be used for degradation of phytate present in the grain to overcome the problem of micronutrient deficiency in humans and monogastric animals. Significant positive correlation has been reported between native phytase activity and phosphorus utilization (Oloffs *et al.*, 2000) in broilers and micronutrient bioavailability (Lopez *et al.*, 2003) in rats. Different strategies could be applied to optimize phytate degradation during food processing and digestion in the human alimentary tract such as adjustment of more favourable conditions during food processing for naturally occurring phytases in the raw material, addition of isolated phytases to the production process and the use of recombinant food-grade microorganisms as carriers for phytate-degrading activity in the human gastrointestinal tract. In humans, phytate breakdown in the stomach and the intestine is influenced mainly by the dietary phytases (Sandberg and Andersson, 1998) which are active in the human stomach. In addition, there are reports indicating improved bread making quality by higher phytase levels during fermentation (Haros *et al.*, 2001). Therefore, it is needed to see the variability in phytase levels occurring naturally

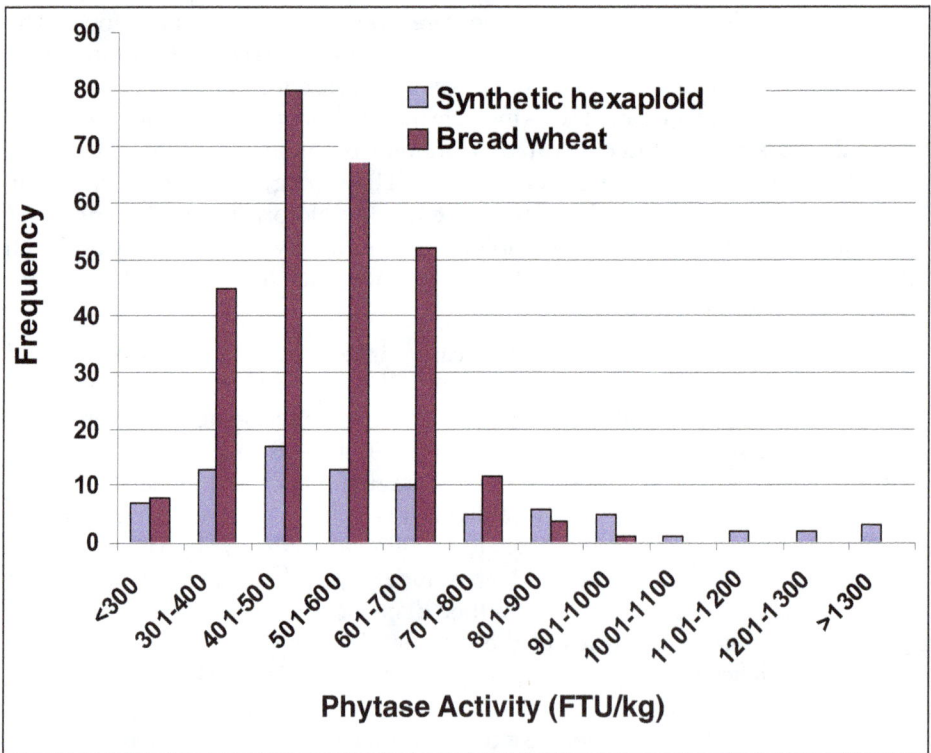

Figure 11.2: Phytase Activity (FTU/kg) of Indian Wheat Varieties, Advanced Lines and Synthetic Hexaploids grown during 2008-2009 (Ram *et al.*, 2011).

in diverse wheat genotypes. In our studies we found 3.4 fold variations in phytase levels among varieties and 5.9 fold among synthetic hexaploids (Figure 11.2). In contrast, lower variability (1.6 fold in varieties and 2.2 fold in synthetic hexaploids) was observed in phytate levels (Ram *et al.*, 2011).

The number of phytases in wheat is still under investigation. Two enzymes, Phy1 and Phy2, have previously been purified from wheat bran and two isozymes with the N-terminal amino acid sequence EPAXTLTGPSRPV have also been purified (Nakano *et al.*, 1999). Based on amino acid sequence and masses of tryptic peptides, a third enzyme with homology to PAP phytases was cDNA cloned from wheat and its homologue from barley (Rasmussen *et al.*, 2007). Recently four cDNAs encoding for MINPPs were also cloned. At least two of these show in vitro phytase activity and they are probably expressed during late seed development and germination (Dionisioet al., 2007). Madsen *et al.* (2013) reported that high mature grain phytase activity in the Triticeae evolved by duplication followed by neofunctionalization of the purple acid phosphatase phytase (*PAPhy*) gene. They characterized the *PAPhy* gene complement from wheat, barley, rye, einkorn, and *Aegilopstauschii* and showed that the Triticeae *PAPhy* genes generally consist of a set of paralogues, *PAPhy_a* and *PAPhy_b* and have been mapped to Triticeae chromosomes 5 and 3, respectively. The promoters share a conserved core but the *PAPhy_a* promoter has acquired a novel *cis*-acting regulatory element for expression during grain filling while the *PAPhy_b* promoter has maintained the archaic function and drives expression during germination.

7. Beta-carotene Contents (Yellow Pigmentation) in Durum Wheat

The genetic improvement of durum wheat involves not only the desirable agronomic traits but also its capability to yield good quality products (Semolina or pasta) with higher nutritional value. Nutritional value of durum wheat depends upon the quantity of beta-carotenes, precursor of vitamin A present in grains. Carotenoids are natural compounds that reduce the oxidative damage to biological membrane by scavenging peroxi-radicals involved in certain human diseases and in the ageing process. Furthermore, carotenoids protect cells and organisms against the harmful effects of light and air. Beta-carotenes have also been implicated in reducing the risk of cancer in human beings. Therefore, an improvement in the endogenous quantity of beta-carotenes should enhance the nutritional value of pasta products. Vitamin A deficiency also increases susceptibility to malaria and diarrhoeal disease, and reduces the bioavailability of micronutrients, including iron, important to the health of all family members (WHO, 2002). Genetically increasing the levels of vitamin A and its precursors in dietary staples is especially relevant for poor families living in isolated rural areas for whom vegetable sources are often in short supply. There are reports in other crops such as rice, maize and sorghum etc. where efforts have been made in manipulating beta-carotene content by molecular techniques; however, little information is available in wheat. Evaluation of thousands of durum wheat germplasm lines from national and international sources as well as advanced entries in coordinated programme indicated the availability of lines with higher beta-carotene content and gluten strength.

Genetic studies showed that wheat endosperm yellow colour is highly heritable (67–90 per cent) and under polygenic control (Parker *et al.*, 1998; Elouafi *et al.*, 2002). Quantitative trait loci (QTL) for endosperm yellow colour accounting for a large proportion of the genetic variation (50 per cent –60 per cent) have been mapped on chromosome arms 7AL and 7BL. Smaller QTLs were detected on chromosomes 2A, 3A, 5A and 5B (Parker *et al.*, 1998; Elouafi *et al.*, 2002). Later BAC clones were identified having genes responsible for carotenoid biosynthesis (Cenci *et al.*, 2004). They reported the isolation of BAC clones containing genes coding for three different enzymes of the carotenoid biosynthesis pathway: phytoene synthase (PSY), phytoene desaturase (PDS) and -Carotene desaturase (ZDS). Primers were designed on the basis of wheat ESTs similar to the sequences of these three genes in other species, and used to screen a BAC library from *Triticum turgidum* var. *durum* ($2n = 28$, genomes AABB). Eight, six and nine 384-well plates containing at least one positive clone were found for PSY, PDS, and ZDS, respectively. BACs selected for each of these genes were then divided in two groups corresponding to the A and B genomes of tetraploid wheat, based on differences in the length of the PCR amplification products, conformation-sensitive gel electrophoresis (CSGE) or cleavage amplification polymorphisms. Positive clones were then assigned to chromosomes using a set of D genome substitution lines in *T. turgidum* var. *durum* 'Langdon'. PSY clones were localized on chromosomes 5A and 5B, PDS on chromosomes 4A and 4B, and ZDS on chromosomes 2A and 2B.

Though wide variability in yellow pigment content has been observed ranging from 3 ppm to >12 ppm in durum wheat, β-carotene content is rather low. β-carotene content can be enhanced specifically using genetic engineering approach by manipulating carotene biosynthetic pathway. This approach has been demonstrated in rice. "Golden Rice 2" has been developed by transforming rice using genes of β-carotene biosynthetic pathway namely *psy* encoding phytoene synthase and *crt1* encoding carotene desaturase taken from maize and *Erwinia uredovora*, respectively. This increased β-carotene content from 0.0 to 37.0 ppm. Recently, phytoene synthase (*psy*) gene has been characterized in wheat and functional marker YP7A closely related to grain yellow pigment content developed. This demonstrates the possibility of enhancing content of carotenoid pigments in wheat grains by both marker assisted breeding as well as by manipulating biosynthetic pathway using genetic engineering.

7.1. Biosynthesis of Yellow Pigments

The process involves the synthesis of a 40 carbon atom chain (C40) backbone, its desaturation, cyclization, and subsequent modifications (*e.g.*, hydroxylation and epoxidation). The first two steps are common for all carotenoid pathways, whereas cyclization can produce βor erings having different properties. Phytoene synthase (PSY) catalyses the reaction making phytoene (C40) from two molecules of geranyl-geranyl pyrophosphate (C20); phytoenedesaturase (PDS) catalyses the double desaturation at the 11–12 and 11'–12'symmetrical positions of phytoene producing z-carotene; and z-carotene desaturase (ZDS) catalyses the reaction producing lycopene by desaturation of 7–8 and 7'–8'symmetrical positions of z-carotene. Double lycopene cyclization can produce β, e or α-carotene, having β – β, e–e, and β –erings, respectively. Subsequent modifications transform α -, β -, and e-carotene into lutein or other xanthophylles.

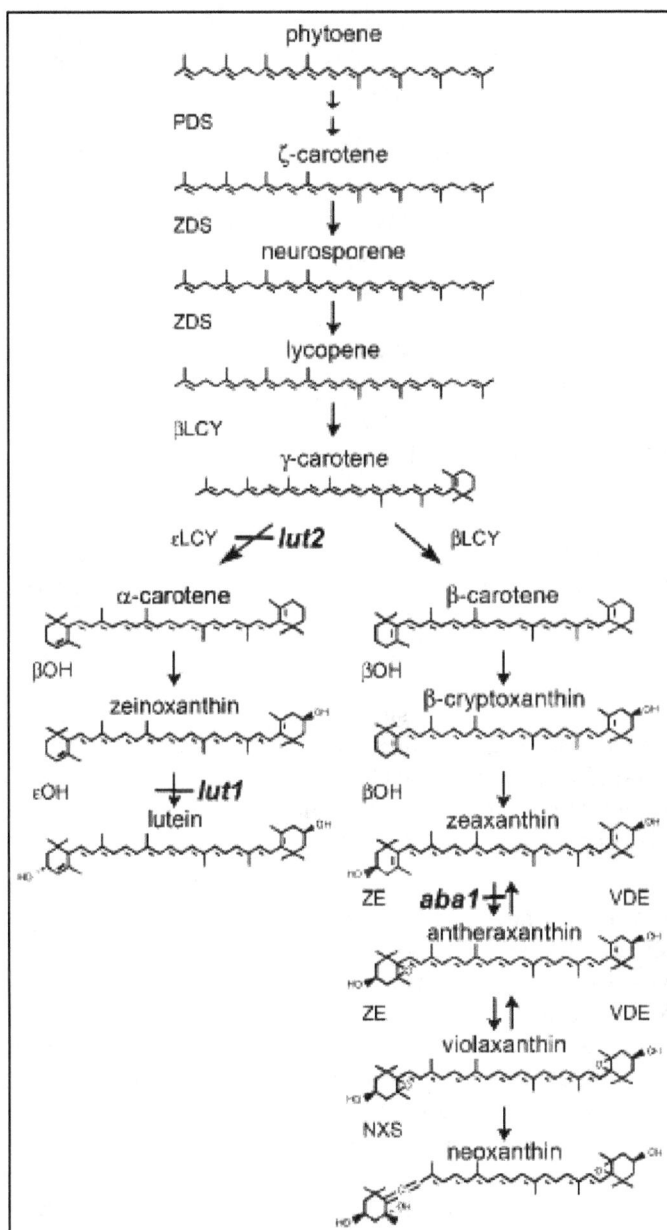

Figure 11.3L Carotenoid Biosynthetic Pathway.

The commonly held view of the carotenoid biosynthetic pathway in plants is a series of four desaturations to form all-*trans*-lycopene from phytoene. Lycopene is subjected to two cyclization reactions to form _α- or _β-carotene, which are modified further to form the various xanthophylls. PDS, phytoenedesaturase; ZDS: Carotene desaturase; βLCY: β-cyclase; LCY: cyclase; βOH: β-hydroxylase; OH: Hydroxylase; ZE: Zeaxanthinepoxidase; NXS: Neoxanthin synthase; VDE: Violaxanthindeepoxidase (Park *et al.*, 2002).

7.2 Lipoxygenase

The content of pigments in the final product depends not only on the natural carotenoid content in the seeds but also by the oxidative degradation by lipoxygenase (LOX). The LOXs are non-heme iron containing dioxygenases that catalyze the oxidation of polyunsaturated fatty acids in plants, animals and microorganisms. Lipoxygenases are involved in lipoperoxidation of membranes and the synthesis of signaling molecules.LOX seems to have role in plant development and senescence, wound responses and pest resistance. In durum wheats, the role of lipoxygenase is due to its involvement in yellow pigment level particularly during the processing of semolina into pasta. Radicals produced during the intermediate steps of this reaction are responsible for oxidative degradation of carotenoid pigments. The optimal pH for hydroperoxidation and bleaching activities are near to dough pH.

Cereal LOX genes have been studied in more detail in barley than in wheat. Two LOX isoenzymes have been purified from germinating barley grains and mapped on barley chromosomes. Wheat LOX isoenzymes were first assigned to chromosomes 4A (*Lpx-A1*), 4B (*Lpx-B1*), 4D (*Lpx-D1*), 5A (*Lpx- A2*), 5B (*Lpx-B2*) and 5D (*Lpx-D2*) using nulli-tetrasomic lines. Mapping studies in wheat showed colinearity of LOX genes between wheat and barley suggesting that they are orthologous. These mapping results indicate that the wheat *Lpx-2* locus corresponds to the barley *LoxC* gene (homoeologous group 5), whereas the wheat *Lpx-1* locus corresponds to barley *LoxA* or *LoxB* locus (homoeologous group 4). Hessler et al. (2002) sequenced several wheat fragments and based on their higher similarity to the barley *LoxA* gene assigned them to the *Lpx-1* locus. Carrera *et al.* (2007) identified wheat sequences on homoeologous group 4 corresponding to the barley *LoxB* gene, which they proposed to designate *Lpx-3*. Two different genes related to the barley *LoxA* on chromosome 4B were also identified and designated as *Lpx-B1.1* and *Lpx-B1.2*.

Verlotta *et al.* (2010) further characterized the *Lpx-B1* gene family in durum wheat germplasm and reported three different haplotypes, I, II and III that are associated with high, intermediate and low LOX activity in mature grains, respectively. Although haplotype I was lost during the breeding of the modern short-straw genotypes, there remains high variability in the modern cultivars for this trait. The findings of this study will allow a precise selection of the genotypes bearing the deleted *Lpx-B1.1* allele, and the fixing of haplotype III in all breeding lines, which will contribute to significant improvements in durum wheat quality. The identification of lines lacking one or more LOX isoenzymes, which can be employed in breeding programmes, could be useful in improving pasta quality. Another approach can be the development of transgenics wheat lines in which the LOX activity is down-regulated by antisense or co-suppression.

8. Summary

Wheat is the main source of energy and nutrition in human diet providing carbohydrates, proteins, vitamins, minerals and fats. Though wheat grain has many components of diet, it is deficient in certain essential nutrients as vitamins A, B12 and vitamin C fats and the amino acid lysine. In addition it has certain anti-nutritional factors which reduce the availability of micronutrients especially Fe and Zn. Since most of the minerals and vitamins are concentrated in outer layers of grain, major part of the nutrients is lost during milling because white endosperm is preferred in baking industry. There are reports of allergy to wheat based products and also enteropathy to wheat gluten. This has led to considerable research in the area of improving wheat nutritional quality by improving lysine content and also reducing anti-nutritional factors. Recently genetic engineering techniques have opened new opportunities for manipulating starch content and structure, lysine content and phytic acid. This may led to enhanced nutritional quality of wheat and making it as complete food.

References

Arruda, P., Kemper, E.L., Papes, F. and Leite, A. (2000). Regulation of lysine catabolism in higher plants. *Trends Plant Sci.* 5, 324–330.

Azevedo, R.A. and Lea P.J. (2001). Lysine metabolism in higher plants. *Amino Acids* 20: 261-279.

Azevedo, R.A., Arruda P, Turner WL, Lea PJ (1997). The biosynthesis and metabolism of the aspartate derived amino acids in higher plants. *Phytochemistry* 46: 395-419.

Betschart, A.A. (1988). Nutritional quality of wheat and wheat foods. In: *Wheat Chemistry and Technology.* Ed. Y. Pomeranz. St. Paul, Minnesota, USA. pp. 91-132.

Bird, A. R., Flory, C., Davies, D. A., Usher, S., Topping, D. L. 2004. A novel barley cultivar (Himalaya 292). with a specific gene mutation in starch synthase IIa raises large bowel starch and short-chain fatty acids in rats. *Journal of Nutrition.* 134, (4): 831-835.

Bohn, L., Josefsen, L., Meyer, A.S. and Rasmussen, S.K. (2007). Quantitative analysis of phytate globoids isolated from wheat bran and characterization of their sequential dephosphorylation by wheat phytase. *Journal of Agricultural and Food Chemistry,* 55(18): 7547-7552.

Bohn, L., Meyer, A.S. and Ramussen, S.K. (2008).Phytate: impact on environment and human nutrition. A challenge for molecular breeding*. *J Zhejiang UnivSci* B. 9(3): 165-191.

Carrera, A., Echenique, V., Zhang, W., Helguera, M., Manthey, F., Schrager, A., Picca, A., Cervigni, G. and Dubcovsky, J. (2007). A deletion at the Lpx-B1 locus is associated with low lipoxigenase activity and improved pasta color in durum wheat (*Triticum turgidum* ssp. *durum*). *Journal of Cereal Science* 45 (2007). 67–77.

Cenci, S. Somma, N. Chantret, J. Dubcovsky, and A. Blanco 2004. PCR identification of durum wheat BAC clones containing genes coding for carotenoid biosynthesis enzymes and their chromosome localization. *Genome* 47: 911–917.

Chhuneja, P, Dhaliwal, H.S., Baines, N.S. and Singh, K. (2006). *Aegil poskotschyi* and *Aegil opstauschii* as sources of higher levels of grain Iron and Zinc. *Plant Breeding*. 125: 529-531.

Dionisio, G., Holm, P.B., Brinch-Pedersen, H. (2007). Wheat (*Triticum aestivum* L.). and barley (*Hordeum vulgare* L.). multiple inositol polyphosphate phosphatases (MINPPs). are phytases expressed during grain filling and germination. *Plant Biotechnology Journal*, 5(2): 325-338.

Elouafi, I., Nachit, M.M., and Martin, L.M. 2002. Identification of a microsatellite on chromosome 7B showing a strong linkage with yellow pigment in durum wheat (*Triticum turgidum* L. var. *durum*). *Hereditas*, 135: 255–261.

Food and Agriculture Organization-World Health Organization (1991). Protein quality evaluation; report of the joint FAO/WHO expert consultation. FAO Food and Nutrition Paper 51, Rome, Italy.

Food and Agriculture Organization-World Health Organization-United Nations University (1985). Energy and protein requirements; report of a joint FAO/WHO/UNU expert consultation. *WHO Tech. Rep. Ser.* 724, WHO, Geneva, Switzerland.

Forsyth, J.L., Beaudoin, F., Halford, N.G., Sessions, R.B., Clarke, A.R., and Shewry, P.R. (2005). Design, expression and characterization of lysine-rich forms of the barley seed protein CI-2. *Biochim.Biophys.Acta* 1747, 221–227.

Gaziola, S.A., Alessi, E.S., Guimarães, P.E.O., Damerval, C. and Azevedo, R.A. (1999). Quality protein maize: a biochemical study of enzymes involved in lysine metabolism. *J. Agric. Food Chem.* 47: 1268-1275.

Gregersen, PL., Brinch-Pedersen, H. and Holm, P.B. (2005). A microarray-based comparative analysis of gene expression profiles during grain development in transgenic and wild type wheat. *Transgenic Research*. 14(6): 887-905.

Greiner, R. and Konietzny, U. (2006). Phytase for food application. *Food Technol. Biotechnol*. 44 (2). 125–140 (2006).

Grusak, M.A. and Penna D.D. (1999). Improving the nutrient composition of plants to enhance human nutrition and health. *Annual Review Plant Physiology and Plant Molecular Biology*. 50; 133-161.

Guttieri, M., Bowen, D., Dorsch, J.A., Raboy, V. and Souza, E. (2004). Identification and Characterization of a Low Phytic Acid Wheat. *Crop Sci*. 44: 418-424.

Harland, B.F., Morris, E.R., 1995. Phytate–A good or a bad component? *Nutrition Research* 15, 733–754.

Haros, M., Rosell, C.M., Benedito, C., 2001. Use of fungal phytase to improve bread making performance of whole wheat bread. *Journal of Agricultural and Food Chemistry* 49, 5450–5454.

Hessler, T.G., Thomson, M.J., Benscher, D., Nachit, M.M., Sorrells, M.E., 2002.Association of a lipoxygenase locus, Lpx-B1, with variation in lipoxygenase activity in durum wheat seeds. *Crop Science* 42, 1695–1700.

Jung, R. and Falco, S.C. (2000).Transgenic corn with an improved amino acid composition. In: *8th International Symposium on Plant Seeds*, Gatersleben, Germany.

Kemper, E.L., Cord-Neto, G., Papes, F., Martinez-Moraes, K.C., Leite, A. and Arruda.P.(1999).The role of opaque-2 on the control of lysine degrading activities in developing maize endosperm. *Plant Cell* 11: 1981-1994.

Lefèvre, A., Consoli, L., Gaziola, S.A., Pellegrino, A.P., Azevedo, R.A. and Damerval.C. (2002).Dissecting the opaque-2 regulatory network using transcriptome and proteome approaches along with enzyme activity measurements.*Scientia Agricola* 59: 407-414.

Lopez, H.W., Krespine, V., Lemaire, A., Coudray, C., Coudray, C.F., Messager, A., Demigne, C., Remesy, C., 2003. Wheat variety has a major influence on mineral bioavailability; studies in rats. *Journal of Cereal Science* 37, 257–266.

Lucca, P., Hurrell, R.F. and Potrykus, I. (2001). Genetic engineering approaches to improve the bioavailability and the level of iron in rice grains. *Theor. Appl. Genet.* 102: 392-397.

Madsen, C.K., Dionisio, G., Holme, I.B., Holm, P.B., and Brinch-Pedersen, M. (2013). High mature grain phytase activity in the Triticeae has evolved by duplication followed by neofunctionalization of the purple acid phosphatase phytase (PAPhy). gene. *Journal of Experimental Botany*, Vol. 64(11), 3111–3123.

Mazur, B., Krebbers, E. and Tingey, S. (1999). Gene discovery and product development for grain quality traits. *Science* 285, 372–375.

Nakano, T., Joh, T., Tokumoto, E. and Hayakawa, T. (1999).Purification and characterization of phytase from bran of *Triticum aestivum* L. cv. *Nourin* #61. *Food Science and Technology Research*, 5(1): 18-23.

Park, H., Kreunen, S.S., Cuttriss, A.J., DellaPenna, D. and Pogson, B.J. (2002). Identification of the carotenoid isomerase provides insight into carotenoid biosynthesis, prolamellar body formation, and photomorphogenesis. *The Plant Cell*, 14: 321–332.

Parker, G.D., Chalmers, K.J., Rathjen, A.J., and Langridge, P. 1998. Mapping loci associated with colour in wheat (*Triticum aestivum* L.). *Theor. Appl. Genet.* 97: 238–245.

Pomeranz, Y. (1998). *Wheat Chemistry and Technology*.American Associaion of Cereal Chemists, St. Paul Minnesota, USA.

Raboy, V. (2002). Progress in breeding low phytate crops. *J. Nutr.* 132: 503S–505S.

Raboy, V., 2007.The ABCs of low-phytate crops. *Nature Biotechnology*, 25(8): 874-875.

Ram, S. and Mishra, B. (2010). *Cereals: Processing and Nutritional Quality*. New India Publishing Agency, New Delhi.

Ram, S., Verma, A.and Sharma, S. (2011). Large variability exists in Phytase levels among Indian wheat varieties and synthetic hexaploids. *Journal of Cereal Science* 52, 486-490.

Rasmussen, S.K., Johansen, K.S., Sørensen, M.B. (2007). Polynucleotides Encoding Phytase Polypeptides. US Patent 10/275,311(7,186,817).

Regina, A., Bird, A., Topping, D., Bowden, S., Freeman, J., Barsby, T., Kosar-Hashemi, B., Li, Z., Rahman, S. and Morell, M. (2006). High-amylose wheat generated by RNA interference improves indices of large-bowel health in rats. *PNAS* 103 (10); 3546-3551.

Sandberg, A.S., Andersson, H., 1998. Effect of dietary phytase on the digestion of phytate in the stomach and small intestine of humans. *Journal of Nutrition* 118, 469-473.

Sarwar G., Peace R.W., Botting H.G. (1985). Corrected relative net protein ratio (CRNPR). method based on differences in rat and human requirements for sulfur amino acids. *J. Assoc. Off. Anal. Chem.* 68: 689-693.

Sarwar, G. (1997). The Protein Digestibility-Corrected Amino Acid Score Method Overestimates Quality of Proteins Containing Antinutritional Factors and of Poorly Digestible Proteins Supplemented with Limiting Amino Acids in Rats. *The Journal of Nutrition*. Vol. 127(5): 758-764.

Shi, J., Wang, H., Schellin, K., Li, B., Faller, M., Stoop, J.M., Meeley, R.B., Ertl, D.S.,Ranch, J.P. and Glassman, K. (2007). Embryo-specific silencing of a transporter reduces phytic acid content of maize and soybean seeds. *Nature Biotechnology*, 25(8): 930-937.

Singh, NK, Donovan, GR, Carpenter, HC, Skerritt, JH and Langridge, P. 1993. Isolation and characterization of wheat triticincDNA revealing a unique lysine-rich repetitive domain. *Plant Molecular Biology*. 22(2): 227-237.

Van den Broeck, H.C., de Jong, H.C., Salentijn, E.M.J., Dekking, L., Bosch, D., Hamer, R.J., Gilissen, L.J.W.J., van der Meer, I.M., Smulders, M.J.M. (2010). Presence of celiac disease epitopes in modern and old hexaploid wheat varieties: wheat breeding may have contributed to increased prevalence of celiac disease. *Theor. Appl. Genet.* 121: 1527–1539.

Verlotta, A., De Simone, V., Mastrangelo, A.M., Cattivelli, L., Papa, R. and Trono, D. (2010). Insight into durum wheat Lpx-B1: a small gene family coding for the lipoxygenase responsible for carotenoid bleaching in mature grains. *BMC Plant Biology*.

Xiang, T., Liu, Q., Deacon, A.M., Koshy, M., Kriksunov, I.A., Lei, X.G., Hao, Q. and Thiel, D.J. (2004).Crystal structure of a heat-resilient phytase from *Aspergillus fumigatus*, carrying a phosphorylated histidine. *Journal of Molecular Biology*, 339(2): 437-445.

Zhu, X and Galili, G. 2004. Lysine metabolism is concurrently regulated by synthesis and catabolism in both reproductive and vegetative tissues. *Plant Physiology*, 135: 129–136.

Zhu, X., Tang, G., Granier, F., Bouchez, D. and Galili, G. 2001. A T-DNA Insertion Knockout of the Bifunctional Lysine-KetoglutarateReductase/Saccharopine Dehydrogenase Gene Elevates Lysine Levels in Arabidopsis Seeds. *Plant Physiology*, 126: 1539–1545.

Recent Advances in Crop Physiology Vol. 2 (2016)   *Pages* **337–356**
Editor: **Dr. Amrit Lal Singh**
Published by: **DAYA PUBLISHING HOUSE, NEW DELHI**

Chapter 12

Nitrogen-use Efficiency and Productivity of Wheat Crop

C. Gireesh[1]*, B.C. Ajay[2], R. Abdul Fiyaz[1],
K.T. Ramya[3] and C. Mahadevaiah[4]

[1]*ICAR-Directorate of Rice Research, Rajendranagar, Hyderabad*
[2]*ICAR-Directorate of Groundnut Research, Junagadh – 362 001, Gujarat*
[3]*ICAR-Directorate of Oilseed Research, Rajendranagar, Hyderabad*

1. Introduction

Nitrogen is the most important nutrient element essential for growth and development of plants. Nitrogen availability is a major constraint on grain production and the main component of the carbon footprint of cereal production. Large amount of nitrogen fertilizers are being used to increase the crop production. Nitrogen fertilizers are used as an important agronomic tool to improve output, quantity as well as quality in all cultivated crops. However, current agricultural and economic environmental concerns require farmers to constantly optimize the application of nitrogen fertilizers throughout the growing season to avoid pollution by nitrates while preserving their economic margin. Over the past four decades, the doubling of agri-cultural food production worldwide has been associated with a seven fold increase in the use of nitrogen fertilizers (Hirel *et al.*, 2007).

Excessive use of fertilizer increases the cost of production and causes ecological risks like nitrate leaching, volatilized ammonia and greenhouse gas emission (nitrous oxides) (Galloway *et al.*, 2008; Hirel *et al.*, 2007). It has been estimated that nitrogen fertilizer accounts for more than 70 per cent of the greenhouse gases (GHGs) associated

* *Corresponding Author:* E-mail: giri09@gmail.com

with the production of wheat (Mortimer *et al.*, 2004). Therefore, efficient use of available nitrogen reduces the cost of production, reduces the harmful impact on environment and increases the yield per unit nitrogen use. Given the cost of N fertilizer and the negative environmental impact of high fertilizer use, efficient utilization of nitrogen resources is vital to sustainable cereal production.

Breeding new crop plant cultivars with improved productivity in low nitrogen environments provided an effective approach to increase nitrogen use efficiency (Rengel and Marsch-ner, 2005). Thus, breeding for cultivars that absorb and metabolize nitrogen most efficiently for grain or silage production is becoming increasingly important. Introduction of semi-dwarf wheat varieties which were responsive to nitrogen fertilizer application during 1970's saw an extensive use of fertilizers (Austin, 1999) but the present agricultural system demands improved productivity and profits with reduced inputs. It is well established fact that semi-dwarf wheat varieties have better yield potential over tall varieties due to better partitioning of nutrients to grain at the expense of straw. Semi-dwarf varieties have better grain yield potential for the same total N-uptake and total dry matter production *i.e.* greater grain-NutE (Fischer and Wall, 1976). Such crops would make better use of nitrogen fertilizer supplies giving higher yields with improved protein contents. Cultivars with increased NUE will be of economic to farmers and help reduce environmental contamination associated with excessive inputs of N fertilizers. There is increasing emphasis worldwide in breeding wheat cultivars with NUE (Ortiz-Monasterio *et al.*, 2001; Guarda *et al.*, 2004; LaPerche *et al.*, 2006; Hirel *et al.*, 2007; Li *et al.*, 2008).

Development of Nitrogen use efficient cultivars requires a better understanding of nitrogen metabolism and its regulation, and identification of target genes to monitor N uptake by either direct gene transfer or marker-assisted breeding. Nitrogen uptake is an essential element in crop improvement, either directly for grain protein content or indirectly for photosynthetic production. Nitrogen movement within the plant life cycle may be conveniently divided in two phase's i.e vegetative and reproductive phases. During vegetative phase young developing roots and leaves act as a sink for assimilation of inorganic N to produce amino acids (Hirel and Lea, 2001). During reproductive stage remobilization of accumulated N takes place and shoots and/or roots behave as a source by providing amino acids which are later exported to reproductive and storage organs represented, such as seeds, bulbs, or trunks (Masclaux *et al.*, 2001). In cereal crops such as rice and wheat 60 to 95 per cent of grain N comes from the remobilization of N stored in roots and shoots before anthesis (Palta and Fillery, 1995; Mae, 1997; Habash *et al.*, 2006; Tabuchi *et al.*, 2007). In oilseed crops amount of N taken up by plant during grain filling is low (Rossato *et al.*, 2001) and as a result large quantities of N absorbed by plant and stored in vegetative organs is not used and lost in early falling leaves (Malagoli *et al.*, 2005)

2. Defining Nitrogen Use Efficiency

Nitrogen use efficiency (NUE) can be defined as the grain dry matter yield divided by the supply of available nitrogen from the soil and fertilizer (Moll *et al.*, 1982). Nitrogen-use efficiency is dependent on efficient uptake of nitrogen by crop per unit nitrogen available from soil and fertilizer (known as nitrogen uptake efficiency) and accumulation of nitrogen by crop in the form of grain and dry matter yield (nitrogen utilization efficiency). Hence, there are two major components of NUE: 1) uptake efficiency (UPE = N in plant/N applied), which is important in low to moderate N environments and 2) utilization efficiency (UTE = N in grain/N in plant), which is most important in high N environments. Environments with moderate N levels appear suited for both components. Therefore, both nitrogen uptake efficiency and nitrogen utilization efficiency play important role in achieving the nitrogen-use efficiency in crop plants.

Identification of metabolic and physiological traits to increase NUE that may be transferred into elite varieties through breeding is therefore imperative. Promising approaches of nitrogen management strategies are also essential to increase NUE. Candidate metabolic and physiological traits with potential for improving NUE and for reducing excessive input of fertilizers, while maintaining an acceptable yield, is essential aspect of breeding for NUE. Improvement of traits through breeding will be most efficiently applied at the farm level when N-efficient cultivars are combined with N management strategies for high NUE.

3. Genetic Characterization of NUE

Characterizing and quantifying recent genetic progress bring meaningful information to researchers. The first decision that breed-ers have to take is to choose the N level for which they want to breed. Indeed, in numerous studies which analysed agronomic traits, significant genotype × N (G × N) inter-actions were detected (*e.g.* Le Gouis *et al.*, 2000; Laperche *et al.*, 2006a; Barraclough *et al.*, 2010), meaning that differential behaviour of variety depends on N treatment. Quanti-fying G × N interactions is, therefore, crucial for efficient selection. The identification of traits to improve NUE in wheat and the characterisation of their variability provide useful direc-tions to breeders (*e.g.* Barraclough *et al.*, 2010; Foulkes *et al.*, 2009; Gaju *et al.*, 2011).

High grain yield with adequate protein content is an important goal in crop improvement, especially for bread wheat (*Triticum aestivum* L.). Unfortunately, it has been shown in various cereals, that these two traits are genetically negatively correlated in extensive (Simmonds, 1995; Oury *et al.*, 2003), although this correlation can be broken down by adequate nitrogen (N) supply late in plant development (Krapp *et al.*, 2005; Laperche *et al.*, 2006). More-over, it is well known that a negative correlation between yield and protein content exists in wheat (Kibite and Evans, 1984; Simmonds 1995, Oury *et al.*, 2003; Oury and Godin, 2007; Bogard *et al.*, 2010). A yield increase may, therefore, lead to a decrease in protein content which could cause lower end-use quality (Ortiz-Monasterio *et al.*, 1997b; Shewry, 2004). Thus, the question of the genetic improve-ment in yield or NUE cannot be assessed independent of quality. Studies conducted at different N lev-els have concluded that genetic progress occurred in both HN and LN conditions, but was higher at HN (Ortiz-Monasterio *et al.*, 1997a; Brancourt-Hulmel *et al.*, 2003; Guarda *et al.*, 2004). Fewer studies have been published on the genetic pro-gress for NUE and its components (Ortiz-Monasterio *et al.*, 1997a; Guarda *et al.*, 2004; Muurinen *et al.*, 2006).

Table 12.1: NUE Definitions and their Units (Foulkes et al., 2009)

Name	Definitions	Units
Nitrogen use efficiency	Kilogram (grain dry mass) at harvest per kilogram available nitrogen (from soil plus fertilizer)	kg kg^{-1}
Nitrogen-uptake efficiency (UPE)	Kilogram (above-ground nitrogen) at harvest per kilogram available nitrogen (from soil plus fertilizer)	kg kg^{-1}
Nitrogen-utilization efficiency	Kilogram (grain dry mass) per kilogram (above-ground N) at harvest	kg kg^{-1}
Nitrogen harvest index (NHI)	Proportion of above-ground nitrogen in the grain at harvest	Unitless
Above-ground nitrogen uptake (AGN)	Gram (above-ground nitrogen) per square metre (ground) at harvest	g m^{-2}
Grain nitrogen concentration (GNC)	Gram (grain nitrogen) per gram (grain dry mass) at harvest X 100	per cent
Root length density (RLD)	Centimetre (root) per cubic centimetre (soil)	cm cm^{-3}
Green area index (GAI)	Square metre (green area) per square metre (ground)	m^2 m^{-2}
Specific leaf nitrogen content (SLN)	Gram (leaf nitrogen) per square metre (leaf area)	g m^{-2}
Radiation-use efficiency (RUE)	Gram (above-ground biomass) per megajoule intercepted global radiation	g MJ^{-1}
Light-saturated CO$_2$ exchange rate (A$_{max}$)	Micromoles (CO$_2$) per square metre (leaf area) per second in light-saturated conditions	μmol m^{-2}s^{-1}
Nitrogen remobilization efficiency (NRE)	Proportion of nitrogen in the whole plant or organ at anthesis which is not recovered in the straw at harvest	Unitless

Various studies worldwide have identified genetic associations between grain yield and NUE components under contrasting conditions of high and low N input supply. In general, these studies indicated that UPE accounts for more of the genetic variation in NUE at low N than at high N supply (Dhugga and Waines, 1989; Ortiz-Monasterio *et al.*, 1997; LeGouis *et al.*, 2000; Muurinen *et al.*, 2006). Few studies have attributed genetic gains in NUE to UTE (Brancourt- Hulmel *et al.*, 2003; Foulkes *et al.*, 1998). Modern cultivars normally had higher yields than old cultivars under low N input conditions (Ortiz-Monasterio *et al.*, 1997; Foulkes *et al.*, 1998; Brancourt-Hulmel *et al.*, 2003). Genetic variation for NUE has been noted in rice (Borrell *et al.*, 1998), wheat (Le Gouis *et al.*, 2000) and barley but improvements in NUE have been limited by expensive and laborious phenotyping (Clark, 1983; Le Gouis *et al.*, 2000; Wang *et al.*, 2011). This variability could help us understand the genetic basis of NUE.

Two major approaches are used to assess genetic pro-gress: (1) historical trial analyses and (2) direct com-parison of old and modern varieties in the same envi-ronment. But these two approaches suffer from some limitations. (1) When historical trials are analysed, as genotypes are tested in different year × environment combinations, there is a need to take into account agro-climatic variation. This may induce bias as elimination of 2yeary effects is often based on variation from year-to-year of common controls leading to inadequate consid-eration of genotype × yeary interactions (Oury *et al.*, 2012; Graybosch and Peterson, 2012). (2) Direct comparisons of old and modern varie-ties are often limited by the experiment size (Bran-court-Hulmel *et al.*, 2003; Guarda *et al.*, 2004; Muurinen *et al.*, 2006; Green *et al.*, 2012) with few genotypes stud-ied in few environments. This can cause sampling errors. Lopez *et al.* (2012) proposed to base genetic progress assessment only on the highest yielding variety per date of release but still with a quite low number of cultivars. Moreover, the period under study is usually spread out and includes major changes in plant height due to intro-duction of dwarfing alleles. Indeed, height decrease is one of the major sign of winter wheat genetic improve-ment between 1946 and 1992 in France (Brancourt-Hul-mel *et al.*, 2003) as well as other countries (Ortiz- Monasterio *et al.*, 1997a; Austin, 1999). It is directly linked to NUE through an increase in lodging resistance and nitrogen partitioning (Hedden, 2003). Plant height is now stabilised; therefore, the question of recent genetic gain can be asked independently of this major physiolog-ical change using a large panel of recent cultivars grown in the same environments.

Quantitative trait locus (QTL) analysis has proved to be an effective approach to dissect a complex quantitative trait into component loci to study their relative effects on the trait (Doerge, 2002). Nowa-days, QTL mapping has become a routine procedure for the identification of genomic regions harboring the genes which control polygenic traits (Saal *et al.*, 2011). Several QTL experiments have been conducted in wheat to study N use efficiency under different N levels in hydroponic cul-ture (An *et al.*, 2006; Guo *et al.*, 2012; Laperche *et al.*, 2006), pot trails (Habash *et al.*, 2007) and field trials (An *et al.*, 2006; Fontaine *et al.*, 2009; Laperche *et al.*, 2007; Quarrie *et al.*, 2005), or to identify QTLs for phosphorus (P) use efficiency in P sufficient and limited conditions (Li *et al.*, 2007b; Su *et al.*, 2006, 2009). Several QTL analyses have been performed during the last two decades on barley (Kjær *et al.*, 1995), maize (Agrama *et al.*, 1999;

Bertin and Gallais, 2001; Hirel *et al.*, 2001), rice (Obara *et al.*, 2001; Lian *et al.*, 2005), wheat (An *et al.*, 2006; Habash *et al.*, 2007; Laperche *et al.*, 2007; Fontaine *et al.*, 2009), and Arabidopsis thaliana (Rauh *et al.*, 2002; Loudet *et al.*, 2003). Xu *et al.*, 2014 reported three major QTLs for NUE in wheat located on chromosomes 2D, 4B and 6A which were coincident with *Rht8*, *Rht-B1b* and *TaGW2*, respec-tively.

Genes encoding major enzymes have been cloned and shown to drive N economy in plants (Miflin and Habash, 2002; Bernard and Habash 2009). Eleven major chromosomal regions controlling NUE in wheat that co-localise with key developmental genes such as Ppd (photoperiod sensitivity), Vrn (vernalization requirement) and Rht (reduced height) have been identified in wheat (Quraishi *et al.*, 2011). Physical mapping, sequencing, annotation and candidate gene validation of an NUE metaQTL on wheat chromosome 3B allowed them to propose that a glutamate synthase (GoGAT) gene that is conserved structurally and functionally at orthologous positions in rice, sorghum and maize genomes may contribute to NUE in wheat and other cereals.

4. Role of Glutamine Synthetase and Glutamate Dehydrogenase in Nitrogen Assimilation

During the growth and development of plants, nitrogen is moved into and out of proteins in the different organs and transported between organs in a limited number of transport compounds. Some of the organic nitrogen is moved between compounds via the activity of transaminases and glutamine-amide transferases, but a significant portion is released as NH_3 and re assimilated via Glutamine synthetase (GS) which plays a central role in nitrogen metabolism. Glutamine synthetase (GS; EC 6.3.1.2) was first purified and characterized from plants in 1956. One particular characteristic is its high affinity for ammonia and thus its ability to incorporate ammonia efficiently into organic combination. However, the discovery of NAD(P)H glutamate synthase in bacteria (Tempest *et al.*, 1970) and later ferredoxin-dependent glutamate synthase in plants (Lea and Miflin, 1974) established a route, the glutamate synthase cycle, for NH_3^{2-} to enter into organic compounds via its assimilation by GS. Evidence based on labelling kinetics, use of inhibitors, in organello studies, and genetics established that this was the major route of primary nitrogen assimilation in plants (Miflin and Lea, 1980).

The GS is the first key enzyme for N metabolism, as it catalyses assimilation of all inorganic nitrogen incorporated into organic compounds, such as proteins and nucleic acids. This reaction is coupled to the formation of glutamate by glutamate synthase (GoGAT) as part of the GS/GoGAT cycle. Although ferredoxin-GoGAT plays a critical role in the re-assimilation of ammonium released by glycine decarboxylase during photo-respiration, NADH-GoGAT assimilates ammonium from both primary and secondary sources during nitrogen remobilization (Lea and Miflin, 2003). Genes coding for these two key enzymes of NH_4 assimilation has been cloned in monocots [rice (Tabuchi *et al.*, 2007; Cai *et al.*, 2009), wheat (Caputo *et al.*, 2009), maize (Valadier *et al.*, 2008)] and Arabidopsis (Ishiyama *et al.*, 2004; Potel *et al.*, 2009), Brassicaceae (Ochs *et al.*, 1999), Medicago truncatula (Lima *et al.*, 2006).

The GS lies at the intersection of C and N metabolic pathways, and its manipulation in wheat plants could potentially raise NUE through more efficient internal recycling of N from older to new leaves. During late leaf senescence NH_3 accumulates in leaves and is liable to be lost from the plant by volatilization and high GS activity reduces these losses (Mattsson *et al.*, 1998). Two major forms of GS exist, comprising: (i) up to five cytosolic isoforms of GS (*GS1*) in root and shoot cells and (ii) a plastidic isoform, GS (*GS2*), in the chloroplasts of photosynthetic tissues and in the plastids of roots and other non-photosynthetic tissue. The *GS1* isoforms are differentially expressed in different plant tissues and specific isoforms appear to play a major role in the synthesis of glutamine in senescing leaves for transport to new tissues. Increases in the activities of *GS1* during leaf senescence have been reported in many plant species (Feller and Fischer, 1994). Masclaux *et al.* (2001) observed a positive relationship between GS1 activity and UTE and grain yield of maize RILs grown under low N. Similarly, Hirel *et al.* (2001) found positive correlations between leaf nitrate content, *GS1* activity and yield in maize RILs under low N, and co-incidental location of QTLs for NUE and a structural gene for GS1. Adding an extra GS gene to wheat altered the balance of *GS1* and *GS2* activity in flag leaves of plants (Habash *et al.*, 2001), and in pot trials one such transformed line showed a fourfold increase in leaf *GS1* activity and had more roots, more grain and higher grain N. Two cytosolic *GS1* isoforms have been shown to have major impacts on grain production in maize (Martin *et al.*, 2006) whilst a deletion in single specific *GS1* isoform had a strong impact on growth and grain filing in rice (Tabuchi *et al.*, 2005). QTLs for grain N but not yield were found to map with GS genes in wheat (Habash *et al.*, 2007).

Most cultivar testing systems worldwide use a relatively high N supply, which means that plant breeders select cultivars that perform well under these conditions. This means that cultivars selected for high yield under high N conditions may not be the highest yielding under lower N conditions. In order to reliably identify which cultivars will yield well at lower N supplies will requires both breeding and testing at low levels of N input (Brancourt-Hulmel *et al.*, 2005) or an understanding of which plant traits are important for yield under these conditions.

One of the key traits that could explain the genetic variability associated with yield variation under low N is leaf senescence kinetics. The capacity of a genotype to retain green leaf area for longer than a standard genotype during grain-filling has been referred to as the 'stay-green' phenotype (Thomas and Smart, 1993). Stay green genotypes may have an increased source : sink ratio during grain filling with an increased fraction of N derived from the soil N. Using stay green genotypes under low N was associated with improved performance in both sorghum (Borrell and Hammer, 2000) and maize (Mi *et al.*, 2003). Genetic variation in functional stay-green lines has been reported in wheat (Silva *et al.*, 2000; Verma *et al.*, 2004; Foulkes *et al.*, 2007), although the underlying physiological mechanisms have not been studied extensively. Christopher *et al.* (2008) found that the stay-green phenotype in spring wheat *SeriM82* was associated with extraction of deep soil water in Australia. In summary, the maintenance of a photosynthetically active canopy in the post-anthesis period under low to moderate N supply should optimize the efficiency of N input use for grain.

5. Traits of NUE

Promising traits and their measurement methodologies for selection to increase NUE are:

☆ Anthesis date (AD)

☆ Crop dry mass and N content

☆ The concentration of N in the straw and ears.

☆ Nitrogen Nutrition Index (NNI) could be estimated according to the ratio of the actual above-ground crop N per cent at anthesis and the critical N per cent (N per cent ct), where N per cent ct was estimated according to the 'critical dilution curve' described by Justes *et al.* (1994).

☆ The above-ground N and DM per unit ground surface area

☆ Number of ears m^{-2} changes

☆ Crop height measured 14 d after the end of anthesis, estimates of grain number per unit of ground surface area (GN), ear number per unit of ground surface area, grain number per ear and DM of straw, chaff and grain per unit of ground surface area were calculated.

☆ The harvest index (HI) and N harvest index (NHI) calculated as the proportion of DM and N in the above-ground plant which is in the grain at harvest, respectively.

☆ The nitrogen-use efficiency (NUE) was calculated by dividing the DM grain yield (kg DM ha^{-1}, 0 per cent moisture content) by the amount of N available to the crop from the soil and fertilizer (kg N ha^{-1}).

☆ N × cultivar interaction

☆ The N-uptake efficiency was calculated by dividing the aboveground N at harvest (kg N ha^{-1}) by the amount of N available to the crop from the soil and fertilizer (kg N ha^{-1}); and the N-utilization efficiency was calculated by dividing the machine-harvested DM grain yield (kg DM ha^{-1}, per cent moisture content) by the above-ground N at harvest (kg N ha^{-1}). The crop N remobilization efficiency (NRE) was calculated as: $\{AGN_H -(AGN_H - GN)\}/ AGNA$, where AGNA is the above-ground N at anthesis (kgNha^{-1}), AGNH (kg N ha^{-1}) is the above-ground N at harvest and GN (kg N ha^{-1}) is the grain N at harvest (kg N ha^{-1}).

☆ Lodging was defined as the stems leaning at an angle of at least 45° from the vertical. Close to the time of harvest the proportion of the plot area that was lodged was visually assessed to the nearest 0.05.

☆ Senescence kinetics of the flag leaf of five main stems at CF or of the whole canopy at other sites were assessed visually by recording the percentage green area senesced using a standard diagnostic key based on a scale of 0–10 (100 per cent senesced).

☆ Increased root length density (RLD) at depth.

☆ High capacity for N accumulation in the stem, is associated with high N-uptake rate.

☆ Low leaf lamina N concentration,

☆ More efficient post-anthesis remobilization of N from stems to grain, but less efficient remobilization of N from leaves to grain, both potentially associated with delayed senescence

☆ Reduced grain N concentration may be of particular value for increasing NUE in wheat cultivars used for feed purpose; and

☆ Physiological traits relevant for nitrogen-use efficiency: Root length density at depth, GS activity, Alanine aminotransferase (AlaAT) activity, RuBisCo CO_2 specificity factor, Introduction of C_4 'Krantz' anatomy into C_3 species, specific leaf N content, vertical N distribution with leaf layer, leaf posture, leaf photosynthetic rate post-anthesis, stem N storage, stay-green, N remobilization, efficiency post-anthesis and post-anthesis N uptake.

6. Association of NUE with yield

In a number of studies there was inverse relationships between grain yield and grain N concentration ((grain N/grain DM)*100, GNC) (Kibite and Evans, 1984; Triboï *et al.*, 2006). Assuming a constant N harvest index (the proportion of above-ground N (AGN) at harvest in the grain; NHI), grain yield will be positively associated with AGN and/or negatively associated with GNC. Various CSR indices show high correlations with barley and wheat grains yield, biomass and N concentration and are already used in N management strategies. The CSR estimates of biomass and N content can be combined with yield and grain protein to estimate UPE and UTE. The association of CSR with biomass is particularly important, as biomass is a key component of NUE and yield, it is expensive to measure, and harvest index is approaching its theoretical limit, restricting future yield gains to increases in biomass. Under low N supply, genetic gains in yield were positively associated with AGN and NHI and negatively associated with GNC (Foulkes *et al.*, 1998; Ortiz-Monasterio *et al.*, 1997a; Brancourt- Hulmel *et al.*, 2003). Whereas under high N supply, genetic gains in yield were positively associated with AGN (Ortiz-Monasterio *et al.*, 1997a; Brancourt-Hulmel *et al.*, 2003; Foulkes *et al.*, 1998), positively (Ortiz-Monasterio *et al.*, 1997a; Brancourt-Hulmel *et al.*, 2003) or neutrally (Foulkes *et al.*, 1998) associated with NHI, and negatively (Ortiz-Monasterio *et al.*, 1997a; Brancourt-Hulmel *et al.*, 2003) or neutrally (Foulkes *et al.*, 1998) associated with GNC. Essentially, under both low and high N supply, wheat crops with higher NUE compared to current cultivars will require an increase in UPE to raise AGN and/or an increase in UTE. The latter may be achieved by raising NHI and/or lowering GNC. There have been relatively few attempts to use 'direct selection' breeding to improve NUE in wheat. CIMMYT in Mexico adopted a strategy to select for grain yield in medium-to-high fertility conditions, since at this fertility level both UPE and UTE contribute to the observed variation in NUE, resulting in lines which were more N-efficient (Ortiz-Monasterio *et al.*, 1997a). More recently, it has been suggested that this method of selection may not be as efficient as selecting lines under alternating high-low N selection regimes commencing with high N in the F_2 (Van Ginkel *et al.*, 2001).

7. Root Traits Influencing N-uptake

The factors affecting the N uptake efficiency are root axis number, rooting depth, rooting density and root longevity. There is a Root membrane N transporter systems and root proliferation which has a direct relation to N status. Two distinct gene families of nitrate transporters, NRT1 and NRT2, have been identified (Crawford and Glass, 1998; Forde, 2000; Forde and Clarkson, 1999; Williams and Miller, 2001; Hawkesford and Miller, 2004) in the Arabidopsis genome. Some members of both NRT1 and NRT2 gene families are nitrate inducible, and are expressed in the root epidermis and in root hairs, and are likely to be responsible for the uptake of nitrate from the soil (*e.g.,* Lauter *et al.,* 1996; Zhuo *et al.,* 1999; Ono *et al.,* 2000; Orsel *et al.,* 2002). Many plant ammonium transporter (*AMT*) genes, which complement a yeast mutant deficient in normal ammonium uptake, have been identified (Von Wire´n *et al.,* 2000). There are six *AMT* genes in Arabidopsis, while rice has 10, with two types being distinguishable on the basis of sequence but both conferring high affinity uptake. Like the nitrate transporters, some *AMT* genes are expressed in root hairs (Lauter *et al.,* 1996; Ludewig *et al.,* 2002). Manipulation of N transporters in wheat plants to give higher rates of nitrate transport across the plasma membranes of root cells, thereby improving nitrate and ammonium capture from the soil. Signalling and root proliferation in relation to N status in Arabidopsis (Zhang *et al.,* 1999; Walch-Liu *et al.,* 2006; Rogg *et al.,* 2001; Casimiro *et al.,* 2003; Remans *et al.,* 2006). There are again prospects for transferring this information to wheat for improving UPE in the long term if the root screens used for Arabidopsis could be adapted to the larger and structurally different root system of wheat.

8. Traits Influencing N-utilization

N-utilization efficiency (NUTE) is dependent on the N efficiency of biomass formation, the effect of N on carbohydrate partitioning, nitrate reduction efficiency and remobilization of N from senescent tissues as well as storage functions (Novoa and Loomis, 1981; Good *et al.,* 2004; Lea and Azevedo, 2007; Hirel *et al.,* 2007).

8.1 Nitrate Assimilation

Absorbed nitrate is reduced to nitrite in the cytosol by the enzyme nitrate reductase (NR). Nitrite is transported to the plastid or chloroplast and reduced to ammonium by the enzyme nitrite reductase (NiR). Ammonium is then converted to glutamine and glutamate in the plastid chloroplast by the glutamine synthetase/glutamate synthase (GS/GOGAT) enzyme system (Miflin, 1980; Ireland and Lea, 1999).

Nitrate reductase was long considered to be the bottleneck in nitrate assimilation and was thought to be pivotal in the growth response to N supply. However, numerous studies altering the expression of NR-encoding genes have resulted in no change to plant growth (Crawford, 1995). Studies of maize (Masclaux *et al.,* 2001) and wheat (Kelly *et al.,* 1995) have also failed to find correlations between leaf NR activity and yield. It appears therefore that N assimilation and yield are not generally limited by the level and activity of the NR/NiR enzymes.

8.2 Manipulating Photochemistry to Improve NUTE

Genetic progress in conversion of N into grain yield could be enhanced by improving the efficiency of fixation of CO_2 (Parry *et al.*, 2007; Long *et al.*, 2006). The key photosynthetic enzyme ribulose-1,5-bisphosphate carboxylase/oxygenase (RuBisCo) typically constitutes up to 30 per cent of the total N in wheat leaves (Lawlor, 2002) and under field conditions leaf photosynthetic rate is highly correlated with RuBisCo content (Hudson *et al.*, 1992; Lauerer *et al.*, 1993). However, the RuBisCo of thermophilic red algae, *e.g. Galderia partita*, is up to three times more efficient than those of C_3 cereals due to greater specificity for CO_2 (Uemura *et al.*, 1997). There are therefore long term prospects of boosting UTE by introducing forms of RuBisCo with greater specificity for CO_2 from red algae into wheat plants, thereby reducing photorespiration.

8.3 Post-anthesis N Remobilization

During grain filling the N stored as protein in the vegetative tissues becomes important as root N uptake falls short of the grain N demand. Bread-making wheat cultivars in particular require efficient translocation of N during grain filling. Genetic variation in N remobilization efficiency of the vegetative tissues has been reported in wheat in the overall range 0.52–0.92 (Cox *et al.*, 1986; Van Sanford and Mackown, 1987; Papakosta and Garianas, 1991; Barbottin *et al.*, 2005; Tahir and Nakata, 2005), and the genotype effect has been shown to depend on year (Przulj and Momcilovic, 2001) and N fertilization level (Cox *et al.*, 1986; Papakosta and Garianas, 1991).

8.4 The Stay-Green Trait

Several investigations have concluded that the genetic control of N remobilization seems likely to be involved in the regulation of leaf senescence (Sinclair and De Wit, 1975; Masclaux *et al.*, 2001), and an improved understanding of N remobilization could be important in developing cultivars with stay-green properties. Although under optimal conditions wheat crops are in general little limited by the assimilate supply during grain filling (Dreccer *et al.*, 1997; Borra´ s *et al.*, 2004; Calderini *et al.*, 2006), under low to moderate N fertilizer levels yields may be more limited by postanthesis assimilate supply. The capacity of a genotype to retain green leaf area for longer than a standard genotype during grain filling has been referred to as the 'stay-green' phenotype (Thomas and Smart, 1993). In the grasses, such as Festuca pratensis, stay green mutant lines have been identified (Thomas *et al.*, 2002) in which the phenotype is non-functional (delayed senescence not associated with extended photosynthesis). One stay-green mutation of this type was located in the phaeophorbide an oxygenase gene (Thomas *et al.*, 2002), increasing the stability of the light harvesting and reaction-centre thylakoid membrane proteins during senescence. Genetic variation in functional stay-green (delayed senescence associated with extended photosynthesis) lines has been reported in bread wheat (Silva *et al.*, 2000; Verma *et al.*, 2004; Foulkes *et al.*, 2007), although the underlying physiological mechanisms have not been studied extensively. Christopher *et al.* (2008) found that the stay-green phenotype in spring wheat SeriM82 was associated with extraction of deep soil water in Australia. More studies have been carried out on the mechanisms underlying genetic variation in stay-green in sorghum. Nitrogen dynamics are an important factor in the maintenance of green leaf area in sorghum, with stay-green in sorghum hybrids linked to changes in the balance between N demand and supply during grain filling resulting in a slower rate of N translocation from the leaves to the grain compared with senescent genotypes (Borrell and Hammer, 2000). Stay-green mutants have also been identified in durum wheat (*Triticum turgidum* spp. durum) (Spano *et al.*, 2003), with delayed senescence being correlated with a higher rate and duration of grain filling. A transcription factor (NAM-B1) accelerates senescence and increases N remobilization from leaves to grains in emmer wheat (an ancient cultivated tetraploid species, *T. turgidum* ssp. dicoccoides), whereas modern durum wheat varieties carry the non-functional NAM-B1 allele (Uauy *et al.*, 2006).

8.5 Optimizing Grain Protein Content and Composition

Grain proteins can be broadly divided into structural/metabolic and storage proteins (Shewry and Halford, 2002). The embryo and outer layers of the grain (including the aleurone) contain about 30 per cent of the total grain N. They are enriched in albumins, globulins and insoluble proteins, most of which are structural and metabolic in function, but both tissues also store a 7S globulin protein. Structural and metabolic proteins are also present in the starchy endosperm cells, but the predominant protein fraction in this tissue is the gluten storage proteins, comprising a mixture of monomeric gliadins and polymeric glutenins. These groups of proteins are present in approximately equal amounts and together account for about 60–70 per cent of the total N in the endosperm tissue. The gluten proteins are crucial for the processing of wheat into bread, other baked food, pasta and noodles, as they confer viscoelastic properties to dough. A precise balance of gliadin and glutenin proteins is also required, as glutenins are predominantly responsible for dough elasticity (strength) and gliadins for dough viscosity and extensibility. Thus, highly elastic (strong) doughs are required for bread making and more extensible doughs for making biscuits and cakes.

Cultivars for bread making are selected for high protein content and strong gluten properties with appropriate levels of N fertilizer being applied to the crop to ensure that the required protein content is achieved. For example, the Chorleywood Bread making Process, which is the predominant process in the UK and a number of other countries, requires wheat with a minimum protein content of 13 per cent on a dry weight basis. Although "high protein" genes have been reported in wheat, there is in fact limited variation between the grain protein contents of elite wheat cultivars grown under similar agronomic conditions.

9. Summary and Conclusion

Nitrogen is the important nutrient element required for crop growth and glutamine sythetase plays an important role in its assimilation in plants. Increased cost of fertilizers and environmental concerns has forced us to use fertilizers more judiciously. Semi-dwarf rice and wheat varieties released during green revolution possessed higher nitrogen utilization efficiency as they had an ability to translocate most of the absorbed nitrogen to the developing grains. Hence, breeding cultivars with that absorb and use nitrogen efficiently is becoming important. Present day cultivars of cereals such as rice and wheat possess better N utilization efficiency than old cultivars and further improvement of N uptake efficiency under low N conditions would enhance NUE of the present cultivars.

References

Agrama, H.A.S., Zakaria, A.G., Said, F.B. and Tuinstra, M.R. (1999). Identication of quantitative trait loci for nitrogen use efciency in maize. *Mol. Breed.* 5, 87–195.

An DG, Su JY, Liu QY, Zhu YG, Tong YP, Li JM, Jing RL, Li B, Li ZS (2006). Mapping QTLs for nitrogen uptake in relation to the early growth of wheat (*Triticum aestivum* L.). *Plant Soil* 284: 73–84.

Austin RB (1999). Yield of wheat in the United Kingdom: recent advances and prospects. *Crop Sci* 39: 1604–1610.

Barbottin, A., Lecomte, C., Bouchard, C., Jeuffroy, M.-H., 2005. Nitrogen remobilisation during grain filling in wheat: genotypic and environmental effects. *Crop Sci.* 45, 1141–1150.

Barraclough PB, Howarth JR, Jones J, Lopez-Bellido R, Parmar S, Shepherd CE, Hawkesford MJ (2010). Nitrogen efficiency of wheat: genotypic and environmental variation and prospects for improvement. *Eur J Agron* 33: 1–11.

Bernard SM. And Habash DZ. 2009. The importance of cytosolic glutamine synthetase in nitrogen assimilation and recycling. *New Phytol* 192: 608-620.

Bertin, P. and Gallais, A. (2001). Physiological and genetic basis of nitrogen use efciency in maize. II. QTL detection and coincidences. *Maydica* 46: 53–68.

Bogard M, Allard V, Brancourt-Hulmel M, Heumez E, Machet JM, Jeuffroy MH, Gate P, Martre P, Le Gouis J (2010). Deviation from the grain protein concentration-grain yield negative relationship is highly correlated to post-anthesis N uptake in winter wheat. *J Exp* Bot 61: 4303–4312.

Borrell, A.K., Garside, A.L., Fukai, S. and Reid, D.J. (1998). Season, nitrogen rate, and plant type affect nitrogen uptake and nitrogen use efciency in rice. *Aust. J. Agric. Res.* 49: 829–843.

Borrell, A.K., Hammer, G.L., 2000. Nitrogen dynamics and the physiological basis of stay-green in sorghum. *Crop Sci.* 40, 1295–1307.

Brancourt-Hulmel M, Doussinaut G, Lecomte C, Berard P, LeBuanec B, Trottet M (2003). Genetics improvement of agronomic traits of winter wheat cultivars released in France from 1946 to 1992. *Crop Sci* 43: 37–45.

Brancourt-Hulmel, M., Heumez, E., Pluchard, P., Béghin, D., Depatureaux, C., Giraud, A., LeGouis, J., 2005. Indirect versus direct selection of winter wheat for low input or high input levels. *Crop Sci.* 45, 1427–1431.

Brisson N, Gate P, Gouache D, Charmet G, Oury FX, Huard F. 2010. Why are wheat yields stagnating in Europe? A comprehensive data analysis for France. Field *Crops Res.* 119: 201–212.

Cai, H., Zhou, Y., Xiao, J., Li, X., Zhang, Q. and Lian, X. (2009). Overexpressed glutamine synthetase gene modies nitrogen metabolism and abiotic stress responses in rice. *Plant Cell Rep.* 28: 527–537.

Caputo, C., Criado, M.V., Roberts, I.N., Gelso, M.A. and Barneix, A.J. (2009). Regulation of glutamine synthetase 1 and amino acids transport in the phloem of young wheat plants. *Plant Physiol. Biochem.* 47: 335–342.

Christopher, J.T., Manschadi, A.M., Hammer, G.L., Borell, A.K., 2008. Developmental and physiological traits associated high yield and stay-green phenotype in wheat. *Aust. J. Agric. Res.* 59, 354–364.

Clárk RB (1983). Plant genotype differences in the uptake, translocation, accumulation, and use of mineral elements required for plant growth. *Plant Soil* 72: 175–196.

Cormier F, Faure S, Dubreuil P, Heumez E, Beauchêne K, Lafarge S, Praud S, Le Gouis J. 2013. A multi environmental study of recent breeding progress on nitrogen use efficiency in wheat (*Triticum aestivum* L.). *Theor Appl Genet* (2013). 126: 3035–3048.

Cox, M.C., Qualset, C.O., Rains, D.W., 1986. Genetic variation for nitrogen assimilation and translocation in wheat. 3. Nitrogen translocation in relation to grain yield and protein. *Crop Sci.* 26, 737–740.

Dhugga, K.S., Waines, J.G., 1989. Analysis of nitrogen accumulation and use in bread and durum wheat. *Crop Sci.* 29, 1232–1239.

Dodig D (2005). A high-density genetic map of hexaploid wheat (*Triticum aestivum* L.). from the cross Chinese Spring × SQ1 and its use to compare QTLs for grain yield across a range of environments. *Theor Appl Genet* 110: 865–880.

Doerge RW (2002). Mapping and analysis of quantitative trait loci in experimental populations. *Nat Rev Genet* 3: 43–52.

Fischer R, Edmeades G (2010). Breeding and cereal yield progress. *Crop Sci* 50: 85–98.

Fischer, R.A., Wall, P.C., 1976. Wheat breeding in Mexico and yield increases. *Journal of the Australian Institute of Agricultural Science* 42, 139–148.

Fontaine J-X, Ravel C, Pageau K, Heumez E, Dubois F, Hirel B, Le Gouis J (2009). A quantitative genetic study for elucidating the contribution of glutamine synthetase, glutamate dehydrogenase and other nitrogen-related physiological traits to the agronomic performance of common wheat. *Theor Appl Genet* 119: 645–662.

Foulkes M, Hawkesford M, Barraclough P, Holdsworth M, Kerr S, Kightley S, Shewry P (2009). Identifying traits to improve the nitrogen economy of wheat: recent advances and future prospects. *Field Crops Res* 114: 329–342.

Foulkes, M.J., Sylvester-Bradley, R., Scott, R.K., 1998. Evidence for differences between winter wheat cultivars in acquisition of soil mineral nitrogen and uptake and utilization of applied fertiliser nitrogen. *J. Agric. Sci. (Camb.).* 130, 29–44.

Foulkes, M.J., Sylvester-Bradley, R., Weightman, R., Snape, J.W., 2007. Identifying physiological traits associated with improved drought resistance in winter wheat. *Field Crop Res.* 103, 11–24.

Gaju O, Allard V, Martre P, Snape JW, Heumez E, Le Gouis J, Moreau D, Bogard M, Griffiths S, Orford S, Hubbart S, Foulkes MJ (2011). Identification of traits to improve the nitrogen-use efficiency of wheat genotypes. *Field Crops Res* 123: 139–152.

Gajua O, V. Allardb, P. Martreb, J.W. Snaped, E. Heumeze, J. LeGouisb, D. Moreaub, M. Bogardb, S. Griffithsd, S. Orfordd, S. Hubbarta, M.J. Foulkesa, 2011. Identification of traits to improve the nitrogen-use efficiency of wheat genotypes. *Field Crops Research* 123 (2011). 139–152.

Gallardo F, Fu J, Canton FR, Garcia-Gutierrez A, Canovas FM, Kirby EG.1999. Expression of a conifer glutamine synthetase gene in transgenic poplar. *Planta* 210: 19–26.

Galloway JN, Townsend AR, Erisman JW, Bekunda M, Cai Z, Freney JR, Martinelli LA, Seitzinger SP, Sutton MA (2008). Transformation of the nitrogen cycle: recent trends, questions, and potential solutions. *Science* 320 (5878): 889–892.

Graybosch R, Peterson C (2012). Specific adaptation and genetic progress for grain yield in Great Plains hard winter wheats from 1987 to 2010. *Crop Sci* 52: 631–643.

Green A, Berger G, Griffey C, Pitman R, Thomason W, Balota M, Ahmed A (2012). Genetic yield improvement in soft red winter wheat in the eastern United States from 1919 to 2009. *Crop Sci* 52: 2097–2108.

Guarda G, Padovan S, Delogu G (2004). Grain yield, nitrogen-use efficiency and baking quality of old and modern Italian bread wheat cultivars grown at different nitrogen levels. *Eur J Agron* 21: 181–192.

Guo Y, Kong FM, Xu YF, Zhao Y, Liang X, Wang YY, An DG, Li SS (2012). QTL mapping for seedling traits in wheat grown under varying concentrations of N, P and K nutrients. *Theor Appl Genet* 124: 851–865.

Habash DZ, Bernard S, Schondelmaier J, Weyen J, Quarrie SA (2007). The genetics of nitrogen use in hexaploid wheat: N utilisation, development and yield. *Theor Appl Genet* 114: 403–419.

Habash DZ, Bernard S, Shondelmaier J, Weyen Y, Quarrie SA. 2006. The genetics of nitrogen use on hexaploid wheat: N utilization, development and yield. *Theoretical and Applied Genetics* 114, 403–419.

Habash DZ, Massiah AJ, Rong HL, Wallsgrove RM, Leigh RA.2001. The role of cytosolic glutamine synthetase in wheat. *Annals of Applied Biology* 138: 83–89.

Harrison J, Brugiere N, Phillipson B, Ferrario-Mery S, Becker T, Limami A, Hirel B.2000. Manipulating the pathway of ammonia assimilation through genetic engineering and breeding: consequences to plant physiology and plant development. *Plant and Soil* 221: 81–93.

Hedden P (2003). The genes of the green revolution. *Trends Genet* 19: 5–9.

Hirel B, Lea PJ. 2001. Ammonium assimilation. In: Lea PJ, Morot- Gaudry JF, eds. *Plant nitrogen*. Berlin: Springer-Verlag, 79–99.

Hirel B, Le-Gouis J, Ney B, Gallais A (2007). The challenge of improving nitrogen use efficiency in crop plants: towards a more central role for genetic variability and quantitative genetics within integrated approaches. *J Exp Bot*. 58 (9): 2369–2387.

Hirel, B., Bertin, P., Quillere´, I. *et al.* (2001). Towards a better understanding of the genetic and physiological basis for nitrogen use efciency in maize. *Plant Physiol*. 125: 1258–1270.

Hudson, G.S., Evans, J.R., von Caemmerer, S., Arvidsson, Y.B.C., Andrews, T.J., 1992. Reduction of ribulose-1,5-bisphosphate carboxylase oxygenase content by antisense RNA reduces photosynthesis in transgenic tobacco plants. *Plant Physiol*. 98, 294–302.

Ishiyama, K., Inoue, E., Watanabe-Takahashi, A., Obara, M., Yamaya, T. and Takahashi, H. (2004). Kinetic properties and ammonium-dependent regu- lation of cytosolic isoenzymes of glutamine synthetase in Arabidopsis. *J. Biol. Chem*. 279: 16598–16605.

Justes, E., Mary, B., Meynard, J.M., Machet, J.M., Thelier-Huche, L., 1994. Determination of a critical nitrogen dilution curve for winter wheat crops. *Ann. Bot*. 74: 397–407.

Kibite S, Evans LE (1984). Cause of negative correlations between grain yield and grain protein concentration in common wheat. *Euphytica* 33: 801–810.

Kjær, B., Jensen, J. and Giese, H. (1995). Quantitative trait loci for heading date and straw characters in barley. *Genome* 38: 1098–1104.

Krapp, A., Saliba-Colombani, V. and Daniel-Vedele, F. (2005). Analysis of C and N metabolisms and of C/N interactions using quantitative genetics. *Photosynth. Res*. 83, 251–263.

Laperche A, Brancourt-Hulmel M, Heumez E, Gardet O, Hanocq E, Devienne-Barret F, Le Gouis J (2007). Using genotype × nitrogen interaction variables to evaluate the QTL involved in wheat tolerance to nitrogen constraints. *Theor Appl Genet* 115: 399–415.

Laperche A, Devienne-Barret F, Maury O, Le Gouis J, Ney B (2006). A simplified conceptual model of carbon/nitrogen functioning for QTL analysis of winter wheat adaptation to nitrogen deficiency. *Theor Appl Genet* 113: 1131–1146.

Lauerer, M., Saftic, D., Quick, W.P., Labate, C., Fichtner, K., Schulze, E.D., Rodermel, S.R., Bogorad, L., Stitt, M., 1993. Decreased ribulose-1,5-bisphosphate carboxylase- oxygenase in transgenic tobacco transformed with antisense rbcs. VI. Effect on photosynthesis in plants grown at different irradiance. *Planta* 190, 332–345.

Lawlor, D.W., 2002. Carbon and nitrogen assimilation in relation to yield: mechanisms are the key to understanding production systems. *J. Exp. Bot*. 53, 773–787.

Le Gouis J, Béghin D, Heumez E, Pluchard P (2000). Genetic differences for nitrogen uptake and nitrogen utilisation efficiencies in winter wheat. *Eur J Agron* 12: 163–173.

Lea PJ, Miflin BJ 1974 An alternative route for nitrogen assimilation in higher plants. *Nature* (Lond.). 251: 614-616.

Lea, P.J. and Miin, B.J. (2003). Glutamate synthase and the synthesis of glutamate in plants. *Plant Physiol. Biochem*. 41: 555–564.

Li ZX, Ni ZF, Peng HR, Liu ZY, Nie XL, Xu SB, Liu G, Sun QX (2007b). Molecular mapping of QTLs for root response to phosphorus deficiency at seedling stage in wheat (Triticum aestivum L.). *Prog Nat Sci* 17: 1177–1184.

Lian, X., Xing, Y., Yan, H., Xu, C., Li, X. and Zhang, Q. (2005). QTLs for low nitrogen tolerance at seedling stage identied using a recombinant inbred line population derived from an elite rice hybrid. *Theor. Appl. Genet*. 112: 85–96.

Lima, L., Seabra, A., Melo, P., Cullimore, J. and Carvalho, H. (2006). Post- translational regulation of cytosolic glutamine synthetase of Medicago truncatula. *J. Exp. Bot*. 57: 2751–2761.

Long, S.P., Zhu, X.-G., Naidu, S.L., Ort, D.R., 2006. Can improvement in photosynthesis increase crop yields? *Plant Cell Environ*. 29, 315–330.

Lopez MS, Reynolds MP, Manes Y, Singh RP, Crossa J, Braun HJ (2012). Genetic yield gains and changes in associated traits of CIMMYT spring bread wheat in a "historic" set representing 30 years of breeding. *Crop Sci* 52: 1123–1131.

Loudet, O., Chaillou, S., Merigout, P., Talbotec, J. and Daniel-Vedele, F. (2003). Quantitative trait loci analysis of nitrogen use efciency in Arabidopsis. *Plant Physiol*. 131: 345–358.

Mae T. 1997. Physiological nitrogen efficiency in rice: nitrogen utilization, photosynthesis, and yield potential. In: Ando T, ed. *Plant Nutrition for Sustainable Food Production and Environment*. Dordrecht, The Netherlands: Kluwer Academic Publishers, 51–60.

Malagoli P, Laine P, Rossato L, Ourry A. 2005. Dynamics of nitrogen uptake and mobilization in field-grown winter oilseed rape (*Brassica napus*). from stem extension to harvest. *Annals of Botany* 95, 853–861.

Masclaux C, Quillere´ I, Gallais A, Hirel B. 2001. The challenge of remobilization in plant nitrogen economy. A survey of physioagronomic and molecular approaches. *Annals of Applied Biology* 138, 69–81.

Mi, C., Liu, J., Chen, F., Zhang, F., Cui, L., Liu, X., 2003. Nitrogen uptake and remobilization in maize hybrids differing in leaf senescence. *J. Plant Nutr*. 26, 447–459.

Miflin BJ, Habash DZ. 2002. The role of glutamine synthetase and glutamate dehydrogenase in nitrogen assimilation and possibilities for improvement in the nitrogen utilization of crops. *Journal of Experimental Botany* 53: 979–987.

Miflin BJ, Lea PJ 1980 Ammonia assimilation. In (BJ Miflin ed). "*The Biochemistry of Plants*" Vol 5, Academic Press, New York, pp 169 - 202.

Moll RH, Kamprath EJ, Jackson WA (1982). Analysis and interpretation of factors which contribute to efficiency of nitrogen utilization. *Agron. J.* 74: 562-564.

Mortimer ND, Elsayed MA, Horne RE, 2004. Energy and Greenhouse Gas Emissions for Bioethanol Production from Wheat Grain and Sugar Beet. Sheffield, UK: Sheffield Hallam University: Final Report no. 23/1.

Muurinen S, Slafer GA, Peltonen Sainio P (2006). Breeding effects on nitrogen use efficiency of spring cereals under northern conditions. *Crop Sci* 46: 561–568.

Obara, M., Kajiura, M., Fukuta, Y., Yano, M., Hayashi, M., Yamaya, T. and Sato, T. (2001). Mapping of QTLs associated with cytosolic glutamine synthetase and NADH-glutamate synthase in rice (*Oryza sativa* L.). *J. Exp. Bot.* 52: 1209–1217.

Ochs, G., Schock, G., Trischler, M., Kosemund, K. and Wild, A. (1999). Com- plexity and expression of the glutamine synthetase multigene family in the amphidiploid crop Brassica napus. *Plant Mol. Biol.* 39: 395–405.

Ortiz-Monasterio I, Pena RJ, Sayre KD, Rajaram S (1997b). CIMMYT's genetic progress in wheat grain quality under four N rates. *Crop Sci* 37: 892–898.

Ortiz-Monasterio I, Sayre KD, Rajaram S, McMahon M (1997a). Genetic progress in wheat yield and nitrogen use efficiency under four N rates. *Crop Sci* 37: 898–904.

Oury FX, Godin C (2007). Yield and grain protein concentration in bread wheat: how to use the negative relationship between the two characters to identify favourable genotypes? *Euphytica* 157: 45–57.

Oury FX, Godin C, Mailliard A, Chassin A, Gardet O, Giraud A, Heumez E, Morlais JY, Rolland B, Rousset M, Trottet M, Charmet G (2012). A study of genetic progress due to selection reveals a negative effect of climate change on bread wheat yield in France. *Eur J Agron* 40: 28–38.

Oury, F.X., Be´rard, P., Brancourt-Hulmel, M.et al. (2003). Yield and grain protein concentration in bread wheat: a review and a study of multi-annual data from a French breeding program. *J. Genet. Breed*.57, 59–68.

Palta JA, Fillery IRP. 1995. N application increases pre-anthesis contribution of dry matter to grain yield in wheat grown on a duplex soil. *Australian Journal of Agricultural Research* 46, 507–518.

Papakosta, D.K., Garianas, A.A., 1991. Nitrogen and dry matter accumulation, remobilisation, and losses for Mediterranean wheat during grain filling. *Agron. J.* 83, 864–870.

Parry, M.A.J., Madgwick, P.J., Carvalho, J.F.C., Andralojc, P.J., 2007. Prospects for increasing photosynthesis by overcoming the limitations of Rubisco. *J. Agric. Sci.* 145, 31–43.

Potel, F., Valadier, M.H., Ferrario-Me´ry, S. *et al.* (2009). Assimilation of excess ammonium into amino acids and nitrogen translocation in *Arabidopsis thaliana*

– roles of glutamate synthases and carbamoylphosphate synthetase in leaves. *FEBS J.* 276: 4061–4076.

Przulj, N., Momcilovic, V., 2001. Genetic variation for dry matter and nitrogen accumulation and translocation in two-rowed spring barley. II. Nitrogen translocation. *Eur. J. Agron.* 15, 255–265.

Quarrie SA, Steed A, Calestani C, Semikhodskii A, Lebreton C, Chinoy C, Steele N, Pljevljakusic D, Waterman E, Weyen J, Schondelmaier J, Habash DZ, Farmer P, Saker L, Clarkson DT, Abugalieva A, Yessimbekova M, Turuspekov Y, Abugalieva S, Tuberosa R, Sanguineti MC, Hollington PA, Aragues R, Royo A,.

Quraishi UM, Abrouk M, Murat F, Pont C, Foucrier S, Desmaizieres G, Confolent C, Rivière N, Charmet G, Paux E, Murigneux A, Guerreiro L, Lafarge S, Le Gouis J, Feuillet C, Salse J. 2011. Cross-genome map based dissection of a nitrogen use efficiency ortho-metaQTL in bread wheat unravels concerted cereal genome evolution. *Plant Journal* 65: 745-756.

Rauh, L., Basten, C. and Buckler, S. (2002). Quantitative trait loci analysis of growth response to varying nitrogen sources in *Arabidopsis thaliana*. *Theor. Appl. Genet.* 104: 743–750.

Raun, W.R., Solie, J.B., Johnson, G.V., Stone, M.L., Mullen, R.W., Freeman, K.W., Thomason,W.E., Lukina, E.V., 2002. Improving nitrogen-use efficiency in cereal grain production with optical sensing and variable rate application. *Agron. J.* 94, 351–815.

Rengel Z, Marschner P (2005). Nutrient availability and management in the rhizosphere: Exploiting genotypic differences. *New Phytol* 168: 305–312.

Rossato L, Laine' P, Ourry A. 2001. Nitrogen storage and remobilization in Brassica napus L. during the growth cycle: nitrogen fluxes within the plant and changes in soluble protein patterns. *Journal of Experimental Botany* 52, 1655–1663.

Saal B, von Korff M, Leon J, Pillen K (2011). Advanced-backcross QTL analysis in spring barley: IV. Localization of QTL × nitrogen interaction effects for yield-related traits. *Euphytica* 177: 223–239.

Shanahan, J.F., Kitchen, N.R., Raun, W.R., Schepers, J.S., 2008. Responsive in-season nitrogen management for cereals. *Comput. Electron. Agric.* 61, 51–62.

Shewry PR (2004). Improving the protein content and quality of temperate cereals: wheat, barley and rye. In Cakmak I, Welch R (eds). *Impacts of agriculture on human health and nutrition*. USDA, ARS, U.S. Plant, Soil and Nutrition Laboratory, Cornell University, USA.

Silva, S.A., de Carvalho, F.I.F., Caetano, V.R., Oliveira, A.C., Coimbra, J.L.M., Vasconcellos, N.J.S., Lorencetti, C., 2000. Genetic basis of stay-green trait in bread wheat. *J. New Seeds* 2, 55–68.

Simmonds, N.W. (1995). The relation between yield and protein in cereal grain. *J. Sci. Food Agric.* 67, 309–315.

Su JY, Zheng Q, Li HW, Li B, Jing RL, Tong YP, Li ZS (2009). Detection of QTLs for phosphorus use efficiency in relation to agronomic performance of wheat grown under phosphorus sufficient and limited conditions. *Plant Sci* 176: 824–836.

Su, J. Y., Xiao, Y., Li, M., Liu, Q., Li, B., Tong, Y., Jia, J., and Li, Z. (2006). Mapping QTLs for phosphorus-deficiency tolerance at wheat seedling stage. *Plant Soil* 281, 25–36.

Tabuchi, M., Abiko, T. and Yamaya, T. (2007). Assimilation of ammonium ions and reutilization of nitrogen in rice (*Oryza sativa* L.). *J. Exp. Bot.* 58: 2319–2327.

Tahir, I.S.A., Nakata, N., 2005. Remobilisation of nitrogen and carbohydrate from stems of bread wheat in response to heat stress during grain filling. *J. Agron. Crop Sci.* 191, 106–115.

Tempest DW, Meers JL, Brown CM 1970 Synthesis of glutamate in Aerobacter aerogenes by a hitherto unknown route. *Biochem. J.* 117: 405-407.

Thomas, H., Smart, C.M., 1993. Crops that stay green. Ann. Appl. Biol. 123, 193–219.

Triboï, E., Martre, P., Girousse, C., Ravel, C., Triboï-Blondel, A.M., 2006. Unravelling environmental and genetic relationships between grain yield and nitrogen concentration for wheat. *Eur. J. Agron.* 25, 108–118.

Uemura, K., Anwaruzzaman, S., Yokota, A., 1997. Ribulose-1, 5-bisphosphate carboxylase/oxygenase from thermophilic red algae with strong specificity for CO_2 fixation. *Biochem. Biophys. Res. Commun.* 233, 568–571.

Valadier, M.H., Yoshida, A., Grandjean, O., Morin, H., Kronenberger, J., Boutet, S., Raballand, A., Hase, T., Yoneyama, T. and Suzuki, A. (2008). Implication of the glutamine synthetase/glutamate synthase pathway in conditioning the amino acid metabolism in bundle sheath and mesophyll cells of maize leaves. *FEBS J.* 275: 3193–3206.

Van Ginkel, M., Ortiz-Monasterio, I., Trethowan, R., Hernandez, E., 2001. Methodology for selecting segregating populations for improved nitrogen use efficiency in bread wheat. *Euphytica* 119, 223–230.

Van Sanford, D.A., Mackown, C.T., 1987. Cultivar differences in nitrogen remobilisation during grain filling in soft red winter wheat. *Crop Sci.* 27, 295–300.

Verma, V., Foulkes, M.J., Caligari, P., Sylvester-Bradley, R., Snape, J., 2004. Mapping quantitative trait loci for flag leaf senescence as a yield determinant in winter wheat under optimal and drought-stressed environments. *Euphytica* 135, 255–263.

Wang RF, An DG, Hu CS, Li LH, Zhang YM, Jia YG, Tong YP (2011). Relationship between nitrogen uptake and use efficiency of winter wheat grown in the North China Plain. *Crop Pasture Sci* 62: 504–514.

Xu Y, Wang R, Tong Y, Zhao H, Xie Q, Liu D, Zhang A, Li B, Xu H, An D. 2014. Mapping QTLs for yield and nitrogenrelated traits in wheat: influence of nitrogen and phosphorus fertilization on QTL expression. *Theor Appl Genet.* DOI 10.1007/s00122-013-2201-y

Recent Advances in Crop Physiology Vol. 2 (2016)
Editor: **Dr. Amrit Lal Singh**
Published by: **DAYA PUBLISHING HOUSE, NEW DELHI**

Pages 357–356

Chapter 13

Quality Seed: A Mega Factor in Enhancing Crop Productivity

J.S. Chauhan[1], A.L. Singh[2], S. Rajendra Prasad[3]
and Satinder Pal[1]*

[1]*Indian Council of Agricultural Research,
Krishi Bhawan, New Delhi – 110 001*
[2]*ICAR-Directorate of Groundnut Research, Junagadh – 362 001, Gujarat*
[3]*ICAR-Directorate of Seed Research, Mau, Uttar Pradesh*

1. Introduction

Agriculture is the mainstay of livelihood worldwide, with majority of rural households and population being directly or indirectly dependent on it. Presently agriculture and food security is the topmost policy agendas of the global development. In a world where 900 million people are undernourished and 2 billion are micronutrient deficient, ensuring access to nutritious foods is critical to lifting more than 1 billion poor people out of poverty (IFPRI, 2014). World cereal production in 2014 is predicted to be 2,532 million tonnes (mt). The forecast for world cereal utilization in 2014-15 is put at 2,465 m t, up by 48 m t (2 per cent) from 2013-14. The anticipated year-to-year increase mainly reflects greater cereal usage by the livestock sectors, supported by falling prices. The volume of cereals destined for food is expected to increase to 1104 m t, 1 per cent up from 2013/14, implying a stable average global per capita consumption of 153 kg. The FAO's, at the close of 2015, forecast 628 m t world cereal stocks, the highest since 2000. As a result, the global cereal stocks-to-use ratio would hit a 13-year high of 25.2 per cent in 2014-15,

* *Corresponding Author:* E-mail: adgseedicar@nic.in

suggesting a generally comfortable supply situation for the 2014-15 marketing season. World cereal trade is forecast to contract by about 17.7 m t in 2014-15, mainly because of wheat and coarse grains reach 339 m t.

The world population, which was approximately 5 million at the dawn of agriculture about 10, 000 years ago (8000 B.C.), grew to 200-600 million with a growth rate of < 0.05 per cent per year over a period of 8,000 year (up to 1 A.D). Further it had taken around 1800 years to reach one billion. But, tremendous growth occurred with the industrial revolution and the second billion was achieved in only 130 years (1930), the third billion in less than 30 years (1959), the fourth billion in 15 years (1974), and the fifth billion in only 13 years (1987). During 20th century alone, the world population has grown from 1.65 billion to 6 billion. In 1970, there were roughly half as many people in the world as of there now (7.29 billion as of Jan 2015). Though, due to declining growth rates, now it will take over 200 years to double it again, the predicted population will be around 10 billion during 2050 and 11 billion by the end of this century. Accordingly, to feed the population, we have to increase the productivity ans the pressure will be more in Asian countries.

In world agriculture the "Green Revolution" during the 1960s and 1970s was the turning point and introduction of high yielding, semi-dwarf and fertilizer responsive varieties of wheat and rice led to a dramatic shift from "food scarce" to "food secure" status in India and many Asian country (Paroda, 2013). The credit goes to the Indian agriculturist who, keeping the pace with such a huge increase in population, made a spectacular advancements in agriculture production during the post-green revolution years, with the country's food grain production increasing from 115 m t in 1965 to 265 m t in 2013-14 (Table 13.1) as a result presently India rank 2nd in the world on the basis of agriculture production and we are exporting several food commodities. As per DGSCIS annual export India's export of various food commodities to different countries during 2013-14 was 5,09,665 t of with a value of 3,18,773 lakh Indian rupees (531 million US$).

However, average productivity of many crops in India is still far below the world's average. Same time the food availability to all at affordable price is a major concern.

Achieving food security for all is the main mission of FAO and to make sure people have regular access to enough high-quality food to lead active and healthy lives it has three main goals (FAO 2015):

☆ The eradication of hunger, food insecurity and malnutrition;

☆ The elimination of poverty and the driving forward of economic and social progress for all and

☆ The sustainable management and utilization of natural resources, including land, water, climate and genetic resources for the benefit of present and future generations

The Indian population is likely to be 1.4 billion by 2030 and the rise in per capita income is expected to raise further the demand for food grains to over 320 mt by 2030 along with the fact that non-agricultural land uses are increasing and natural resources are shrinking and degrading rapidly. Hence, there is an urgent need to increase productivity and quality in all these crops and devise mechanism to make it available to all to have both food and nutritional security. This can only be possible by optimising the yield and quality through improved production technologies and integration of natural resources for producing healthy food. Use of quality seed is the most important production factor (Chauhan *et al.*, 2013).

Table 13.1: Food Production in India

Year	Food Grain Production (million tonnes)	Year	Food Grain Production (million tonnes)
1950-51	51	1960-61	82
1970-71	108	1980-81	129
1990-81	176	2000-01	197
2002-03	175	2003-04	213
2004-05	198	2005-06	209
2006-07	217	2007-08	231
2008-09	235	2009-10	218
2010-11	245	2011-12	259
2012-13	257	2013-14	265

Source: Directorate of Economics and Statistics, Ministry of Agriculture, GOI
http://dacnet.nic.in/eands).

Seed is a critical input for enhancing productivity of all agricultural and horticultural crops and plays a vital role in ensuring food security. It also offers to integrate production, protection and quality enhancement technologies in a single entity, in a cost effective way as use of quality seeds alone increases productivity by 15-20 per cent highlighting its importance in agriculture. The response of all other inputs depends on quality of seeds to a large extent and it can be further raised up to 45 per cent with efficient management of other inputs. Therefore, any attempt to enhance agricultural productivity will largely depend on higher replacement rate of quality seeds of high yielding varieties/hybrids of agri-horticultural crops (Chauhan *et al.*, 2013). In this chapter an effort has been made to highlight the role of quality seed, its production and availability in enhancing the production of Indian agriculture.

2. Varietal Improvement

Sustained increase in agricultural production and productivity necessarily requires continuous development of new and improved varieties as well as hybrids of crops and efficient production system and supply of seeds to farmers. Though domestication of most of the crops is only a few thousand years old, origin and speciation of most crops is millions of years old (Chaudhary, 2000; Damania, and Valkoun, 1997). The conventional breeding is about a hundred years old while the youngest, yet more powerful of all biotechnology is three decade old. Improved varieties have made significant impact (Khush, 1995) and the increase is more significant in and almost all of it has been through the public research.

The Indian Seed Improvement Programme is backed by a strong crop improvement programme in both the public and private sectors and now the industry is highly active and well recognized internationally and several developing and neighbouring countries were benefited from quality seed from India. India's Seed Programme has a strong seed production base in terms of diverse and ideal agro-climates spread throughout the country for producing high quality seeds of several tropical, temperate and sub-tropical plant varieties in enough quantities at competitive prices. Over the years, several seed crop zones have evolved with extreme levels of specialization. There are more than 20,000 seed dealers and seed marketing distributors in the business.

Genetic yield enhancement is the single most significant technological intervention introduced and supported by the National Agriculture Research System (NARS). From 1960, the NARS has been continuously developing new varieties suitable for different agro-climatic regions and changing production conditions. Presently 26 crop science and 23 horticulture institutes, 31 crop science and 12 horticulture project coordinators of ICAR and all the state agricultural universities (SAU's) are engaged in developing of new varieties/hybrids of different crops and production of breeder and basic seed to ensure continued varietal improvement. As a result a total of 3801 varieties of field crops and 718 varieties of horticultural crops have been developed till September 2013 (Gautam, 2013). The crop-wise number of varieties and hybrids released during 2009-2013 are given in Table 13.2.

Table 13.2: Varieties and Hybrids of different Crops Released during 2009-2013

Crop Groups	Number of Varieties and Hybrids Released during Various Years						
	2009	2010	2011	2012	2013	2014	Total
Cereals	63	68	31	75	70	47	354
Oilseeds	24	29	10	19	15	8	105
Pulses	21	29	12	19	8	12	101
Fiber crops	5	15	1	10	3	0	34
Forage crops	7	5	4	1	4	5	26
Sugar crops	0	5	2	5	6	2	20
Total	**120**	**151**	**60**	**129**	**106**	**74**	**640**

The NARS has released 706 and 648 varieties of field crops during 10th and 11th plan period, respectively with the largest number of varieties (188) released during the year 2006 (Figure 13.1). Of the total 1354 varieties, 44 per cent were released by Central Sub-Committee on Crop Standards, Notification and Release of Varieties for Agricultural Crops. During 2009-2013, 354 varieties of cereals,105 oilseeds, 101 pulses, 34 fibres, 26 forages, 20 sugarcane and 2 of underutilized crops were released. These varieties are input-responsive, high yielding and show tolerance to the biotic and abiotic stresses and have shown their impact on the crop production by increasing their productivity (Table 13.3).

Table 13.3: Production and Productivity of Selected Crops of India during the Last Decade

Years	Production (Million tonnes)					Productivity (kg/ha)				
	Rice	Wheat	Maize	Oilseeds	Pulses	Rice	Wheat	Maize	Oilseeds	Pulses
2002-03	71.8	65.8	11.2	14.8	11.1	1744	2612	1681	691	543
2003-04	88.5	72.2	15.0	25.2	14.9	2078	2713	2041	1064	635
2004-05	83.1	68.6	14.2	24.4	13.1	1984	2602	1907	885	577
2005-06	91.8	69.4	14.7	28.0	13.4	2102	2619	1938	1004	598
2006-07	93.4	75.8	15.1	24.3	14.2	2131	2708	1912	916	612
2007-08	96.7	78.6	19.0	29.8	14.8	2202	2802	2335	1115	625
2008-09	99.2	80.7	19.7	27.7	14.6	2178	2907	2414	1006	659
2009-10	89.1	80.8	16.7	24.9	14.7	2125	2839	2024	958	630
2010-11	96.0	86.9	21.7	32.5	18.2	2239	2989	2540	1193	691
2011-12	105.3	94.9	21.6	30.0	19.1	2393	3177	2476	1133	699
2012-13	105.2	93.5	22.2	30.9	18.3	2462	3117	2566	1168	789
2013-14	106.5	95.9	24.3	32.9	19.3	–	–	–	–	–

No. of Varieties	2003	2004	2005	2006	2007	2008	2009	2010	2011	2012
Central varieties	37	78	23	73	54	95	46	78	38	75
State varieties	66	84	87	115	89	88	75	74	22	57

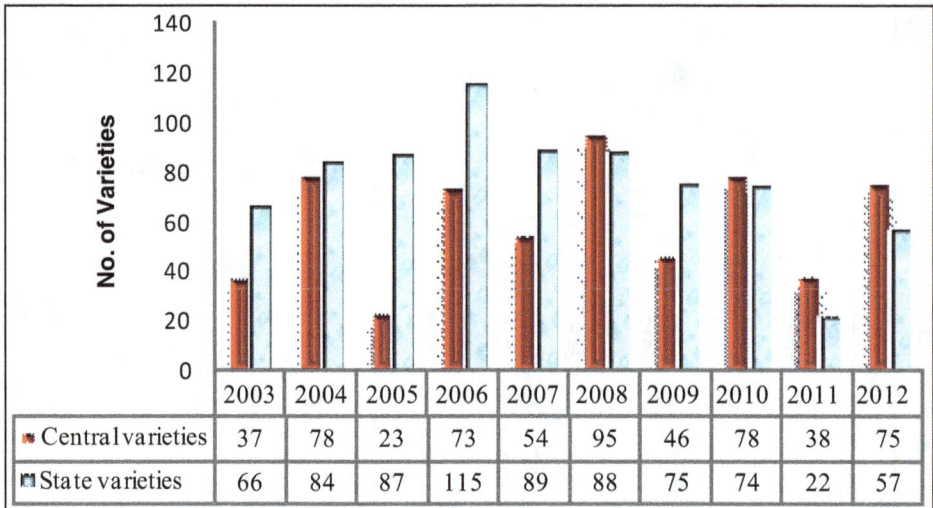

Figure 13.1: Varieties of Field Crops Released in India during 2003-2012 (Chauhan *et al.*, 2013).

Seed security is most important prerequisite for food security and in past few years, a significant stride in this aspect through effective implementation of ICAR Seed Project and AICRP-NSP (Crops) was made. It is proved that the increase in quality seed availability has a huge bearing on food grain production. The gene banks were established for safe keeping the "heritage of human kind", which is functioning to large extent.

3. Seed Sectors and Research in India

Indian seed programme includes the participation of Central and State governments, Indian Council of Agricultural (ICAR), State Agricultural Universities (SAU) system, Public sector, cooperative sector and private sector institutions. Seed sector in India consists of two national level corporations *i.e.* National Seeds Corporation (NSC) and State Farms Corporation of India (SFCI now merged with NSC), 15 State Seed Corporations (SSCs) and about 100 major seed companies. The private sector has started to play a significant role in the production and distribution of seeds. However, the organized seed sector particularly for food crops cereals continues to be dominated by the public sector.

Though the seed sector and seed production in India started long back, maximum development of the Indian seed industry took place only during the last 30 years. The major re-structuring of the seed industry done by Government of India (GOI) through the National Seed Project Phase-I (1977-78), Phase-II (1978-79) and Phase-III (1990-1991) was a turning point in shaping of an organized seed industry in India which has strengthened the most needed seed infrastructure. Introduction of New Seed Development Policy (1988-1989) was another mile stone which transformed the Indian Seed Industry and gave access to Indian farmers of the best of seed and planting material available anywhere in the world. This policy stimulated appreciable investments by private individuals, Indian corporate and multinational companies (MNCs) in the Indian seed sector with strong R and D base for product development in each of the seed companies with more emphasis on high value hybrids of cereals and vegetables and hi-tech products such as *Bt.* Cotton. Today, farmer has a wide product choice and seed industry is set to work with a 'farmer centric' approach and is market driven.

Basically, the seed sectors in India, comprises of public sector institutions and private seed companies. The main focus of private seed companies has been on the high value low volume seeds and market for low value high volume seeds of cereals, pulses and oilseeds is still dominated by the public sector seed corporations. Public seed sector includes various organizations, *viz.*, NARS comprising of several crop research institutes of ICAR, SAUs and deemed universities dealing with crops, All India Coordinated and Network Projects on each crop and eight zonal project directorates, NSC established in 1963, SSCs, SFCI. The public sector seed companies/ state governments forecast seed demand for various crops three years in advance and place requirement for breeder seeds with the Ministry of Agriculture, Government of India. Using the breeder seeds provided by NARS, the public sector seed companies produce foundation seeds on government farms or reliable, well-trained contract farms. The foundation seed is further multiplied in contract farmers' fields as certified seed for commercial distribution.

Looking to the demand and importance of quality seed, ICAR upgraded the National Seed Project into a Directorate of Seed Research as well as launched Mega Seed project during 10th plan which led to double the breeder seed production in 11th plan. The ICAR seed project also aimed at creation of infrastructure for quality seed production, processing, storage and also competence enhancement of the personnel involved in public seed sector. For seed technology research, presently India has a national level Directorate at Mau and separate departments in several institute under ICAR as well as in the State Agricultural Universities. Among them most prominent one are seed technology division of ICAR-IARI at New Delhi, GBPUAT at Pantnagar and TNAU Coimbatore. In seed education, nearly 10 SAUs and traditional universities offers post graduation in seed technology leading to M.Sc. and/Ph.D degree. The seed industry has three well reputed national level associations apart from several provincial level groups to take care of the interests of the industry.

Private seed sector also plays equally important role in seed production, but usually of high value low volume crops such as vegetables and also hybrid corn, cotton, pearl millet, sunflower, etc. (Hanchinal, 2013). Private seed sector experienced rapid growth under liberalized government policy particularly under New Policy on Seed Development (1988) which resulted in establishment of around 500 seed companies engaged in seed production or seed trade across the country. Private sector companies have a significant place mainly in the case of maize and sunflower and cotton. However, in the case of vegetable seeds and planting materials of horticultural crops, the private sector is the dominant player.

As the private sectors are not interested in entering into seed production of high volume low margin crops of wheat, paddy, other cereals, oilseeds and pulses, the public sector seed corporations will continue to remain dominant in cereals, pulses and oilseeds for many more years to come. Besides, significant quantities of seeds are also produced by the State Departments of Agriculture, where the State Seeds corporations are not in existence. There is an urgent need for the SSCs also to transform themselves in tune with the industry in terms of infrastructure, technologies, approach and the management culture to be able to survive in the competitive market and to enhance their contribution in the national endeavour of increasing food production to attain food and nutritional security.

Most countries increased their crop productivity during the last two decades, but there is very poor representations of private sectors. In case of rice till 1996, there were no known multi-national company and production trend, growth rate for production and productivity were high enough taking Asia-Pacific region in general over decades (Chaudhary, 1998, 1999). It is generally argued that IPR and patenting will assure return in research investment by providing product secrecy. Thus attract private investment for agricultural researches which are mostly funded by public sector. Privatisation of public research appears feasible and on the surface offers bright possibilities. Looking to the importance of seed and its direct relevance to the farmer, the Govt of India through it DAC upload all the informations on seed under the report and publications to its web: http://agricoop.nic.in/documentreport.html

Some of the current informations are:

1. Agenda Items for National Conference on Agriculture for *Kharif* Campaign-2013
2. Annual Report of Gujarat State Seeds Corporation Ltd
3. Annual Report of National Seeds Cooperation
4. Annual Report of PPV and FRA (English)/(Hindi)
5. Annual Report of State Farms Cooperation of India Limited
6. Calendar of Operation for Breeder Seed
7. Details of Tissue culture labs
8. Final breeder seed indent for *kharif*-2014
9. Government Review on Gujarat State Seeds Corporation
10. Govt Review on SFCI
11. Govt. Review on National Seeds Corporation
12. Govt. Review on PPV and FRA
13. India Minimum Seed Certification Standards : For Comments/Suggestions from various Stake Holder
14. List of Seed Testing Laboratory in India
15. Minutes of EXIM Committee for export and import of seeds and planting materials
16. Physical Targets of Certified Seeds in Seed Bank during 2012-13
17. Potato Allocation for *Rabi* 2013-14
18. Requirement and Availability of Certified/Quality Seed For *Kharif*-2013

19. Seed Replacement Rates
20. State Seed Rolling Plan 2013-2014 to 2016-2017

4. Seed Production Systems in India

Seeds are critical basic input for enhancing productivity of agricultural crops. In India after release of a variety, its seeds have to be multiplied in sufficient quantity which takes 3-4 generations before it reaches to the farmers for commercial use. During multiplication cycles, care is taken to maintain the original characteristics of the variety and avoid degeneration. The Indian seed programme largely adheres to the limited generations' system for seed multiplication in a phased manner. The system recognizes three generations namely breeder, foundation and certified seeds and provides adequate safeguards for quality assurance in the seed multiplication chain to maintain the purity of the variety as it flows from the breeder to the farmer.

Breeder seed is the progeny of nucleus seed of a variety and is produced by the originating breeder or by a sponsored breeder. Foundation seed is the progeny of breeder seed and is required to be produced from breeder seed or from foundation seed which can be clearly traced to breeder seed.

Certified seed is the progeny of foundation seed and must meet the standards of seed certification prescribed in the Indian minimum seeds certification standards, 1988. In case of self pollinated crops, certified seeds can also be produced from certified seeds provided it does not go beyond three generations from foundation seed stage-I. For quality control and certification, there are 22 State Seed Certification Agencies (SSCAs) and 104 State Seed Testing Laboratories (SSTLs) in India (listed in the Annexure 1).

The seed production programme is an exhaustive task requiring high technical skills, financial investments and proper methodology and care. It is expected that the nucleus/breeder seed should be of high purity, because the genetic purity in subsequent generation will largely depend upon the quality of nucleus/breeder seed. Hence, the nucleus and breeder seed are the back-bone of the seed programme.

Nucleus/breeder seed production of crop varieties is coordinated by All India Coordinated Research Project on the crop through State Agricultural Universities as per Department of Agriculture and Cooperation (DAC) indent under the supervision of qualified breeders.

In Indian agriculture the seed programme, with its well poised for continued growth, is now occupying a pivotal place. The NSC is the largest single seed organization in the country with such a wide product range, pioneered the growth and development of a sound industry in India. The NSC, SFCI, SSCS and other seed producing agencies are continuously expanding their activities and products, volume and value of seeds, level of seed distribution to the remotest areas. During the past five decades, these seed producing agencies have built up as a competent and experienced seed producers and seed dealers in various parts of the country and have adequate level of specialization and competence in handling and managing various segments of seed improvement on scientifically sound and commercially viable terms.

Presently, the Indian seed improvement programme is highly vibrant and energetic, backed up by a strong crop improvement programme in both the public and private sectors and is well recognized in the international seed arena. Several developing and neighbouring countries were benefited from quality seed imports from India. India's Seed Programme has a strong seed production base in terms of diverse and ideal agro-climates spread throughout the country for producing high quality seeds of several tropical, temperate and sub-tropical plant varieties in enough quantities at competitive prices.

5. Seed Quality Health, Drying, Storage and Viability

It is essential to properly store the seed of the crop varieties to maintain its health, a particular percent of viability and assured field germination, especially during rainy season. The drying and storage methods affect the germinability, seedling vigour and field emergence, thus require special attention during drying and storage. The seed of most of the crops after drying is generally stored at ambient condition in air tight bags. This enables adequate seed germination after a season in most of the crops. However, in several crops the loss of viability, due to poor storage condition, is very fast which enables the seed unfit for next season. These seeds need special care to avoid deterioration. The high relative humidity (RH) and favourable temperature during rainy season increases seed moisture and activate seed causing further spoilage. There are many cost affecting methods to store the seed commercially in various packing material and in cold storage, but the farmer need cheap and on farm storage technologies. Covering the cereals bags with straw and use of cheap desiccants such as calcium chloride and silica gel takes care of moisture and high humidity (RH) during rainy season and well known adopted method by farmers.

The tetrazolium chloride (TTC) test is a well-known method for testing the viability of most of the seed and it also gives an indication of seed deterioration (Singh *et al.*, 1987). Groundnut pods with < 7 per cent moisture content stored in a polyethylene-lined gunny bag with the desiccant $CaCl_2$ maintained seed viability even after eight months of storage with more than 75 per cent germination due to low moisture percentage of the seeds during the storage period (Nautiyal and Joshi, 1991; Nautiyal and Ravindra,1996). The drying by a method developed by Directorate of Oilseeds Research (DOR) system (a process of heaping and inverting heaps over a 6-d period), drying in artificial shade and windrow (sun) drying when compared on viability vigour and the leakage of solutes in 6 Spanish groundnut cultivars, pods dried in windrows showed a faster rate of moisture loss during drying period as temperature reached up to 44° C which resulted in rapid loss of viability, while in pods dried under shade or the DOR system the temp remained below 39°C, dried more uniformly, had a slower rate of moisture loss thus maintaining better seed viability for a longer period (Nautiyal and Zala, 1991). Damage of the seed membrane was inferred to be greater in seeds dried in windrows because almost double the amount of solutes was leached out during imbibition compared with seed dried by the other methods.

Further studies showed that drying of groundnut pods by the DOR method and storage in polythene-lined gunny bag with a desiccant (CaCl$_2$ or silica gel) maintained germinability up to 95 per cent at Bhubaneshwar and 92 per cent at Bargarh, Odisha even after 6 months of storage (*i.e.* until sowing during the next summer season). On the other hand, the seeds dried and stored by the conventional method lost germinability completely within 6 months of storage (Nautiyal *et al.*, 2004). The drying of pods under natural field conditions in windrows at high temperatures (about 45° C) in summer (May-June) adversely affected germination (74 per cent) and lost about 50 per cent after 3 months of storage. However, pods dried by NRCG method, which protects the pods from the direct exposure of sun by the haulm of the plant in a tripod structure, retained >80 per cent germination, even after 9 months of storage and also helped in maintaining the seedling vigour, and seed-coat colouration (Nautiyal and Zala, 2004).

The field emergence in most of the crop seeds, due to invasion by soil fungus, is less than laboratory germination. In groundnut this reduction was more pronounced with large size seeds due to more reserve food and highly prone to infection by soil borne fungus. High seed weight, efficient utilization of reserve food material, development of secondary roots and lower SLA are desirable agronomic traits in crop cultivation, but it varies with crops and genotypes. The graded small (< 6.8 mm diameter), medium (6.8 to 7.8 mm dia), large (> 7.9 mm dia) and ungraded (mixed) seeds of six Spanish bunch J 11, X 1720, Girnar 1, SB XI, GG 2, and ICGS 44) and two Virginia runner GAUG 10 and M 13 groundnut cultivars when studied, germination was greater with small and medium size seeds than large seeds in most of the cultivars. Large seeds produced more vigorous plants having more shoot and root biomass at initial growth stages and more large seeds at harvest than those produced from small and medium seeds. However, at maturity the plant produced by various seed categories did not differ in height, pod yields, 100-seed weight and shelling. These suggest that small and medium seeds, which germinate better and require 50 and 25 per cent lesser amount of seeds, respectively, than those of large one, should be used for sowing, the large and handpicked seeds should be used as food or other edible purposes (Singh *et al.*, 1997; Singh *et al.*, 1998).

However, studies on utilisation of reserve food material in 10 cultivars of different seed weight indicated that both medium and higher seed-weight groups are efficient in utilization of reserve food material from cotyledons to establish vigorous seedlings than that of lower seed-weight group. An inverse relationship (r=-0.71, n=15) between specific leaf area (SLA) and total seedling biomass was established, *i.e.*, lower the SLA (thicker leaf) higher the biomass, and higher the SLA (thinner leaf) lower the biomass.

The soil moisture-deficit stress during different phenophases alters the quality of seed in term of composition of total sugars, phenolics, protein and fatty acid (Nautiyal *et al.*, 1991; Chakraborty *et al.*, 2013), but varies with the crops and varieties. Stress during pod development increased palmitic acid, stearic acid and oleic acid and protein content in groundnut, but stress during vegetative and flowering phases reduced the sugar content, however, these changes are governed more by cultivar and its interaction with the environmental conditions time an intensity of imposed soil moisture-deficit stress. Water stress at pod initiation/development phase reduced germinability, vigour, seed membrane integrity and also affected chlorophyll synthesis and dehydrogenase activity in cotyledons during germination (Nautiyal *et al.*, 1991). However, fertilizers and organic manure increases the quality of groundnut seed (Khatediya *et al.*, 2013).

During storage after a particular time period most of the seed losses their viability due to loss of membrane integrity. The storage condition further can enhance or reduce the time period of viability. The loss of viability in groundnut seed is associated with cell membrane integrity that causes increase in leakage of electrical conductivity, K+ and sugar contents and leachate (adsorption at 260 nm) taken 48 h after seed immersion (Nautiyal *et al.*, 1988).

Seedling vigour index (SVI), calculated from germination and root length, provides the correct interpretation of quality seed for germination. Though it varies with crops, the crop varieties being grown round the year the germination and SVI were better in material harvested in Oct-Dec than that in May-June. If the seeds are not stored properly, high humidity (70-99 per cent) and temperature (30-38°C) during rainy season causes quick deterioration and loss of viability (Singh *et al.*, 1987, 1992) Drying temperatures had a significant effect on both germination immediately after drying and its subsequent loss of viability during storage. Studies on physiological and biochemical changes during accelerated ageing in the seeds indicate that accelerated ageing was associated with a decline in germinability and SVI.

The dormancy in seed is an inherent protection from sprouting and further deterioration of seed under favourable condition immediately after harvest. However the seed dormancy varies with crops and genotypes. Seeds of the dormant type maintain higher germinability, root length and hypocotyl length, and had lower electrical conductivity of the seed leachate than that of non-dormant type. The genetic control of seed dormancy in groundnut is considered to be quantitative in nature and it is regulated mainly by the testa (a maternal tissue) in the Spanish type, but by cotyledons, and embryonic axis (both zygotic tissue) as well as testa in Virginia types (Bandyopadhyay *et al.*, 1999).

Fresh-seed dormancy is a desirable seed character in most of the crop and need to be introduced in the non-dormant crops and varieties to avoid sprouting of seed in the field due to rain at harvest. If seed has to be used immediately after harvest the ethrel and ABA are the well known chemical for breaking seed dormancy and achieving maximum germination. There is remarkable differences in seed dormancy of many crops varieties such as groundnut. The groundnut cultivars TAG 24, Girnar 1, J 11 and KRG 1 of spanish type are suitable for sowing in drought prone areas and germplasm accessions *viz.* NRCG 12752 and 12642 were tolerant to high temperature during germination and early seedling growth (Nautiyal, 2009). The amount of K and Ca present in seed also decides the germination quality of several crops as low seed Ca and K leads to poor germinations (Singh 2002, Singh *et al.*, 2004).

Seed dressing with fungicides is most important and essential to assure better seed germination and initial crop establishment as there are plenty of soil organisms which eat or deteriorate seed once it come into contact with soil (Singh *et al.*, 1992,

1993). Being low in quantity the biofertilizers, biopesticides and fertilizers are often used to increase the efficacy of these, however some of these fertilizers are toxic to certain seeds (. In groundnut with soft testa, various micronutrient salts (borax, boric acid, copper sulphate, copper acetate, copper chloride, ferrous sulphate, ferric chloride, manganese sulphate, manganese chloride, zinc sulphate and zinc chloride) at 5 kg ha^{-1} and macronutrients salts (Calcium chloride, calcium nitrate, potassium dihydrogen orthophosphate and potassium chloride) at 10 kg ha^{-1} when compared as seed dressing and also in soil furrows revealed that, in general, most of the micronutrient salts and $CaCl_2$ showed their positive response and increased pod yield and seed size of groundnut. However as seed dressing, only $CuSO_4$, Cu $(CH_3COO)_2$, $FeSO_4$, $MnSO_4$ and $CaCl_2$ could increase the yield and other parameters. Interestingly, H_3BO_3, $CuCl_2$, $MnCl_2$, $ZnCl_2$, $ZnSO_4$, $FeCl_3$, Ca $(NO_3)_2$, KH_2PO_4 and KCl as seed dressing caused damage to seed and reduced field emergence hence should not be used (Singh, 2001 Singh, 2002).

It is desirable to have seed and crop genotypes with high micronutrient density as the micronutrient malnutrition is spread worldwide and eating micronutrient dense crop is the best way to solve the problem. Under harvestplus programme there is emphasis on the identification and breeding the crops for high micronutrient particularly of Fe and Zn. Looking to the urgency our top priority should be to strengthen research on these micronutrient rich varieties with incentive to grow the same even on the little bit compromise of yield by the farmers. There is an urgent need to identify high Fe and Zn containing cultivars and germplasm in each and every crops more particularly of the crop being consumed with less preparation to minimize the loss of micronutrient.

The seventy groundnut varieties were sorted based on their Zn concentrations in seed and categorized as low (below 30 mg kg^{-1}), medium (31-50 mg kg^{-1}) and high (51 mg kg^{-1} Zn and above) zinc density genotypes. The Zn concentration in seed of various groundnut genotypes ranged from 11-77 mg kg^{-1} with a mean value of 45 mg kg^{-1}. However, the Fe concentration in these ranged from 35-150 mg kg^{-1} and mean 60 mg kg^{-1}. Of these 12 groundnut cultivars (Tirupati 4, UF 70-103, GG 2, GG 5, GG 7, ICGV 86590, CSMG 84-1, JL 24, JL 220, CO 1, CO 2 and TMV 2) with more than 45 mg kg^{-1} Zn in their seeds, were identified as high Zn density groundnut and 10 cultivars (M 145, M 335, Tirupati 4, UF 70-103, ICGS (FDRS) 4, GG 2, GG 5, GG 7, Jawan and ICGV 86590) with more than 85 mg kg^{-1} Fe in their seeds, were categorized as high Fe-density groundnut. The GG 2, GG 7, JL 24 and ICGV 86590 are high yielding commercial cultivars and also good source of Fe and Zn (Singh *et al.*, 2011, 2013). The identified, high Zn density groundnut genotypes are being recommended for their cultivation and incorporation in human food to combat the Zn malnutrition in India (Singh and Chuni Lal, 2007).

There is a now good network of quality control and infrastructure in India as seed testing has been recognized as an essential aspect of seed quality. Accordingly, Section 4(2) of the Seeds Act, 1966 empowers the State Government to establish one or more State Seed Testing Laboratories in the state. There are many state laboratories and 2 central laboratories in the country. The central laboratories include National

Seed Research and Training Centre (NSRTC) at Varanasi and the laboratory at Central Institute for Cotton Research, Nagpur (for GM cotton only)

The National Seed Research and Training Centre, Government of India, Ministry of Agriculture Department of Agriculture and Cooperation, Varanasi (U.P.), periodically organise NSRTC training courses to Indian officials dealing with seed on 'Seed Quality Regulation and Seed Health Testings' and also on 'Varietal Identification through in vitro Techniques'. The training are basically designed for the officials engaged in the seed production, seed quality regulation and testing programme at Central and State Govt Organizations/Institutions, SAU's, ICAR Institutes, public and private seed sectors at national level, with an objective to provide hands-on experience and up-to-date the knowledge and skills of Seed Analysts, Seed Testing Officers, Seed Technologists and seed quality regulation personnels of various agencies and straightened cooperation, uniformity and information exchange in the field of seed quality control and testing at national level. The main course contents of training covers:

1. Seed development programme and quality regulation.
2. Seed production, pre and post harvest seed handling techniques.
3. Mixing and dividing techniques for obtaining homogenous representative sample.
4. Seed sampling and methods.
5. Seed quality determination tests- purity, moisture, germination etc.
6. Germination testing, methods, media and evaluation.
7. Use of tolerance tables, calculation and reporting results.
8. Genetic purity test : GOT, ODV and Molecular marker.
9. Varietal identification through molecular/various *in vitro* techniques
10. Tetrazolium testing for quick evaluation.
11. Seed Health Test.
12. Seed law enforcement for GM Crops.
13. Seed Law enforcement and seed industry management.
14. ISTA accreditation standard, quality assurance in the laboratory, staff and infrastructure.

These courses provide general information and hands-on experience in various techniques for quality regulation and seed health, varietal identification and gives an overview on the role of seed quality control and seed testing in the seed sector, rules for sampling and seed testing, principles and practice on seed production quality, pre and post harvest handling techniques for seeds, the ISTA accreditation standard establishment and the infrastructure and management of a seed testing laboratory.

The International Seed Testing Association (ISTA) founded in 1924 during the 4th International Seed Testing Congress held in Cambridge, UK is well known International body for its service to seed industry globally for lab validation, accreditation, trainings and research on tests. With its vision on 'Uniformity in seed quality evaluation worldwide' the ISTA aimed at developing and publishing standard procedures in the field of seed testing as it is inextricably linked with the history of seed testing with 202 member laboratories in 77 countries worldwide. The 120 of the ISTA Member Laboratories are accredited by ISTA and entitled to issue ISTA International Seed Analysis Certificates. The latest ISTA (2015) International rule for seed testing has been released and will be effective from 2015.

In hot and humid areas maintenance of seed viability and vigour in most of the crops, during storage, is a serious problem as there is rapid seed deterioration which result into poor crop establishment and productivity. Seed priming, under sub-optimal conditions, invigorate and improve seed vigour by reducing water soluble sugars and enhanced activities of dehydrogenase, peroxidase, catalase and super oxide dismutase (Nautiyal *et al.*, 2013). Seeds of chickpea varieties kabuli (Pusa 1108) and desi (Pusa 5028) treated with turmeric or garlic extract exhibited higher speed of germination, thiram inhibited nodule formation, while ascorbic acid enhanced seedling vigour and nodule number. Osmopriming with PEG (-1.5MPa) under sub-optimal conditions (10-12°C) in okra varieties, A-4 and Varsha uphar, improved field emergence and crop stand. Hydro priming (17 h 25°C) and magneto priming (1000 GI 2 h) enhanced seed germination and field emergence in low vigour seed lots of HQPM-1 and sweet corn.

Seed coating with commercial polymers Seed coat Red' (5ml kg⁻¹ seed) in low vigour seed lots of maize showed early vegetative growth and higher seed yield. Paddy seed (PRH-10) treated with pulse electromagenetic field (EMF: 50Hz) for five days daily for 15 hours showed higher grain yield (4.3 t ha⁻¹). The cereal crops have tremendous potential to re-mobilize pre-anthesis non-structural carbohydrates (NSC) stored in stem to the developing grains and wide genetic variation was recorded in maize, wheat and paddy. The seedling vigour of wheat recombinant inbred lines (RILs) for drought tolerance traits showed wide variation in initial and final germination percentage, root and shoot length, and 1000-seed weight and the pre-anthesis shoot biomass was associated (r=0.50) with root biomass (<10 cm depth). In efficient maize line Prakash the re-mobilization of NSC enhanced seed quality, harvest index and 1000-seed wt. Studies are on at Seed Technology Lab of ICAR-IARI to utilize seed priming potential for enhancing both seed quality and productivity of all crops and vegetables.

6. Breeder Seeds Production

Breeder seed production is basically the mandate of the ICAR and is being undertaken with the help of i) ICAR Research Institutions, National Research Centres and All India Coordinated Research Project of different crops; ii) State Agricultural Universities (SAUs) with 14 centres established in different states; iii) Sponsored breeders recognized by selected State Seed Corporations, and iv) Non-government organizations. ICAR also promotes sponsored breeder seed production programme through the NSC, SFCI, SSCs, KVKs etc.

The production of breeder seed requires several steps right from placement of indent of breeder seed with Director of Agriculture by the State Government and State

Public Seed Producing Agencies, Communication of the screened and compiled indents by Director of Agriculture of the State to Seed Division of Ministry of Agriculture, GOI. Seed Association of India forwards the indents of private parties to Seed Division of this Ministry. Central Agencies such as NSC, SFCI etc. would place their indents directly with Seed Division of Ministry of Agriculture, New Delhi. Forwarding of compiled indents by Seed Development section, Ministry of Agriculture (MOA), Government of India (GOI) to the ICAR. Communication of Breeder Seed Production Plan in BSP-1 by Project Coordinator (Crops) to Seed Development Section, MOA and ADG (Seeds), ICAR. Communication of the BSP-2 and BSP-3 by the concerned Breeder to the Seed Development Section of MOA and ADG (Seeds), ICAR. Communication of the final production figures of breeder seed by the ICAR in BSP-5 to the Seed Development Section, MOA, GOI and Allocation of Breeder seed by Seed Development section, MOA, GOI to Director of Agriculture to the concerned indenters.

The indents from various seeds producing agencies are collected by the State Departments of Agriculture and submitted to the DAC, which compiles the whole information crop wise and sends it to the Project Coordinator/Director of the respective crops in ICAR for final allocation of production responsibility to different SAUs/ ICAR institutions. The allocation of responsibility for production of breeder seed is discussed in the workshop in respect of the particular crop and is made to various centres as per the facilities and capabilities available at the centres and the availability of nucleus seed of a particular variety. It may be noted that indents are compiled and forwarded to ICAR at least 18 months in advance. The details of these and deadlines being regulated are given as a calendar of operation in Table 13.4. These informations are updated regularly on the web site of DAC (*http://dacnet.nic.in/eands*).

The monitoring team, consisting of breeder of the variety, the concerned Project Director or his nominee, representative of NSC have been constituted and reporting proformas have been devised to make the programme systematic, and for proper evaluation of the breeder seed production programme. The production of breeder seed is reviewed every year by ICAR-DAC in the annual seed review meeting.

There has been a steady increase in the production of breeder seed over the years. The actual production of breeder seed by different centres is intimated to DAC by ICAR. On receipt of information from ICAR, the available breeder seed is allocated to all the indenters in an equitable manner. In the case of varieties which are relevant only to a particular state, the indents for breeder seed are placed by the concerned Director of Agriculture with the SAUs/ICAR institutions located in the state. The breeder seed produced is lifted directly by the Director of Agriculture or foundation seed producing agencies authorized by him.

The breeder seed production increased during the last decade from 30,671 q during 2002-03 to 1,04,784 q during 2011-12 (242 per cent) (Figure 13.2). During that period, the indents for breeder seed also rose from 26,140 q during 2002-03 to 94,220 q during 2011-12 (260 per cent increase). The breeder seed indent and production of various crops during 11th plan was higher by 143 per cent and 101 per cent, respectively over that of 10th plan. However, the breeder seed production reduced and during 2012-13 and 2013-14 was 89,437 and 70,704 q, respectively (Table 13.5).

Table 13.4: Calendar of Operations for Production and Distribution of Breeder Seed

Sl.No.	Steps	Last Date of Action — Kharif	Last Date of Action — Rabi
1.	Placement of Indent of breeder seed with Director of Agriculture by the State Government and State Public Seed Producing Agencies.	15th December of previous year	31st May of year
2.	Communication of the screened and compiled indents by Director of Agriculture of the State to Seed Division of Ministry of Agriculture (MOA), Government of India (GOI). Seed Association of India would forward the indents of private parties to Seed Division of this Ministry. Central Agencies such as NSC, SFCI etc. would place their indents directly with Seed Division of MOA, New Delhi.	1st week of January	7th June
3.	Communication of compiled indents by Seed Development section, MOA, GOI to ICAR.	28th February	15th July
4.	Communication of Breeder Seed Production Plan in BSP-1 by Project Coordinator (Crop) to Seed Development Section, MOA and ADG (Seeds), ICAR.	15th May	15th October
5.	Communication of the BSP-2 by the concerned Breeder to the Seed Development Section of Ministry of Agriculture and ADG (Seeds), ICAR	After 15 days of the planting breeder seed	After 15 days of the planting breeder seed
6.	Communication of the BSP-3 by the concerned breeder to the Seed Development Section, Ministry of Agriculture, Govt. of India and ADG (Seeds), ICAR.	After 15 days of inspection of breeder seed by the team	After 15 days of inspection by the monitoring team
7.	Communication of the final production figures of breeder seed by the ICAR in BSP-5 to the Seed Development Section, MOA, GOI.	15th February	15th July
	Groundnut compensatory production.	15th April	
	Cotton 1. North Zone 2. Central and South Zone	15th February 1st March	
	Pigeonpea 1. Early and medium duration varieties 2. Long duration varieties 3. Toria and Rape seed	7th March 15th April 31st May	

Contd...

Table 13.4–*Contd....*

Sl.No.	Steps	Last Date of Action	
		Kharif	Rabi
8.	Allocation of breeder seed by Seed Development section, Ministry of Agriculture, GOI to Director of Agriculture and concerned indentors.	31st March	15th September
9.	Communication of the details of lifting of breeder seed against the GOI allotment to MOA by Director of Agriculture in the performa 'A'.	After 15 days of the cut-off-date	After 15 days of the cut-off-date
10.	Communication of details of seed supply to the allottees by the breeder to MOA and ICAR in Performa 'B' enclosed with supply plan.	After 15 days of the cut-off-date	After 15 days of the cut-off-date

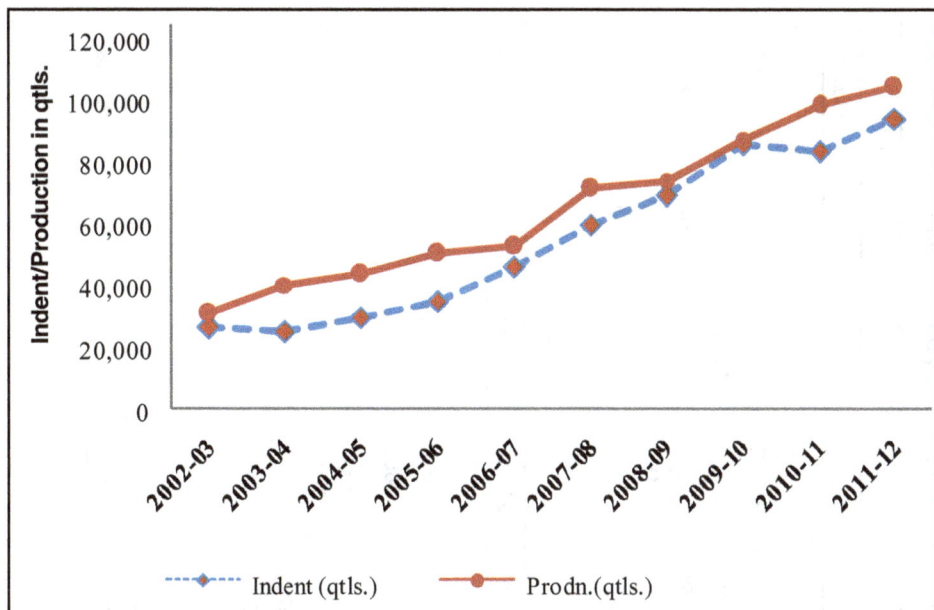

Figure 13.2: Breeder Seed Indent and Production of Field Crops during 2003-12.

6.1 Crop-wise Breeder Seed Production

The crop-wise breeder seed indent and the production for the last five year are given in Table 13.5. During 10th plan the breeder seed indent of cereal crops increased from 9,253 in 2002-03 to17,006 q (83.8 per cent increase) in 2006-07. During 11th plan it was 25,903q in the first year (2007-08) and 36,831q in 5th year (2011-12) (Figures 13.3a and b) with an increase of 42.2 per cent. But, further it was declined to 27,106 q in 2012-13 and 25,836 q in 2013-14. Production of breeder seed during 10th plan rose from 11,126 in 2002-03 to 20,489 q in 2006-07 (84.2 per cent increase). During 11th plan there was an increase of 37 per cent from 32,759 q in 2007-08 to 44,889 q in 2011-12. During 2012-13 and 2013-14 production was 40,135q and 35,436 q, respectively (Table 13.5). Overall, an increase of about 146.4 per cent in breeder seed production during 11th plan was recorded over that of 10th plan.

The breeder seed indent of oilseed crops increased consistently in both 10th and 11th plans (Figures 13.3c and d) from 11,702 in 2002-03 to 21,195q (81.1 per cent increase), in 2006-07 during 10th plan, while in 11th plan an increase of 83.4 per cent was observed from 22,577 (2007-08) to 41,404 q (2011-12). The breeder seed production of oilseed crops increased from 12,431 (2002-03) to 21,767 q (2006-07) and 26,037 (2007-08) to 41,446 q (2011-12), respectively, during 10th and 11th plan. During last five years the production increased from 29,417 q (2009-10) to 33,235 q (2012-13) with an increase of 12 per cent while it was 22,398 in 2013-14 (Table 13.4). Further, the indents for breeder seed production for the year 2008-09 and 2009-10 were higher than the actual production. Though the breeder seed production of oilseeds in 11th plan increased by 64.9 per cent over that of 10th plan there was a short fall of 18 per cent and 15.2 per cent during 2008-09 and 2009-10, respectively.

Table 13.5: Breeder Seeds Indent and Production (q) of Various Crops during the Last Five Years

Crops	2009-10		2010-11		2011-12		2012-13		2013-14	
	Indent	Production	Indent	Production	Indent	Production	Indent	Production	Indent	Production
Cereals										
Wheat	32,330	35,049	29,692	38,469	28,860	35,745	20,542	27,502	20,050	22,492
Paddy	3,880	5,387	4,604	6,095	5,772	6,828	5,267	11,455	4,745	10,586
Barley	2,496	3,053	1,778	2,900	1,842	1,906	1,029	698	843	1,820
Sorghum	55	221	36	167	113	158	115	375	83	305
Smallmillet	5	24	22	42	18	47	37	109	44	116
Maize	179	243	178	232	211	173	99	109	60	89
Pearlmillet	8	8	10	28	15	32	17	67	11	28
Total	**38,954**	**43,985**	**36,320**	**47,934**	**36,831**	**44,889**	**27,106**	**40,315**	**25,836**	**35,436**
Oilseeds										
Groundnut	22,897	16,407	11,423	15,092	18,115	20,076	13,075	12,014	11,027	12,996
Soybean	11,625	12,517	22,293	18,327	22,973	20,853	24,688	20,718	19,509	8,660
Safflower	10	138	20	51	27	53	28	65	21	331
RM	76	138	75	150	49	151	108	212	95	213
Linseed	38	68	49	97	145	157	96	139	41	99
Sesame	3	9	28	49	42	67	32	41	26	59
Sunflower	14	36	9	36	32	48	5	16	3	15
Castor	17	88	24	202	11	28	9	15	4	15
Niger	6	17	16	10	11	15	10	15	8	10
Total	**34,685**	**29,417**	**33,937**	**34,015**	**41,404**	**41,446**	**38,051**	**33,235**	**30,734**	**22,398**

Contd...

Table 13.5–*Contd...*

Crops	2009-10		2010-11		2011-12		2012-13		2013-14	
	Indent	Production	Indent	Production	Indent	Production	Indent	Production	Indent	Production
Pulses										
Chickpea	9,381	8,850	9,889	10,787	9,915	11,141	9,944	10,452	9,433	8,768
Mungbean	798	1,169	1,059	1,073	1,244	1,343	1,168	703	799	679
Pigeonpea	276	499	475	975	537	1,317	646	787	391	674
Fieldpea	178	1,304	332	997	838	959	774	863	588	631
Lentil	347	516	431	433	644	718	622	916	470	614
Urdbean	501	617	508	805	846	1,031	799	606	518	533
Rajmash	2	5	–	–	–	–	-	-	–	–
Cowpea	66	82	30	28	54	42	55	37	39	53
Mothbean	151	113	221	262	213	95	140	63	95	40
Horsegram	1	2	–	–	12	11	7	3	8	–
Total	**11,700**	**13,155**	**12,944**	**15,360**	**14,303**	**16,656**	**14,153**	**14,429**	**12,341**	**11,992**
Forage Crops										
Oats	224	371	202	305	1,082	890	1278	611	402	398
Guar	480	389	248	520	277	575	289	431	344	206
Maize	54	72	63	93	75	77	99	109	89	138
Berseem	45	60	69	50	94	84	87	77	35	37
Sorghum	55	221	23	29	34	53	33	74	34	19
Teosnite	4	4	5	10	–	–	–	–	–	–
Cowpea	8	17	9	16	43	12	29	14	20	10
Bajra	1	2	2	9	6	6	5	9	3	6

Contd...

Table 13.5—Contd...

Crops	2009-10		2010-11		2011-12		2012-13		2013-14	
	Indent	*Production*	*Indent*	*Production*	*Indent*	*Production*	*Indent*	*Production*	*Indent*	*Production*
Rice Bean	3	3			3	3	3	2	3	4
Lucerene	8	8	6	6	13	7	7	7	6	5
Gobhisarson	–	–	0	1	1	2	0	2	0.32	0.43
Total	**882**	**1,145**	**627**	**1,039**	**1,627**	**1,708**	**1,832**	**1,336**	**936**	**823**
Fibre Crops										
Cotton	37	102	44	58	32	59	40	107	26	36
Jute	6	8	8	13	13	15	11	14	17	19
Sunhemp	–	–	–	–	10	11	–	–	–	–
Total	**42**	**110**	**52**	**71**	**55**	**85**	**51**	**121**	**43**	**55**
Grand Total	**86,264**	**87,812**	**83,880**	**98,419**	**94,220**	**1,04,784**	**81,193**	**89,437**	**69,890**	**70,704**

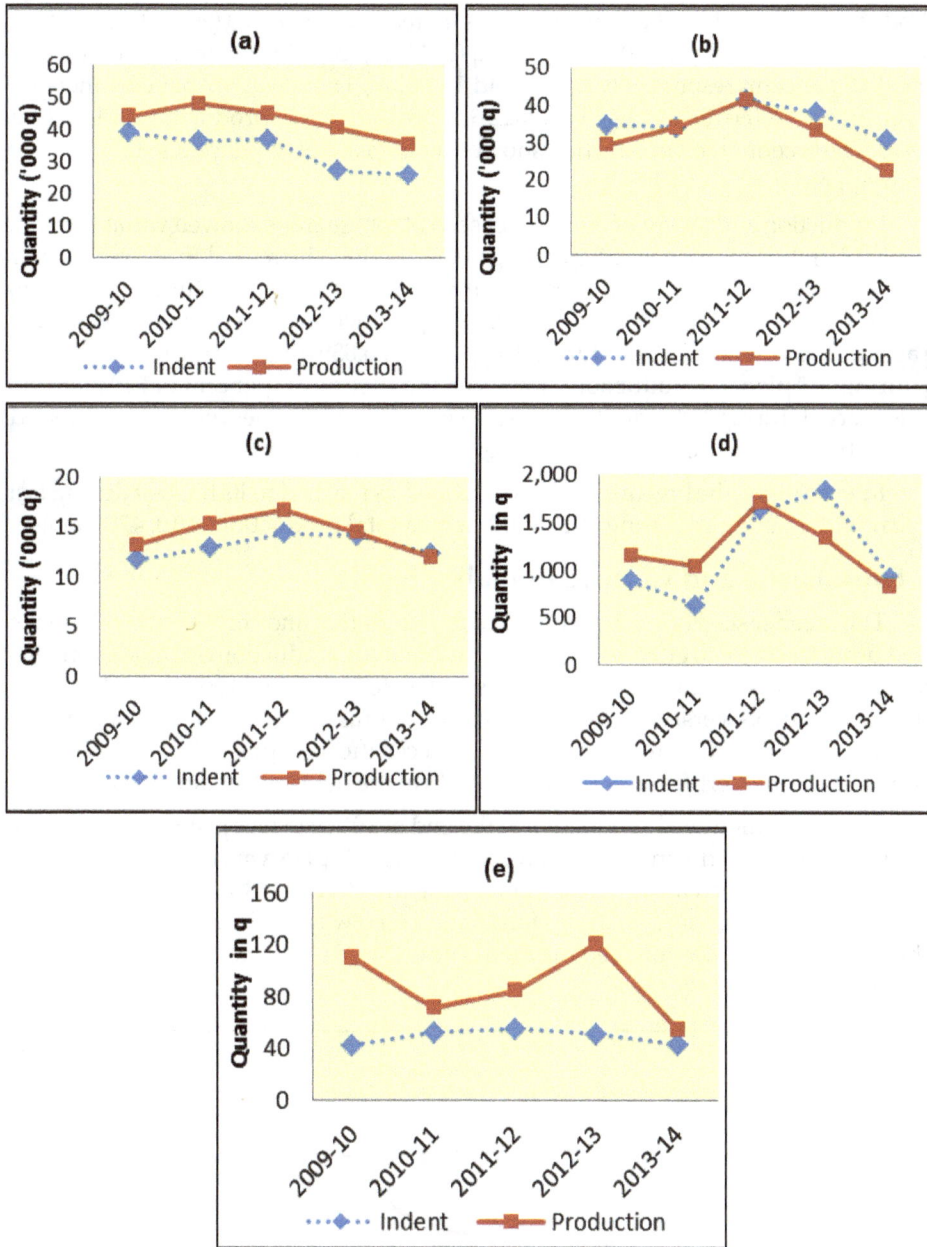

**Figure 13.3: Breeder Seed Indent and Production
(a) Cereals; (b) Oilseeds; (c) Pulses, (d) Forage crops
and (e) Fibre Crops during Last Five Years.**

The indent for breeder seeds for pulse crops enhanced from 4,696 q in 2002-03 to 7,348 q (56.5 per cent increase) in 2006-07 during 10th plan, while during 11th plan 9,948 q in 2007-08 to 14,303 q (43.8 per cent increase) in 2011-12 (Figures 13.3 e).

Production of breeder seeds of pulse crops increased from 6,573 (2002-03) to 9,382 q (2006-07) and 11,234 (2007-08) to 16,656 q (2011-12), registering an increase of 42.8 and 48.3 per cent, respectively in 10th and 11th plan. There was 9.7 per cent increase from 11,700 (2009-10) to 14,429 q (2012-13) breeder seeds production in last five years. The breeder seed production showed an increase of 91 per cent in the 11th plan as compared to that of 10th plan.

Production and indent of the breeder seed of forage crops showed variable trend. In 10th plan the indent was 385 q during 2003-04 which increased to 456 q (18.5 per cent), during 2006-07 while in 11th plan it increased from 931 q during 2007-08 to 1,627 q (75 per cent) during 2011-12. The total production of breeder seed was 3,259 q against the indent of 2,012 q during 10th plan and 6,839 q against the indent of 5,610 q during 11th plan. Its production showed an increase of 159 per cent in 10th plan and 21.8 per cent during 11th plan. There was an increase of 109.8 per cent in breeder seed production of forage crops in 11th plan over that of 10th plan.

Interestingly, the breeder seed production has reached to its highest during last 5 year with maximum during 2011-2012, with a total production of 10, 478 tonne.

7. Foundation and Certified/Quality Seeds

The breeder seed has to be multiplied to foundation and certified seeds to make seed production chain effective. The responsibility for production of foundation seed has been entrusted to the NSC, SFCI, SSCs and State Departments of Agriculture and private seed producers, who have the necessary infrastructure facilities. Foundation seed is required to meet the standards of seed certification prescribed in the Indian Minimum Seeds Certification Standards, both at the field and laboratory testing.

The foundation seed production at the end of 10th plan was just 80,000 q (Figure 13.4), which showed consistent increase during 11th plan varying from 85,000 to 2,23000 q with an increase of 9.7 per cent during the 1st year and 178.9 per cent during the final year. A total of 7.7 lakh q of foundation seed was produced during 11th plan. The increase in foundation seed production during 2011-12 was 162.3 per cent over that of the year 2007-08.

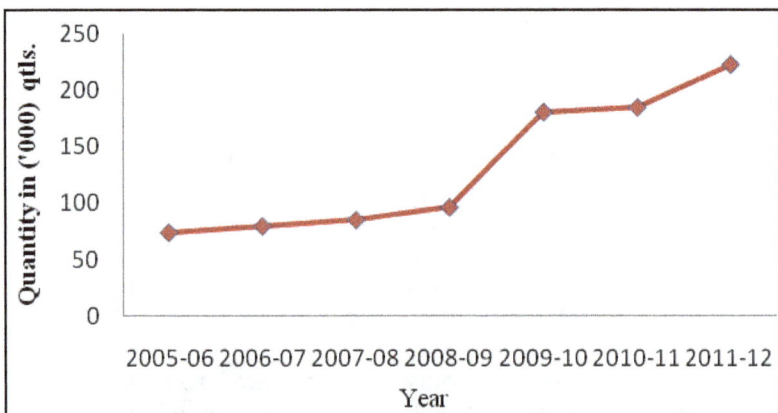

Figure 13.4: Trend in Foundation Seed Production in India.

Total quality seed production by public sector during 2003-04 to 2007-08 was 8.09 million tonnes which rose to 15.35 million tonnes during the next 5 years, with an increase of 89.7 per cent. Similarly, there was an increase of 31.6 per cent during the 11th plan period (Table 13.5). The percent increase during 10th plan was 47 per cent during 2007-08 as compared to that of the year 2003-04. The seed production by the private sector increased from 0.63 million tonnes during 2003-04 to 1.67 million tonnes during 2012-13 showing an increase of 165.1 per cent during the last 10 years. The contribution of public sector reached up to 61.1 per cent in the year 2009-10 but declined thereafter and during 2012-13 it was only 49.1 per cent (Table 13.6). During the same period, the contribution of private sector showed marginal decline from 42.6 per cent during 2007-08 until 2009-10 but registered an increase since then and reached up to 50.9 per cent. Private seed sector dominates in quality seed production of hybrids of maize, pearl millet, sorghum, sunflower and cotton and 90 per cent of vegetable seed production comes from this sector only.

Table 13.6: Quality Seed Production by Public and Private Sectors and their Contributions

Year	Certified/Quality Seed Production/ Availability (million tonnes, m t)	Seed Produced by Public Sector		Seed Produced by Private Sector	
		Quantity (m t)	Share (per cent)	Quantity (m t)	Share (per cent)
2003-04	1.32	0.70	52.5	0.63	47.5
2004-05	1.41	0.77	55.0	0.63	45.0
2005-06	1.48	0.79	53.2	0.69	46.8
2006-07	1.94	1.15	59.0	0.80	41.0
2007-08	1.94	1.12	57.4	0.83	42.6
2008-09	2.50	1.51	60.2	1.0	39.8
2009-10	2.80	1.71	61.1	1.09	38.9
2010-11	3.22	1.66	51.6	1.56	48.4
2011-12	3.54	1.81	51.1	1.73	48.9
2012-13	3.29	1.61	49.1	1.67	50.9

During the last 10 years availability of certified/quality seed showed a consistent increase till 2011-12, but declined marginally (8 per cent) during 2012-13 over the highest value achieved in 2011-12. Nevertheless, the production of quality/certified seeds was always higher than the required quantity (Table 13.7). The requirement for quality seed consistently increased from 1.8 million tonnes (m t) during 2007-08 to 3.32 m t in 2013-14 showing an increase of 84 per cent. The requirement of quality seed was marginally reduced, by 4.5 per cent, during the year 2012-13 over that of the year 2011-12. But availability of quality seeds varied from 1.94 (2007-08) to 3.44 m t (2013-14), registering an increase of 77 per cent. However, the highest increase of 82 per cent in availability of quality seeds was achieved during 2011-12 over that of 2007-08 which narrowed down the gap between demand and availability up to 3.6

per cent. Therefore, there is a need for strengthening the quality seed production programme in the country.

Table 13.7: Demand and Availability of Certified/Quality Seeds in India

Year	Quantity (million tonnes)		
	Availability	Demand	Difference (per cent)
2007-2008	1.94	1.80	7.7
2008-2009	2.50	2.07	20.8
2009-2010	2.79	2.49	12.0
2010-2011	3.21	2.90	10.7
2011-2012	3.54	3.30	7.3
2012-2013	3.29	3.15	4.4
2013-2014	3.44	3.32	3.6

The production and distribution of quality/certified seeds is primarily the responsibility of the State Governments. Certified seed production is organized through SSCs, Departmental Agricultural Farms, Cooperatives etc. The distribution of seeds is undertaken through a number of channels such as departmental outlets at block and village level, cooperatives, outlets of seed corporations, private dealers etc. The efforts of the State Governments are being supplemented by NSC and SFCI which produce varieties of national importance. NSC markets its seeds through its own marketing network and also through its dealer network. SFCI markets its seeds mainly through the State Departments of Agriculture and the SSCs. The production of certified seed by NSC and SSCs is mainly organized through contract growing arrangements with progressive farmers. The SFCI undertakes seed production in its own farms which is now as it is merged with NSC. The private sector has also started playing an important role in the supply of quality seeds of vegetables and crops like hybrid maize, sorghum, bajra, cotton, castor, sunflower, paddy etc. Preparation of state seed rolling plan and strategy are tied up with seed production agencies (Selvaraj 2013).

State Governments assess requirement of certified/quality seeds, on the basis of the area sown under different crop varieties, area covered by hybrid and self-pollinated crop varieties as well as the seed replacement rate achieved. On the basis of the production of seed in government farms, SSCs and other agencies, the availability of seed is ascertained by the State Departments of Agriculture. The Government of India periodically assesses the requirement and availability of seeds through detailed interaction with State Governments and seed producing agencies in the bi-annual Zonal Seed Review Meetings and the National *Kharif* and *Rabi* Conferences. The Department of Agriculture and Cooperation facilitates tie-up arrangements with seed producing agencies to ensure that the requirement of seeds is met to the maximum extent possible. As an example the Zonal seed review meeting for *Kharif* 2013 reported all India total availability of certified/quality seed for *Kharif* 15.4 lakh t against the requirement of 13.99 lakh t (Crop-wise and state-wise availability of certified/quality seed for *Kharif* 2013 against the requirement are given in Tables 13.8 and 13.9).

Table 13.8: Crop-wise Requirements and Availabilities of Certified/Quality Seeds in India

Crops	Quantity (Quintals) for Kharif 2013		
	Requirement	Availability	Deficit/Surplus
Paddy	6254264	6923320	669056
Maize	779459	742633	–36826
Bajra	231119	317380	86261
Jowar	164610	242243	77633
Ragi	29984	31163	1179
Little Millet	40	49	9
Banyard Millet	315	25	–290
Kodo Millet	472	1149	677
Italian Millet	3300	3300	0
Total Cerals	7463563	8261262	797699
Arhar	255623	246598	–9025
Urd	147057	249314	102257
Moong	168470	189145	20675
Cowpea	22079	18781	–3298
Moth Bean	20900	16648	–4252
Horse Gram	4785	4789	4
Rajma	510	510	0
Peas/Others	75	75	0
Total Pulses	619498	725859	106361
Groundnut	2086881	2176047	89166
Soybean	3299968	3694733	394765
Sesame	20737	22334	1597
Sunflower	17269	19538	2269
Castor	62576	72782	10206
Niger	3104	2803	–301
Total Oilseeds	5490535	5988237	497702
Cotton	215158	239885	24727
Jute	32208	14451	–17757
Roselle	8	14	6
Total Fibre	247374	254350	6977
Bajra Napier	115	115	0
Ricebean	150	150	0
Dhaincha	71350	70850	–500
Sunnhemp	20622	24850	4228
Guar	74700	68616	–6084
Total others	166937	164581	–2356
Grand Total	**13987907**	**15394290**	**1406383**

**Table 13.9: State-wise Requirement and Availability of
Certified/Quality Seeds in India**

States	Quantity (Quintals) for Kharif 2013		
	Requirement	Availability	Deficit/Surplus
Andhra Pradesh	2832712	3243129	410417
Karnataka	1061998	1038580	−23418
Kerala	40000	40000	0
Tamilnadu	375141	400586	25445
Pondicherry	1215	1253	38
Maharashtra	1999076	2022818	23742
Gujarat	566064	581943	15879
Chhattisgarh	707040	902907	195867
Madhya Pradesh	1721344	1951462	230118
Rajasthan	681179	828770	147591
Goa	4622	4622	0
Punjab	214405	265329	50924
Haryana	104700	225595	120895
Uttar Pradesh	957810	698908	−258902
Uttrakhand	26726	54411	27685
Himachal Pradesh	44427	44427	0
Jammu & Kashmir	57647	57647	0
Bihar	503910	597366	93456
Jharkhand	198751	229129	30378
Odisha	687056	718050	30994
West Bengal	508230	793238	285008
Sikkim	3094	3094	0
Nagaland	46056	46056	
Arunachal Pradesh	8402	8402	0
Assam	583644	583644	0
Mizoram	13818	13818	0
Manipur	6130	6130	0
Meghalaya	14870	14870	0
Tripura	17840	18105	265
Grand total	**13987907**	**15394290**	**1406383**

All State Governments and seed producing agencies assess all crops seed production in that particular season and take remedial measures to increase the seed yield per unit area including organizing contingency seed production in *Rabi* and *Summer* seasons (Maize for Bihar, Jharkhand and U.P., paddy, urd, moong, cowpea and groundnut for TN, Karnataka, Maharashtra, Gujarat, Rajasthan and Odisha) and advance tie up arrangement to procure required seeds so as to enable the state to meet seed demand. The State Governments implement seed village, strengthening the seed infrastructure facilities and seed components available under various Crop Development Programmes/schemes effectively. The total quality seed production increased from 2.4 lakh quintals at the time of inception of the mega seed project (2005-06) to 6.9 lakh quintals during 2011-12. Total quality seed availability in India during 2013-14 was 3.44 million tonnes (m t) as compared to 2.50 m t in 2008-09 which increased availability of quality seeds to the farmers.

8. Seed Multiplication and Replacement Rates and Ratio

The seed multiplication ratio (SMR) is basically the number of seeds to be produced from a single seed when it is sown and harvested. But in agricultural crops it is difficult to measure such ratio as the seed rates vary with seed size of particular crop variety and cropping season. However, keeping in view constant seed wt for both produce and seed sown, the ratio of quantity of yield achieved to the quantity of seed sown is taken as the criterion of SMR (SMR = Seed Yield/Seed Rate). Thus the SMR, is the ratio of quantity of yield to the seed sown and increasing in most of the crops since last one decade. Taking into consideration the yield level of 2010-11 and constant seed rate the SMR of various crops has been calculated (Table 13.10). However the SMR of hybrid seed in foundation and certified seed vary (Table 13.11).

The Seed Replacement Rate (SRR) is basically percent of old seed replaced with the new one It is calculated as the percentage of area sown out of total area of crop planted in the season by using certified/quality seeds other than the farm saved seeds.

SRR = (X/Y)x 100

where,

X is quantity of farmer saved seed and Y is quantity of quality seeds of a particular variety reported to cover a given area.

The SRR of important crops in India during the decade from 2001-2011 are given in Table 13.12. The SRR in 2010 as compared to that of 2007, increased from 25.2 to 32.6 per cent in wheat, 25.9 to 37.5 per cent in paddy, 44.2 to 54.1 per cent in maize, 19.9 to 25.9 per cent in sorghum, 48.5 to 61.4 per cent in pearl millet, 21.8 to 26.7 per cent in green gram and 14.3 to 24.5 per cent in groundnut (Table 13.12).

This is essential for maintaining genetic purity and quality seed production. The seed replacement rate gives an idea about the quantity of the quality seeds used by the farmers.

The state-wise, 5 yearly seed replacement rate (per cent) of important crops from 2001-2011 are given in Table 13.13. Based on the SRR the physical targets are fixed for quality seed production for achieving maximum SRR (Table 13.14). Also the price of breeder seed are fixed to encourage promotion of new varieties with quality seeds. Presently, availability of quality seed is more than the requirement due to coordinated efforts of entire seed sector resulted into high SRR in major cereal, oilseed and pulse crops. Analysis of data in rice revealed that quality seed availability with the present

Table 13.10: Crop-wise Seed Rate, Yield (during 2010-11) and their Multiplication Rate in India

Sl.No.	Crops	Seed Rate (kg/ha)	Seed Yield (kg/ha)	Multiplication Ratio 1:
1.	Jute	5	2329	466
2.	Mesta	12.5	1115	90
3.	Sunhemp	25	750	30
4.	Cotton mcu-5	12.5	499	40
5.	Arhar	10	655	66
6.	Moong	15	600	40
7.	Urd	12.5	500	40
8.	Cowpea	15	600	40
9.	Castor	12.5	1534	123
10.	Soybean	62.5	1327	41
11.	Groundnut	100-150	1411	14
12.	Til	5	429	86
13.	Guar	12.5	500	40
14.	Cowpea fodder	12.5	500	40
15.	Sorghum	12	949	80
16.	Foragze sorghum			
	MP chari	5	500	100
	Pc- 6	10	1000	100
	Pc-23	5	500	100
17.	Berseem	20	200	10
18.	Lucern	7.5	200	27
19.	Oat	62.5	900	15
20.	Lentil	25	800	32
21.	Peas	100	3517	35
22.	Gram I-550	58	895	16
23.	L-144	75	750	10
24.	Mustard/toria	5	1185	237
25.	Lathurus	25	500	20
26.	Paddy	30	2239	75
27.	Wheat	100	2989	30
28.	Maize	20	2540	127
29.	Bajra	5	1079	216
30.	Potato	3000	42339	14
31.	Safflower	12	617	52
32.	Moth	15	600	40
33.	Horse gram	12.5	500	40
34.	French bean	60	500	9
35.	Linseed	25	408	17
36.	Sunflower	20	701	36

Table 13.11: Seed Rate, Yield and their Multiplication Ratio in Foundation and Certified Seeds of Hybrid Crops in India

Sl.No.	Crops and Seeds	Seed Rate (kg/ha)	Seed Yield (kg/ha)	Multiplication Ratio 1:
1.	**MAIZE**			
	1. Foundation II to Certified			
	a) Seed Parent	11	2200 (1100)	200 (100)
	b) Male Parent	5	N.A	100
	c) Composite	18	1800	100
	2. Foundation I to Foundation II			
	a) Female Line	10	1000	100
	b) Male Line	5	N.A	200
	c) Line Increase	15	500	33
2.	**SORGHUM**			
	1. Foundation II to Certified			
	a) Seed Parent	7.5	750	100
	b) Male Parent	5	N.A.	150
	c) Varieties	10	1000	100
	2. Foundation I to Foundation II			
	a) Female Line	7.50	750	100
	b) Male Line (B)	5	N.A.	150
	c) Restorer Line	10	1000	100
	d) Varieties	10	1000	100
3.	**BAJRA**			
	1. Foundation II to Certified			
	a) Seed Parent	2.5	500	200
	b) Male Parent	1.25	N.A	400
	2. Foundation Seed Production			
	a) Female Line	2.5	500	200
	b) Male Line (B)maintenance	1.25	N.A.	400
	c) Restorer Line	3.75	750	200
	d) Fline	3.75	750	200

rate of basic seed production is sufficient for even 100 per cent SRR, provided breeder seed conversion in seed multiplication chain is streamlined. Similarly, concerning wheat, the present level of basic seed can suffice for the gross cropped area at 50 per cent SRR. But the current SRR figures in rice and wheat are 37.5 per cent and 32.6 per cent, respectively. Main reasons for low to moderate seed replacement rates in majority of the crops are due to problems in seed multiplication chain and outreach activities.

The SRR has a strong positive relationship with the crop productivity, hence to achieve desired productivity levels and for attaining sustained food security levels,

Table 13.12: Seed Replacement Rate (per cent) of Important Crops in India during Decade 2001-2011

Year	Wheat	Paddy	Maize Hyb	Jowar Hyb	Bajra Hyb	Gram	Urd	Moong	Arhar	G.Nut	R/M	Soya-bean	Sunflower Hyb	Cotton	Jute
2001	13.0	19.2	20.9	18.4	45.9	4.2	16.6	13.5	8.7	5.2	38.4	12.4	13.7	21.2	29
2002	13	19.3	21.4	18.8	48.5	4.2	17.1	13.8	8.8	5.5	44.6	12.5	15.6	21.8	30
2003	13	19.2	24.4	26.7	51.0	7.1	20.5	19.5	13.6	11	66.9	15.6	19.6	19.8	27
2004	16.5	16.3	31.5	19.3	44.9	9.9	17.2	12.3	9.8	7.1	58.5	27	60.2	20.7	26
2005	17.6	21.3	35.4	19.0	55.4	9.4	15.7	12.5	10.5	6.9	55.4	28.9	67.7	21.8	26
2006	21.8	22.4	43.8	19.4	55.1	9.0	13.7	19.9	11.6	9.8	60.7	28.4	66.9	19.8	35
2007	25.2	25.9	44.2	19.9	48.5	11.9	23.9	21.8	16.1	14.3	58.6	33.4	62.9	15.3	33
2008	26.8	30.1	48.5	26.2	62.9	14.4	26.3	21.9	16.0	17.0	52.7	35.1	43.6	12.1	35
2009	31.8	33.6	46.8	26.4	48.9	22.0	30.9	23.0	27.8	22.9	74.8	38.9	51.5	11.7	33
2010	32.6	37.5	54.1	25.9	61.4	18.4	29.2	26.7	17.5	24.5	63.6	35.9	61.2	10.4	36
2011	32.6	40.4	56.6	23.9	60.4	19.4	34.4	30.3	22.2	22.5	78.9	52.8	32.5	33	42

Table 13.13: State-wise, 5 Yearly Seed Replacement Rate (per cent) of Important Crops from 2001-2011

States	Year	Wheat	Paddy	Maize Hyb	Jowar Hyb	Bajra Hyb	Gram	Urd	Moong	Arhar	G.Nut	R/M	Soya-bean	Sunflo-wer Hyb	Cotton Hyb	Jute
A. Pradesh	2001	–	42.0	48	91	44.0	3.00	18.0	22.7	12.5	6.00	–	77.0	81	13	–
	2006		60.0	87	63	67	49.0	27.0	32.0	37.0	25.0	–	100	0	0	–
	2011		87.2	–	62.6	–	85.0	59.2	50.9	78.0	62.1		100	0	–	–
Karnataka	2001	6	22.0	–	13	26.0	5.00	7.00	7.00	8.00	2.50	–	16.0	18.0	6.0	–
	2006	26	34.0	–	26	29	16.0	20.0	15.0	15.0	13.0	–	48.0	25	22	
	2011		41.0	–	29.5	41.9	31.4	36.3	21.0	13.0	28.8		89.0	30.4	36.4	
Tamil Nadu	2001	–	17.0	8.0	6.0	6.0	0.5	17.5	13.7	6.00	5.00	–	20.0	50.0	15.0	–
	2006		56.0	2	6		5.5	11.9	8.40	5.00	3.18	–	–	8.25	15	–
	2011		68.0	–	16.4	3.7	89.7	46.7	21.5	94.8	12.0	–	–	36.6	87.9	–
Kerala	2001	–	10.0	–	–	–	–	–	–	20.0	10.0	–	–	–	–	–
	2006		31.2	–	–	–	–	–	–	20.0	10.0	–	–	–	–	–
	2011		74.7	–	–	–	–	–	–	–	–	–	–	–	–	–
Gujarat	2001	20	18.2	–	33	–	3.42	14.4	22.2	10.4	1.16	71.4	–	–	31.0	–
	2006	24.8	21.6	–	–	–	16.1	34.2	18.8	21.8	1.88	60.3	–	–	23.4	–
	2011	26.5	38.1	–		–	25.2	38.8	35.3	30.8	4.07	100	–	–	65.0	–
Maharashtra	2001	25.0	18.0	53	15	72.0	6.00	44.0	26.0	13.0	2.00	–	33.0	28.0	80.0	–
	2006	41.0	24.0	75	10	75	9.00	45.0	3.00	15.0	3.00	–	45.0	30	85	–
	2011		45.9	93	13.5	98.6		50.8	35.1	30.7	2.69		57.7	–	60.9	–
Rajasthan	2001	11.2	4.41	1.70	1.43	32.9	6.64	3.08	8.67	14.3	0.75	68.9	4.37	–	61.6	–
	2006	19.1	7.61	19.8	7.50	46.2	3.91	5.16	9.12	8.8	2.15	60.0	10.9	–	3.8	–
	2011	30.5	7.34	52.6	21.8	57.2	12.5	6.94	18.3	21.7	6.05	85.1	28.3	–	58.3	–

Contd...

Table 13.13—Contd...

States	Year	Wheat	Paddy	Maize Hyb	Jowar Hyb	Bajra Hyb	Gram	Urd	Moong	Arhar	G.Nut	R/M	Soya-bean	Sunflo-wer Hyb	Cotton Hyb	Jute
MP	2001	5.64	3.3	8.05	5.99	35.1	1.29	1.39	2.47	2.78	0.03	7.15	6.35	15.0	3.40	–
	2006	13.1	3.96	12.9	14.2	55.3	2.15	1.17	7.58	6	0.33	21.4	15.8		0.35	
	2011	30.3	16.9	47.5	20.9	73.2	9.94	10.3	21.2	17.2	0.97	42.8	30.2			
UP	2001	15.3	14.3	6.57	5.35	13.8	4.13	7.24	13.8	12.0	1.04	26.8	7.11	36.0	71	–
	2006	24.1	22.7	19.8	11.6	51.8	14.3	12.2	20.7	18.8	4.41	58.3	41.3	61.9	66	
	2011	40.9	31.6	30.4	24.2	71.0	16.5		20.8	25.4	18.5	63.8	45.1	73	73	
Haryana	2001	17.9	11.3	5.00	–	64.3	6.3	46.6		48.6	–	46.4		–		
	2006	23.3	17.5			57.5	10.5	69				72				
	2011	33.8	30.5												42	
Punjab	2001	7	11	42.0	–	–	22	11		26	–	26	–	–		
	2006	13	21	95			56	38				58			95	
	2011	38.8	52.8	99.1			77.0					63.8			106	
HP	2001	16.7	11.6	0.3	–	–	3.47	18.3				–	20			–
	2006	13.7	14.3	0.07			3.47	16.6					25			–
	2011	26.9	96.9	33.7			95.2	93.7	77.5			83.3	28.5			–
J&K	2001	10.0	3.53	2.03	–	–	19.1		15.6			–				–
	2006	10.7	6.81	5.66			3.07		0.07			40.8				–
	2011	26.1	22.4	18.1	0.66			42.8				41.8				–
Odisha	2001	40.4	9.59	8.04			7.72	2.3	1.52		20.2	18.5		54.2	42.2	98
	2011	39.0	21.7	13.2			5.68	3.52	2.41	4.75	32.4	32.4		64.0	65.6	28

Contd...

Table 13.13—*Contd...*

States	Year	Wheat	Paddy	Maize Hyb	Jowar Hyb	Bajra Hyb	Gram	Urd	Moong	Arhar	G.Nut	R/M	Soya-bean	Sunflo-wer Hyb	Cotton Hyb	Jute
W. Bengal	2001	30.0	22.0	15.0	–	–	15.0	24.0	24.0	33.0	30.0	30.0	–	–	–	70
	2006	37	26	22	–	–	22	28.5	29	40	35	38	–	–	–	72
	2011	43.4	33.7	29.0	–	–	26.9	35.6	33.6	45	40.3	41.4	–	–	–	79
Bihar	2001	8.43	6.33	21.1	–	–	–	–	–	–	–	–	–	–	–	–
	2006	11	12	60	–	–	8	–	–	–	–	40	–	–	–	–
	2011	34.8	38.0	100	–	–	15.7	18.5	20.2	11.2	–	47.3	–	–	–	–
Chattisgarh	2006	9	8.50	11	–	–	7.8	5	6.75	12	2	9.8	26	42	–	–
	2011	33.7	34.3	21.2	–	–	15.9	8.05	3.65	20.7	3.37	24.7	64.4	–	–	–
Assam	2006	34.8	10.7	0.62								15.7				
	2011		46.8	22.1				79.7	107.1	35.5		26.1				51
Uttarakhand	2006	20	15	3.35									9.7			
	2010	36.9	10.8	10.4			43.6	71.4				42.1	36.1			
Jharkhand	2006	9		7			3					7				
Tripura	2011	31.9	17.1	7.27			1.29	6.98	21.2	23.5		20.5				
	2006	100	35.0	40.0				39.0	30.0	29.0		96.0				
	2010	69.5	33.8	58.0				67.2	38.5	43.3	56.1	69.3				34
Manipur	2006						35.7					64.8				
Sikkim	2010		59.5	10.0				99.8	99.5	100	35.0		23.6			
	2006	39.7	11.9	15.4				9.58				63.1	16.3			
	2011	40.0	20.0	14.0								75.0	6.00			

Contd...

Table 13.13–*Contd...*

States	Year	Wheat	Paddy	Maize Hyb	Jowar Hyb	Bajra Hyb	Gram	Urd	Moong	Arhar	G.Nut	R/M	Soya-bean	Sunflo-wer Hyb	Cotton Hyb	Jute
Mizoram	2006		10.8	76.5								85.7	42.6			
	2010		28.0	44.0							80.0		30.0			
Meghalaya	2006	4.26	11.3	6.12							63.3	18.0	28.4			19
	2010		17.0	53.4									97.2			
Nagaland	2006		13.8	1.00			25.0		18.0	5.00	8.50		2.00			
	2010		17.7	37.0					83.3	20.8	45.2		21.9			3.7
Arunachl Prad.	2006	12.0	6.8	12.8				44.0	44.0	65.0	47.0	33.0	54.4	74.0		
	2011	28.4	7.4	12.9				38.9	51.4	79.7	70.0	36.3	67.5	93.3		

anomalies, *viz.*, skewed SRR and varietal replacement rates (VRR) should be addressed appropriately. With the adoption of quality seed of improved varieties coupled with appropriate production technologies and enabling environment, there was appreciable increase in the production of oilseeds (21.4 per cent), rice (12.7 per cent), wheat (25.1 per cent) and pulses (20.4 per cent) in 2011-12 over that of 2006-07 (Table 13.3) and India achieved an all time high food production of 265 million tonnes during 2013-14. The yield increased during 2011-12 over that of 2002-03 were 37.3 per cent in rice, 21.6 per cent in wheat, 47.3 per cent in maize, 64 per cent in oilseeds and 28.7 per cent in pulses in India (Table 13.3).

Guidelines to Improve the SMR and SRR

All the states prepare seed plan for the specific plan period (presently for 2013-14 to 2016-17) season and crop wise depicting targeted SRR, variety-wise requirement of certified seeds, for agriculture production under normal as well as contingent situations like droughts floods, cyclones etc. State wise seed plan are discussed under the chairmanship of Secretary (A and C) and are implemented as per the following guidelines:

☆ The State Department of Agriculture provide variety-wise seed production plan based on the seed plan prepared by them to all the seed producers, viz, State Department of Agriculture, SSF, SSC, SAUs, NSC, SFCI, Co-operative and private seed industries etc. to take up certified seed production as per the plan (see Table 13.14 as an example).

☆ States execute MOUs with seed producers for supply of the required amount of certified seeds crop, season and year wise 12th plan *i.e.* 2016-17, so as to ensure their timely availability.

☆ States chalk out breeder seed and foundation seed requirements as per the seed plan and place requisitions to the concerned organisation, accordingly.

☆ States should ensure that the seed produced under the MOUs is purchased by them, subject to fulfilment of quality standards.

☆ All the states should monitor and periodically review seed production at the field level and take follow up action.

☆ States must review contingent situations arising in the state objectively and accordingly plan seed requirements in the seed plan and seed bank. Due cognizance may be given for fodder and green manure crops in contingent planning.

☆ All states must make efforts to produce seed within the state through credible seed producers having technical competence, sufficient seed infrastructure facilities and experience in seed production.

☆ States should ensure increase in SRR and VRR of crops.

☆ States which do not take up seed production as per the seed plan and which do not purchase the contracted quantity of seed from the seed producers as per MOU will face reduction in allocation/release of fund under various crop development programme.

Table 13.14: Physical Targets of Certified Seeds in Seed Bank during 2012-13 (Qty. in qtls.)

Crops	APS SDC	BR BN	GS SC	HS DC	KS SC	MS SC	MPS SDC	NSC	SF CI	OS SC	PS SC	RS SC	UPS DC	UKS and TDC	WBS SC	KSS DA (K+R)	Jhar-khand	Chan-di-garh	TN	Total
Kharif																				
Paddy	8200	1200	500	2000	6000	8500	1430	30000	7000	5100	500		4000	2500	3300	4000	1000	5000	4300	94530
Sorghum					350	1000	20	1000												2370
Bajra			400	200		500		200				150								1450
Maize		550					120	100									200			970
Urd	1400		500	50	450		60	3000	1500	700		500	500	100	220		200		500	9680
Moong	500	400	1000	500				2000	1200	500		500	100				200		200	7100
Arhar			550	100	300	500	145	4000	600				100				150			6445
G.Nut	3500		700					2000	4000	1600		100			800				600	13300
Soyabean			500			4000	6000	2000				1000		150						13650
Sunflower					50															50
Moth												200								200
Cowpea					150							125								275
Ragi					1000															1000
Guar				200				100	300			200								800
Castor	300		1000			100														1400
Sesamum	100		500			50		250					150		350					1400
T. Targets	14000	2650	5650	3050	8300	14650	7775	44650	14600	7900	500	2275	4850	2750	4670	4000	1750	5000	5600	154620

Contd...

Table 13.14—*Contd...*

Crops	APS SDC	BR BN	GS SC	HS DC	KS SC	MS SC	MPS SDC	NSC	SF CI	OS SC	PS SC	RS SC	UPS DC	UKS and TDC	WBS SC	KSS DA (K+R)	Jhar-khand	Chan-di-garh	TN	Total
Rabi																				
Wheat		2200	5650	4000	1500	6500	5000	10500	15000		2000	3250	4850	5000			700			66150
Sorghum					1700	500														2200
Gram	1000	250	1100	1000	4500	4500	2500	4000	14000			500	200				300			33850
Barley				1000			50	100	5000			300								6450
Lentil		330		200			145	2000	300				200	25	300					3500
Cowpea								500												500
Pea							110	1000	200				200							1510
Mustard/Toria		1100	250	1250			45	500	1900	100		500	200	150	800		500			7295
Lin seed							30													30
safflower						1000		100												1100
Taramira		550							100											650
T. Targets	1000	4430	7000	7450	7700	12500	7880	18700	36500	100	2000	4550	5650	5175	1100	0	1500			123235
G.T. (R+Kh)	15000	7080	12650	10500	16000	27150	15655	63350	51100	8000	2500	6825	10500	7925	5770	4000	3250	5000	5600	277855

Table 13.15: Sale Price of Breeder Seed during 2005-06 to 2014-15

Sl.No. Crop Variety/Hybrid/Line	2005-06	2006-07	2007-08	2008-09	2009-10	2010-11	2011-12	2012-13	2013-14	2014-15
Cereals										
1. Paddy										
Coarse varieties	2000	2000	2000	2400	2400	3400	3570	3700	3800	3800
Medium varieties	2400	2400	2400	2800	2800	3600	3780	3900	4000	4100
Pusa-44	2400	2400	2400	2800	2800	6000				
Basmati Varieties	3800	3800	3800	4500	5000		6300	6500	6500	6700
Paddy hybrid										
A Line	12500	12500	12500	13750	14000	17500	17500	19000	19000	20000
B Line	3000	3000	3000	3300	3300	3600	3600	3900	3900	4000
R Line	3000	3000	3000	3300	3300	3600	3600	3900	3900	4000
2. Maize										
Inbred Lines	9000	9000	9000	9900	10500	13000	13650	14300	14300	14700
Varieties and Compos	3000	3000	3000	3300	3300	3500	3670	3800	3800	4000
3. Sorghum hybrid										
A Line	9000	9000	9000	9900	10000	13000	13650	14000	14500	14500
B Line	7500	7500	7500	8200	8200	10000	10000	10000	10000	10000
R Line	6000	6000	6000	8250	8250	10000	10000	10000	10000	10000
Varieties and Compos	3800	3800	3800	5062	5062	6000	6300	6500	6500	6500
4. Bajra hybrid										
A Line	13000	13000	13000	14300	14300	16000	16800	19000	19000	23000
B Line	6600	6600	6600	7260	7260	8000	8000	8000	8000	8000
R Line	5500	5500	5500	6050	6050	8000	8000	8000	8000	8000
Varieties and Compos	5500	5500	5500	6050	6050	7000	7350	8000	8000	8000

Contd...

Table 13.5—*Contd...*

Sl.No. Crop Variety/Hybrid/Line	2005-06	2006-07	2007-08	2008-09	2009-10	2010-11	2011-12	2012-13	2013-14	2014-15
5. Wheat										
Varieties-Bread Wheat	2000	2000	2000	3000	3000	3800	4000	4100	4200	4200
Durum/Dicoccum	2150	2150	2150	3150	3150	3800	4000	4500	4600	4700
Desi		2500	2500	3500	3500	4200	4400	4500	4600	4700
6. Barley										
Barley	1600	1600	1600	1760	1760	2200	4000	4100	4100	4100
Malt Barley	1800	1800	1800	1980	1980	2400	4200	4600	4600	4600
7. Small Millets										
Ragi	1800	1800	1800	1980	1980	2400	2520	3000	3500	3600
Foxtail Millet	1800	1800	1800	1980	1980	2400	2520	3000	3000	3000
Kodo Millet	1800	1800	1800	1980	1980	2400	2520	3000	3000	3000
Proso Millet	1800	1800	1800	1980	1980	2400	2520	3000	3000	3000
Little Millet	1800	1800	1800	1980	1980	2400	2520	3000	3000	3000
8. Pulses										
Moong	5000	5000	5000	6000	6000	9000	9900	11000	12000	12000
Urd	5000	5000	5000	6000	6000	9000	9900	11000	12000	12000
Arhar (Pigeon pea)	5500	5500	5500	6500	6500	9000	9000	10000	11000	11000
Arhar A Line						13000	13000	13000	13000	13000
Arhar B and R line						9000	9000	9000	9000	9000
Cowpea	4500	4500	4500	4950	4950	7000	7000	7500	7500	7500
Gram (Kabuli)	5000	5000	5000	6000	6100	9000	9000	11000	11000	11000
Gram (Desi)	4100	4100	4100	5100	5100	6500	7000	8000	8000	8000

Contd...

Table 13.5–Contd...

Sl.No.	Crop Variety/Hybrid/Line	2005-06	2006-07	2007-08	2008-09	2009-10	2010-11	2011-12	2012-13	2013-14	2014-15
	Lentil	4100	4100	4100	5100	5100	6500	7150	8500	8500	8500
	Peas	3600	3600	3600	4600	4600	5000	5000	5500	5500	5600
	Moth	3500	3500	4000	5000	5000	6000	8000	8000	8000	8500
	Rajmash	5400	5400	5400	6400	6400	7000	7000	8000	8000	8000
	Horse gram								4000	4000	4000
9.	Fibre crops										
	Jute	7000	7000	7000	7500	8000	10000	10500	12500	12500	12700
	Mesta	4700	4700	4700	5170	5200	7000	5000	6000	6000	6000
	Sunhemp	2000	2000	2000	2200	2300	3000	1850	3500	3500	6000
10.	Cotton hybrid										
	Female Parent	40000*	40000*	40000*	44000	44000	48000	48000	53000	53000	55000
	Male Parent	40000*	40000*	40000*	44000	44000	48000	48000	53000	53000	55000
	Varieties	12500*	12500*	12500*	13750	13750	15000	15000	15000	15000	15000
	G. Hirsutam								16500	16500	17000
	G. Barbadens								20000	20000	22000
	Male Sterlity Syst A Line	50000*	50000*	50000*	55000	55000	60000	60000	66000	66000	60000
	Based Hybrid R Line B Line	50000*	50000*	50000*	55000	55000	60000	60000	66000	66000	60000
11.	Fodder crops										
	Guar	3000	3000	3200	3520	3520	4500	4700	35000	14500	12000
	Teosinite	1800	1800	1800	1980	1980	2400	2500	2500	2500	2500
	Lucerne	20000	20000	20000	25000	25000	27000	27000	30000	30000	31000

Contd...

Table 13.5—Contd...

Sl.No.	Crop Variety/Hybrid/Line	2005-06	2006-07	2007-08	2008-09	2009-10	2010-11	2011-12	2012-13	2013-14	2014-15
	Berseem	18000	18000	18000	22000	22000	24000	24000	26000	26000	26000
	Oat	2500	2500	2500	3500	3500	3850	4000	4200	4200	4200
	Cowpea	4500	4500	4500	4950	4950	7000	7000	7500	7500	7500
	Maize	3000	3000	3000	3300	3300	3500	3670	4000	4000	4000
	Jowar	3800	3800	3800	4180	4180	6000	6300	8000	8000	8000[a] 12000[b]
	Bajra fodder	5500	5500	5500	6050	6050	7000	7350	7500	7500	7500
	Dhaincha									5000	5000
12.	Oilseeds										
	Groundnut	4500	4500	4500	5000	5000	6000	6300	7200	9000	10000
	Soyabean	4500	4500	4500	5000	5000	6000	6300	6800	7000	7500
	Sunflower										
	A Line	20000	21000	21000	22000	23000	27000	29700	30000	30000	31000
	B Line	10000	10000	10000	11000	11000	13000	14850	15000	15000	15000
	R Line	10000	10000	10000	11000	11000	13000	14850	15000	15000	15000
	Varieties	6000	6000	6000	7000	7000	8000	8800	9000	9000	9000
	Castor										
	Female Parent(A Line)	20000	20000	21000	22000	22000	25000	27500	30000	30000	30000
	Male Parent(B and R Line)	7500	7500	7500	8000	8000	9000	10000	11000	11000	11000
	Varieties	5000	5000	5000	6000	6000	7000	7700	8400	8400	8500
	Sesamum	10000	10000	10000	11000	11000	12000	12000	12000	12000	13500
	Niger	6000	6000	6000	6600	6600	6600	6600	7000	7000	8000

Contd...

Table 13.5—Contd...

Sl.No. Crop Variety/Hybrid/Line	2005-06	2006-07	2007-08	2008-09	2009-10	2010-11	2011-12	2012-13	2013-14	2014-15
Rapeseed/Mustard										
A Line								15000	15000	15000
B Line								7000	7000	7000
R Line								7000	7000	7000
Varieties	4600	4600	4600	5400	5400	6500	6800	7000	7000	7500
Toria								7000	7000	7000
Taramira								7000	7000	7000
Safflower										
Varieties	4000	4000	4000	4400	4400	5000	5250	5500	5500	5500
Female Parent(H)	10000	10000	10000	11000	11000	12000	12000	12600	12600	12600
Male Parent (H)	4000	4000	4000	4400	4400	5000	5000	5500	5500	5500
Linseed	3500	3500	3500	3850	3850	4000	4200	4600	4600	4600

☆ States should take up certified seed production of the latest varieties/hybrids as and when released and notified.

☆ All states are advised to initiate action for procuring seeds immediately after harvest of the crop to avoid non-supply of seeds by the producing farmers/organisations in the event of increase in the price of the crop in the market.

☆ States which are not availing the facilities provided under Seed Village Programme may make full use of the programme.

☆ NFSM, RKVY, Accelerated Pulses Production Programme (A3P) clusters may be used for seed production to meet the seed requirement of the designated crop under the scheme.

☆ During the last five years, certified/quality seed production has doubled but the certification staff has not increased, for which all the states may take appropriate action to make good the shortages in staff designated for seed certification.

☆ States should also focus on crops, in which production was impacted during the current year due to deficiency in rainfall as demand for seed of such crop in the succeeding year could be high.

7. Seed Certification

Seed certification is a legally sanctioned system for quality control of seed multiplication and production. It is a process designed to maintain and make available to the general public continuous supply of high quality seeds and propagating materials of notified kinds and varieties of crops, so grown and distributed to ensure the physical identity and genetic purity.

The main objective of the Seed Certification is to ensure the acceptable standards of seed viability, vigour, purity and seed health. The concept of seed certification dates back to the earlier part of the 20th century which grew out of the increased concern for the rapid loss of identity of varieties during production cycles. The Swedish were the first to initiate the process of field evaluation of the seed crops, with the visits of agronomists and plant breeders to the fields of progressive farmers who grew seeds of new varieties. This was primarily to educate them on seed production, followed by field inspection and later on found to be very helpful in keeping varieties pure in the production chain. But, in process other problems appeared, to overcome which scientists from USA and Canada met in Chicago, Illinois in 1919 and formed an International Crop Improvement Association (ICIA), which in 1969, by paving the way for modern day seed certification, changed its name to Association of Official Seed Certifying Agencies (AOSCA).

Some of the milestones in the Indian seed sectors are:

☆ Establishment of Central Seed Testing Laboratory at IARI, New Delhi (1955).

☆ Establishment of the National Seed Corporation (NSC) in 1963, the main agency for production, distribution and certification of foundation and certified seeds.

☆ State Seed Corporations and State Seed Certification Agencies were also established.

☆ World bank assisted National Seeds Project (NSP) paved the way for further growth of seed sector of the country (1976).

☆ New Policy on the Seed Development (1988) and the New Industrial Policy (1991) that opened doors for the foreign investors in the Indian seed industry. Since then, a number of multinational companies have entered in the seed market.

The seed certification in India, through field evaluation, started with the establishment of National Seeds Corporation in 1963 and legal status was given to seed certification with the enactment of first Indian Seed Act in the year 1966 and formulation of Seed Rules in 1968. The Seed Act of 1966 provided the required impetus for the establishment of official Seed Certification Agencies (SCAs) by the states. Accordingly, Maharashtra first established an official SCAs during 1970 as a part of the Department of Agriculture, however, Karnataka established the SCA as an autonomous body during 1974. At present 22 states in the country have their own SCAs established under the Seed Act, 1966. For India a comprehensive list of the minimum standard of seed certification for various crops has been prepared by Trivedi and Gunasekharan (2013) and put on the wed of the DAC under seed section for reference, use, modification and suggestion. However all these standards are subject to change based on the requirements of the farmers.

In great majority of the countries in the world, including India, seed certification is voluntary and labelling is compulsory. The organizational set up of the certification agency includes board of directors, technical and other staff for operating the programme. Seed certification agency may have its own seed testing laboratory or it may get its seed samples tested through seed testing laboratories. A well organized seed certification should help in accomplishing the following three primary objectives:

☆ The systematic increase of superior varieties.

☆ The identification of new varieties and their rapid increase under appropriate and generally accepted names.

☆ Provision for continuous supply of comparable material by careful maintenance.

Any variety to become eligible for seed certification should meet the following requirements:

☆ Should be a notified variety under Section-5 of the Indian Seed Act, 1966.

☆ Should be in the production chain and its pedigree should be traceable.

☆ Should meet the field standards (selection of site, isolation requirements, spacing, planting ratio, border rows etc).

☆ Presence of off-types in any seed crop, pollen-shedders in sorghum, bajra, sunflower etc., shedding tassels in maize crosses, disease affected plants, objectionable weed plants etc., should be within the maximum permissible levels.

Broad Principles of Seed Certification Agencies

The seed certification agencies are essential and established in each state as per the Seeds Act, 1966 and function on the following guidelines:

☆ Seed certification agency should be an autonomous body and should not involve itself in the production and marketing of seeds.

☆ Uniform seed certification standards and procedures throughout the country.

☆ Close linkage with the technical and other related institutions.

☆ Its long-term objective should be to operate on no-profit no-loss basis.

☆ Adequate staff trained in seed certification should be maintained.

☆ Provision for creating adequate facilities for ensuring timely and thorough inspections.

☆ It should serve the interests of seed producers, farmers and users.

Seed certification is carried out in following six broad phases:

☆ Verification of seed source, class and other requirements of the seed used for raising the seed crop.

☆ Receipt and scrutiny of application.

☆ Inspection of the seed crop in the field to verify its conformity to the prescribed field standards.

☆ Supervision at post-harvest stages including processing and packing.

☆ Drawing of samples and arranging for analysis to verify conformity to the seed standards.

☆ Grant of certificate, issue of certification tags, labelling, sealing etc.

Control Measures

Organization and establishment of a seed certification agency need careful planning taking into consideration the anticipated acreage for certification of various crops and varieties, area of operation, farm sizes etc. Following control measures are taken by seed certification agencies:

Origin of the propagating material: The first step of seed certification programme is to verify source of seed, and unless the seed is from approved source and of designated class, certification agency will not accept the seed field for certification, thereby ensuring the use of high quality true type seed for sowing of seed crops.

Field inspection : The crop variety grown in the field is evaluated for varietal purity, isolation of seed crop to prevent out-cross, physical admixtures, disease dissemination and also ensure crop condition as regards to the spread of designated diseases and the presence of objectionable weed plants etc.

Sample inspection: The certification agency draws representative samples from the seeds produced under certification programme and assess the planting value of the seeds by germination and other purity tests required for conforming to varietal purity.

Bulk inspection: To know the genuinity of lot and sample, provision has been made for bulk inspection under certification programme. The lot is evaluated for checking homogeneity of the bulk seed produced as compared with the standard sample.

Control plot testing: To compare the samples drawn from the source and final seed produced are grown side by side along with the standard samples of the variety in question and determined whether the varietal purity and health of the produced seed are equal to the results based on field inspection.

Grow-out test: The samples drawn from the lots are grown in the field along with the standard checks and evaluated for their genuineness to species or varieties or seed borne infection as it helps in the varietal purity and elimination of the sub-standard seed lots.

Seed Quality Control

There are adequate provisions under existing seed legislations (Seed Act, 1966 and Seed Rules, 1968) to regulate the quality of seeds keeping in view the federal structure of the country. State Governments have powers to appoint Seed Analysts and Seed Inspectors. Seed Inspectors are vested with adequate powers for quality control *viz.* to draw the sample; enter and search; examine records, registers and documents; seize the stock and issue 'Stop Sale' order in case the commodities under reference contravene provisions of law. Inspectors are authorized to take punitive action/launch proceedings against dealers found to be selling sub-standard seeds. The seed in respect of which the contravention has been committed can be forfeited under Section 20 of the Seeds Act. Penalties are provided under Section 19 of the Act.

State Governments have the powers of enforcement and implementations, however, uniformity and consistency is maintained through standards for certification, labelling etc. prescribed by Government of India (GOI). The Joint Secretary (Seeds), GOI is the Controller of seeds for the country as a whole. Also, the Government of India can give directions under Section 23 of the Act to the States. The Central Seed Certification Board and the Central Seed Committee advise the Central and State Governments on various aspects of seed.

The Seeds Act/Rules are applicable to notified seeds while Seeds (Control) order (1983) to both notified non-notified varieties. Seed Act (1966) under Section 5 prescribes the notification of varieties, but it is voluntary. When notified variety is sold in the market, it should be labelled as prescribed under section 6 (a) of the Seeds Act, 1966 that deals with the standard of germination and physical purity and section 6(b) that deals with color, content and size of the label. Export and import of seed of any notified kind or variety is subject to conditions such as conforming to the minimum limits of germination and physical purity etc. Exemption under Section 24 from the provisions of the Act covers farmers' seeds. The Seeds (Control) Order, 1983 brought seed under the ambit of Essential Commodities Act, 1955 and made it mandatory that the business of selling, exporting and importing seeds can be carried out only under a license issued by the State government. The dealers can also be directed to distribute seeds in specified manner in public interest. Seed dealers are required to maintain books and accounts and display the stock position and its price. A dealer's license is liable to be suspended/cancelled for contravention and the penalties are imposed as per the Essential commodities Act, 1955. Seed testing has been recognized as an essential aspect of seed quality. Accordingly, Section 4(2) of the Seeds Act, 1966 empowers the State Government to establish one or more State Seed Testing Laboratories in the State.

8. Export and Import of Seed

The Export and Import (EXIM) Policy, 2002-07 (revised 2009-14) framed by Govt. of India governs the export and import of seeds and planting materials. To encourage export of seeds in the interest of farmers, the procedure for export of seeds has been simplified. Seeds of various crops have been placed under the Open General License (OGL) except the seeds of wild varieties, germplasms, breeder seeds and upon seeds which are on restricted list under the new Export and Import Policy 2002-07 and 2009-2014. There is no restrictions on export of seeds of cultivated varieties w.e.f. 01.04.2002 except a few. The export of the seeds of following is restricted and is only allowed on case-to-case basis under licence issued by Director General, Foreign Trade on the basis of the recommendations of Department of Agriculture and Cooperation:

- ☆ Breeder or foundation seeds or wild varieties
- ☆ Onion, berseem, cashew, nux vomica, rubber, pepper cuttings, sandalwood, saffron, neem, forestry species and wild ornamental plants such as Red sanders, Russa Grass, tufts and Seeds of tufts
- ☆ Export of niger is canalized through TRIFED, NAFED, etc.
- ☆ Export of groundnuts is subjected to compulsory registration of contract with APEDA

The provisions regarding import of seeds and planting material are as under:

- ☆ Import of seeds/tubers/bulbs/cuttings/saplings of vegetables, flowers and fruits is allowed without a licence in accordance with import permit granted under Plant Quarantine (Order), 2003 and amendment made therein.
- ☆ Import of seeds, planting materials and living plants by ICAR, etc. is allowed without a licence in accordance with conditions specified by the Ministry of Agriculture.
- ☆ Import of seeds/tubers of potato, garlic, fennel, coriander, cumin, etc. is allowed in accordance with import permit granted under PQ Order, 2003.
- ☆ Import of seeds of wheat, rye, barley, oat, maize, rice, millet, jowar, bajra, ragi, other cereals, soybean, groundnut, linseed, palm nut, cotton, castor, sesamum, mustard, safflower, clover, jojoba, etc. is allowed without licence subject to the New Policy on seed development, 1988 and in accordance with import permit granted under PQ Order, 2003.

All imports of seeds and planting material is regulated under the Plant Quarantine Order 2003. Import licences are granted by DGFT only on the recommendations of DAC. On receipt of applications for commercial import, DAC considers the trial/evaluation report on the performance of the seed and their resistance to seed/soil borne diseases. DAC is required to either reject or recommend the application to DGFT for grant of import licence within 30 days of receipt. All importers have to make available a small specified quantity of the imported seeds to the ICAR at cost price for testing/accession to the gene bank of National Bureau of Plant Genetic Resources (NBPGR). The import of seeds has to be cleared/rejected by Plant Protection Adviser (PPA) after quarantine checks within three weeks and the rejected consignment has to be destroyed. During quarantine, the imported consignment is kept in a bonded warehouse at the cost of the importer. While importing seeds and plating material, there is absolutely no compromise on plant quarantine procedures and each and every effort is made to prevent the entry of exotic pests, diseases and weeds into India, that are detrimental to the interests of the farmers.

An EXIM Committee constituted in the seeds division deals with application for exports/imports of seeds and planting materials in accordance with the New Policy on Seed Development and EXIM regulations. Exporters/importers are required to submit 20 copies of applications for export/import in the prescribed formats. The Committee meets every month, subject to tendency of proposals analyzes applications and gives recommendations to PPA/DGFT for issuing of the licence for import/ export of seeds and planting material. The minutes of the EXIM Committee are posted on the Seednet portal (http://seednet.gov.in).

The statistics of world seed trade indicate that India has 6th largest size of domestic seed market of about 1300 million dollars in the world but, it's share in global trade in seeds (import and export) is only about 37 million dollars.

Under the EXIM policy 2009-14 the new initiatives are:

☆ Scheme for Promotion of Seed Export in sub-Mission of Seed and Planting Material under National Mission on Agricultural Extension and Technology (NMAET).

☆ Encourage the export of seed by providing incentives to the exporters

The new policy for import of seeds and planting materials are aimed at:

☆ To provide the best planting materials available anywhere in the world to Indian farmers to increase the farm productivity.

☆ All imports of seeds and planting materials etc. are allowed freely subject to EXIM Policy, 2009-2014, New Policy on Seed Development, 1988 and Plant Quarantine (Regulation of import into India) Order, 2003.

The New Policy on Seed Development (NPSD), of 1988 contains :

☆ The NPSD 1988 heralded a new era of private enterprise.

☆ Vegetable, flower and ornamental seeds could be imported freely under OGL.

☆ Seeds of oilseeds, pulses, fodder and coarse cereals like maize, sorghum and other millet could be imported for two years by companies which had technical and financial collaboration agreements for production of seed with companies abroad and also subject to the condition that initially a small quantity of seeds will be imported and tested in ICAR's multilocation trials.

☆ The bulk import was also allowed subject to the provision that the foreign supplier agreed to supply parent line seed or breeder seeds to the Indian company within two years of the date of first commercial consignment.

☆ Import of items on the restricted list is allowed on case to case basis under import permit issued by Plant Protection Advisor to the Government of India, who issues import permits on the basis of recommendations from the Department of Agriculture and Cooperation.

The Policy on Seed Development, 1988 was revised on 27.6.2011 to allow import of specifie quantity of seeds of wheat and paddy initially for trial and evaluation purpose. Based on the results of trial for one crop season, the company may be allowed to import bulk quantity of seeds of wheat and paddy for a period not exceeding two years subject to the conditions. The trial and evaluation of imported seeds will be conducted by ICAR in their Research Stations or in the farms which are accredited by them.

FDI Policy on Seed Sector 2011

☆ FDI is permitted up to 100 per cent under the automatic route in development and production of seeds and planting materials subject to certain conditions.

☆ The permission for FDI up to 100 per cent would encourage infusion of foreign investment into the seed sector and would also facilitate indigenous seed companies for strengthening of Research and Development activities for development of seeds of better varieties.

☆ About 38 companies have been cleared for FDI for seed related activities.

OECD Varietal Certification

The OECD seed scheme is one of the International frameworks available for certification of agricultural seeds moving in international trade with an objective to encourage use of seeds of consistently high quality in participating countries. The schemes were established in 1958 driven by a combination of factors including a fast-growing seed trade, regulatory harmonization in Europe, the development of off-season production, the seed breeding and production potential of large exporting countries in America (North and South) and Europe and the support of private industry. Membership of the schemes is voluntary and participation varies.

To give a boost to seed export, India is participating in OECD seed schemes for the following categories of crops:

☆ Grasses and legumes
☆ Crucifers and other oil or fiber species
☆ Cereals
☆ Maize and sorghum
☆ Vegetables

The OECD seed scheme authorizes use of labels and certificates for seed produced and processed for international trade according to agreed principles. The Joint Secretary (Seeds) in the DAC is the National Designated Authority. Further, heads of seed certification agencies in Karnataka, Andhra Pradesh, Tamil Nadu, Maharashtra, Rajasthan, Uttaranchal, Uttar Pradesh, Haryana, Bihar and Assam have been nominated as the Designated Authorities under the scheme to undertake certification work under OECD Seed Schemes.

9. Policies, Regulations and Legislative Framework

The Seeds Act, 1966 provides for the legislative framework for regulation of quality control of seeds sold in the country. The Central Seed Committee (CSC) and the Central Seed Certification Board (CSCB) are apex agencies set up under the Act to deal with all matters relating to administration of the Act and quality control of seeds.

The Government of India has taken following important policy initiatives in seed sector:

☆ Enactment of the Seeds Act (1966)

☆ Seed Review Team-SRT (1968)

☆ National Commission on Agriculture's Seed Group (1972)

☆ Launching of the world bank aided National Seeds Programme (1975-85) in three phases leading to the creation of State Seeds Corporations, State Seed Certification Agencies, State Seed Testing Laboratories, Breeder Seed Programmes etc

☆ Seed Control Order (1983)

☆ Creation of the Technology Mission on Oilseeds and Pulses (TMOP) in 1986, lateron Integrated Scheme of Oilseeds, Pulses, Oil Palm and Maize (ISOPOM) and National Mission on Oilseeds and Oil Palm (NMOOP)

☆ Production and Distribution Subsidy

☆ Distribution of Seed Mini-kits

☆ Seed Transport Subsidy Scheme (1987)

☆ New Policy on Seed Development (1988)

☆ Seed Bank Scheme (2000)

- National Seeds Policy (2002)
- The Seeds Bill (2004)
- Formulation of National Seed Plan (2005)
- National Food Security Mission (2007)
- Rashtriya Krishi Vikas Yojna (2007)

Some of the highlights of the Seeds Act, 1966 and the Seed Rules, 1968 are:

☆ The Seeds Act and rules are applicable to notified seeds

☆ Notification of kind/varieties of seeds under Section 5 of the Act

☆ Minimum limits for germination, physical and genetic purity of varieties/ hybrids for crops have been prescribed.

☆ Labelling of seed is compulsory as per the Seeds Act.

☆ State Governments appoints seed Analysts and seed Inspectors.

☆ Seed Inspectors are vested with powers for quality control *viz.* to draw the sample; enter and search; examine records, registers, and documents; seize the stock and issue 'Stop Sale' order in case the commodities under reference contravene provisions of law.

☆ Inspectors are authorized to take punitive action and launch proceedings against dealers found to be selling sub-standard seeds.

☆ The seed in respect of which the contravention has been committed can be forfeited under Section 20 of the Seeds Act.

☆ The notified variety sold in the market should be labelled with the standard of germination and physical purity (under section 6 a) colour, content and size of the label (section 6 b) and here is penal provision if a person breaching these, the Seed Inspector is empowered to prosecute him (section 13) with a fine of Rs. 50/- the first offence (upto Rs. 1000/if repeated) or imprisonment for six months or with both.

The highlights of Seeds (Control) Order, 1983 are:

☆ Applicable to notified and non-notified seeds.

☆ Issued in exercise of the powers conferred by section 3 of the Essential Commodities Act, 1955.

☆ State Governments appoints Seed Inspectors who are vested with powers for quality control *viz.* to draw the sample; enter and search; examine records, registers, and documents; seize the stock and issue 'Stop Sale' order in case the commodities under reference breach provisions of law.

☆ Inspectors are authorized to take punitive action and launch proceedings against dealers found to be selling sub-standard seeds.

☆ The business of selling, exporting and importing seeds can be carried out only under a license issued by the State government.

☆ A dealer's license is liable to be suspended/cancelled for contravention.

☆ Seed dealers are required to maintain books and accounts and display the stock position and its price.

☆ The dealers can also be directed to distribute seeds in specified manner in public interest.

☆ The penalties are provided under Essential commodities Act, 1955.

☆ The standards notified under section 6 of the Seeds Act are equally applicable.

☆ Seed inspectors (Clause 12) are empowered to draw the samples of seed, whether it is notified or non-notified under Clause 13 of the said order, meant for sale, export and import and to send the same to a laboratory notified under the Seeds Act.

☆ GM crops which are mostly non-notified are covered under this provisions.

The National Seeds Policy, 2002 was created:

☆ To achieve the food production targets and enhances the SRR of various crops.

☆ To create conducive climate for growth of the competitive and localized seed industry, encouragement of import of useful germless and boosting of export.

☆ To evolve a long term policy for export of seed with a view to raise India's share of global seed export from the present level of less than 1 per cent to 10 per cent by the year 2020.

☆ To establish and strengthen seeds export promotion Zones and Customise Production of seed for export.

☆ Create a data bank to provide information on International market for Indian varieties in different parts of the world and establishment of seed testing/ certification facilities in conformity with International requirements.

The major thrust areas of **National Seeds Policy, 2002** are: variety development, plant variety protection, seed production, quality assurance, seed distribution and marketing, infrastructure facilities, transgenic plant varieties, import of seeds and planting materials, seed exports, promotion of domestic private sector seed industry and strengthening of the monitoring systems.

The salient features of **Seeds Bill, 2004** are:

☆ Registration of kinds and varieties of seeds etc.

- Evaluation of performance
- Compensation to farmers
- Registration of Seed producers and Processing units
- Seed dealers to be registered

☆ Regulation of seed certification and its sale

☆ Seed analysis and testing

☆ Export and import of seeds and planting material

☆ Offences and punishment.

From the year 2005-06, the Seed Division department has launched a Central Sector Scheme "Development and Strengthening of Infrastructure Facilities for Production and Distribution of Quality Seeds". The main components of which are quality control arrangements on seeds, transport subsidy on movement of seeds to North-East and other hilly areas, establishment and maintenance of Seed Bank, Seed Village Scheme, assistance for creation of infrastructure facilities, assistance for boosting seed production in private sector, Human Resources Development, assistance for seed export, propagation of application of biotechnology in agriculture, promotion of use of hybrid seeds of rice and evaluation/review.

Variety Registration Procedures

In India each variety has to pass through three phases of evaluation (local, IET/ PYT and AVT). The best entries of the Breeders on the basis of their evaluation in local programmes are contributed for testing in the Initial Yield Evaluation Trial (IET) or Preliminary Yield Trial (PYT) organized by project coordinator (crops) in selected number of places in each crop zone. The Pathologists and Entomologist simultaneously also study their reaction to important diseases and pests. The entries qualifying from yield, disease and quality point of view in IET/PYT are further tested in the Uniform Regional Trials (URT) which are also called Advanced Varietal Trials (AVT) or Coordinated Varietal Trials (CVT) which are organized at a very large number of locations in each zone and the plot size is larger than that in IET. During the tests, reaction to various diseases, pests and quality traits are also studied. Entries found suitable in the second phase are again evaluated in the URT and simultaneously supplied to Pathologists, Entomologists, Nematologists, Agronomists and Quality Evaluation Groups to study the entries comprehensively for important factors. Actual measurements are also made on other parameters. Agronomy group evaluates these entries for their adaptability to varied range of agronomic variables such as sowing dates, levels of fertilizers and number of irrigations etc. These tests are followed by a critical discussion in a crop workshop.

A special multi-disciplinary committee of scientists is constituted at the workshop to consider the proposals for identification of the varieties for release. Varieties evolved by the SAUs and Government Research Institutes are tested within the concerned states at limited locations. Central Seed Committee (CSC) pointed out in 1982 that varieties of state importance might also be tested in the concerned All India Crop Improvement Project. Simultaneous testing of all state varieties along with the Central varieties provides exposure to the state varieties to a wide range of environments. This will help to identify the varieties which are highly prone to diseases and pests and release of which may cause problems in some other states.

To derive the benefit of identified superior genotypes by public, it is essential to maintain a system to have quantities of promising genotypes made available for commercial production. This process is referred to as release of the varieties. The purpose of release system is to introduce newly evolved varieties to the public for general cultivation in the suitable regions. It serves as a guideline in the choice of varieties for cultivation in any region. The practice of official release of varieties started in October, 1964 with the formation of the Central Variety Release Committee (CVRC) at the central level and State Variety Release Committee (SARC) at state level. The CVRC functioned up to November, 1969 when its functions were taken over by the CSC established Seeds Act, 1966. The CSC constituted a Central Sub-committee on crop Standards, Notification and Release of Varieties (CSC on CS, N and RV). The sub-committee discharges the functions of release and notification of varieties at Central level, while State Seed Sub-Committees (SSSCs) discharge similar function at state level. The CSC and its Sub-committee have due representation for all the agencies involved in seed research, production and quality control namely State Governments, SCAs, SAUs, ICAR Institutes, Seed producing agencies in public and private sectors and seed farmers.

The PPVFRA suggested that nomenclature of varieties is very critical as denominations of the varieties as per notification have to be maintained in protection of varieties as per PPV and FRA rules and guidelines. However, in the 70th meeting of CVRC in 2014, guidelines were issued for the nomenclature of all centrally released varieties ccording to the recently approved guidelines of AICRPs in following manner:

☆ For all the released varieties by CVRC the prefix of each variety would be central variety followed by the name of the crop. For example : Central Wheat

☆ The proposed name by the university, institute/university centres will be maintained along with above mentioned prefix. For example: Central Wheat PBW 343

☆ Nomenclature of the varieties on name of individuals, god/goddess and political personalities is not acceptable.

☆ PCs/PDs will ensure following these guidelines on nomenclature at the time of VIC Meeting.

However, there are still debates going on these proposed names of centrally released varieties.

Varietal Protection in India

The World Trade Organization (WTO) has at least half a dozen of intergovernmental agreements that directly affect agriculture. The WTO member countries must provide patent protection for any invention, whether product or process, in all fields of technology and India is the signatory of WTO. In order to fulfil the obligations under Article 27(3) b of Trade Related Aspects of the Intellectual Property Rights (TRIPS) Agreement of the WTO, India has ratified, the Legislation for Protection of Plant Varieties and Farmers' Rights (PPV and FR) and enacted in year 2001. Barton and Siebeck (1992) has clearly highlighted the prospects and consequences of IPR issues. *Sui Generis* is not a system but a French adjective meaning "distinctive" or "specific" and hence its meaning depend on place, people, and time. A number of forms and their combinations will include *sui generis* forms of IPR. So India will be an osterich if she hides behind *sui generis*, without being specific. A number of systems could pass as *sui generis*, as alternative to IPR as many argue (RAFI-UNDP 1999).

As per TRIPS the member states are allowed to "exclude from patentability *inter alia* plants and animals other than micro-organisms, and essentially biological process for production of plants and animals other than non-biological and microbiological process; however, members are required to provide protection to plant varieties either by patents or by an effective *sui generis* system or by any combination thereof". Developing economies and least developed economies are allowed a grace period of four and ten years, respectively to implement these provisions failing that conventional IPR will be enforced, which may.

☆ Undermine the efforts to implement the convention's new regime on control over genetic resources;

☆ Would not ensure benefit sharing; and

☆ Limit *in situ* conservation of plant genetic resources (PGR) as it restricts local and informal exchange of germplasm.

Under the TRIPS Agreement Article 27(3) (b), resulted from the negotiations of the Uruguay Round, it is essential to protect plant varieties either by patents or by an effective *'sui generis'* system of protection or by a combination of both these systems. The legislation provides for the establishment of an effective system for protection of plant varieties, the rights of farmers and plant breeders and to encourage the development of new varieties of plants. Finally India established Protection of Plant Varieties and Farmers Rights (PPV and FR) Authority, under the Protection of Plant Varieties and Farmers Rights Act, 2001, which become operative from 11th Nov. 2005 with the following objectives:

☆ Establishment of an effective system for protection of plant varieties, the rights of farmers and plant breeders and to encourage development of new varieties of plants.

☆ Recognition and protection of the rights of farmers for their contribution in conserving, improving and making the available plant genetic resources for development of new varieties.

☆ Accelerated agricultural development by stimulation of investment for research and development both in public and private sectors.

☆ Facilitate growth of seed industry to ensure availability of quality seeds and planting material to the farmers.

The authority is in process of registration of plant varieties of 14 selected crops. The national draft guidelines for the conduct of tests for distinctiveness, uniformity and stability (DUS) for selected 35 crops have been finalized. The scheme provides financial support for functioning of PPV and FR Authority and for developing DUS test guidelines for crops and strengthening and equipping the DUS centres and identified institutions. In 11th plan, the scheme had 12 components with an outlay of Rs.120 crores for implementation of PPV and FR Act with a target of setting up of two branch offices of the authority and establishment of Plant Variety Protection (PVP) Appellate Tribunal besides other projections during the plan.

Under this provision, any of the following can make an application to the PPV and FRA for registration of a variety:

☆ Any person claimed to be a breeder of a variety.

☆ Any person being the assignee of the breeder of a variety.

☆ Any farmer or group of farmers or community claiming to be the breeder of a variety.

☆ Any university or publicly funded agricultural institution claiming to be breeder of a variety.

The Certificate of Registration shall confer an exclusive right on the breeder, his successor, his agent or licensee the right to produce, sell, market, distribute, import or export the variety. A National Register of plant varieties will be maintain by the PPV and FR Authority. The Certificate of Registration shall be valid for 9 years in the case of trees and vines and 6 years in case of other crops. The total period of variety shall not exceed 18 years for trees and vines and 15 years for extant varieties notified under Seeds Act and for other crops. The PPV and FR Authority shall invite claims for beneficiary of any registered variety on the basis of following:

☆ The extent and nature of the use of genetic material of the claimant.

☆ Commercial utility and demand in market of the variety relating to which benefit has been claimed.

The benefit determined by the PPV and FR Authority shall be deposited by the breeder with the 'National Gene Fund' and the amount of benefit sharing shall be recoverable as area of land revenue.

Farmer who has developed or bred a new variety shall be entitled for registration as a breeder of a variety. Farmer shall be deemed to be entitled to save, use, sow, re-sow, exchange, share or sell his farm produce including seed of a variety protected under this Act in the same manner as he was entitled before coming into force of this Act provided that the farmer shall not be entitled to sell branded seed of a variety protected under this Act. Farmers' variety shall be entitled for registration.

Farmer who is engaged in the conservation of genetic resources of land basis and wild relatives of economic plans and their improvement and preservation shall be entitled to recognition and reward from the Gene Fund provided the material so selected and preserved has been used as a donor of genes in varieties registerable under the PPV and FR Act. Any person or group of persons (whether actively engaged in farming or not) or any other government or non-government organization may stake a claim on behalf of the village or local community.

There is a provision for compulsory licensing to meet the reasonable requirement of the public for seed or other propagating materials. The PPV and FR Authority has official website: www.plantauthority.gov.in for further informations and update.

7. Summary and Conclusions

Seed, with its inherent genetic potential, is an important component of production technology of any crop and variety. Now-a days, seed security is a prerequisite for achieving food security as use of quality seeds alone increases 15-20 per cent crop productivity besides enhancing the efficacy of all other inputs. Successful quality seed programme encompasses production of sufficient quantity of seed with appropriate research back-up on seed production, maintenance, certification, quality assurance, processing, storage, seed protection and quality enhancement. A systematic, strong and vibrant seed production system is essential for future agriculture growth of the country and finally food security.

The Indian seed improvement programme is backed by a strong crop improvement programme in both the public and private sectors and now the industry is highly active and well recognized internationally and several developing and neighbouring countries are benefited from quality seed from India. Indian seed sector consists of public sector institutions such as ICAR institutes, SAU's, NSC, SSCs and SFCIs and private seed companies. The ICAR institutes and SAU's are engaged in developing new varieties and hybrids of different crops and production of breeder and basic seeds.

The Seed Programme, which is now occupying a pivotal place in Indian agriculture and is well poised for continued growth, has evolved, seed quality specifications of international standards over the years and adopted by both public and private sectors. India has a strong rigorous mechanism for seed quality control through voluntary seed certificate and compulsory labelling monitored by provincial level seed law enforcement agencies. The Indian seed processing/conditioning industry has also perfected the post harvest handing techniques of quality up-gradation and maintenance to ensure high standards of physical condition and quality. By virtue of the diverse agro-climates several geographical zones in the country emerged as ideal seed storage locations under ambient conditions.

India's Seed Programme has a strong seed production base in terms of diverse and ideal agro-climates spread throughout the country for producing high quality seeds of several tropical, temperate and sub-tropical plant varieties in enough quantities at competitive prices. The seed industry has three well reputed national level associations apart from several provincial level groups to take care of the interests of the industry. The NSC is the single largest seed organization in the country with a wide product range, pioneered the growth and development of a sound industry in India. The NSC, SFCI, SSCs and other seed producing agencies are continuously and gradually expanding all their activities, product range, volume and value of seed handled and level of seed distribution to the un-reached areas, etc. Over the years, several seed crop zones have evolved with extreme levels of specialization. There are more than 20,000 seed dealers and seed marketing distributors in the business.

For seed technology research, presently India has a national level Directorate at Mau and separate departments in several ICAR institutions as well as SAUs with seed technology division of ICAR-IARI at New Delhi the most prominent one. In seed education, nearly 10 state agricultural and traditional universities offer post graduation in seed technology.

A total of 706 and 648 varieties of field crops were released during X and XI plan, respectively. The seed supply system in India can be divided into formal and informal systems. The formal seed supply is from public sector organizations and private seed companies. Informal seed supply deals with farmer saved seed and exchange/sharing within farming community which comprises about 60 per cent of seed availability. Public sector produces bulk of self pollinated crops/high volume low value crops such as rice and wheat accounting for about 60 per cent. Private seed sector deals mostly in seed production of hybrids, vegetables and flowers contributing almost 50 per cent to the total quality seed production.

During 10th plan with launch of a mega seed project and upgradation of National Seed Project into a Directorate of Seed Research, ICAR led double the breeder seed production from 2.2 lakhs q in 10th plan to 4.4 lakh q in 11th plan. Creation of infrastructures for seed production, processing, storage by ICAR seed project resulted in competence enhancement of the personnel involved in public seed sector and finally resulted in increase of the quality seed production from 2.4 lakh q in 2005-06 (at the inception of project) to 6.9 lakh qu during 2011-12. The availability of total quality seed during 2013-14 to the farmers in India was 3.44 m t as compared to 2.50 m t in 2008-09.

There has been a tremendous increase in the SRR of various crops during the last decade which improved production and India achieved an all time high food production of 265 million tonnes during 2013 whic led to increased export of several crop commodities. During 2011 the achieved SRR were 33 per cent in wheat, 40 per cent in paddy, 57 per cent in hybrid maize, 24 per cent in sorghum, 60 per cent in pearl millet, 30 per cent in moong and 24 per cent in groundnut. The adoption of quality seeds of improved varieties and hybrids appreciably increased the production of oilseeds by 21 per cent, rice by 13 per cent, wheat by 25 per cent and pulses by 20 per cent in 2011-12 over that of 2006-07.

Sustained increase in agricultural production and productivity necessarily requires continuous development of new and improved varieties as well as hybrids of crops and efficient system of production and supply of seeds to farmers. As seed is a critical input for enhancing productivity of all agricultural and horticultural crops, some of the priority areas which require research and developments are listed below:

☆ All the states should give more attention for production and distribution of Quality seed of various crops.

☆ All the varieties of field crops released during X, XI and XII plans may be multiplied.

☆ All the crop varieties released recently may be got registered under PPV and FR act.

☆ Release of crop varieties for specific nitches *i.e.* acid, calcareous and saline soils.

☆ Rapid multiplication of recently released varieties from breeder to foundation seeds.

☆ Encourage potential varieties through quality seed production.

☆ Discourage multiplication of low yielding old varieties, but encourage multiplication of nutrient efficient potential old varieties with consistent yield.

☆ The quality seed may be determined based on germination and survival, not on the size.

☆ Systematic and meticulous field evaluation and seed certification.

☆ Seed policy may be implemented to the best of its benefit to farmer.

☆ The emphasis needed on the release of nutrient efficient crop varieties.

☆ There is an urgent need for the SSCs to transform themselves in tune with the industry in terms of infrastructure, technologies, approach and the management culture to be able to survive in the competitive market and to enhance their contribution in the national endeavour of increasing food production to attain food and nutritional security.

☆ Enabling the resource poor farmers with quality seed and its production technology is still an imminent challenge and needs to be focused upon.

☆ Streamline seed research system to deliver commercially viable technologies on production, storage and processing, seed quality enhancement and control etc.

☆ As 60 per cent of farmers use farm saved seeds, technological intervention for up-gradation of such seed requires immediate attention. Village-based seed banks may serve as an alternative to help farmers become self-reliant.

☆ Participatory seed production involving farmers, seed village scheme, community seed banks is effective strategy to make available quality seed of improved varieties hybrids at appropriate time and affordable price enabling partnership with private sector; self help groups, non-government organizations and community based organizations.

☆ Capacity building with focus on skill intensification, technology dissemination needs to be further strengthened. This would require organized communities, institutional technical backstopping and continued interaction between various institutions, policymakers and stakeholders to strengthen local seed systems to enhance seed productivity and availabilty thereby enabling food security.

☆ Similarly, for post harvest handing, the Indian seed processing/ conditioning industry has perfected the techniques of quality up-gradation and maintenance to ensure high standards of physical condition and quality. By virtue of the diverse agro-climates several geographical zones in the country have emerged as ideal seed storage locations under ambient conditions. In terms of seed marketing and distribution, more than about 20000 seed dealers and distributors are in the business.

☆ Organization and establishment of a seed certification agency need careful planning taking into consideration the anticipated acreage for certification of various crops and varieties, area of operation, farm sizes etc.

☆ There is a need for strengthening the quality seed production programme in the country.

References

Barton, J.H., and Siebeck, W.E. 1992. Intellectual property issues for the international agricultural research centres. Issue in Agriculture 4, CGIAR Secretariat, Washington DC, USA 52 pp.

Chakraborty, K., Bishi, S.K., Singh, A.L., Kalariya, K.A. and , K. 2013. Moisture stress affects yields and quality of groundnut seeds. *Indian J Plant Physiology*, 18: 136-141.

Chaudhary, R. C. 1999. Ethics and mechanisms to internationalise elite germplasm while still honouring IPR and farmers' right. IN: Proc. 11th Australian Plant Breeding Conf. Vol 2. (Eds. Langrdige *et al.*) Adelaide, Australia, 119 pp.

Chaudhary, R.C. 1998. The rights for germplasm and wrongs for crop production sustainability: Perspectives and retrospectives from GATT. *Asian Crop Science* 1998, Taichung, Taiwan, pp. 139-159.

Chaudhary, R.C. 2000. Exploitation of plant genetic resources: Many tools, many rules In: Proc International Conference On Managing Natural Resources For Sustainable Agricultural Production in the 21st Century, 14-18 Feb 2000 (Eds. A.K. Singh *et al.*), Vol. 1 pp. 1-20. *Indian Soc. Soil Sci.*, IARI, New Delhi, India.

Chauhan, J.S., S. Rajendra Prasad and Satinder Pal, 2013. Quality seed and productivity enhancement in major crops in India. In: *Current Trends in Plant Biology Research* (Ed A.L. Singh *et al.*), National Conference of Plant Physiology, 13-16th Dec 2013, DGR, Junagadh, India. pp. 160-168. www.ncpp13.nrcg.res.in.

DAC 2015. Department of agriculture and cooperation, ministry of agriculture, Government of India, 23 January 2015 http: //agricoop.nic.in/ documentreport.html.

Damania, A. B. and J. Valkoun (Ed.). 1997. The Origins of Agriculture and Domestication of Crop Plants in the Near East-The Harlan Symposium, 10-14 May 1997, ICARDA, Aleppo, Syria, 70 pp.

Directorate of Economics and Statistics, Department of Agriculture and Cooperation, Ministry of Agriculture, Govt. of India, New Delhi (*http: //dacnet.nic.in/eands*).

FAOSTAT 2015 | © FAO Statistics Division 23 January 2015 http: //faostat.fao.org/ site/567/default.aspx#ancor.

Gautam Mukesh (2013). Strategies for quick spread of new varieties among the farmers. p. 49-53. In. *6th National Seed Congress on Advancement in agriculture through quality seeds*. Lucknow. Sept. 12-14, 2013.

Hanchinal R.R. (2012). An overview of developments in Indian seed sector and future challenges. p.1-12.In. *National Seed Congress on Welfare and economic prosperity of the Indian farmers through seeds*. Raipur, Chhattisgarh. Dec. 21-23, 2012.

ISTA 2015. ISTA Rule for 2015 http: //www.seedtest.org/en/international-rules-for-seed-testing-_content—1—1083—895.html.

IFPRI 2014. Annual report 2013, International Food Policy Research Institute, Washington DC. IFPRI releases the 2014 Global Hunger Index NSSP News and Notes nssp.ifpri.info/2014/10/15/ifpri.2014-global-hunger-index.

Khatediya N., Mahatma M.K, Lokesh Kumar, Singh A.L. and Chaudhari V. 2013. Fertilizers and organic manure increases the quality of groundnut seed. In: *Current Trends in Plant Biology Research* (Ed A.L. Singh *et al.*), National Conference of Plant Physiology, 13-16th Dec 2013, DGR, Junagadh, India. pp. 928-929. www.ncpp13.nrcg.res.in.

Khush, G. S. 1995. Modern varieties – their real contribution to food supply and equity. *Geojournal* 35 (3): 275 – 284.

Nautiyal, P.C. 2009. Seed and seedling vigour traits in groundnut (*Arachis hypogaea* L.). *Seed Science and Technology*. 37(3): 721-735.

Nautiyal, P.C. and Ravindra, V. 1996. Drying and storage method to prolong seed viability and seedling vigour of rabi-summer-produced groundnut. *Journal of Agronomy and Crop Science*. 177: 2, 123-128.

Nautiyal, P.C. and Zala, P.V. 1991. Effect of drying methods on seed viability and seedling vigour in Spanish groundnut (*Arachis hypogaea* L.). *Seed Sci. Tech*; 19(2): 451-459.

Nautiyal, P.C., Bandyopadhyay, A. and Misra, R.C. 2004. Drying and storage methods to prolong seed viability of summer groundnut (*Arachis hypogaea*) in Odisha. *Indian Journal of Agricultural Sciences*. 74(6): 316-320.

Nautiyal, P.C., Bandyopadhyay, A. and Zala, P.V. 2001. In situ sprouting and regulation of fresh-seed dormancy in Spanish groundnut (*Arachis hypogaea* L.). *Field Crops Res* 70: 233-241.

Nautiyal, P.C., Ravindra, V. and Joshi,Y.C. 1990. Varietal and seasonal variation in seed viability in Spanish groundnut (*Arachis hypogaea*). *Indian Journal of Agricultural Sciences*. 60(2): 143-145.

Nautiyal, P.C., Vasantha, S., Suneja, S.K. and Thakkar, A.N. 1988. Physiological and biochemical attributes associated with the loss of seed viability and vigour in groundnut (*Arachis hypogaea* L.). *Oleagineux*. 43: 12, 459-463; 28.

Nautiyal, P.C., Sudipta Basu, Sandeep Kumar Lal and Neeta Dwevedi. 2013. Seed priming and quality enhancement in field crops. In: *Current Trends in Plant Biology Research* (Ed A.L. Singh *et al.*), National Conference of Plant Physiology, 13-16th Dec 2013, DGR, Junagadh, India. p 224. www.ncpp13.nrcg.res.in.

NSC, 2014. Annual report 2013-14, National Seeds Corporation Ltd, New Delhi, A govt of India undertaking- "Mini Ratna" company http: // www.indiaseeds.com/seeds_avablt.html.

NSRTC, 2014. National Seed Research and Training Centre, Varanasi, www.nsrtc.nic.in.

Paroda R.S. (2013). Indian seed sector: The way forward. Special lecture delivered at Indian Seed Congress. Hotel Leela Kempinsky, Gurgaon, Haryana, Feb. 8, 2013. pp.16.

RAFI-UNDP 1999. Conserving Indigenous Knowledge: Integrating Two Systems of Innovation. UNDP, New York, USA, 79 pp.

Seednet India Portal, http: //seednet.gov.in/.

Selvaraj S. (2013). Preparation of state seed rolling plan and strategy to tie up seed production with different seed agencies. p.37-48. In. *6th National Seed Congress on Advancement in Agriculture through Quality Seeds*. Lucknow. Sept. 12-14, 2013.

SFCI 2014. Annual report 2013-14. State farms corporation of India, Govt of India Undertaking.

Singh, A.L. 2002. Potassium influences kernel filling of large–seeded groundnut in calcareous soil. *J. Potassium Research*, 18: 47-52.

Singh A.L. 2012. High zinc density groundnut a boon to alleviate zinc malnutrition in India. In: *Proc UGC-Sponsored Natl Seminar on New Frontiers of Plant Science for Sustainable Development*, pp 271-275 (Ed S.K. Nayak) 25-26 Feb 2012, PN college Khurda, Odisha.

Singh, A.L. 2014. Micronutrient biofortification in crop plants to alleviate its malnutrition in India. In: 3rd International Conference on Agriculture and Horticulture, October 27-29, 2014 at Hyderabad International Convention Centre, Novotel, Hyderabad, India. Agrotechnol 2014, 2: 4 p 43. http://dx.doi.org/10.4172/2168-9881.1.011.

Singh A.L. and Vidya Chaudhari, 2015. Zinc biofortification in sixty groundnut cultivars through foliar application of zinc sulphate. *J. Plant Nutrition* (in press).

Singh, A.L., P.C. Nautiyal and P.V. Zala 1998. Growth and yield of groundnut (*Arachis hypogeae* L.) varieties as influenced by seed size. *Tropical Science*, 38: 48-56.

Singh, A.L., V. Chaudhari and C.B. Patel 2011. Identification of High Zinc Density Groundnut Cultivars to Combat Zinc Malnutrition in India. In Zinc Crops 2011 (3rd International Zinc Symposium 2011) "*Improving Crop Production and Human Health*" Hyderabad, India, 10-14 October, 2011. http://www.zinccrops2011.com or.

Singh, P.L., M.N. Gupta and A.L. Singh 1993. Screening of fungicides and fumigants for their use in storage of chilgoza (*Pinus gerardiana* Wall.) seed. *J. Mycology and Plant Pathology*, 23(1): 58-63.

Singh, P.L., M.N. Gupta, and A.L. Singh 1992. Deterioration of physico-chemical properties of chilgoza (*Pinus gerardiana* Wall.) seed during storage. *Indian Journal of Plant Physiology* 35(3): 231-237.

Singh, Pushp Lata, M.N. Gupta and A.L. Singh 1987. A rapid method for detection of mycotoxins in chilgoza seed. *Seed Research*, 15 (2): 195-200.

Trivedi, R.K. and M. Gunasekaran, 2013. Indian Minimum Seed Certification Standards, The Central Seed Certification Board, Department of Agriculture and Co-operation, Ministry of Agriculture, Government of India, New Delhi, India.

Annexure 1

State-wise List of Seed Testing Laboratories and Certification Agencies in India

Andhra Pradesh

1. Seed Testing Laboratory, 1561-A, Smith Road, Cuddapha.
2. A. P. State Seed Development Corporation, Quality Control Laboratory, Hyderabad.
3. Seed Testing Laboratory, Dist. West Godavari, Tadapaligudam – 534101.
4. National Seeds Corporation Ltd., Quality Control Lab (South), 17-11, Tukaramgate, North Lallaguda, Secunderabad – 500017
5. Seed Testing Laboratory, Vijaywada.
6. A.P. State Seed Certifn Agy, Mini Seed Testing Lab, Govt. Farm, Amravati, Guntur.
7. A.P. State Seed Certifn Agency, Seed Testing Lab, Rajendra Nagar, Hyderabad 500030.

Bihar

8. Seed Testing Laboratory, Purvi Champaran.
9. Regional Seed Testing Laboratory, Combind Building, Agriculture Department, Dumka.
10. Regional Seed Testing Laboratory, Bharat Krishi Mahavidyalya, Dholo, Dist. Muzaffarnagar.
11. Regional Seed Testing Laboratory, Krishi Bhawan, Sahabaganj.
12. Seed Testing Laboratory, Mithapur Farm, Patna.
13. Regional Seed Testing Laboratory, Laharia Sarai, Dist. Darbhanga.
14. Seed Testing Laboratory, Patna Seed Certification, Krishi Bhawan, Mithapur, Patna.

Delhi

15. National Seeds Corporation Ltd., Quality Control Laboratory, Beej Bhawan, Pusa. Complex, New Delhi – 110012.
16. Seed Testing Laboratory, Development Department, Govt. of Delhi, Barawala Complex, Barawala, New Delhi – 110039.
17. Seed Testing Laboratory, Mori Gate, Delhi.

Assam

18. Seed Testing Laboratory, Chandmari Road, Silchar, Distt. Cachar.
19. Seed Testing Laboratory, Assam State Seed Certification Agency, Ulubari, Guwahati – 781 007.

20. Seed Testing Lab, Assam State Seed Certification Agency, Jail Road, Barbheta, Jorhat.

21. Seed Testing Officer, Assam State Seed Certification Agency, Barapura Main Road, Bonganigaon – 783380.

Gujarat

22. Seed Testing Laboratory, Mini Krishi Bhawan, Nilam Baug, Junagadh – 362 001.

23. Seed Testing Laboratory, Sector-15, Gandhinagar – 361 001.

24. Seed Testing Laboratory, University Campus, Navsari, Dist. Valsad.

Goa

25. Seed Testing Laboratory, Agriculture and Horticulture Research Station, Ela, Old Goa.

Haryana

26. Seed Testing Laboratory, IADP, C/o Deputy Director of Agriculture, Uchani, Karnal.

27. Seed Testing Unit, Department of Plant Breeding, HAU, Hissar.

28. Seed Testing Laboratory, Haryana State Seed Certification Agency, Beej Pramanikaran, Bhawan, Bay 11-12, Sector – 14, Panchkula – 134109.

Himachal Pradesh

29. Seed Testing Lab, Holta at Palampur, O/O DDA, Palampur-176 061, Himachal Pradesh.

30. Seed Testing Lab, Solan at Chambaghat, O/O DDA, Solan – 173 213, Himachal Pradesh.

Jammu and Kashmir

31. Seed Testing Laboratory, Department of Agriculture, Lalmandi, Srinagar.

32. Seed Testing Laboratory, Department of Agriculture, Talab Tilloo, Jammu – 180 002.

Jharkhand

33. Regional Seed Testing Laboratory, Krishi Golpahari, Tatanagar, Jamshedpur.

34. Seed Testing Laboratory, Ranchi.

Karnataka

35. Seed Testing Laboratory, Dy. Director of Horticulture, Lalbagh, Bangalore – 560 004.

36. Seed Testing Laboratory, Opposite Pepsi Factory, Belgaum Road, Dharwad – 580008.

37. Seed Testing laboratory,Karnataka State Seed Certification Agency,Opposite Baptist Hospital, KAIC Compound, Bellary Road, Hebbal, Bangalore – 560024.

38. Seed Testing Laboratory, Davanagere

39. Seed Testing Laboratory, Gangavathi

40. Seed Testing Laboratory, UAS, Bangalore – 560 024.

41. Seed Testing Laboratory, UAS, Dharwad – 580 008.

Kerala

42. Seed Testing Laboratory, Parattukonam, Thiruvannthapuram.

43. Seed Testing Laboratory, Kalarcode, P.O. Alleppyy – 688003.

44. Seed Testing Laboratory, Pattambi.

Chhattisgarh

45. Seed Testing Laboratory, Krishak Nagar, Raipur.

Maharashtra

46. Seed Testing Laboratory, Opposite Institute of Science, Maharaj bagh Square, Civil Lines, Nagpur – 440001.

47. Seed Testing Laboratory, Department of Agriculture, New Modha, Parbhani – 431401.

48. Quality Control laboratory, Maharashtra State Seeds Corporation Ltd., Parbhani.

49. Seed Testing Lab, Dept of Agriculture, Ashirwad Building, Rautwadi, Akola – 444005.

50. Seed Testing Lab, Department of Agriculture, Agri. Engg. Workshop Premises, Dargah Road, Aurangabad – 431005.

51. Seed Testing Lab, Dept of Agriculture, Krishi Bhawan, Shivaji Nagar, Pune – 411 005.

52. Quality Control Lab, Maharashtra State Seed Corpn., Plot No. 8, Shastri Nagar, Akola.

Madhya Pradesh

53. Seed Testing Lab, M.P. State Seed Certification Agency, Krishi Nagar, Adhartal, Jabalpur.

54. Seed Testing Laboratory, Veterinary Campus, M.O.G. Lines, Mhow Naka,Indore.

55. M.P. Fodder Seed Production and Distribution Proj.19-M.P. Nagar, Zone-II, Bhopal 462001.

NEH

56. Seed Testing Laboratory, Department of Agriculture, Manipur, Mantirpukhri, Imphal – 795 003.

57. Seed Testing Laboratory, Dept of Agr, Fruit Garden, East Khasi Hills, Shillong- 793 003.

58. Seed Testing Laboratory, Department of Agriculture, Mizoram, Aizawal.

59. Seed Testing Laboratory, Department of Agriculture, Majitar– 737 135. Sikkim

60. Seed Testing laboratory, Department of Agriculture, Arundherti Nagar, Agartala, Tripura

Odisha

59. Seed Testing Laboratory, Rayadada, (DAO, Office Campus), Dist. Rayagada – 765 001.

60. Seed Testing Laboratory, Bargarh, At/Ekamra Chhack, Post Sarasara, bargarh – 768 028.

61. Seed Testing Laboratory, At – P.O. Balia, Distt. Balasore.

62. Seed Testing Laboratory, Odisha State Seed Certification Agency, At Samantarapur, PO – Bhubaneswar, District – Khurdha – 751 002.

63. Seed Testing Laboratory, Odisha State Seed Certification Agency, Palaapalli

64. Seed Testing Laboratory, Odisha State Seed Certification Agency, Sambalpur

Pudducherry

65. Seed Testing Laboratory, Krishi Vigyan Kendra Campus, Lyyankutti-palayam, Pondicherry- 605 009.

Punjab

66. Seed Testing Laboratory, Punjab Agricultural University Campus, Ludhiana.

67. Seed Testing Laboratory, Gurdaspur.

68. Seed Testing Laboratory, Chief Agricultural Officer, Faridkot.

Rajasthan

69. Seed Testing Laboratory, Department of Agriculture, Durgapura, Jaipur.

70. Seed Testing Laboratory, Department of Agriculture, Sriganganagar.

71. Seed Testing Laboratory, Department of Agriculture, Karkhaw Bagh, Kota.

72. Seed Testing Laboratory, Department of Agriculture, 24, Paota, Jodhapur.

73. Seed Testing Laboratory, Department of Agriculture, Alwar.

74. Seed Testing Laboratory, Department of Agriculture, Chittorgarh.

Tamil Nadu

76. Seed Testing Laboratory, Directorate of Seed certification, Coimbatore – 641003.

77. Seed Testing Laboratory, Tamil Nadu Agricultural University, Dept of Seed Science and Technology, Coimbatore – 641 003.

78. Seed Testing Laboratory, Department of Seed Certification, Alwar Nagar, Nagamalai Pundu Kottai, Madurai – 625 019.

79. Seed Testing Laboratory, Department of Seed Certification, Dharampuri – 636 705.

80. Seed Testing Lab, Dept. of Seed Cert, South Street, Manarpuram, Trichirapalli – 620 020.

81. Seed Testing Lab, Dept. of Seed Cert, Kattihottam Mariamman, Kali, Thanjavaur- 631 001.

82. Seed Testing Laboratory, Nirubar Colony, Palayam Kattai, Tirunelveli – 627 002.

83. Seed Testing Lab, Dept. of Seed Certification, Panjepettai, Kancheepuram – 613 502.

84. Seed Testing Laboratory, Department of Seed Certification, Erode.

85. Seed Testing Laboratory, Department of Seed Certification, Salem.

Uttar Pradesh

87. Regional Agriculture Seed Testing and Demonstration Station, Dept of Agric, Barabanki.

88. Regional Agriculture Seed Testing and Demonstration Station, Department of Agriculture, 9515, Civil Line, Jhansi, U.P.

89. Regional Agriculture Seed Testing and Demonstration Station, Dept of Agric, Meerut.

90. Seed Testing Laboratory, U.P. Seed Development Corporation C-973/74 B, Faizabad Road, Mahanagar, Lucknow – 226 006.

91. Regional Agriculture Seed Testing and Demonstration Station, Dept. of Agric, Azamgarh.

92. Seed Testing Lab, U.P. State Seed Certification Agency, 35-C/6, Rampur Bagh, Bareilly.

93. Regional Agricultural Testing and Demonstration Station (R.A.T.D.S.), Post – Industrial Estate, Varanasi – 221 106.

94. Regional Agriculture Seed Testing and Demonstration Center, Department of Agriculture, 32/8, Civil Line, Mathura.

95. Regional Agriculture Seed Testing and Demonstration Center, Department of Agriculture, Station Road, Bardoi.

96. Seed Testing Laboratory, Department of Seed Technology, C.S. Azad University of Agriculture and Technology, Kanpur.

Uttarakhand

97. Regional Agric Seed Testing and Demonstration Center, Dept. Agri Haldwani, Nainital.

98. Seed Testing Laboratory, Department of Genetics and Plant Breeding, College of Agriculture, G.B. Pant University of Agriculture and Technology, Pantnagar – 263 145.

99. Seed Testing Laboratory, Uttaranchal Seeds and Tarai Development Corporation Ltd., Pantnagar, P.O. Haldi, District- Udham Singh Nagar – 263 146.

100. Seed Testing Laboratory, Uttarakhand State Seed Certification Agency, 12/II, Vasant Vihar, Dehradun – 248 006.

101. Seed Testing Laboratory, Rudrapur.

West Bengal

102. Seed Testing Lab, Dept of Agric, Malda Gaur Road, P.O. – Mukdumpur, Dist Malda.

103. Seed Testing Laboratory, State Agriculture Research Institute (SARI), Department of Agriculture. 230-A, Netaji Subhas Chandra Bose Road, Tollygunge, Kolkata –700 040.

104. Seed Testing Laboratory, Department of Agriculture, Kolkata

105. Seed Testing Lab, Dept of Agri, Makdoompur.

Central Seed Testing Laboratory

106. Central Seed Testing Laboratory, (For testing of all seeds except Bt. Cotton Seed), National Seed Research and Training Centre, G.T. Road, Collectry Farm, Varanasi – 221 106, U.P.

107. Central Seed Testing Laboratory, (Only for testing Bt. Cotton Seed) Central Institute of Cotton Research, (ICAR), Nagpur.

List of ISTA Accredited Lab

1. Bejo Sheetal Seeds Pvt. Ltd. Bejo Sheetal Corner, Mantha Road Jalna – 431203, MS.

2. Indo-American Hybrid Seeds (India) Pvt Ltd., Seed Lab, 7th km, Banashankari, Kengeri Link Rd, Uttarahalli Hobli, Channasandra, SubramanyaPura, Bangalore 560 061.

3. Maharashtra Hybrid Seeds Company Ltd. Quality Assurance Laboratory Jalna- Aurangabad Road, Dawalwadi, Badnapur Tq P.O.Box 76, Jalna – 431 203 (M.S.)

4. Namdhari Seeds Pvt. Ltd., Seed Production and Testing 119, 9th Main, Ideal Homes Townships, Rajarajeshwari Nagar 560 098, Bangalore.

Previous Volume Content

— Volume 1 —

2015, xii+445p., col. plts., figs., tabls., ind., 25 cm   Rs. 2800

ISBN 978-93-5124-276-5

Index